Praise for EDISON

"[A] singular accomplishment of capturing a quicksilver intellect and conveying, in often luminous language, what it was like to be Thomas Edison . . . The sobering realization that these are among the last words we will have from Edmund Morris's pen only heightens our gratitude for this Edisonian portrait that, in intimacy and insight, constitutes its own Eureka moment."
—*The Wall Street Journal*

"Morris deploys those extraordinary talents again to sculpt a staggeringly grand likeness of the American genius Thomas Alva Edison. . . . Fortunately, both Edison and Morris were eccentric and brilliant enough to make even a life told in reverse a compelling experience."
—*The Washington Post*

"The delight of Edmund Morris's *Edison* is that, instead of arguing with earlier writers or debating the terms of genius, it focusses on the phenomenological impact of Edison's work."
—*The New Yorker*

"Exhaustive in scope but paced like a novel, *Edison* is a definitive biography by one of the finest practitioners of the craft."
—*The StarTribune*

"Morris skillfully weaves this lesser-known personal history, replete with friction and domestic dramas, into his epic account of experiments, setbacks, triumphs, and corporate empire-building. The result is a fully formed, engrossing portrait of one of the world's most important and influential figures."
—*The Chicago Review of Books*

"[An] outstanding biography . . . Morris rivetingly describes the personalities, business details, and practical uses of Edison's inventions as well as the massive technical details of years of research and trial and error for both his triumphs and his failures."
—*Kirkus Reviews*

"Morris vividly fleshes out Edison's extraordinary intellect and industry. . . . The result is an engrossing study of a larger-than-life figure who embodied a heroic age of technology."
—*Publishers Weekly*

EDISON

EDISON

EDMUND MORRIS

RANDOM HOUSE
NEW YORK

Published in the United States by Random House, an imprint and division
of Penguin Random House LLC, New York.

RANDOM HOUSE and the HOUSE colophon are registered trademarks of
Penguin Random House LLC.

Originally published in hardcover in the United States by Random House, an
imprint and division of Penguin Random House LLC, in 2019.

LIBRARY OF CONGRESS CATALOGING-IN-PUBLICATION DATA
Names: Morris, Edmund, author.
Title: Edison / by Edmund Morris.
Description: New York: Random House [2019] | Includes bibliographical
references and index.
Identifiers: LCCN 2019005173 | ISBN 9780812983210 |
ISBN 978067964465 1 (ebook)
Subjects: LCSH: Edison, Thomas A. (Thomas Alva), 1847–1931. | Inventors—
United States—Biography. | Electrical engineers—United States—Biography.
Classification: LCC TK140.E3 M685 2019 | DDC 621.3092 [B]—dc23
LC record available at https://lccn.loc.gov/2019005173

Printed in the United States of America on acid-free paper

randomhousebooks.com

246897531

Book design by Simon M. Sullivan

For Pauline

I do not find it so easy to talk about my Father. . . . I have yet to find a biography of him that satisfies me as a picture of the whole man. The emphasis is so much on what he did that few people know what he was.

I have been astonished at times to find the general impression is that he was a sort of superhuman lightning rod, pulling down inventions from heaven at will—a miraculous robot who never got tired—a disembodied brain whose success, bringing fabulous riches, was effortless and assured, in spite of a background of abject poverty, almost total lack of education, and no personal life at all. I may say that the picture is not quite accurate.

—MADELEINE EDISON SLOANE

CONTENTS

EDISON

PROLOGUE

1931

TOWARD THE END, as at the beginning, he lived only on milk.
When he turned eighty-four in February, and pretended to be
able to hear the congratulations of the townspeople of Fort
Myers, and let twenty schoolgirls in white dresses escort him under
the palms to the dedication of a new bridge in his name, and shook
his head at being called a "genius" by the governor of Florida, and
gave a feeble whoop as he untied the green-and-orange ribbon, and
retreated with waves and smiles to the riverside estate he and Mina
co-owned with the Henry Fords, he declined a slice of double-iced
birthday cake and instead drank the fourth of the seven pints of milk,
warmed to nursing temperature, that daily soothed his abdominal
pain.[1]

From earliest youth he had half-starved himself, faithful to the dic-
tum of the temperance philosopher Luigi Cornaro (1467–1566) that
a man should rise from the table hungry. It was not always a matter
of choice. At times during his teenage years as a gypsy telegrapher, he
had wandered the streets of strange cities, unable to afford a cheekful
of tobacco. But even in early middle age, while earning big money and
enabling two successive wives to fatten on haute cuisine, he would eat
no more than six ounces a meal—generally only four—and drink
nothing except milk and flavored water. "A man can't think clearly
when he's tanking up." His one indulgence was cheap Corona cigars,
which he smoked, or rather chewed, by the boxful and liked not for
their price but for their strong, coarse taste.[2] These "long-toms"
jazzed his already hyperactive metabolism to the point that he could
work fifty-four hours at a stretch. Until about two years ago, he had

habitually run up flights of stairs, and could swing a spry leg over his desk. Long before that, his stomach had shrunk so much that anything more than a lamb chop or a couple of fishballs made him feel sluggish. At seventy-seven he reduced his daily diet to a slice of toast, a tablespoonful of porridge, another of spinach, a sardine, and four Uneeda biscuits, washed down with pint after pint of milk. At eighty-one he switched to milk entirely, except for a quarter of an orange at either end of the day. Now he was afflicted by a toxic mix of renal failure and diabetes. Famously indestructible, having near-blinded himself with the study of incandescence, suffered countless acid burns and electrical shocks, bombarded his arms and face with roentgens, and breathed enough mine dust to give a lesser man pneumoconiosis, he seemed at last to be in final decline—along with the national economy, about $15 billion of which derived from his inventions.[3]

"My message to you," he advised his fellow citizens in a valedictory radio broadcast from his botanical library, "is to be courageous. I have lived a long time. I have seen history repeat itself time and again. I have seen many depressions in business. Always America has come back stronger and more prosperous."[4]

Before returning home at winter's end to New Jersey, he prayed to some power other than God (whose existence he denied) to be spared long enough to finish his current round of botanical experiments. "Give me five more years, and the United States will have a rubber crop that can be utilized in twelve months' time."[5]

It was clear, however, when he arrived at the station in Newark, that he would not see another spring. He was frail and stooped under his thick fall of white hair, and needed help to walk. Three of his six children were on hand to greet him. Outside, a warm thunderstorm was pounding down. Mina threw a protective rug over her husband's shoulders as he tottered toward a waiting automobile for the short drive to West Orange.[6]

Next morning, employees at his vast laboratory complex up Main Street waited for the Old Man—as he had been known since his twenties—to punch in early as usual. But for the rest of June and all of July, he uncharacteristically remained at Glenmont, his mansion in the gated confines of Llewellyn Park. On the first day of August he appeared at the front door, dressed for a country excursion, only to collapse and be carried upstairs to bed. Three physicians arrived in a hurry, one of them by chartered plane. That night they announced that their patient was "in failing health," afflicted by chronic nephritis on top of his metabolic disorder. Aware that Wall Street would react negatively to this news, they added, "The diabetic condition now is under control, and the kidney condition seems improved."[7]

NEWSROOMS AROUND THE world hastened to update the obituary of Thomas Alva Edison. They had been doing so for fifty-three years, ever since his self-proclaimed greatest invention, the phonograph, won him overnight fame.[8] Then and now, journalists marveled that such an acoustic revolution, adding a whole new dimension to human memory, could have been accomplished by a man half deaf in one ear and wholly deaf in the other.

Even the most text-heavy periodicals lacked enough column inches to summarize the one thousand and ninety-three machines, systems, processes, and phenomena patented by Edison.* (Not to mention an invention impossible to protect, yet as seminal as any—his establishment of history's first industrial research and development facility, at Menlo Park, New Jersey.) Although his disability was progressive—"I haven't heard a bird sing since I was twelve years old"[9]—he had invented two hundred and fifty sonic devices: diaphragms of varnished silk, mica, copper foil, or thin French glass, flexing in semifluid gaskets; dolls that talked and sang; a carbon telephone transmitter; paraphenylene cylinders of extraordinary fidelity; duplicators that molded and smoothed and swaged; a pointer-polisher for diamond splints; a centrifugal speed governor for disk players; a miniature loudspeaker utilizing a quartz cylinder and ultraviolet light; a dictating machine; audio mail; a violin amplifier; an acoustic clock; a radio-telephone receiver; a device that enabled him to listen to the eruptions of sun-

* Edison averaged one patent for every ten to twelve days of his adult life. The complete list, arranged by number and execution date, is available online at edison.rutgers.edu/patents. It does not include inventions, such as the X-ray fluoroscope, that he chose to leave patent-free.

spots; a recording horn so long it had to be buttressed between two buildings; bone earbuds that could be shared by two or more listeners, and a voice-activated flywheel.

He was even more legendary for his creation of the long-burning incandescent lightbulb, accompanied by two hundred and sixty-three other patents in illumination technology. That number could be increased by one, had he not made his X-ray fluoroscope available without license to all medical practitioners. Most spectacularly, Edison had designed, manufactured, powered, and built the world's first incandescent electric lighting system. At the flick of a switch, one September evening in 1882, he had transformed the First District of lower Manhattan from a dimly gaslit warren into a great spread of glowing jewels.

Out of his teeming brain and ever-mobile hands (the rest of him rigid with concentration, as he hunched over his tools and flasks) came the universal stock ticker, the electric meter, the jumbo dynamo, the alkaline reversible battery, the miner's safety lamp, slick candy wrappers, a cream for facial neuralgia, a submarine blinding device, a night telescope, an electrographic vote recorder, a rotor-lift flying machine, a sensor capable of registering the heat of starlight, fruit preservers, machines that drew wire and plated glass and addressed mail, a metallic flake maker, a method of extracting gold from sulfide ore, an electric cigar lighter, a cable hoist for inclined-plane cars, a self-starter for combustion engines, microthin foil rollers, a sap extractor, a calcining furnace, a fabric waterguard, an electric pen, a sound-operated horse clipper, a moving-sphere typewriter, gummed tape, the Kinetograph movie camera, the Kinetoscope projector, and moving pictures with sound and color. He built the world's first film studio, the world's biggest rock crusher, tornado-proof concrete houses, scores of power plants, and an electromagnetic railway complete with locomotive, trolleys, brakes, and turntable. He dreamed up a Goldbergian set of variations on the theme of telegraphy, including duplex, quadruplex, and octoplex devices that transmitted multiple messages simultaneously along a single wire, "grasshopper" signals that leaped from speeding trains, and receivers that chattered out facsimiles or turned dots and dashes into roman type. If he had not been so busy inventing other things in the early 1880s, he could have combined his discoveries of etheric sparking, thermionic emission, extended induction, and rectifying reception into the wireless technology of radio.

His lifelong policy (adopted at age fourteen, when he wrote, printed, and published an onboard train newspaper) had been to create only what was practical and profitable. But in aspiring to be primarily an entrepreneur, with over a hundred start-up companies to his credit, he did not have to admit that his need to invent was as compulsive as lust. Each of his honeymoons had triggered a concurrent flood of technological ideas. On a single day, when he was forty and full of innovative fire, he had jotted down a hundred and twelve ideas for "new things," among them a mechanical cotton picker, a snow compressor, an electrical piano, artificial silk, a platinum-wire ice slicer, a system of penetrative photography (presaging radiology by twelve years), and a product unlikely to occur to anyone else, except perhaps Lewis Carroll: "Ink for the Blind."

At fifty-nine, he solved in two hours a hygroscopic problem that had baffled a professional chemist for eleven months.[10]

Only when old age advanced upon him did his shafts of perception slow. He executed a mere 134 patents in his sixties, less than half that number in his seventies. He filed just two in 1928—a year more memorable for the award of a Congressional Gold Medal to the "Father of Light"—and none at all in 1929 or 1930. His final successful application—a mount for the electroplating of precious stones—had come in the early days of this, the last year he would see.[11]

AFTER A WEEK in bed, Edison rallied enough to read a textbook on insulin therapy, as if erudition might help him fix the workings of his pancreas. Although he did not claim to be a pure scientist, he had always kept abreast of the latest professional literature, arguing that expertise should precede experiment. The doctors dispersed. But the chief of them, his personal physician, Dr. Hubert S. Howe, was only guardedly optimistic. "I do not think he will ever be out of danger."[12]

By mid-August Edison was ambulatory and talking, with little conviction, of returning to his laboratory. He had an old rolltop desk there, in a library filled with a lifetime's worth of scientific and technological literature. Throughout his career he had demonstrated an almost dissociative ability to function in different disciplines, moving on a typical day between chemistry, radiography, mineralogy, and electrical engineering. For the last eight years he had been obsessed with botany, struggling to produce rubber from domestic laticiferous plants, including *Solidago edisonia,* a variety of goldenrod developed by himself. It was a project financed by his good friends Henry Ford and Harvey Firestone, both of them wholly dependent on foreign rubber. After testing seventeen thousand native plant species, ranging from tropical ficus to desert shrubs, Edison had fixed on goldenrod as the most promising source, and been encouraged by Maj. Dwight Eisenhower, U.S. Army, to develop it as a strategic war reserve. However, impurities in the weed's watery latex kept frustrating his attempts to concentrate its polyisoprenic particles. Now at last, four chunks of springy coagulum vulcanized by his Florida research team were pressed into his hand.[13]

Charles Edison, president of Thomas A. Edison, Inc., announced that the Old Man was "very happy" to receive them.[14]

THERE HAD BEEN a time when Thomas A. Edison, Jr., hoped for Charles's title. As Edison's eldest son by his first wife, Mary, Tom claimed it by right of primogeniture—only to be slapped down as unworthy. William, Tom's brother, also nourished a sense of early rejection, its sting sharpened now, in middle age, by what he took to be his father's "intense dislike."[15] Marion, the elder sister of both men, was only slightly less starved for paternal affection. For a while, after Mary's mysterious death in 1884, she and Edison had been of comfort to each other. But that intimacy had not lasted much longer than the year and half it took him to marry a girl straight out of finishing school.

The three children he proceeded to have with Mina—Madeleine, Charles, and Theodore—were better mothered if not better fathered. "He is so shut away from us," Madeleine complained. When Edison took enough time off his work to notice them, he felt they were an improvement on Mary's brood. Theodore in particular was a scientist of considerable brilliance. But to varying extents, all six siblings were crushed by the weight of their sire's overpowering celebrity. Only Madeleine had given him any grandchildren—four sons, who bore his name secondarily.[16]

Not that it was likely to be forgotten. Edison had always, with fanatical thoroughness, identified himself with every business he founded, from 1869 on: Pope, Edison & Co., Edison's Electric Pen and Duplicating Press Company, Edison Ore-Milling Company, Edison Telephone Company of London, Ltd., Edison Machine Works, Thomas A. Edison Central Station Construction Department, Edison Phonoplex System, Edison Wiring Company, Edison Phonograph Company, Edison Iron Concentrating Company, Edison Manufacturing Company, Edison Industrial Works, Edison Ore-Milling Syndicate, Ltd., Edisonia, Ltd., Edison Portland Cement Company, Edison Storage Battery Company, Edison Crushing Roll Company, Edison Kinetophone Company, and Thomas A. Edison, Inc.—not to mention such polysyllabic affiliates as Compañía chilena de teléfonos de Edison, Société industrielle et commerciale Edison, Société Kinetophon Edison, and Deutsche Edison-Gesellschaft.[17]

A separate constellation of lighting firms blazed his name around the world, some in characters too strange for Western eyes to read.[18]

IN THE SECOND week of September his health began to fail again. He sensed that he was dying and said goodbye to his wife and children. Dr. Howe issued daily pessimistic bulletins. One stated that Edison had Bright's disease, and stomach ulcers complicating his uremia and diabetes. He was having dizzy spells and losing his sight as well as the last of his hearing. The only voice he seemed to recognize was that of Mina yelling "Dear, how are you?" into his right ear, her hand cupped against his cheekbone. By early October he was ingesting only milk, although one morning Dr. Howe got him to swallow a few spoonfuls of stewed pear. After that he lay inert, except for the obstinate beating of his pulse.[19]

Word spread that he could die at any moment. President Hoover asked to be kept informed. Pope Pius XI cabled twice to express his concern. A woman in Kansas offered her own blood, if it would keep the old inventor alive. Newsmen began an around-the-clock vigil in a press room set up over Glenmont's garage. Others hung around the laboratory downtown, as if half-expecting its founder might still emerge in the small hours, silver-stubbled, reeking of chemicals, spattered from collar to cuffs with tobacco juice and beads of wax, and saying with a wink that he had to go home to save his marriage.[20]

The mansion filled up with family. Notwithstanding the ancient split between Mary's and Mina's children, they clung together in the den downstairs. The sickroom upstairs glowed through dawn, as Dr. Howe and a relay of nurses kept watch over their patient. The gates of Llewellyn Park were closed to motor traffic. Neighbors refrained from entertainments, forgetting that Edison had never been aware of outside noise.

Howe gave up hope on the fifteenth, when his patient briefly opened his eyes—large, blue, and blind—then slipped into a final coma. From time to time his hands made kneading movements, as if he were still testing the malleability of rubber. "Father can't last much longer," Charles told reporters. An urgent call came from Henry Ford, asking for the great man's last breath to be preserved in a test tube.[21]

Mina and all the children were at Edison's bedside when he died at 3:24 A.M. on Sunday 18 October.

TWO MINUTES LATER, the high wall clock in his laboratory library stopped ticking.[22] Its pointers maintained their acute angle for the next three days while Edison, clad in an old-fashioned frock coat, lay beneath in an open coffin. Ten thousand mourners filed past to stare at his waxen profile. "A marvelous, powerful face," the sculptor James Earle Fraser remarked. "The beautiful, full forehead, the nose, the mouth, the chin . . . The hands, too, are wonderful. Delicate, sensitive nails and fingertips, yet withal they show great power."[23]

To gawkers less fixated on flesh, the surrounding gallery could be seen as a sort of wooden cranium, packed with evidence of Edison's searching intellect. *Si monumentum requiris, circumspice.* Staired book stacks, rising to triplex height above the floor, held thousands of scientific and technological tomes, along with runs of periodicals alphabetically devoted to aeronautics, automobiles, chemistry, construction trades, drugs, electrical engineering, hydraulic power, mechanics,

metallurgy, mining, music, philosophy, railroads, telegraphy, and theater. (He left unthumbed those on mathematics, one of the few disciplines that bored him.) A corner pedestal mysteriously supported a
486-pound cube of solid polished copper. Panels and vitrines glittered
with mechanical models, crystals, chunks of ore, medallions, and gold-
stamped awards, along with a framed misquotation: "There is no expedient to which a man will not resort to avoid the real labor of
thinking."[24]

All the library's lamps were dimmed except for the soft radiance
cast by a globe-bearing marble figurine, Aurelio Bordiga's *Genius of
Electricity*. The shabby old rolltop desk that Edison had insisted on
using, in defiance of the splendor of the room, stood against one wall,
temporarily shunted aside to make way for his bier. One of its pigeon-
holes was stuffed with memoranda for inventions he had meant to get
around to. A shadowy alcove half-concealed the blue-covered cot that
Mina had installed for his catnaps—even though he had always been
happy to stretch out on a workbench, with one arm for a pillow, deaf
to the conversations around him.

Now his head rested on silk. An honor guard of veteran employees
kept watch at each corner of the catafalque. The library's normal bookish mustiness was made fragrant by strewn red oak leaves and floral
wreaths. Adding to the aura of sanctity were prayers intoned every few
hours by the minister of the West Orange Methodist Episcopal Church.
They were read at Mina's request, in defiance of her husband's oft-
stated, vehement agnosticism. Dr. Howe tried to make reporters believe that Edison had expressed religious sentiments toward the end.
But the only mystical remark he could recall was "If there is life hereafter, or if there is none, it does not matter."[25]

A private funeral was scheduled for Wednesday at an unstated
hour. Meanwhile an international avalanche of tributes poured in,
attesting to the fact that Edison had done more to irradiate the planet
than any agent save the sun. "An inventive spirit," Albert Einstein
cabled from Berlin, "has filled his own life and all our existence with
bright light." Henry Ford declared that the dead man's achievement
was "etched in light and sound on the daily and hourly life of the
world." Even President Hoover was moved to eloquence: "He multiplied light and dissolved darkness."[26]

Unquoted among all the panegyrics was the frankest of Edison's self-appraisals, recorded some twenty years before: "Everything on earth depends on will. I never had an idea in my life. I've got no imagination. I never dream. My so-called inventions already existed in the environment—I took them out. I've created nothing. Nobody does. There's no such thing as an idea being brain-born; everything comes from the outside. The industrious one coaxes it from the environment; the drone lets it lie there while he goes off to the baseball game. The 'genius' hangs around his laboratory day and night. If anything happens he's there to catch it; if he wasn't, it might happen just the same, only it would never be his."[27]

Botany

1920–1929

AT SEVENTY-THREE, WITH his wartime career as president of the Naval Consulting Board behind him, Edison tried to make sense of a new intellectual order that challenged everything he had learned of Newtonian theory. Abstract thought did not come easily to him. "My line of sorrow," he wrote, "lies in the realm of technical science." He needed to feel things come together under his hands, see the filament glow, smell the carbolic acid, and—as far as possible for a near-deaf man—hear the "molecular concussions" of music.[1]

Laws such as those of Faraday's electromagnetic induction and Ohm's relation of current, voltage, and resistance he understood, having applied them himself in the laboratory. But now, if only to slow as much as possible the entropy of his own particles (the fate of all systems, according to Lord Kelvin), Edison studied Einstein's general theory of relativity.[2] The recent solar eclipse had persuaded him, along with the academic scientists he mocked as "the bulge-headed fraternity," that the theory was valid—even if it failed to suggest any correlation between his attempt to measure the total eclipse of 1878 and his subsequent perfection of incandescent electric light.[3]

The urtext of the theory, as translated by Robert Lawson, defeated him after only eleven pages. "Einstein like every other mathematical mind," he scrawled in the margin of his copy, "has not the slightest capacity to impart to the lay mind even an inkling of the subject he tries to explain." He turned for help to an interpretive essay—Georges de Bothezat's "The Einstein Theory of Relativity: A Glance into the Nature of the Question"—and filled thirty-one notebook pages with scrawled paraphrases of its main points.[4]

Gravitation is due to the retardation in velocity of the ultimate par-
ticle in passing through the fixed aggregates of matter. Ultimate
particles fill the whole of space and proceed in every direction. . . .

He could imagine that at least in terms of his own observation,
forty years before, of the thermionic emission of carbon electrons in a
lightbulb after evacuation—a mysterious darkening since known as
the "Edison Effect." It was about as far as he ever got in his search for
a "new force" in electrochemistry. Disparaged at the time by his
peers, he now knew that he had discovered, if not recognized, the
phenomenon of radio waves eight years before Heinrich Hertz.

Wireless waves cannot proceed thru space but thru Matter in com-
bination with the ultimate particle. . . . From this, if true, all matter
is formed of the same material.

Edison had once teased a science fiction writer with the notion of
interchanging atoms of himself with those of a rose. He noted that
Einstein envisaged particles in space with common axes converging
into solidly constituted "rings," while others remained ethereal.*
Hence the "primal ring" of the solar system, with its interplanetary
nothingness.

We now have matter in a form which is polar & capable of produc-
ing what we call Magnetism & Electricity.

The religion boys, of course, would protest that what drew particles together was the will of God. Edison was as ready as Einstein to
believe in a "Supreme Intelligence" made manifest by the order and
beauty of the stars, and equally reluctant to personalize it: "I cannot
conceive such a thing as a spirit." The furthest he would go in the
direction of metaphysics was to imagine the subcellular particles of a
human being as "infinitesimally small individuals, each itself a unit of
life."[5]

---

* Around this time Edison also exhaustively studied analytical atomic spectroscopy, as
described in his annotated copy of *The Nature of Matter and Electricity,* by Daniel Comstock and Leonard Troland (1919).

*These units work in squads—or swarms, as I prefer to call them—
and . . . live for ever. When we "die" these swarms of units, like a
swarm of bees, so to speak, betake themselves elsewhere and go on
functioning in some other form or environment. If the units of life
which compose an individual's memory hold together after that individual's death, is it not within the range of possibility . . . that
these memory swarms could retain what we call the individual's
personality after the dissolution of the body?*

Having thus anticipated by more than a century both swarm intelligence and DNA inheritance theory, Edison gave up trying to understand relativity and returned to the more tangible universe he
preferred.

A BIG BUMP FOR COOKIES

As he saw it, his first order of business in the new decade was to reim-
pose his own—highly individual—personality upon Thomas A. Edison, Inc., the sprawling industrial conglomerate that he had been
forced to neglect during the war. He chose not to notice that it had
thereby done much better than it had in earlier years, when he had
run its manifold activities—phonograph and record production,
movie making, cement milling, storage battery development, and laboratory research—with such autocratic willfulness as to make his executives despair of ever influencing him.

Edison was not an easy man to advise, being a combination of twinkling charm and bruising imperiousness. In his youth the charm had
prevailed, but now that he was a septuagenarian and almost unreach-
ably deaf, the urge to overbear had become a compulsion, and he had
lost much of the bonhomie that had kept thousands of men working for
him, and worshiping him, over the past half-century. Long gone was the
perpetual hint of a smile flickering around the corners of his mouth, as
if he were about to break into thigh-slapping laughter. The artist Richard Outcault remembered its radiance back in '89, when "the boys"
presented "the Old Man" with a gold and silver phonograph for his
birthday. "Edison's smile! [It] sweetened up the atmosphere of the
whole building. . . . As long as I live the sweet spirit that pervaded the
atmosphere of the laboratory will always remain with me."[6]

Edison still moved with the jerky energy that kept him awake, and acting more decisively, than young men unable to match his eighteen-hour-a-day schedule. He regarded exercise as a waste of time, and sleep even more so. Since he was twenty, he had maintained his 175-pound, five-foot-nine-and-a-half-inch frame with only a few lapses, quickly corrected. ("I do believe I have a big bump for cookies.") The most remarkable thing about his appearance, apart from the brilliance of the blue-gray eyes, was the largeness of his head, amplified by its thick mop of snowy hair. He wore custom-made size eight-and-a-half straw hats, and slashed the bands of his caps for comfort. His handshake was perfunctory and surprisingly cold. Monomaniacally focused on whatever current project interested him, he strode at a forward angle, hands in vest pockets, aware only of his destination and completely unconscious of time. He never wore a watch, and made no distinction between day and night, nodding off when he felt like it and expecting his assistants to follow suit. The same went for waking up. If two hours of rest was enough for him, he did not see why anyone else should want more.[7]

Lovable as he was—or had been in the past—Edison did not return affection, beyond the occasional beaming familiarity, in which there was often a note of tease. He thought hurtful practical jokes—electrified washbasins, a wad of chewing tobacco spat onto a white summer suit, firecrackers tossed at the bare feet of children—were funny. Having made money easily all his life, thanks to phenomenal energy and the mysterious gift of imagination (his personal wealth, at latest calculation, was almost $10 million),* he was unmoved by the lesser luck or ill fortune of others, and casual about the loneliness of his wives. Now, returning to his laboratory desk in 1920, he was determined to teach Charles Edison a thing or two about running a large corporation.

NOTHING'S RIGHT AND ALL IS CHANGE

For four years Charles had been under the impression that he, not his father, was the chief executive of Thomas A. Edison, Inc. His formal

---

* The equivalent of $125 million in 2018, according to the "Purchasing Power Calculator" at measuringworth.com.

titles were chairman of the board and general manager, but now that the Old Man had come home from the navy, reasserting command and firing off orders like grapeshot, he felt demoted. There was little he could do about it, since Edison had never relinquished the title of president.[8]

Charles Edison, circa 1920.

Charles was nearing thirty, married but childless, an oddly divided personality. At work he was the quintessential businessman, cautious, courteous, efficient, and fair. The patrician manners of Hotchkiss and MIT sat easily on his sober-suited shoulders. Small and wiry (Edison called him "Toughie"), he was a handsome man, with heavy-browed eyes of the palest blue. In later life he would develop a startling resemblance to his father.

At home or in the Greenwich Village cafés he loved to frequent, Charles was a bohemian. For two years he had helped run an avant-garde theater off Washington Square, commuting back nightly to West Orange on the "owl" train. He spoke fluent French, composed songs with titles like "Wicky Wacky Woo," attracted squads of young women, and wrote quantities of light poetry under the nom de plume "Tom Sleeper."[9]

He had displayed all the forceful spirit of extreme youth when he became chairman in June 1916. Until that moment, Edison's skinflint, union-busting management style had made the West Orange complex "the last place at which men desired to work." Charles had taken advantage of his father's naval appointment to bring in some younger, more progressive executives, while decentralizing Edison Industries into a web of largely independent divisions, serviced by an administration in charge of communal interests. He prided himself on having "put the business on a little more humane basis," and expanding it so judiciously that by 1920 Thomas A. Edison, Inc., with eleven thousand employees, was admired for its generous pay, medical, and social policies.[10]

Charles's dread was that the returned Commodore, already harrumphing that the company was too large and too loose, would move to dismantle his "beautiful organization" and reestablish totalitarian control. If so, there was bound to be blood on the boardroom floor. The prospect was enough to make Charles, whose health tended to be psychosomatic, sick with apprehension. He revered Edison as "Father, Boss & Hero," and half-welcomed his reassumption of power at the plant.[11]

"When he is here," he wrote Mina, "I always feel that there is a safe harbor to go into if the weather gets too rough for me on the open sea."[12]

WE ARE ALONE

March found Edison, as usual, at Seminole Lodge, his winter estate in Fort Myers, Florida. What was less usual was the absence of any children to stay. "Papa and I are sitting under the trees, just where Charles was married," Mina wrote her second son, Theodore, a freshman at MIT. "It is blissfully quiet and we are alone. . . . It has never happened in 34 years."[13]

To her pleased surprise, Edison showed no inclination to start another of his countless experimental notebooks. The failure of Washington paper-pushers to adopt a single one of his forty-five inventions and plans as the navy's top defense adviser seemed to have crushed his creativity—for how long, she could not tell. He was particularly hurt by their transformation of his pet project, a naval research laboratory

Aerial photograph of Thomas A. Edison, Inc., 1920s.

to be located far from Washington and staffed by civilian scientists, into a service facility just downriver from the capital, where "mentally inbred" career officers were sure to suppress any innovative ideas.[14]

Edison had come to despise government bureaucrats, seeing them as a blight on democracy. In his disgust he had just turned down a medal for his defense work, arguing that he deserved it no more than any other member of the Naval Consulting Board. He said he had lost interest in weapons of war—not to mention respect for the patent and copyright clause of the U.S. Constitution, "and the other 27,946 books filled with laws."[15]

Mina luxuriated in her husband's company. Normally, unless guests were around, she had to settle for birdsong. A passionate amateur ornithologist ("My dream is a natural aviary"), she rarely went out of doors without binoculars. Her curiosity about all feathered things extended to their habitat, and she was as schooled in the Latin names of trees and shrubs as she was with those of birds. Early on, it had been a frustration to her that biology was the one natural science Edison ignored—beyond scouring the world for bamboo fibers to carbonize in his lamps, or rare resins to bake into his phonograph records. But in later life he had begun to study botany, collecting and identifying specimens on rural jaunts and taking pleasure in the variety of plantings around Seminole Lodge. He talked now of setting out some groves of red and black mango and Louisiana cup oak, for a possible sideline in veneer cutting. To prepare, Edison read academic papers on the vegetation of Florida, and made sure that his estate manager understood the fine art of squashing the "slippery lumps" in wet humus.[16]

A LOOSE PIECE OF LEATHER

With the insensitivity that characterized his dealings with all his children, he upstaged Charles at a conference of "Ediphone" dictating-machine distributors in West Orange that summer.[17] Nobody expected Edison to address the audience personally, since it was well known that he never spoke in public. Instead, he gave Charles a speech he had written and asked him to read it from the podium.

During the discourse, Edison sat unhearing and apparently unaware that he was the focus of all eyes. His own attention became fixed on his right shoe. He bent down to unlace it, then, in the words of a reporter present, "took it off and pruned a loose piece of leather from the sole with a jackknife." Discovering that the sole itself was detached, he peered and poked at it as if he were back at his workbench in the laboratory.

By now the only person in the room not fascinated by the shoe was Charles, still gamely speaking. Edison sensed the stare of the crowd, looked up, and received an amused ovation. He felt obliged to explain, in a voice overriding his son's, "I went over to New York to buy a pair of shoes, and found they were asking $17 and $18 a pair—"

Charles had no choice but to let him proceed.

Edison said he would not pay that kind of money for pointy-toe footwear. Instead, he had gone to a bargain basement and bought a pair of Cortlands for six dollars. He then launched into a harangue on extortion by haberdashers that segued somehow into a demand for greater productivity from his employees.

By the time he allowed Charles to go on reading, it was evident to the audience that the Old Man was back in charge.

HIRIN' AND FIRIN'

Edison's complaint about inflated prices was not entirely the affectation of a rich man. The shoe he held in his hand may have represented inventory that the Cortland Company was desperate to unload.[18]

Overproduction during the postwar boom, stimulated by rapacious consumption, easy credit, and addictive speculation, had caused such a rise in the cost of living that men of his age, remembering the panics of '73 and '93, could see that the American economy was again a bubble close to bursting. In fact, it had burst already, manifesting itself in millions of canceled orders and a recent 25 percent increase in railroad rates that made cash-poor farmers slaughter their horses for hog feed. Salaried city dwellers felt the inrushing cold air of a major depression, and reacted with a halt to optional purchases. Luxuries like phonographs (until now the topmost item on the Edison profit sheet) stacked up unsold. Shabbiness became the new chic. Women recycled last year's dresses, and men had their suits "turned," shiny side in. William McAdoo, President Wilson's former treasury secretary, publicly sported trouser patches. For once in his life, even Edison began to look fashionable.[19]

On 16 September a wagon bomb packed with shrapnel exploded opposite the headquarters of J. P. Morgan & Co. on Wall Street, killing thirty-two pedestrians and injuring hundreds of others. Investigators blamed the disaster on anarchists. But to financiers, a coincidental sharp drop in the Dow Jones industrial was an even louder inducement to panic. Henry Ford slashed the price of his basic Model T, hitherto hard to keep in the showroom, from $575 to $440. General Motors followed suit. The Chicago billionaire Samuel Insull—Edison's former private secretary—had to borrow $12 million in personal funds to keep his web of power companies together. Deflation set in, at a rate unparalleled in American history.[20]

Edison waited no longer than October to initiate a purge of most of the employees his son had hired during the war. He believed that the slump left him no choice but to trim the payroll and increase automation—in both cases, if necessary, by half. He did not scruple to fire some of his own long-serving aides as well. "Poor Charles I fear is pretty much crushed," Mina wrote Theodore.[21]

As diplomatically as she could, she tried to persuade her husband to give up his lifelong habit of command. She had to do so in writing, rather than shout in his right ear:

My darling—

It is beautiful *to see you a* tower *among the young men—Charles, John, Fagan, Mambert, Maxwell, etc —and I do* love *to see you* quietly *counselling with them, giving them the benefit of your wisdom and experience.*

I and all *have so admired your giving the work over to Charles and backing him up in his efforts. . . .*

*You have made a success of your life—built up tremendous industries successfully so you have nothing more to prove to the world that you are capable—*All know it*—Can't you be happy in just letting the boys struggle along,* with you to guide them. . . . *Charles is all for you—He stands by you at all times and is with you,* wanting to please you in every way. *He always puts up your side and will never let any one say a thing contrary to your praise—*Don't misjudge him.

Success makes success—and if you will only let Charles feel that you do appreciate him you will make him and all happier. Forget a little bit that you are Charles's manager and be a father—a big father![22]

Mina might have shouted into Edison's other deaf ear, for all the notice he took of her letter. Charles, he declared, needed to have the "conceit" knocked out of him. The tension between father and son grew to the point that Mina forbade them to talk business during a family lunch. As a result, she told Theodore, "Papa never opened his mouth during the whole meal."[23]

There was a temporary truce in November after Edison and Charles both voted, as Republicans, to send Warren Gamaliel Harding to the White House. Harding's huge win over James L. Cox (announced that night by a tiny startup station calling itself "8ZZ" in Pittsburgh, Pennsylvania) repudiated the cloudy idealism of the war years. But for as long as the stroke-enfeebled Woodrow Wilson remained in office, the election did nothing to bolster consumer confidence. By late December bank presidents were committing suicide, homeowners losing their all to sheriffs (Edison knew what *that* felt like), and "Billy" Durant, the founder of General Motors, was out of a job.[24]

Edison had no intention of sharing Durant's fate. Working eighteen hours a day and often not returning home until dawn, he increased the savagery of his purge, dismissing the whole of Charles's personnel department before Christmas ("Hell, I'm doin' the hirin' and firin' round here") and laying off 1,650 employees of the Phonograph Works. He jettisoned five-sixths of the engineering force and a like proportion of bookkeepers, clerks, artists, copywriters, salesmen, and talent scouts. Those who survived had their wages slashed and were told to forget about Christmas bonuses. In the process, Edison destroyed his old image as a benevolent autocrat, and Charles lapsed into despair.[25]

A TRIO OF THORNS

Among those caught short by the depression were Edison's children by his first marriage. Disdained by Mina as genetically inferior to her own brood, they had been for more than thirty years a trio of thorns in their father's side. Marion at least had done him the favor of settling in Europe and marrying a German army officer. But now, at forty-seven, she wrote to complain that Oberst Oscar Öser was an unfaithful, abusive husband. She was hiding from him in Switzerland, and if Edison did not send the money she needed for a divorce, she might throw herself into the Rhine.[26]

Thomas Alva Edison, Jr., forty-four, was a sad ne'er-do-well, perpetually broke and ailing. Although he ran a mushroom farm, he had long tried to market inventions under his famous name. The latest was a fuel-saving automotive device that he wanted his father to sponsor. Earlier in the year Mina had been terrified that Tom's wife might give birth to a Thomas Alva Edison III. "Poor papa and poor us!" The pregnancy, like others of Beatrice's, had mysteriously evaporated. She claimed to be a nurse, but there was reason to believe that she had once practiced a much older profession.[27]

William, forty-two, was a jock turned clubman, large, loud, defensively jovial. Like Tom, he was a would-be inventor who settled for a malodorous variety of farming—in his case, poultry. William admitted to his father's secretary, Richard Kellow, that he owed Edison $8,347.36 for a tractor and other items of machinery. "Tell him to cheer up, all is not lost, that I'm not dead yet." In the meantime he needed further funds: his wife Blanche was facing a $500 medical procedure.[28]

Edison gave Marion a monthly allowance of $200, agreed to test but not endorse Tom's Ecometer, and told Kellow to deny William's appeal. "Find out why he don't sell the tractor."[29]

HAIL THE MASTER

In the new year of 1921 Edison, alarmed by a free fall in phonograph sales, went on a rampage of additional firings that had even well-wishers questioning his stability. "The Old Man is certainly out of his mind," Miller Reese Hutchison, the company's former chief engineer, wrote in his diary. "Breaking up his organization and seems pointing to a 'bust up.' "[30]

Edison showed no sympathy for dismissed employees who had failed to save for hard times. "I do not believe in unemployment insurance." Mina reported the new purge to Theodore in anguished letters, sometimes two a day. "What can we do to have father dear see that he is crushing all the spirit throughout the plant? . . . I wish he would calm down and let Charles manage things." A few days later: "Papa is tired to death and Charles is just about at the end of his string."[31]

She did not know how near Charles was to resigning over the closure of another of his creations, the Power Service Division. He wrote a bitter poem on the theme of one of Edison's favorite maxims, "Nothing is permanent but change."[32]

> *Changes bring but other changes;*
> *Progress runs in Error's ring;*
> *Plans are made, but Change deranges;*
> *Hail the master; Change is king.*[33]

Charles later admitted to wanting to leave the company rather than tolerate the humiliations his father heaped on him. One of these was Edison's public remark that Thomas A. Edison, Inc., had lost efficiency during the war "due to the negligence of those who were supposed to be watching it."[34]

Charles could not deny that the company's profit sheet, substantial in 1919 and 1920, was reddening toward a loss of more than $1 million this year. But the depression, not his own management, was at fault: nationwide, corporate profits plummeted by 92 percent. One of the Phonograph Division's biggest competitors, the Columbia Company, had to float a $7.5 million bond issue, at ruinous interest, just to pay for a forest's worth of cabinets it could not sell. U.S. Steel, the nation's first billion-dollar trust, was in the process of firing one hundred thousand workers.[35]

Edison saw, with eyes older and colder than his son's, the necessity of similar action at a time when industrial wages were draining eighty-five cents out of every budgeted dollar. He kept pointing out that he had started out in business at age eleven. "I've been through half a dozen of these depressions. I know how they work, and it's got to be this way or we'll go broke." By February Charles's protests had weakened into second guesses that Edison, who often made a convenience of being deaf, ignored.[36]

One night, brooding in bed, Charles heard himself say, "There's a possible chance that he may be right and I may be wrong."[37]

A SHAKER LIKE RAPPOLD

Edison's preoccupation with staff and wage cuts did nothing to assuage his inventive drought.[38] The only patent applications he had filed since 1919 were for improvements to his elegant alkaline storage battery of a decade before. Now, revisiting another old technology, he spent every available hour in the experimental recording studio Charles had built on Columbia Street across from the plant, trying for the fifth time in his career to perfect the sonics of Edison music products.

To most ears, the Phonograph Division's new take of Marie Rappold and Carolina Lazzari singing Puccini's "Tutti i fior" had remarkable fidelity, with flutes and tinkling percussion complementing the tessitura. When the two women went into duet, their voices seemed to shimmer. Edison could not stand it. Deaf as he was, he persisted in thinking he heard perfectly if he jammed the right side of his head close to the amplifier. "How could anyone who pretends to understand Music record such a Record," he scribbled in his notebook. "All out of balance too loud wrong instruments, 2 singers can't sing together & putting a shaker like Rappold in."[39]

It was too late for him to prevent the disk's release, but he could at

Edison listening to phonograph records at home, 1920s.

least wage war on what he saw as lapsed standards throughout the division, from studio to point of sale. He ordered fresh rosin to be applied to the horsehair of string instruments for every four hours of playing time. This would prevent the ribbons from wearing "square," a phenomenon he had detected under the microscope. Plastic dust adhering to the grooves of any pressing should be whisked out with a sweep of the finest white Chinese bristles (an idea that came to him when he was brushing his teeth), and the phenolic varnish glossed with stearin for extra slickness under the reproducer.[40]

The therapy of working with sound again revived Edison's spirits, if not those of the technicians he bullied. "Of all the children of his brain, the phonograph seems to be the one he loves most," his personal assistant, William Meadowcroft, remarked. Mina rejoiced to see her husband becoming his old jocular self. At such times he affectionately called her "Billy," the boyish name he had given her in the early days of their marriage. "It puts a bright hue on everything when he is happy and makes love to me as he is doing now," she wrote Theodore.[41]

FREE FALL

Warren Harding was sworn in as president on 4 March 1921. A placid, middlebrow, middle-of-the-road midwesterner, he famously personified everything that was "normal" in America. Harding objected to extreme behavior, whether it was too emotional a reaction to the current state of economic affairs, or too precipitous an action to combat it.

His inaugural speech echoed what Edison had been saying to Charles for the last five months. Citing the "delirium of expenditures" that had brought the depression on, Harding declared, "We must face a condition of grim reality, charge off our losses, and start afresh."[42]

If this sounded like a warning of governmental intervention, Harding soon made clear that by *we* he meant the 62 million adult Americans whose buying and selling influenced the economy. He waited for the invisible hand of the market to reassert itself, doing little more than appoint a distinguished group of aides to monitor it. They included Andrew Mellon as Secretary of the Treasury, and Herbert Hoover as Secretary of Commerce. Prices continued their free fall.

WHAT IS COPRA?

Edison congratulated himself that spring on having gotten rid of thousands of "untrained and careless workers"—by one estimate, nearly a third of his eleven-thousand-man payroll—with further pink slips yet to be issued before Edison Industries was, in his opinion, slim and trim again. "You're going to learn a big lesson out of this depression," he said to Charles.[43]

Apparently not caring that he had become the most hated man in West Orange, he worked on a new plan to replace highly paid executives with young men willing to work for less money. This meant a risky investment in recent college graduates. To ensure he got the best out of hundreds of desperate job seekers with degrees, he devised a questionnaire to bring out their general knowledge. Only 4 percent of his initial batch of applicants struck him as worth hiring. "The results of the test are surprisingly disappointing," he announced in May. "Men who have gone through college I find to be amazingly ignorant."[44]

The contempt for higher education implicit in that remark was nothing new for Edison. It betrayed a prejudice much more complex than the anti-intellectualism of a small-town boy who had clawed his way to success with minimal schooling. Although his mother was his primary teacher, at home in Port Huron, Michigan, she had been a woman of enough culture to introduce him to Gibbon and Hume, even as he mastered R. G. Parker's *A School Compendium of Natural and Experimental Philosophy* by himself. And his father—radical, randy, secessionist Sam—had "larned" him the complete works of Thomas Paine when he was still a newsboy on the Grand Trunk Railroad.

Edison's reading in the sixty years since embraced few of the humanities but most of the sciences, as well as a wide range of magazines and newspapers. He now claimed to study twenty-seven periodicals, ranging from the *Police Gazette* and "the liberal weeklies" to the *Journal of Experimental Medicine,* plus five papers a day and "about forty pounds of books a month." He was able to maintain this consumption because of his ability to flip pages fast and memorize whatever data appealed to him. "Nearly all my books are transcripts of scientific societies, which will never be republished."[45]

He was an energetic margin-scribbler, forever endorsing—or more often disagreeing with—passages that struck him. "This is young metaphysics over a pound of platinum," he wrote above a chapter of Oliver Lodge's *Ether and Reality,* and "Why lug bible sayings in" next to a passage on maternal love in Sherwood Eddy's *New Challenges to Faith.* Quotations came easily to him, and he had a transatlantic sense of irony: "As La Rochefoucauld said, our virtues increase as our capacity for sin diminishes." His erudition was beyond that of many university professors, let alone their graduate students. "From my experience," the electrical theoretician George Steinmetz remarked, "I consider Edison today as the man best informed in all fields of human knowledge."[46]

Hence the frustration of a Cornell man who publicized seventy-seven Edisonian questions that he thought had unfairly disqualified him from a job at West Orange, such as "How is leather tanned?" "Who was Danton?" and "What is copra?" Another rejectee complained that he failed to see any useful connection "between the thyroid gland and selling incandescent bulbs, or between gypsies and talking machines, or attar of roses and sales production."[47]

Edison had not meant his questions to be leaked. He was obliged to draft another 113, but they too ended up in newspapers across the country, under such headlines as "IF YOU CANNOT ANSWER THESE YOU'RE IGNORANT, EDISON SAYS."[48]

Harper's Magazine accused him of indulging in "philallatopism," or pedantic pleasure in exposing the ignorance of other people. But the questions, though difficult, were not condescending:

Which country drank the most tea before the war?
What is the first line of *The Aeneid*?
Where is the live center of a lathe?
Name two locks on the Panama Canal.
What is the weight of air in a room 20 × 30 × 10?
Who invented logarithms?
What state is the name of a famous violin maker?
How fast does sound travel per foot per second?[49]

The last item was too much even for Albert Einstein. Sounding defensive when it was put to him, the father of relativity said through an interpreter that he saw no point in cluttering his mind with data obtainable from any encyclopedia. "The value of a college education," Einstein huffed, "is not the learning of many facts but the training of the mind to think."[50]

Nicola Tesla, Einstein's rival in popular "genius" rankings, agreed. "Edison attaches too great a value to mere memory." A professor of psychology at Boston University wrote Edison to suggest that all college students were intelligent, to the extent that they had qualified for higher education. Any questionnaire designed to contradict this must therefore be incorrectly framed—if not an exercise in personal vanity. "Are you not perhaps setting a standard for others by means of your own accomplishments, and yet we have but one Edison in the United States?"[51]

It was a shrewd thrust, to which Edison could reply only that his questionnaire was "in the nature of a rough test" to bring out the executive quality he prized most—curiosity. In a public statement, he added that he was not trying to measure "intelligence, logic, or power of reasoning." He merely wanted to hire young men* who displayed "alertness of mind . . . power of observation, and interest in the life of the world."[52]

This protestation did nothing to quell the delight with which humorists, professional and amateur, satirized his "Ignoramometer." The length of a short circuit, the number of stripes on a zebra, and the provenance of "jazz" bow ties were urgently discussed, as was the etymology of the Mephistopheles mosquito. One cartoonist lampooned Edison as Diogenes, making tiny ignoramuses scurry from the glare of his intellectual flashlight. A group of Wellesley girls sent him a five-foot-long list of their own questions, including "What are the chemical properties of catnip?" and "When you turn off the electric light, where does the light go?"[53]

Edison groused that the newspapers "have balled me all up," and threatened lawsuits if any more of his questions were published. Yet part of him—the attention-loving side—relished the sensation he had provoked. *The New York Times* published almost forty articles on the subject of "the Edison brainmeter," while magazines of the caliber of *Literary Digest, Harper's,* and *The New Republic* began a debate on intelligence tests that promised to continue for years. Edison's multiphasic questionnaire was not the first such probe—in 1917 a War Department aptitude test had alarmingly suggested that almost half of America's white population was "feeble-minded"—but it was deliberately unscientific and sought to illuminate character over cognition.[54]

As such, it was discounted, even mocked, by most professionals, and when it eventually proved ineffective, he abandoned it. But in time it would be seen as a reproof to the nonverbal, overquantified tests that thousands of corporations adopted in the age of Babbitt. *The World* remarked that at a time rendered dismal by depression and

---

* Although women constituted a number of the Edison Industries workforce in 1920, their jobs were either menial or secretarial.

Prohibition, "Mr. Edison with his questionnaire has contributed to the gaiety of life but also to the dissemination of knowledge."[55]

WHO'S GOT THE JULEP?

"Things look dark as far as business goes and Papa seems quite worried," Mina wrote Theodore at the beginning of July. "There is a strangeness about everything—It seems like something sinister in the air. I wonder what is to happen."[56]

What was, in fact, about to happen was an upturn in the national economy, thanks to President Harding's willingness to let the depression run its precipitous course. Prices were at last so low that money had regained its fair weight in gold. But the recovery was not yet apparent to Edison—nor for that matter to Harding, who on 12 July made an appeal to Congress to vote down a popular bill awarding bonuses to veterans. In words that could have been uttered in the boardroom at West Orange, the president spoke of "the unavoidable readjustment, the inevitable charge-off" consequent to any period of overexpansion. Cost cutting was "the only sure way to normalcy." Harding earned a standing ovation and widespread praise for his courage. *The New York Times* declared that he had risen above patronage politics and proved himself to be "President of the whole people."[57]

Two weeks later Edison could judge this for himself in a meadow in the Blue Ridge Mountains. Henry Ford and Harvey Firestone corralled him, as they did almost every summer, into joining an automobile camping trip that purported to be recreational, but served as excellent advertising for Ford cars and Firestone tires. Since 1918 these "Vagabond" excursions had become more and more elaborate, the line of tourers and supply wagons lengthening and the two magnates looking ever sleeker—in contrast to Edison, whom they paraded as a shabby, overworked genius in need of fresh air. This year Firestone supposed that because Harding and Edison were, like himself, native Buckeyes, they would get on well.[58] If a meeting between them could be arranged at some location convenient to the president, the Vagabonds would score their greatest publicity coup yet.

Harding was pleased to get out of Washington, if only for a couple of days. Congress was still in extraordinary session, debating economic policy. Apart from occasional workouts in a White House closet, he had enjoyed few diversions from affairs of state since his inauguration. His acceptance of Firestone's invitation to camp out on the weekend of 23–24 July near Peckville, Maryland, caused the motorcade to swell to its largest size yet, with wives, children, about seventy servants, and even a Methodist Episcopal bishop, the Rev. William Anderson, in attendance. Firestone rounded up six thoroughbred horses in case Harding wanted to ride, Ford provided a refrigerator truck with three hundred dressed chickens, and Edison, ever the technologist, set up a "wireless" radio telephone for communications with the capital.[59]

The president arrived at noon on Saturday morning, trailed by bodyguards, aides, and reporters. Edison seemed determined not to be seduced by Harding's good-natured charm and declined his offer of a cigar. "No, thank you, I don't smoke."

This was so patently untrue that Firestone boggled. But Harding took no offense. "I think I can accommodate you," he said, pulling a big plug from his pocket.

Edison helped himself to a large cheekful. Later Firestone heard him say, "Harding is all right. Any man who chews tobacco is all right."[60]

Ford's cooks prepared lunch. Soon the humid air was fragrant with the fumes of roasted Virginia ham, lamb chops, and sweet corn. Edison ambled off into the woods and returned with a fistful of mint. "Who's got the julep?"[61]

When the company sat down to eat, at a round table whose inner hub rotated for condiment delivery, Harding found it impossible to talk into Edison's deaf left ear. He had no better luck later, when the men adjourned to a "smoking parlor" of camp chairs beneath a giant sycamore. Reporters cordoned off thirty yards away heard the president's stentorian attempts at conversation:

Q. What do you do for recreation?
A. Oh, I eat and think.
Q. Ever take up golf? [*Louder*] Ever take up golf?
A. No. I'm not old enough.[62]

Harding gave up after that and retreated behind a newspaper. Edison elected to take one of his famed on-the-spot naps. Careless

Edison napping in front of Harvey Firestone and President Harding at Vagabond camp, 23 July 1921.

of his white linen suit, he flopped down on the grass and slept like a child. Harding continued reading, and then, in an oddly tender gesture, rose and laid the newspaper over the old man's face. "We can't let the gnats eat him up, now can we?" he said to a little girl watching.[63]

A KALEIDOSCOPE OF RUDE AWAKENING

By the fall of 1921 it was clear that the United States was in a roaring economic recovery. Housing starts doubled, automobile production cranked up by almost two-thirds, and inventory bloat sweated away. But for a reason not yet clear to Edison, the phonograph industry remained stagnant. Cabinet and records sales had always been the most profitable part of his business. So why was Thomas A. Edison, Inc., still encumbered with $2.3 million of recessionary debt? Encouraged all the same by an order from the builders of New York's Yankee Stadium for forty-five thousand barrels of his patented portland cement, he began to rehire factory and office workers. Charles responded with a letter that came close to grovelling.[64]

Yours truly has experienced a kaleidoscope of rude awakening. There were times when I felt you had stuck the spurs in so deep that I'd surely bleed to death. [But] since about last January I have not opposed in principle one solitary thing you have wanted to do. . . .

What I want you to believe is that for some time past any pride in the air castle organization I helped to construct during the past few years is gone—completely, absolutely, unequivocally gone. . . .

Also that I look to you for and only to you for leadership.[65]

"POOR DEARIE, HE is hurt clean thru," Mina told Theodore. "Papa does not realize how deep a hurt he has made."[66]

Edison realized only that his company had come near to bankruptcy.[67] If it had, he as the single largest shareholder would have been wiped out. His personal cash reserve at the end of 1921 was just $84,504.* Although that was more than the average American earned in a lifetime, it still represented a 50-percent loss over the last two years. Admittedly, some of that could be ascribed to the constant appeals of his older children for money (Marion's Swiss exile allowance; Tom's medical bills for a two-month recurrence of chronic "brain spasms"; William's dollar-devouring poultry business; even Madeleine leaning on him to get her car fixed).[68]

The trappings of wealth meant nothing to Edison. Were he not married to a woman who had been brought up rich and wished to stay that way, he would have been ready to plow every spare cent back into his business and live like a laborer. For a while in the 1890s he had done just that, crashing through more than $2 million, and he looked back on it as a period of acute happiness. His most urgent task now was to return Thomas A. Edison, Inc., to solvency. He had fired some seven thousand employees, and needed to cajole the rest into keeping it at the forefront of chemical and electronic technology.[69]

Or more precisely, what he perceived to be the forefront, in an age of change that was fast leaving him behind. "Everything is becoming

* The equivalent of $1.2 million in 2018.

so complex," he complained, ". . . so intricate, so involved, so mixed up."[70]

CAN YOU GET ME AN OLD INDIAN CYCLE

Edison's perception excluded the new phenomenon of commercial radio, which by the new year of 1922 was booming nationwide. Station 8ZZ in Pittsburgh had found, after its pioneer coverage of Harding's election, that its tiny, headphone-hugging audience liked to hear music between news reports. Renaming itself KDKA, it stepped up its biweekly "broadcasts" to an hour every night and greatly strengthened its signal, wildly exciting a schoolboy who picked it up as far away as Dixon, Illinois.*[71] A new magazine, *Radio Broadcasting*, hailed the "almost incomprehensible" increase in the number of people who spent at least part of their evening listening in. Purchasers of "wireless" equipment, it reported, were standing five feet deep in radio stores, while ready-made sets sold "before the varnish was scarcely dry."[72]

Edison could not long ignore the phonograph-threatening radio craze, with station WJZ-Newark starting up just eight miles east of his laboratory. But he discounted the appeal of a medium hampered by a burring, crackly "static" that was the aural equivalent of cataract vision. "Music is considerably mutilated, and always will be on Radio apparatus," he scoffed. "Can never be a true reproduction." He was prepared to bet that in the end, music lovers would prefer to hear their own selections in their own time, *sans* "atmospherics" and through a fine speaker.[73]

Accordingly, he redoubled his efforts to perfect the acoustic recording process, confident that his infallible sonic and musical instincts would put the Edison Phonograph Division in the black again. The work was bound to be exhausting, given the boneheadedness of everyone he dealt with, and the fact that he was about to turn seventy-five. But he had been through these developmental frenzies before, and invariably succeeded. All that was required was to work harder, sleep less, and starve himself more.

A stream of eccentric record-producer notes began to emit from his office:

---

* The boy's name was Ronald Reagan.

Benny
Can you get me an old Indian cycle using one cylinder controlled
by a knob
  I want the exhaust to beat time in a Jazz Orchestra
  Hitting wood is not loud or sharp enough—see me

Edison[74]

I AM STILL A BOY

"Papa has just come in after working all night," Mina informed Theodore a few days before Edison's birthday. "This is the second time in a week, and the trouble seems to be with the [disk] presses. It is just too bad for him to do it as he looks ashen this morning. He has not the strength for that kind of work any more."[75]

And a couple of months later, "He is getting to be so very deaf."[*][76]

Edison acknowledged that his disability was now almost total, but he was less bothered by it than those who struggled to communicate with him. "Do not mind it in the least," he scribbled on a letter of

Thomas and Mina Edison on his seventy-fifth birthday, 11 February 1922.

---

* A silent documentary filmed that summer, *A Day with Mr. Edison,* can be seen on YouTube at https://www.youtube.com/watch?v=ep5NGVOi6QE. It poignantly conveys his energy, tetchy decisiveness, and extreme deafness.

inquiry, "in fact I consider it an advantage as it has preserved me from the distractions of a noisy world."[77]

The "trouble" Mina cited was caused by his determination to maintain his thick, ten-inch Diamond Disc as the standard by which all other records were measured. This was in spite of the fact that most buyers were more interested in "hit" tunes than in high fidelity. (To Edison's horror, the younger ones often cranked up the r.p.m. of their turntables to make tunes sound zippier.) By quixotically trying to improve sound quality and reduce manufacturing costs at the same time, he succeeded only in slowing production down. He replaced the resinous wood-powder core of his blanks with unwarpable China clay, and then—after glossing them with four coats of varnish that needed time to dry naturally—grooved each under incremental steam pressure of one thousand pounds per square inch. He executed a pair of new patents to reduce blunting of originals in duplication, and water-cooled mold frames so that they would eject every "round" of twelve disks smoothly. The result was a per-press output of only 250 disks a day.[78]

Edison also tried to stave off the collapse of his Blue Amberol cylinder record business by eliminating its dependence on celluloid, an expensive compound due to the rarity of camphor. Retiring to his private chemistry laboratory, he searched for a new varnish formula. At once he became happy. It occurred to him that this was how he had started out at age ten: juggling flasks and retorts and breathing pungent vapors. "I am still a boy," he wrote a correspondent, "and still experimenting."[79]

A TOUGH RUBBER BAND

In October the British Colonial Office, presided over by Winston Churchill, announced that it would henceforth restrict the supply, and drastically raise the price, of rubber—a commodity over which it enjoyed a near-monopoly worldwide. This move, known as the Stevenson Scheme, was a reaction to a postwar production glut that made such rubbers as Malayan ribbed smoked-sheet crude cheaper than canned figs from California. Edison, with his phonograph and battery factories, was such a large purchaser of the polymer that he was planning to open his own rubber factory in Bloomfield, New

Jersey. But his need for pressed *jelutong* sheeting did not compare with the voracity of Harvey Firestone, whose plant consumed 10 million pounds of crude a month. And the two million Model Ts that Henry Ford expected to sell next year would need four rubber tires apiece just to roll out of the showroom.

Other manufacturers of tires and automobiles were just as dependent on the sap of Britain's East Indian plantations. Indeed, it was hard to think of an American industry, from transportation to textiles, that did not utilize rubber in some form. The United States consumed more than three-fourths of the world's entire output. What petroleum would one day be to developed nations, rubber presently was: a raw material essential enough to provoke armed conflict.

Germany's rubber famine had contributed much to the stasis of the Great War. In 1917 Bernard Baruch of the War Industries Board had rated rubber the most vital commodity that the government should stockpile in an emergency. Ever since then, various Cassandras had repeatedly warned Congress against being lulled into a false sense of security by the renewed abundance of foreign rubber. It was the word *foreign,* they said, that should give any thinking American pause. As one of them pointed out, "It is the only important commodity to modern warfare which we have not yet learned to produce."[80]

The principal alarmist was Firestone. He recognized, as did his fellow moguls at Goodyear, Geneva Tire, and B.F. Goodrich, that if rubber prices fell below seven cents a pound, it would bankrupt many British plantation owners and likely cause an industrial catastrophe. Firestone was prepared to pay any reasonable rate that would prevent *that.* But he was not optimistic that Britain would always be able to control its centers of rubber production. If the Russian Empire could be toppled by a small gang of Bolsheviks, how stable was Whitehall's loose clutch of colonies? What if Japan, which already had naval dominance in the Pacific, conquered all of Southeast Asia one day and wound a tough rubber band round the neck of the American economy?

These and similar questions possessed Wall Street and Washington— not to mention West Orange—as the spot market price of rubber tripled to twenty-three cents, and British authorities warned that they would halve the supply, if necessary, to increase it still further. Herbert Hoover, President Harding's hyperactive commerce secretary and "undersecretary of everything else," undertook to work through diplomatic channels for repeal of the Stevenson Scheme. Knowing his chances were slim (Britain needed every export shilling it could get, to repay its huge war debts), Hoover backed a legislative proposal by Senator Medill McCormick that the United States should consider establishing its own plantations abroad, and also research the possibility of growing rubber at home.[81]

Firestone did not altogether trust Hoover, believing him to be more interested in building power than in protecting an American industry. But he agreed that a serious attempt should be made to develop a domestic rubber plant, and he thought he knew the right man to undertake it. Thomas Alva Edison had recently been named "the greatest living American," in a *New York Times* poll. Here was a challenge to his legendary powers of discovery.[82]

COLD ON THE CONVEX

Edison soon received from Akron a large, leather-bound book entitled *Rubber: Its History and Development*. "I hope it will prove interesting to you," Firestone wrote in a covering letter.[83]

The hint was unnecessary, because he knew Edison had always been interested in rubber. During a camping excursion some years before, Firestone had been amazed at his friend's rubber expertise.[84] But until the current crisis, that knowledge was primarily technological. Edison still thought of rubber as something manufactured, rather than seeping from trees. It was the elastomer that protected the conveyor belts of his rock-crusher in mining days, and in its softest, solute form had kept his early records slick and durable. Hardened into lattices, it insulated the electrodes of his storage batteries. A "dark box" of ebonite, polished internally, flashed with the "etheric force" he discovered back in 1875. If he took a sheet of the same material and bent it in his hands, it became cold on the convex and warm in the concave.[85]

Firestone's lavishly illustrated volume left Edison little the wiser.[86] His laboratory work on a cheaper recording medium had already taught him much about the chemical properties of raw rubber. He knew how to vulcanize it by the Peachey process of double saturation with sulfur dioxide and hydrogen sulfide, and how to chlorinate it by predissolving crepe chunks in benzol. He could melt rubber in naphthalene and analyze it down to its most residual particles of manganese and copper. But how to produce it himself, and from what homegrown source of supply?[87]

Unaware that he was embarking on the last great quest of his career, he read some botanical studies of rubber milkweed, wild lettuce, and hemp, and marked up a monograph by H. M. Hall and F. L. Long, *Rubber Content of North American Plants*, underscoring one passage in particular: "If natural rubber is ever produced in commercial quantities in the United States, it will be taken from a plant which will give large yields on cheap land, and one which can be handled almost entirely by machinery."[88]

"SEVENTY-FIVE SQUARES"

The first indication that Edison's experiments with rubber were maturing into biochemistry came around the turn of the year, when he jotted an entry in his pocket notebook, after many pages of record production data:

Edison had no sooner jotted some preliminary biochemical ideas (*Slice ⅛ inch Milkweed into water with HCL or sol that prevents Coagulation Stirring all the time then 150 mesh screen used to seperate the milk from debris of the weed*) than the radio craze diverted his attention—so urgently, Firestone could only conclude that he had lost interest in rubber research.* Edison would not return to rubber research for another eight months, although occasional entries in his pocket notebooks indicated that he held it in mind.[89] He needed meanwhile to fortify his entertainment business against collapse. Sales of Edison Diamond Discs were slipping almost as fast as those of Edison record cylinders. The only way of competing he could think of was to invent something so dramatically new, in the way of recording and reproduction, that even young flappers, with their preference for rhythm over melody, would be seduced back to the phonograph.[90]

---

* In Washington, on 27 February, Firestone publicly called for American rubber independence. The House voted a $500,000 appropriation to research the project, and President Harding signed the McCormick Bill in early March.

"I have set my heart on reproducing perfectly Beethoven's Ninth Symphony with seventy-five people in the orchestra," he said at a ceremony to commemorate the forty-fifth anniversary of his invention. "When I have done that, I'll quit."[91]

Edison had a theory, more imaginative than informed, that sound waves remained turbulent and unresolved until they traveled 125 feet. To test it, he ordered his machine shop to cast a brass recording horn of that length, confident that it would capture the instrumental timbres of a full orchestra. A Brobdingnagian monster slowly took shape, section by conical section.[92] As each was finished, it was carted across to Columbia Street, where the Phonograph Division had a recording facility, for sequential assembly. The longer the horn got, the more it separated the performance studio—a two-story, barnlike structure, muffled with cowhide—from the adjoining "lathe shack," where wax masters were engrooved. Fortunately the surrounding lot was spacious enough for the shack to be moved east as far as necessary. The horn's great weight, increased by a total of thirty thousand rivets, was supported on a horizontal scaffold, and roofing and siding sheltered it from the elements. A telephone wire was strung up for communications between the two buildings, although talking proved simpler through the horn itself.

Its small end, which connected to a diaphragm and cutting apparatus in the lathe shack, was only three inches in diameter. The other orifice gaped so large in the studio wall that a six-foot man could stand and stretch in it, Leonardo-style. On first trial, the horn proved dismayingly directional. It picked up some instruments much better than it did others, depending on their position and proximity.

When Edison, who could not hear these imbalances, was told about them, he mapped out and numbered the studio's floor tiles, as if he were plotting a giant game of snakes and ladders.

"Have the saxophonist start on one and play 'Leave Me With A Smile,'" he said to his music director, Ernest L. Stevens. "Have him go through those seventy-five squares, and I'll go take a nap."

Stevens woke him up an hour or two later. Edison listened to all the takes and selected the one he thought best represented sax timbre. "Now take every instrument of the orchestra and go through the same thing."[93]

It took several weeks before he thought he found the ideal spot for every player. But when he channeled more than two or three threads of sound into the horn, they came out so ill-balanced as to torment any listener with normal ears. A test take of Saint-Saëns's contrapuntal *Prélude du Déluge,* played by the Haydn Orchestra of New Jersey, registered in the lathe shack as if all the string instruments had been stuffed with cotton. Only woodwinds floated clear. It did not seem to occur to Edison that the last remnant of his hearing was monaural, disqualifying him from any sense of sonic space. The purest tones he derived from his giant tube were those of Stevens playing solo piano, and a wind duo exquisitely performing Stephen Foster's "Old Folks at Home." *[94]

Even then Edison fussed over false harmonics that may have pinged only in his imagination. He tried to eliminate an echo that developed in the horn at certain frequencies, first by packing ice around the tube, then by warming it electrically. One day Theodore, who helped out at the plant during vacations from MIT, found him puzzling over a volume of acoustical mathematics.

"There's an easier way to do this," he said, knowing his father's difficulties with numerical theory. But Edison pushed him off. "I'll do it my way."[95]

SUGAR IN HIS SYSTEM

Overwork, stress, and compulsive fasting took their toll on Edison that winter, and his departure for Florida in mid-March came none too soon for anyone in the family. "Father certainly must plan to stay more than the four weeks he was talking about before he left," Theodore wrote Mina.[96]

She replied that Edison was "miserable" with an acute attack of diabetes. "His stomach bothers him considerable and the doctor says that it is the sugar in his system that makes a tingling in his fingers."[97]

Edison remained ill through the end of April, by which time his walk had stiffened alarmingly and he had to be monitored for pneu-

---

* Ernest L. Stevens survived into the age of stereophonic recording and remained proud of the Diamond Discs he recorded for Edison. "They sound exactly as if you're listening to a piano in the room," he said in 1973. "No overtones, no vibration or anything—the best piano record on the market, in tone quality."

monia. But he still limped daily to his garden laboratory for rubber-related experiments with milkweed and guayule, a southwestern desert plant that seemed likely to adapt to scientific cultivation. He was cheered to receive a report from Charles that battery sales were surging. "I am going to stop right here and enjoy it overnight," he said. Mina made him promise that when he returned north, he would pace himself and let Charles handle most of the general management of Edison Industries.[98]

He did, if only because it took him another month to clear his urine and regain some strength. Relations between him and Charles warmed to the extent they could resume their old exchange of "negro jokes." Then an old enemy, neuritis, struck. It nearly prevented him from attending Theodore's graduation in Cambridge at the beginning of June. "He's a good boy," Edison told a reporter, "but his forte is mathematics. I am a little afraid of that for he may go flying into the clouds with that fellow Einstein. And if he does . . . he won't work for me."[99]

Theodore Edison, 1924.

The remark, coming at a time of personal frailty, was the first public hint of Edison's desire to have Charles and Theodore inherit his business as chief executive and chief engineer, respectively. It also betrayed his concern that Theodore was an intellectual loner who might have different ideas. After the graduation ceremony he told Samuel Stratton, president of MIT, that he expected the young man to report for work in West Orange immediately.[100]

Tall, thin, peppery, and garrulous, a chess player and lover of classical music, Theodore hankered for the very career Edison disdained: that of a pure scientist. But he quailed at the thought of disappointing his father. As gently as possible, he wrote him a letter promising to work at the plant "for a good long time," if he could only return to college for a year of postgraduate study, and take the flanking summers off for travel. "I figure that this year and a half will mean a whole lot more to me . . . than will a year and a half of my services in the shop mean to you."[101]

Edison discovered that his youngest son was also his toughest, and Theodore got what he wanted.

"I SHALL NOT GO INTO RADIO"

By early summer, Edison felt he had perfected the Diamond Disc, after thirteen years of obsessive tinkering. "I sincerely hope that it is settled this time for good," Mina wrote her sister Grace. "The surface is better than ever & Thomas is happy."[102]

Actually, he had succeeded only in perfecting an obsolescent technology. Acoustic recording—his proudest achievement—could not advance beyond certain mechanical limits. No matter how responsive a diaphragm might be to the force of sound waves, and warm wax to their transferral, there remained higher and lower frequencies that could not be reproduced mechanically. For some time Edison had been hearing from Walter Miller, the general manager of his Recording Division, that scientists at Bell Laboratories were at work on a new method of electrical recording that registered a frequency range of 50 to 6,000 cycles, well over twice the reach of any acoustic system. AT&T, Bell's corporate owner, was looking to lease this innovation to phonograph companies when it became commercially available.[103]

Ironically, Edison had pioneered two of the three devices that made electrical recording possible (the vacuum tube encapsulating thermionic emission, and the microphone translating sound waves into signals), as well as inventing the phonograph itself. In further irony, his long-ago discovery of "etheric force" was powering the radio boom[104]—a word more apt than ever, thanks to the bass-heavy magnetic speakers now installed in commercial receivers. Radio was no longer an unamplified, earphones-only medium. On the contrary, the new sets produced, for free, such astonishing volumes of sound that the Big Four phonograph companies—Victor, Brunswick, Columbia, and Edison—were at a loss as how to compete. Apart from Edison's clearly doomed attempt to keep his thick disks and rosewood players at the top of the luxury market, there were only two choices: to cram radio receivers awkwardly into record players, or to unload an enormous inventory of acoustic machines and records in favor of a new generation of all-electric phonographs. Either way, there would be the cost of having to pay a per-record royalty to AT&T, unless some way could be found to avoid the numerous patents involved.[105]

Edison was not alone among phonograph executives in refusing to acknowledge that acoustic recording (with its unbalanced intake and uncontrollable output) was going the way of the dodo. Columbia and Brunswick also ignored AT&T's initial advances. Eldridge Johnson, who had built Victor up into a cash cornucopia with sales of $51 million a year, suffered a nervous breakdown over the threat of the electrical system. A year and a half of desperate strategizing began in Big Four boardrooms, while a new entertainment behemoth, the Radio Corporation of America, doubled its earnings to $55 million, and the nation's broadcast frequencies filled with sounds and sweet airs that muffled an old man's cry in West Orange, "I shall not go into radio."[106]

GUAYULE

That June Edison issued a new standing order at the laboratory for "Experimental work on Extraction of Rubber-like sap from various plants." It looked strange in a progress report that otherwise tracked the activities of men on machines.[107] Even stranger—except to his eyes—were the stream-of-consciousness jottings in his pocket notebook, with song titles jostling botanical, chemical, and other data:

Get lot thistle leaves Milkweed & Experiment
Strip 1st—wash off Latex & filler
When the Swallows Homeward Fly
My pretty pretty primrose
Guylue Plant Dry Weight 8 individual plants 3.638 lbs[108]

He soon learned how to spell *guayule* correctly, and focused on it, in preference to milkweed, as a researchable source of domestic rubber. It grew wild in arid conditions north and south of the Rio Grande, could be uprooted easily, and regenerated at once. By late summer, with the help of his beloved set of *Watts's Dictionary of Chemistry*, he had developed a method of extracting its huge polyisoprenoid molecules.[109] Working on some samples from Mexico, he calculated that he could get seven and a half grams of rubber out of every plant. "They occupy a horizontal space of about one foot, and probably grow 40,000 bushes to an acre," he wrote Henry Ford, who was as nervous as Firestone about the effects of the Stevenson Scheme. "This would give 680 pounds of rubber per acre, worth $183."[110]

If Edison's arithmetic was correct (not always the case), home-grown guayule rubber could conceivably sell at twenty-seven cents a pound—about the current price of foreign crude. But he warned Ford that extensive work was necessary to determine whether guayule, *Parthenium argentatum*, would adapt to plantation culture in the United States as well as *Hevea brasiliensis* had in the East Indies. He intended to try growing *Hevea*, a prodigious latex bleeder, on his Florida estate, although it had never flourished in North America. In the meantime, he was planting guayule seeds both there and in the greenhouse at Glenmont.[111]

That was normally the domain of Mina's roses and orchids, but since she had grown up in Akron, Ohio—"Rubber Capital of the World"—she should be tolerant of the invasion of a rubber-bearing species.

I REMEMBER, I REMEMBER

Edison's own half-forgotten Buckeye background thrust itself back into memory when Warren G. Harding died of a heart attack on 2

August 1923. The shock loss of the president, eighteen years his junior and to all recent appearance a superb physical specimen, was a reminder that life could be short. Accompanied by Mina, the Fords, and the Firestones, Edison attended the funeral ceremony eight days later in Marion, Ohio, then took the opportunity to show them his birthplace in Milan, only sixty miles north. The property belonged to him, but he had seen it only a couple of times since moving to Port Huron, Michigan, in 1854. A distant cousin, Miss Metta Wadsworth lived in it as caretaker.

The party set off along Route 4 in three new Lincoln touring cars, rolling on enormous, low-pressure "balloon" tires that partially explained Firestone's greed for rubber.*[112] Twin branches of the Huron River flowed in the same direction, as if escorting Edison to their confluence near where he was born. He had to brace for an onslaught of primary memories. For him, this was a journey both linear and circular, weirdly connecting Harding's grave in Marion to that of Marion, his eldest sister, in Milan—even as Marion, his elder daughter, planned to end her exile in Europe.

Edison liked to joke to geologists and paleontologists that he was too interested in the future to bother about the past. This was certainly true of his first seven years, only three or four of which he could recall.[113] Although they had been flush years for his father—and golden ones for Milan, with its canal basin and warehouses full of wheat—he associated the town with painful events: a public whipping, the loss of a fingertip to an ax, a swimming buddy left to drown, a teacher complaining that young Al Edison seemed "addled." By contrast, he had always regarded Port Huron as a place of ecstatic self-discovery.

Late in the afternoon, some beautiful hills ahead parted to reveal Milan.[114] Two or three thousand citizens were waiting to greet him in the public square, the scene of his chastisement some seventy years before. But the convoy proceeded without stopping to the highest point of the hogsback overlooking the basin. There stood the elegant

* Edison's personal dark green Lincoln is now an exhibit at the Henry Ford Museum in Dearborn.

little seven-room house Sam Edison had built with his own hands in 1841, its redbrick walls and tall, stone-linteled windows solid and straight as ever.*

From the front, it looked like a simple, one-story Federal structure. Only when Edison stepped down into the back garden did lower and upper floors disclose themselves above the slope of the bluff.

"Does the old home look familiar to you?" Henry Ford shouted into his right ear.

Reporters in the street had no difficulty hearing Edison's reply. "Yes, it does."[115]

He seemed more interested in the view downtown than in the house itself. But where a boy, seventy years before, had been able to survey a freshwater harbor jostling with barges, all his white-haired self could see now was a depressed townscape. Wild scrub and weeds traced the line of the old canal, long drained and silted up. A shabby cannery squatted in the curve of the road once jammed with wagons waiting to unload their grain. Milan and its waterborne economy had never recovered from the advent of the railroad.[116]

When at last Edison took his party into the house, he was astonished to find that Cousin Metta still used kerosene lamps. So much for his success at giving the rest of the world electric light. He led the way to the little northeast bedroom, feeling whatever a man feels when he contemplates the first walls he ever saw. Then he went out onto the porch and posed for a photograph, standing alone with the setting sun on his face.[117]

Metta said, "Tom, you'll have to go to the square and make a speech now, or the town will be heartbroken."[118]

There followed the usual ceremony saluting him as "the greatest inventor in all the world," and the usual disappointment when he politely refused to reply. "I'm too deaf to speak," he explained, and emphasized the point by cupping his ear when the brass band struck up "The Star-Spangled Banner."

Although Mina could see that he was tiring fast, he endured a long hand-shaking ritual afterward, saying over and over, "I remember, I remember, yes, I remember."[119]

---

* The Thomas A. Edison Birthplace Museum is now a National Historic Landmark.

THE DIVERSITY OF THINGS

After leaving Milan, Edison joined his fellow Vagabonds on a camping trip to the upper peninsula of Michigan. He remained in a pensive mood, and taught Firestone's twenty-five-year-old son, Harvey Jr., his theory of "memory swarms" that perpetuated human character. "The microscope cannot find them at all. . . . When these entities leave the body, the body is like a ship without a rudder—deserted, motionless, and dead."[120]

Edison flattered himself that he was talking metaphysics, but the tortuous lengths he went to to avoid using the word *God* betrayed, more than concealed, an aging man's need for some sort of divine reassurance that death was not final. Two summers before, at the Vagabond camp in Maryland, he startled Bishop William Anderson with the question, "Tell me what is to become of us and where are we to be when this short life ends?"[121]

That was, of course, one of the basic questions of human existence, and science could not answer it any more than reason. For most of his not-short life, Edison had been a disciple of Thomas Paine, about whom, around this time, he wrote:

> I have always been interested in this man. My father had a set of Tom Paine's books on the shelf at home. I must have opened the covers about the time I was thirteen. And I can still remember the flash of enlightenment which shone from his pages. It was a revelation, indeed, to encounter his views on political and religious matters, so different from the views of many people around us. . . .
>
> Many a person who could not understand Rousseau, and would be puzzled by Montesquieu, could understand Paine as an open book. He wrote with a clarity, a sharpness of outline and exactness of speech that even a schoolboy should be able to grasp. . . .
>
> He has been called an atheist, but atheist he was not. Paine believed in a supreme intelligence, as representing the idea by which other men often express the name of deity.[122]

Edison's self-identification with the great rationalist showed when he praised Paine the inventor. "He conceived and designed the iron bridge and the hollow candle, the principle of the modern central draught burner. The man had a sort of universal genius. He was interested in the diversity of things."[123]

If that suggested they shared a purely mechanistic pantheism, Edison now found himself going beyond the artificiality of manufactured "things" and studying natural ones more. *There* was a diversity more awe-inspiring than anything in Paine's purview. "The Book of Nature never lies; in it may be found lessons concerning almost every fact of life, death, and perhaps immortality."[124]

THE PROTOPLASM OF THE OLEANDER

Returning to West Orange in September, Edison tried again to save the acoustic phonograph—tinkering endlessly with his long horn, putting Theodore to work polishing diamond needles, and resisting the desire of both brothers to enter the radio business. But he devoted increasing amounts of time to studying the mysteries of biochemistry, often spending sleepless nights in his laboratory library and private experimental "Room 12" on the floor above. He authorized a bifurcated rubber research project that would operate simultaneously in West Orange and Fort Myers, and had no qualms about charging it to the corporate budget. Thanks to robust cement and storage battery sales, Thomas A. Edison, Inc., was once again in the black.[125]

"Until a man duplicates a blade of grass, nature can laugh at his so-called scientific knowledge," he assured a former employee, as he realized how much botany he had to learn.[126] The breadth of his erudition in other sciences was extraordinary, but it was also linear, in the sense that a common force—electricity—had linked his experiments in telegraphy, telephony, sound and light technology, magnetic mining, movies, and battery design. Now he needed to embrace systems of growth, morphology, and propagation that were in no way electrical.

It meant that he must saturate himself in the technical literature and work harder than ever before in his life, until comprehension came, and with it success. Surely in time he would be able to coax rubber from plant tissue, as he had once coaxed music out of tinfoil.

Edison secluded himself in his laboratory library, studying Alfred Allen's *Commercial Organic Analysis* to see which alcoholic solvents would give him the most viscous rubber extracts, and William H. Johnson's *The Cultivation and Preparation of Pará Rubber* for advice on the coagulation and purification of latex. He also read and heavily annotated Kurner von Marilaun's *Natural History of Plants*. Soon he felt knowledgeable enough to dismiss Frank Braham's *Rubber Planter's Handbook* as "a hash . . . untechnical," and challenge many of the conclusions in William Wicherley's *The Whole Art of Rubber-Growing*. Brailsford Robertson's *The Chemical Basis of Growth and Senescence* impressed him with its scholarship, which muted some of the things the book had to say about his own bodily decline.[127]

MILKY VARIETIES

For most of 1924 and 1925 the pattern of Edison's supplementary research in his private laboratories, and what came to be known as "the hay fever room" at the West Orange plant, was haphazard, depending on the availability of seeds and specimens. He seemed unable to drive past a New Jersey weed patch without jumping out in search of milky varieties. To eyes other than his own, his pocket notebooks for the period were a manic collage of disparate data: Latin plant names, lists of organic solvents, sizings of the pores in sponge rubber, mechanical drawings, geographical and climatological statistics, an acoustical analysis of the theme of Beethoven's *Moonlight Sonata*, urinalysis results, and (since he could not shake his interest in defense technology) deception systems to deploy in warfare.[128]

Although he told Henry Ford, early on, that he had devised "a very good method of extracting the rubber" from guayule, that was just his habitual way of imagining success far in advance, as a goad to himself.[129] From time to time he penciled the word PHENOMENON into his notebook, but often as not it denoted a hard-to-understand failure. He began to infer that finding an appreciable amount of good rubber in any plant other than *Hevea brasiliensis* was the botanical equivalent of getting blood out of stone.

Guayule was hopefully classified by the USDA as a domestic source of the polymer in an emergency. Edison sowed some beds with it at Fort Myers, but the seedlings came up so slowly that he calculated the shrub's reproductive cycle at four to five years—far too long for practical cultivation. Nor did he like the fact that guayule's rubber molecules were dispersed colloidally in the parenchyma of root, stem, and branch bark, as well as in the more crushable leaves. Unless he could devise a better extraction method than flotation (the whole shrub pulverized, then steeped in dilute sodium hydroxide until its woody dirt sank and rubbery "worms" swam up for skimming), he doubted that the soft, sticky end product could ever compare with, say, fine pure plantation Pará, or even the mats of Pontianak crude he imported for vulcanization in his battery division—tough and dark, slicing white with a moist sour reek.

For these reasons, he decided that guayule could never be grown profitably.[130] The inventor in him, forever wanting to be original, tested other polyisoprenic varieties less favored by the government. He was excited by the potential of *Cryptostegia grandiflora*, a fast-growing vine with exceptionally virile seeds. Its latex rubber content averaged only 3 percent, but he believed he could triple that by judicious breeding. The plant had one major disadvantage: its sprawling habit militated against a mechanical harvester. Next he considered the rubber fig tree *Ficus elasticus*, a white-sapped banyan that milked as easily as *Hevea*. However, it too was antimechanical, since it spread by sending out flying buttresses that swooped downward and rooted themselves, creating aisles and transepts irresistible to children but not to any tapping device Edison could conceive. That did not stop his planting a specimen of the giant variety, *Ficus benghalensis,* at Fort Myers, unaware that in another century it would become a green cathedral covering almost an acre of his estate.*

ONE GOOD ONE

To Mina's consternation, an ambitious, cigarette-smoking, "flapperish" Modern Girl appeared in West Orange in the summer of 1924, engaged to her favorite son. Miss Ann Osterhout was a twenty-three-year-old medical student from Massachusetts. "I love Ted to the limit," she assured Mina, although she had been reluctant to give up her dream of becoming a doctor for married life in New Jersey. Mina, who thought housekeeping (with plenty of servants) and motherhood were the twin peaks of femininity, half-hoped Ann would return Theodore's ring—being confessedly in love with him herself.[131]

Edison, in contrast, looked favorably upon Miss Osterhout, if only

---

* The banyan tree at Edison-Ford Winter Estates is now the largest in North America.

because her father was a Harvard biochemist.[132] He was sufficiently impressed with the young woman's own scientific bent (she had an avid interest in the new subject of colloidal behavior) to allow her to work with Theodore in the research department of his laboratory, hitherto exclusively male. It was about time, he told Mina, that he had "one good one" among his children's disappointing set of spouses.[133]

Marion's husband, Oscar Öser, was a case in point, his infidelity and post-Versailles hatred of Woodrow Wilson driving her home to America in the fall. Rusty-tongued after thirty years of speaking German, she thought of settling somewhere near Tom and William in New Jersey, but was not sure her father would approve. "I have been hungry for years for some sign of affection from you," she told him, adding that she regretted her teenage rebellion against him for marrying Mina. "If I had not loved you so much I would not have been so jealous."[134]

Mina fought, as she had done many times before, with aversion for Edison's "other family." They had never made it easy for her—Marion least of all, as the eldest and most rebellious of the three. Mina tried to make the refugee welcome in West Orange. Marion was grateful, but eventually chose to settle in a Manhattan residential hotel.[135]

Theodore and Ann married in the spring of 1925 and took an apartment not far from the laboratory. This confirmed, for the time being, Theodore's commitment to work for his father. Edison was so relieved not to lose him to the airless world of academe that he drafted a telegram of thanks in his most elegant calligraphy, forgetting that it would be transmitted as dots and dashes. Mina was touched by "the look of pride on his face" as he inked each letter.[136]

Her own feelings were less triumphant. For as long as her sons had remained single, she had felt of some motherly use. But now, with yet another bedroom at Glenmont empty, she lapsed into despair. "All my life," she wrote Theodore, "I had had love, attention, admiration without any effort on my part but now my attractions are diminishing, which were mostly looks, and I find myself floundering. Readjusting! It comes hard."[137]

Every now and again, goaded by some imagined slight, Mina would lash out, and then for weeks afterward be overcome with remorse. "I am terribly spoiled and it behooves me to take on Sack-cloth and ashes." Charles and Theodore were sympathetic, knowing that a large part of her problem was their father's genial absenteeism. Whenever she barged in on him—even in the midst of an experiment—he would melt her heart with his ready smile. But there was something exclusionary in his willingness to be interrupted, as if he had all the time in the world to wait for her departure.[138]

When Edison, in turn, barged in on an interview Mina was having with a pair of reporters over lunch, they were struck by the fact that though present, he seemed to be elsewhere.

"Mr. Edison has few friends," Mina told her guests after he abruptly left the room. "Because of his work he has had to live a great deal by himself and in himself." Admitting that the "intensity of his application" excluded her from what mattered most to him, she said she felt fulfilled all the same. "I have had a definite life job—the intimate service of Thomas A. Edison. And it has been worth everything I could give."[139]

"BEING THE FIRST"

With his ninth decade looming, Edison took steps to free himself from all responsibilities not directly connected to rubber research. On 1 February 1926 he executed a last will and testament, awarding his corporate holdings—the vast bulk of his estate—to Theodore and Charles. "My dear wife, Mina M. Edison, is already adequately provided for through gifts from me or otherwise." He bequeathed small cash amounts to three longtime employees, and directed that his remaining assets be put in a trust, the proceeds of which "shall be divided equally among my six children."[140] Simultaneously, he sold all the patents he still held to Thomas A. Edison, Inc., for $78,200.

It was a token sum, but he was enriching himself anyway as a major shareholder in his own company, which—along with most American industry—was piling up profits at a record rate. This encouraged him, at last, to hand over the title of president to Charles. His mood on finding himself a laboratory man again, after fifty-six years of executive responsibility, was ebullient. "The secret of staying afloat, Jimmie," he said to James Newton, the young manager of an estate adjoining his in Fort Myers, "is to create something that people will pay for. I didn't work at inventions unless I saw a market demand for them. I wasn't interested in making money so much as in being the first to invent something society needed. But if you do that, the money comes in."[141]

The distinguished astronomer and Nobel laureate Albert A. Michelson did not have to overhear this boast to block his nomination for membership at the spring meeting of the National Academy of Sciences. Prejudice against Edison as a mercenary, publicity-seeking technologist was strong in the organization, as it had been ever since he invented a theoretically "impossible" dynamo in 1880.[142] Nevertheless Robert Millikan, the chairman of Caltech, had the courage to stand up and—nervously balancing up and down on his toes—suggest that it was time for the great inventor to be recognized. "I am sure that no physicist would wish to oppose Mr. Edison's nomination."

Michelson rose from the front row and said quietly, "I am that physicist."[143]

That was enough for Millikan's move to be defeated. Edison was not surprised by the rejection, having invited it with his many jibes against "lead-pencil" theorists.[144] "An inventor is essentially practical." But he felt that he had made at least five genuine scientific discoveries in his career, and listed them in response to an inquiry from *Electrical World* magazine. They were the "Edison Effect" of electronic transmission, "now used in Radio bulbs"; the motograph principle, which smoothed the passage of a stylus over a charged electrolytic surface; the "etheric force" spark, subsequently credited to Hertz; the reversible nickel-iron galvanic cell; and the phenomenon of variable resistance of substances under pressure, embodied in his carbon button telephone transmitter.[145]

Of these claims, the distinguished physical chemist Michael Pupin accepted only the motograph as an original discovery. The others, Pupin said, had been either anticipated or were not purely scientific. "There is no doubt that Mr. Edison is a most resourceful genius in eliminating technical difficulties in the course of technical development of a scientific idea. I do not think that in this respect he has ever had an equal. [But] his true distinction lies in the field of applied, rather than true science."[146]

Edison, stung, dashed off another list of his scientific "firsts," in-

cluding the X-ray fluoroscope, mica insulation in dynamo commutator bars, and the electrochemical receptivity of tellurium in telegraph
recorders. In his haste he forgot to mention the tasimeter, and the
papers he had written in youth on magnetic conductivity and the pyromagnetic dynamo. But academic opinion denied him the honor he
thought he deserved.[147]

ONE HAS TO MOVE

Charles and Theodore were now at liberty to publicize some Jazz Age
initiatives that they had kept secret for fear of annoying their father.
These were a switch from acoustic to electrical recording (which they
neatly branded as "Edisonic" technology), an experimental issue of
twelve-inch long-playing records, and plans to enter the radio market
both as manufacturers and as producers—ultimately, perhaps, creating the company's own broadcasting network.

Charles felt that innovation was the only way to compete with
sonic rivals. Victor had adjusted to the radio boom by introducing its
hugely popular Orthophonic Victrola with built-in RCA receiver;
Brunswick was marketing an all-electric radio-phonograph, the Pan-
atrope; and Columbia was profitable after combining with its British
namesake and signing up for Western Electric technology. Alone
among the Big Four recording companies, the Edison Phonograph
Division was languishing, its instruments too expensive, its vertical-
cut disks and cylinders unplayable on other models, its sales force
unable to persuade young customers that Frank Lucas "the Accor-
dion King" was a better entertainer than Al Jolson. Only the Ediphone dictating-machine department was doing well—which was
ironic, because Edison had originally conceived the phonograph for
just that purpose.[148]

As far as he was concerned, his sons were trying to do too much,
too late. But Mina agreed with them that "one has to move with the
times." She risked divorce by installing a five-tube radio set at Glenmont, and was soon addicted to its coverage of political events. Edison could hardly object to her having something to fill the emptiness
of the big house when he was at work, and his deafness prevented him
from being bothered by the noise. But he was sure Charles would regret investing in the new medium. "In three years," he warned him,

"it'll be such a cutthroat business that nobody will make any money."[149]

With that valedictory, he let the long horn on Columbia Street go quiet, and returned to his experiments in polyisoprenic chemistry. In Mina's words, from the moment her husband turned eighty on 11 February 1927, "Everything turned to rubber in the family. We talked rubber, thought rubber, and dreamed rubber."[150]

PRIME FOR GREAT THINGS

Edison's obsession had burgeoned over the last three years in inverse ratio to the flow of imported rubber into the United States. Except for a freak, temporary shortage in 1924—more of a hunger pang than a famine—the Stevenson Scheme had proved a failure, due to the inability of the British Colonial Office to control competitive Dutch competition in the Far East.[151] Neverthless, Harvey Firestone was as evangelical as ever in trumpeting the slogan "America Should Grow Its Own Rubber," and Henry Ford, as well as Commerce Secretary Hoover, joined him in encouraging Edison to proceed with deep research.

Before announcing that he was now a full-time botanist, Edison had to submit to a birthday luncheon in Newark, attended by well over a hundred "Edison Pioneers"—grizzled veterans of the great days in the '70s and '80s, when their boss was inventing something new every two weeks. The seven-course menu, featuring cream of asparagus soup, shad stuffed with its own roe, and sweetbread patties, was notably easy on the gums.[152]

Such occasions were torture for Edison. He was repulsed by the overeating, tired of being told that he was an intellectual superman.[153] His disclaimer that genius was "one percent inspiration and 99 percent perspiration," had become a cliché, yet the Pioneers clung to it—except perhaps the few who could remember him when he had been young, and they even younger: Francis Jehl and William Hammer, witnesses to the night his first viable lightbulb had burned and burned and burned; Charles Clarke and John Lieb, who had helped him power up that first square mile of Manhattan in '82; and Sammy Insull, his former factotum, richer now than everyone else in the room, with the exception of Henry Ford.[154]

The latter had become so besotted with Edison that a newspaper publishing a photograph of them in conversation, mouth to ear, felt obliged to explain to its readers, "Ford isn't kissing his aged but still vigorous chum." Ford planned to establish a $5 million Edison Institute of Technology in Dearborn, Michigan.[155] It would feature a re-creation of his hero's first laboratory, stocked as authentically as possible, down to the last jar of *aqua regia*. One of the first items he solicited was a testimonial, suitable for framing, that the phonograph had been invented in that shed half a century before. Edison obliged with a letter showing he could still wield a pen beautifully at age eighty.[156]

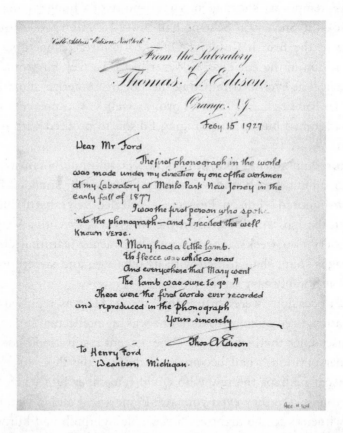

Ford began by purchasing the scrubby hamlet that had once been Menlo Park, New Jersey. Next, he carried off every brick and board of the original complex, and foraged the soil for experimental detritus. To Mina's mounting irritation, he also became a snapper-up of unconsidered trifles around Seminole Lodge, where his immaculate trouser seat was seen protruding from a barrel of old lamps. He proposed appropriating Edison's laboratory there too, in exchange for a modern facility more tailored to rubber research.[157]

Mina threatened to contribute her own dead body to this scheme. She had never been able to understand her husband's goodwill toward Ford or Firestone, and considered them both to be hucksters, capitalizing on his fame. It did not occur to her that maybe Edison was, in turn, capitalizing on their wealth and willingness to finance his work.[158]

Since 1912 he had tolerated Ford's obsequiousness as interest on the corporate loans and battery orders that flowed his way from Dearborn. It had been less easy for him, in recent years, to accept leather-bound volumes of *The International Jew*, a series of antisemitic newspaper articles in which Ford felt impelled to warn Aryans against such threats as the "Jewish Plan to Split Society" and "Jewish Jazz—Moron Music." Edison avoided embarrassment by having his staff noncommitally acknowledge receipt of the books for him. "I know very little about Mr. Ford's efforts. I do not want to get into any controversy about the English Irish Germans or Jews—even Yankees."[159]

Feeling himself thus unsullied, he agreed to accept Ford and Firestone as partners in establishing an official "Edison Botanic Research Corporation" to seriously address the issue of American dependency on foreign rubber.

WHERE LIGHT CAN'T STRIKE

Ford would have preferred to delay publicity for the new venture until all details of financing and staffing had been worked out. But Edison at eighty was as incapable of withholding big news as he had been at thirty. He had no sooner transferred to Fort Myers than he granted a series of "exclusive" interviews to various reporters and press agencies.

"Thomas A. Edison," *The New York Times* announced, "is working way past midnights in his laboratory here on an experiment which he believes will revolutionize the world's rubber trade and change the South from the land of cotton to the rubber planting center of the

United States." Other dispatches described the great inventor's dream of a patchwork of plantations, spreading north as far as Savannah and west into guayule terrain, that would supply all the nation's rubber needs in time of war. Right now Edison was reportedly designing a machine that would reap, crush, press, and suck the rubber globules out of a plant he might breed himself—some milksappy vine or weed or shrub that would grow fast, with minimal maintenance, and reproduce easily. His current focus of interest was a variety of *Cryptostegia* native to Madagascar. A shipment of rare seeds from that island was on its way, paid for by Henry Ford.[160]

Edison cleared ground for the new plantings across the avenue from Seminole Lodge, where his four-year-old grove of fig and rubber trees already stood tall and deep-shadowed in the humid riverside climate. "I am of the earth, earthy," he exulted. By late spring he had sixteen species of laticiferous plants under cultivation across nine acres, including one hundred *Ficus elastica* and 350 *Cryptostegia madagascariensis*. The latter plant thrived to an almost predatory extent, making it difficult to control and impossible to harvest mechanically. Herbaceous things, Edison realized, were less tolerant

Henry Ford, Edison, and Harvey Firestone in Florida, circa 1928.

of automation than the inorganic materials he had dealt with in the past. Beyond this problem were the paradoxes of nature. He puzzled over the green cambium beneath the dead bark of *Hevea*. "Why does the plant place chlorophyll where light can't strike it?" Rubber-bearing plants grown in greenhouses, where there were no insects, secreted less latex and more resin. Was latex some sort of bug repellent?[161]

He made full use of his dusty Model T, exploring the wilds of central Florida for specimens. He learned how to pull a leaf apart and examine the "gossamer threads" that dangled from split capillaries. ("If there is some rubber they will not sag but will stretch out one-quarter to one-half inch.") He anointed a freshly slashed *Ficus* with glycerine and found that it doubled the latex flow. Unfortunately the polyol also retarded coagulation. Pine trees were not laticiferous, but he tapped one anyway, to see how fast it dripped gum: in this case, one bead every eighty-two seconds.[162] By way of relaxation, at night, he studied rubber-industry periodicals, or sat at his desk doodling botanical sketches.

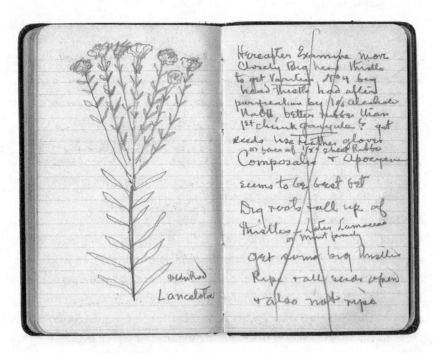

Edison botanical sketch, 1920s.

THICK WHITE SAP

Edison was unaware till he returned to New Jersey for the summer that the National Academy of Sciences had voted to honor him after all. At its latest meeting, an unidentified advocate more eloquent than Robert Millikan had shamed the membership by quoting a French academician's epitaph to Molière. "We cannot afford to say when Mr. Edison dies, 'Nothing can add to his glory, we can only regret that he does not add to ours.'"[163]

He accepted the diploma in writing with an especially graceful signature but otherwise showed no interest in it. By now he was so fixated on polymers that when Mina put a carnation in his buttonhole one morning, he asked if she had tested its stem for latex. He was soon seen at the New York Botanical Garden, abusing various species of Euphorbia "by cutting plant stems, catching the thick white sap in his hand and rubbing it to test its elasticity." Staff at the garden were honored to have so distinguished a vandal on the premises. John K. Small, the head curator, schooled him in the cataloging, preservation, and labeling of specimens.[164]

On 29 July Ford and Firestone officially established the Edison Botanic Research Corporation, with parallel field and laboratory operations ongoing in West Orange and Fort Myers. Its initial capitalization was $93,000, with the two magnates each putting in $25,000 and Edison insisting—over their objections—on contributing an equal amount. He hired fourteen field botanists and gave them each a Ford car and a tent, with orders to fan out across America and "cut every plant in sight" that might suit his purposes. Within a month, he was receiving dozens of express-mail specimens daily, each labeled by genus, finding date, and location soil type. He asked the agents of western railroads to check their rights of way for laticiferous-looking shrubs. Frank Stout, his estate manager in Florida, was told to add more varieties to the plantation there—*Ceara* trees from Brazil, *Landolphia* vines from Liberia, Indian figs, guayules, poinsettias, and scores of other specimens. Meanwhile Edison himself analyzed the enzymes and proteins of as many as fifty plants a day. Aware that most species required two to five years to maximize their rubber cell inclusions, he wrote his doctor that he was embarked on "a race with the Angel of Death."[165]

Mina noticed that he was losing weight, and worried that he was trying too hard to live up to what Ford and Firestone expected of him. But she understood that he was by temperament and disability insulated from praise or gratitude, "simply being impelled to do." At least he was no longer fretting about the Phonograph Division. "He is happy and busy with this rubber research. Just thinks of nothing else."[166]

That made Charles happy too. In a note addressed to "Father: Dept. of Rubberology, Edison Laboratories," he wrote, "I have come to the conclusion you really do want to concentrate on rubber and not bother much with the details of the business."[167]

It spoke much for Charles's confidence that he got away with such a tease. But Edison was, as Mina saw, back in the single-minded mode that had preceded his major accomplishments in the past. If he could live just as long as it took for a guayule shrub to grow, he might surprise the world yet again.

Or not. "I have worked too hard," he said to Marion when she came upon him one day, stone deaf and weary at his desk. In his youth, he had found the difficulties of electric light technology addictively challenging. Now those of botany often bewildered him. He told *Popular Science Monthly* that rubber research was "the most complicated problem I have ever tackled."[168]

At least he could confirm that guayule was not the plant he was looking for. After purifying a large quantity of its secretions and sending the coagulum to Akron for molding, all he got back was a set of fragile tires that cracked and split. Losing interest in desert species, he spent the rest of the year working his experimental way through the *Euphorbiaceae, Asclepiadaceae,* and *Apocynaceae* families, and assuring reporters, "We have only just begun."[169]

IMPALPABLE PULP

That winter Edison executed his one thousand ninetieth patent and his first ever in botanical technology, "Extraction of Rubber from Plants." It claimed to be unique in that it was designed (like his magnetic ore separator of forty years before) to precipitate what was valuable in material that was largely valueless.[170]

He described a two-stage process by which small, air-dried, rubber-bearing plants were first passed through heavy metal rollers "so as to open up the pith seams and break the bark," as well as the woodier stems, branches, and roots. The half-crushed mass was chopped into short strips and soaked until the bark and pith softened, then poured into a water-filled pebble mill, in which tumbling balls pounded the remaining solids, gently separating wood from pulp. In an hour or so the resultant slurry could be decanted from the mill through a fine screen and washed. "The woody material thus retained by the screen," Edison wrote, "is very clean and almost snow-white and in the case of some plants . . . can probably be advantageously used for making paper."[171]

The second stage of his process amounted to a refinement of the first, producing an "impalpable pulp" that slowly liberated and agglomerated all the rubber particles in the mix.[172]

Edison's principal claim of uniqueness for his invention was that it enabled the harvesting and concentration of laticiferous plants containing less than 1 percent of the polymer.[173] Two years later the Patent Office approved this claim, but by then he had lost interest in low yielders and fallen in love with one of the most ignored weeds on the American roadside. For the moment, he kept its name to himself.

HOLIER RED CLAY

"Won't you let us go into radio?" Charles pleaded, as he saw his father off to Florida early in the new year of 1928. Farther down the platform, six Botanic Research Corporation aides were loading a hundred boxes of biochemical equipment aboard the train.

"Well if you want to be a damn fool, go ahead," Edison replied. "You've got my permission, but I'm telling you it's no good."[174]

SEMINOLE LODGE WAS at its most beautiful that January. Mina rejoiced to see her flowers and orchids blossoming, and orange and mango trees bearing almost summerlike loads of fruit. Across McGregor Boulevard the rubber plantation looked equally lush, but the view was spoiled for her by the sight of ground being cleared for a new chemical laboratory, courtesy of Henry Ford. She braced for the disappearance, plank by plank, of the dim old studio Edison had built in the first year of their marriage. He himself was unsentimental about it, pointing out that he needed more space for his burgeoning team of "rubberologists."[175]

William Benney, one of the patient, bull-strong helpers Edison relied on for eclectic duties around the clock, was appointed laboratory superintendent. Francis S. Schimerka, an Austrian-born chemist, headed the analysis and extraction efforts, assisted by a professional botanist, a machinist, and five or six other functionaries of varying usefulness. These ranged from a teenage specimen collector to old Frederick Ott, whose ability to sneeze on cue had made him the first of Edison's film stars.[176]

Mina warmed toward Ford when he arrived in midmonth, full of plans for his projected Edison Museum. She could not help being touched to see the world's richest man worshipping the ground her husband walked on—literally, because Ford was adamant that when the old laboratory was transported, it should take a one-foot depth of Florida soil with it. He intended to do the same with the even holier red clay of Menlo Park, together with whatever fragments of Edisonia remained embedded in it.*[177]

Ford had no idea that he was giving birth to a science which would one day be called industrial archaeology. He knew only that his museum would not be complete without the acquisition of a collection jealously held by the Pioneers—models and machines from all periods of Edison's career, including a magnificent chronological run of his lightbulbs, assembled by William J. Hammer. The Pioneers wanted to display these treasures in a gallery of their own, possibly at the Smithsonian Institution in Washington, D.C.[178]

On another front, the Ford Motor Company faced a challenge from General Electric as to which firm should sponsor the anniversary on 21 October 1929 of Edison's breakthrough incandescent lamp. Given Ford's ability to lobby the Old Man by simply stepping across the lawn that separated their winter villas, it seemed an unequal contest. Yet GE had the powerful support of the Pioneers, and was prepared to reward them and the Smithsonian in return if "Light's

---

* According to Francis Jehl, Ford also managed to salvage "nearly all the timber of the old laboratory, with the doors and most of the window stiles."

Golden Jubilee" could be staged in its hometown of Schenectady, New York.

Edison appeared not to care less about the location of the festival or the enshrinement of his memorabilia. Ford therefore courted Mina, telling her that he was prepared to spend $5 million, if necessary, to achieve both ends at the Edison Institute in Dearborn. She undertook to hold a family conference on the question when she and her husband returned north for the summer. It was unlikely that Edison himself would participate. He no longer had any appetite for public honors. When told that he had been voted one of the three greatest men alive, along with Ford and Benito Mussolini of Italy, he waggishly imitated a Jewish pawnbroker, saying, "Vell, dey vas great men, yes, but de man vot invented interest vasn't no slouch."[179]

NO SUCH FIFTH ACT

The German writer Emil Ludwig visited Seminole Lodge in late February. He had just published a best-selling biography of Napoleon, but if he was thinking of Edison as another heroic subject, he found only an abstracted old man who had reacquired some of the wonder of childhood:

> I saw him step from the door of his flower-covered workshop. . . . He was wearing his white suit, his head was bowed. In his right hand he held a small plant, and his face was filled with joy. For the plant had yielded a good percentage of rubber.
>
> He led us to a rubber tree, which he pierced with a knife, then collected the white liquid that dripped from the cut, meanwhile talking to us in terms of figures and percentages. Then he led us back to his shop, where he showed us preparations he had made from the juice of all gum-producing plants, from oleander to honeysuckle. They had all been weighed and distilled. Lovingly he picked up a tube that contained sap from the leaves.
>
> "Here is the main thing—chlorophyll," he said.
>
> What a drama in the life of this man! Since Goethe's last years there has been no such fifth act![180]

By early summer Edison was able to note, "I have tested 2250 wild plants in Florida, of which 545 have rubber." He also designed a variety of crop-handling machines, including a leaf stripper that could denude twenty thousand oleanders a day. That was still too slow for him: "Must have 160,000 in 8 hours. 2 acres per man."[181]

Meanwhile his efforts to develop a more sophisticated extraction technology were blocked by the difficulty of finding a coagulating agent that precipitated anything less tacky than globules, impossible to vulcanize. But he typically regarded every failure as a step toward success, and told Mina that the past five months had been the happiest he had ever spent.[182]

She could not say the same, feeling again and again the loneliness of a wife waking up nights to find the bed beside her empty— sometimes not even slept in. "Father dear is certainly pushing the rubber idea for all it is worth," she wrote Theodore and Ann one sleepless morning. "He is over at the laboratory now working on his solvents, etc. and it is 2:30 A.M."[183]

Mina's desolation was augmented by letters from Theodore and Charles, full of their enjoyment of life, marriage, and work. She would have preferred family news of a more intimate sort, although she did not expect it from Charles. He was now thirty-eight, and Carolyn considerably older, despite a policy of celebrating her birthdays in reverse order. Mina's hopes focused on Ann, but that purposeful young woman seemed more interested in studying economics than stitching baby clothes.*[184]

Innocent or uncaring of her angst ("Ask me nothing about women—I don't understand them"), Edison continued to ponder the seed, the wind, the sapling, the tree, the branch, and the leaf.[185]

NO NEW "PUPS"

In early June the spacious new green-painted laboratory across McGregor Boulevard began to fill with staff and equipment. It was sequentially laid out, with the crushing and drying rooms servicing the chemical processing tables, and machine and glassblowing shops set alongside. Some visitors assumed it was modeled on Ford's famous

---

* In an unguarded moment, Mina voiced concern to Emil Ludwig over her husband's lack of a grandson. "After marrying twice and producing six children, none of them have perpetuated his name."

production line at River Rouge in Detroit. They did not realize that Ford himself had been inspired by Edison's "beltway" mining complex in the 1890s. And that, in turn, had owed much to the workbench layout at Menlo Park a quarter-century before—so much so that Fred Ott, looking around the long, two-bay-by-four room with its twelve double bays of tables and tubes, could be excused a pang of nostalgia. Except that this laboratory's cabinets were crammed with seed banks, solvents, slicers, grinders, percolators, Büchner funnels, screens, pans, and porcelain balls. Soxhlet extractors sprouted like glass reeds from the farthest tables, their bulbs refracting the tall windows that overlooked the plantation.

Thomas Edison brooding in chem lab.

Mina forbade her husband to work there just yet. The new roof was not yet covered with creepers, and she worried about summer heat beating down on him. It was time, she said, to return to New Jersey for the summer.

"I don't want to leave, but she makes me," Edison joked, as they boarded the northbound train on the twelfth.[186]

After ten months of organization, the Edison Botanic Research Corporation was now a bipolar but smoothly functioning unit.

Around its northern and southern ends swirled many institutions interested in "war rubber," such as the New York Botanical Garden, the U.S. Departments of Agriculture and Commerce, the Army-Navy Munitions Board, and the Ford and Firestone companies. Their various force fields were held together by Edison. No experiment could be undertaken, no new "pups" planted or seeds solicited, without his approval. Even senior scientists were expected to obey his constant stream of orders, oral, written, or telegrammed. To their relief, his disability prevented him from using the telephone.[187]

He made the twenty-four-man Florida campus responsible for analysis and domestication of foreign rubber plants. West Orange handled an ever-swelling inflow of specimens from Botanic Research Corporation collectors in the field. As though this were not enough, Edison spent $8,000 to turn his garden at Glenmont into a Florida-style plantation, growing over five hundred varieties of herbs and weeds in straight rows, to the modified pleasure of neighbors. It was no wonder that his assistant William Meadowcroft, overcoming two decades of deference, complained, "This rubber business seems to stretch out to infinity."[188]

A VERY BIG THING

Mina held her "important" family conclave about Jubilee planning at Glenmont on 20 August. Edison was represented, for once, by his eldest son. Tom had been brought back to the plant as an Edicraft engineer, testing electric toasters, amid general sympathy for his fragile health and marital distress. (Beatrice was cuckolding him with a handyman.) Charles, Theodore, and Ann Edison attended, along with Mina's brother John V. Miller. The only outsider was John Lieb, sitting in on behalf of the Edison Pioneers.[189]

Mina dominated the discussion. As far as she was concerned, General Electric had forfeited any right to celebrate her husband in 1892, when it dropped his name from its title. Lieb, pushing for the Pioneers, said that the company would compensate him now, to the tune of $100,000 a year, and build an Edison museum in Schenectady if he would give his blessing to a festival there. But Mina, still seething over a betrayal she blamed on Samuel Insull, rejected the offer. She confirmed that Henry Ford was prepared to spend as much as $15 million on his proposed Edison Institute, and approvingly cited the magnate's "desire to make this a very big thing—a national affair."[190]

A vote was taken, unanimously favoring Ford and Dearborn. Lieb mangaged to negotiate an agreement whereby General Electric would still sponsor "Light's Golden Jubilee," albeit in Michigan, while Ford simultaneously publicized the opening of the Edison Institute. In time the latter's museum would house the bulk of the Pioneer collection.[191]

Edison thus had to brace for apotheosis a year hence, when all he wanted to do was produce some homegrown rubber that did not stick to his fingers.[192] Already other Greeks sought to ply him with gifts. At the suggestion of Treasury Secretary Mellon, Congress awarded Edison its Gold Medal for "illuminating the path of progress." He said he was too busy to visit Washington to receive it, so on 20 October Mellon came north with an official party to pin it on him in his laboratory.[193]

A radio audience of 30 million heard the secretary praise him as "one of the few men who have changed the current of modern life and set it flowing in new channels." Edison thanked him for the medal, but sounded more pleased when presented with an artifact of duller metal: his first phonograph of 1877, deaccessioned with the utmost reluctance by the Science Museum in London.[194]

HURRAH!!!

Seventeen days later Herbert Hoover was elected president of the United States. One of his earliest votes came from Edison, who could not be sure that Hoover would support domestic rubber research in the White House.[195] Whatever the case, *he* himself meant to continue his botanical quest for as long as it took to succeed—or until either his body or brain failed.

The latter organ showed no decline as far as curiosity and retention of complex information were concerned.[196] But its tolerance of other points of view, never remarkable, was almost gone. Henry Ford's occasional snits were nothing to the spectacle of Edison roaring like a blast furnace when he heard—or misheard—something not to his liking. The bristling brows would contort, the always-jerky gestures become spasmodic, and the voice hoarsen, as if he were convinced that everyone around him was mentally deficient.

Mary Childs Nerney, a cataloguer hired by Charles to organize the papers of Thomas A. Edison, Inc., had just begun to work in the upper stacks of the laboratory library when she became aware that the founder of the company was below.[197]

> Never shall I forget my first sight, or rather sound of him. High voltage invectives winged their way past in mass formation and in solo flight. . . . I looked over the railing of the gallery.
>
> A man of medium height and stocky build stood by the inventor's desk. He was slightly stooped. He had a magnificent head to which his snow white hair gave a venerable look. His fine eyes flashed as he let loose his amazing vocabulary. Could it be—it was—the Old Man himself.[198]

The impression of stockiness she got was caused by Edison's quaint belief that any clothes that fit too well bruised the microvessels of his skin and caused internal damage. So he wore the largest and lightest possible suits, left his high collar loose, and scuffed around in shoes two sizes too big. In winter he declined to wear an overcoat, on the antithermal theory that stiff sleeves let cold air run up his arms. Instead he kept himself warm with two or three layers of underwear—even four in blizzard conditions. Despite the almost invariable shabbiness of his vests and trousers, his shirts were spotless. This was probably due to Mina, yet Edison had an odd love of fine linen, in the form of black satin string ties, Indian silk handkerchiefs one foot square, and enormous pongee nightgowns that billowed around him. It did not stop him from bespattering them with tobacco juice, or from rolling up his jackets to serve as pillows when he napped after lunch.[199]

That meal consisted, these days, of nothing more than a few crackers washed down with warm milk.[200] Dinner, when he bothered to eat it all, was equally frugal. He insisted that solid food dulled the brain, that he needed all his wits to adapt his extraction techniques to the biochemistry of thousands of specimens.

"I am always defeated by the tenacity of the solvents remaining in the rubber extract," Edison complained, despairing of ever getting a precipitate that would toughen enough to vulcanize.[201] Notwithstanding his successful patent for extraction by aqueous flotation, he found the "dry chemistry" of the Soxhlet extractor more efficient. In its tall, teetery, glass-tubed intricacy, it looked not unlike the Sprengel-Böhm pumps he had used in his early lightbulb experiments, striving for a perfect vacuum. Both devices used gravity and airlocks.

The Soxhlet stood on a hot plate that warmed a flat-bottomed flask of solvent (Edison tried ninety different formulas) enough to vaporize the liquid and send it up to a top-mounted, water-cooled condenser. As the vapor reliquefied, it trickled down into a cylinder stuffed with plant pulver and plugged with a thimble of porous paper. The solvent soaked through the pulver, absorbing rubber molecules as it went, forming a slightly syrupy filtrate that was then siphoned back into the warming flask. There the entire cycle of vaporization, condensation, dissolution, and dribble was repeated until virtually all the rubber had been leached out. Decanted into a porcelain drying dish, the syrup solidified into a "stiff tremulous jelly" that was never dry enough, or elastic enough, to please him. Physics kept violating the purity of his residue. He got different results according to how he stirred, kneaded, or washed the coagulum. Even weather, or the kind of light that played on the Soxhlet during extraction, seemed to affect its molecular structure.[202] But then, on 7 November, after applying some dilute sulfuric acid to the powdered leaves of a black mangrove, *Avicennia germinans,* Edison joyfully reached for his pencil and wrote

THE PLANT OF PLANTS

Success with one solvent on one species of plant, however, did not bring Edison appreciably close to the river of domestic crude that he dreamed of diverting into the nation's strategic reserve. At the beginning of 1929 he claimed to have examined fifteen thousand plants and gotten nothing better than a 6.91 percent yield from the milkiest. "I may say that the patience of Job has been considerably overrated," he told a reporter from *The Saturday Evening Post.*[203]

As his eighty-second birthday approached, he could identify with

Job's sufferings, both mental ("*Where shall wisdom be found? and where is the place of understanding?*") and physical. Exhaustion, self-starvation, and a restlessness akin to panic whenever his studies were interrupted took a toll on his health. He was racked with stomach cramps and suddenly began to look frail.[204]

There was no lightening of his workload when he went south. The laboratory in Fort Myers was as busy as the one in West Orange. "He just works and nothing else," Mina wrote to Theodore. "Leaves here about nine and crosses over there until six and comes home exhausted. After a spasm of pain lasting the last three nights about two hours he falls off to sleep . . . until 11:30 and then up again reading until 1:30 or two." Edison blamed his pain on indigestion. But as both he and his doctor were aware, he was suffering from chronic diabetic gastroparesis.[205]

Since life was short, and the regenerative cycle of most plants long, he decided to postpone any further attempts to improve the quality of his extracts, and breed for quantity instead. On 25 January Edison selected about forty plants that he believed might produce double-digit rubber. When he listed them by their Latin names, the genus *Solidago* appeared more often than any other. As a boy among millions of other midwestern children, he had known it as goldenrod—the wild plant whose yellow bloom, every August, warned that schooldays were about to resume. (Might the charged curvature of its tiny bulblike buds, balanced on their stems before exploding into flower, have reminded him of a similar configuration bursting into light, fifty "golden" years ago?) In terms of yield, Prairie goldenrod led the other species—Mexican, Tall, Sweet, and Pine Barren—by more than a full percentage point, save for one Florida specimen so anonymous that Edison simply cataloged it as "Fla. 201." All he noted now was that at 4.15 percent, it was perennial and a "good plant." He would need a few more months to identify it as *Solidago leavenworthii,* and conclude it was the plant of plants that most excited him.[206]

NO APPARENT ROSETTE

Seventeen days later Edison held his usual birthday press conference, this time in the charming hideaway that Mina had built for him, as a surprise, on the estate. Questions—mostly fatuous—were presented to him in writing. He could have answered aloud, for the benefit of a new breed of reporters, the "talking newsreel men." But his mood was subdued, and for most of the time he merely took each slip of paper and penciled a terse reply on it, as if dealing with nuisance mail. When asked what was his recipe for "a happy life," he scrawled, "I am not acquainted with anyone who is happy."[207]

An hour later, however, he was. President-elect Hoover arrived by yacht at the Seminole Lodge dock. Big and calm, tanned from a cruise in quest of marlin, Hoover was the picture of American success, at a moment when the polity he represented—ultracapitalistic, giddily speculative, awash in dividends—was at its apogee. For the next few hours, as he toured Edison's estate and rode with him in a motorcade through Fort Myers, he gave off such waves of pleasure and good nature as to belie his reputation for dourness.[208]

All the same, his visit indicated a concern among public figures that the "Father of Light" might die before his Golden Jubilee. Henry Ford used the occasion to announce his endowment of the Edison Institute of Technology. He released an architectural drawing that projected a five-gallery industrial museum bigger than Versailles and the Kremlin combined, centering on an overscaled replica of Independence Hall in Philadelphia. His reconstructions of Menlo Park and Fort Myers laboratories were not included. Subsequent news reports, issued by the publicist Edward L. Bernays, confirmed that they were to be the core of an adjoining "Greenfield Village," an attempt by Ford to re-create the small-town America his automobiles had done so much to despoil. The entire complex covered 542 acres.[209]

Edison was more interested, that spring, in the nine acres he had under cultivation across McGregor Boulevard—particularly a bed of *Solidago leavenworthii* root cuttings. After six weeks of potting and two more in the ground, they had sprouted as high as fourteen inches. At the end of March he calculated that if they maintained their vigorous growth rate, "possibly and probably [they] will give one ton of leaves for an acre."[210]

The more he looked at this species of goldenrod, the better he liked it. By May he concluded it had the best potential of all the plants he had tested. "No apparent rosette, fast grower, occupies smallest area, 6 x 6, without crowding . . . and mostly forty inches high though not yet in flower." The rubber was concentrated in its leaves, and they were luxuriant all the way to the ground, so if he bred a sufficiently straight stem, they could be stripped by machine. At an extraction rate of 6.9 percent, that projected a per acre rubber yield of 138 pounds, meaning that his entire plantation might generate about fifty-four Firestone balloon tires a year. This was a slight yield by *Hevea* or even guayule standards, but Edison was sure that by massive propagation of the best rootstocks, he could squeeze an additional 2 or 3 percent of rubber molecules into their leaf tissue and greatly increase plant size.[211]

Having taught himself how to crossbreed, he sat all day on his swivel seat in the laboratory, surrounded by dozens of goldenrods in flower. He delicately washed the anthers of some specimens, waited for them to dry, then brushed on the pollen of others, working with the patience of a miniaturist in watercolors.[212]

In June he drafted a detailed sequence of goldenrod treatment for the instruction of his Florida staff. Seven extraction processes were to be performed with linked dispatch. First, low-temperature drying of the leaves to prevent oxidization of the rubber cells; then prompt powdering and purification with acetone; resaturation with benzol for rubber removal; partial distillation of the resultant solution; application of a coagulating agent to the concentrate; and finally, after the stiffened crude was run through a creping roller, hydration to wash out its chlorophyllous resins.[213]

Edison was confident that the rubber of *Solidago leavenworthii*, being "not in the least sticky," could be vulcanized, despite a troublesome "X compound" that affected all his percolations and weakened the springiness of the resultant elastomer.[214] He said nothing publicly about his choice for the time being, knowing that if he did, the press would at once fantasize that he had discovered a miracle plant.*

I AM WRITING THIS

His stomach continued to bother him into the summer, and he found the only relief he could get from abdominal contractions was to adopt

---

* In May 2013 one of Edison's rubber notebooks from this period, featuring his drawing of a goldenrod in flower, was offered by the Paul Fraser auction gallery for sale at an estimated price of $120,000.

an all-liquid diet. Henry Ford flew in a quantity of iced pasteurized milk from Detroit, but Edison preferred the fresh product of his estate brown cow—high in sugar, low in butterfat, still warm from the udder. She was not always cooperative, so he arranged with the Dobbins dairy in Fort Myers for backup supplies of Jersey milk, ordering twenty-four pints whenever he left town on a plant-collecting trip.[215]

Mina noticed that his stomach was always worse when he was distracted from botanical work.[216] She dreaded the day he would discover that Charles and Theodore were failing in their efforts to modernize entertainment technology at Edison Industries. Exactly as he had predicted, they had come to radio, electrical recording, and the long-playing disk too late and at too great a cost. They were now risking their father's wrath by ending production of Blue Amberol cylinders, and the time was fast approaching when the Phonograph Division would have to shut down. Ironically, the one sonic instrument doing well—amid abundant profits elsewhere in the company— was the Ediphone dictating machine. His sons might have forgotten (though Edison had not) that he had visualized the phonograph as a business device from the moment it first spoke to him.[217]

"Father is a little worried and upset over things just now," Mina warned Theodore on the eve of her husband's return to West Orange. "So just let him get settled and realize that it is not your work really that is annoying to him . . . but his experiments in rubber. It might make him irritable and critical so just understand and if anything does seem amiss be patient and know that it will pass."[218]

Evidently it did not. Edison, for whatever alimentary or executive reasons, behaved so tyrannically over Charles's decision to manufacture a line of green "neutrodyne" radios—while also building a majestic stone mansion in Llewellyn Park—that Charles went to the extraordinary length of drafting a proxy suicide note.[219] Ostensibly coming from a friend identified only as "Williams," it read:

> Your son Charles is no longer a boy. Although not yet forty, he
> has literally worn himself gray in your service. His unswerving
> loyalty to you through the blackest days has been a rare and
> admirable thing. . . . He has handled a difficult job with
> imagination and judgement.

*The radio situation is dangerous and no one knows it better
than Charles. . . . If you force him to obey you, he is through. He
is a condition of such despair that I am actually afraid of suicide.*

*You are too great a man to fail him now at this critical moment
in his independent career. I urge you from my soul to let him fight
out his battle if it means sweeping away all you have done.*[220]

Before signing off with his friend's surname, Charles wrote,
"Charles does not know I am writing this, and will never learn about
it from me." His despair was not so great as to prevent him sharing
four versions of the draft with his mother. She in turn showed a copy
to Theodore and Ann. They did not take it seriously. Ann said nothing, and Theodore's only comment was that the author had written
"like a man" and sounded "very earnest." The final version, in Williams's handwriting, had the suicide line deleted.*[221]

THEY'RE ALL HERE

In late August Edison was felled by an attack of pneumonia that had
doctors worrying for his life. But he rallied by Labor Day, only to suc-
cumb once more to gastroparesis. This time his recovery was slower.
When he emerged, snow-haired and alarmingly thin, from Henry
Ford's private railroad car at a depot near Dearborn early on Saturday 19 October, it was as if he were stepping down onto the last
platform of old age.[222]

The Golden Jubilee was two days away, and Ford wanted to famil-
iarize—or refamiliarize—him with certain structures that had arisen,
like brick or clapboard phoenixes, among the lawns and new trees of
Greenfield Village.[223] The latter complex was still under construction,
and would be for years to come. But the Edison Institute's centerpiece,
"Independence Hall," stood complete to the last detail, pristine in the
Indian summer sunshine, ready to receive President Hoover and half
a thousand other VIPs on Monday.

Edison was already somewhat befuddled by the experience of ar-

---

* It is possible the letter was never sent. Edison would easily have identified its purported author as Charles Sumner Williams, a vice-president of Edison Industries, and
fired him for apparent interference in family affairs. Williams was known in the company
as "Charles Edison's right hand man."

riving at a depot that called itself "Smith's Creek" and exactly resembled—in fact was—the station where he had been dumped at age twelve by a conductor infuriated by his onboard chemical experiments. Memory, however, insisted that in those days the depot had been a stop on the Grand Trunk Railroad, sixty miles to the northeast. But this paradox was nothing to the experience of being led by Ford through a barbed-wire fence, with Mina at his side, and seeing at some distance ahead the six buildings that had once comprised Menlo Park—the dominant one white, double-storied, and many-windowed. Was he in Michigan or New Jersey? The very earth he

Menlo Park reconstructed at Greenfield Village, Dearborn, Michigan, 1929.

now trod—seven carloads worth, trucked in by train—was the same eastern clay he had walked on fifty years before.[224]

Edison's cognition flickered back and forth between place and time. When Mina tried to button his overcoat, he pulled away. "I'm all right, I can take care of myself. I'm just as young as I was when I worked there in the old laboratory." He nodded at the white building, which everyone else could see was new. "There's the old boarding house, just like it stood." In this case, he was correct—the hostelry where his research team used to live had survived and been transported intact. "And by golly if Henry hasn't moved in the stump of that old elm tree. I tell you, it's exactly as it was, every bit of it."[225]

At first Edison was animated by what he saw. But when he entered the laboratory and climbed a stair that took him back half a century into the past, he stopped talking. Ford had assumed, with all the naïveté of a surprise party planner, that the restorations he had paid for, whether authentic or duplicated with fanatical fidelity, would evoke nothing but delight in his hero. He was unaware that the effects of sudden déjà vu on an octogenarian might be more complex, not to say depressing.

Again, as in 1879, a long light-filled room opened out, its tables strewn with hundreds of tools and machines collected by the Edison Pioneers. Chemical cabinets glittered against the walls, and gas fixtures—not yet wired for electricity!—spiked down from the ceiling. At the far end stood the pipe organ that Hilborne Roosevelt had built for the entertainment of "the boys" during midnight "lunch." Edison gazed about him with an abstracted half-smile. He pointed at the three volumes of his youthful bible, Faraday's *Experimental Researches in Electricity,* and said in satisfaction, "In their old place."[226] A few straight-backed chairs, designed to discourage sleepiness, stood about. He crossed over to one and sat down.

A silence descended, as by some instinctive scruple, the rest of the party refrained from joining him. For several minutes Edison gazed around, his arms folded and his eyes dimming.[227] At last he became aware of a short man, almost as white-haired as he, waiting deferentially farther down the room. It was Francis Jehl, whom he had hired as a muscular twenty-year-old to help with the hard labor of operating mercury pumps. Jehl was now Ford's resident archivist and the last living witness of the night Edison's first viable electric lamp held its incandescence.

Edison had not seen him in eighteen years, but showed no more awareness of the man's half-reverent, half-hostile attitude now than he had then. He merely rose and led Jehl to a cabinet full of pharmaceuticals, asking, "Where d'you suppose they got 'em all? They're all here, every one of the chemicals I had at Menlo Park."[228]

So they were. Obedient to Ford's fanatical quest for faithfulness, Jehl had ordered them from the laboratory's former supplier, Eimer &

Amend, still doing business in New York. For a while Edison opened random jars, sniffing powders and licking crystals off the palm of his hand. Then he dawdled along the tables, picking up many implements that he recognized. "I could sit right down here and go to work with my old tools." At the request of a photographer, he did just that, scooping up some carbon paste with a spatula to impregnate several raw cotton threads, then kneading and rolling them between his palms until they were stiff and shiny, ready for baking into light filaments.[229]

"Francis, give me the kerosene," he said, assuming that a jar of the hand cleaner would be where it had stood fifty years before. It was, and so was the towel that Jehl brought him to dry his fingers.[230]

He was as cooperative as ever with the playacting of publicity, but at several unguarded moments, his eyes filled with tears.[231]

LIGHT 'ER UP, FRANCIS

In Chicago that weekend, electrical engineers employed by Samuel Insull readied a "sky writing gun," intended to flash Edison's name in capitals fifty feet high on the night of the Jubilee. For the letters to register, there had to be a screen of cloud; but if there was not, a smoke bomb projector was ready to provide one. Countless other technicians in seven continents prepared to celebrate the birth of electric light, amid general agreement that Edison's system had been the most seminal technological advance since the invention of printing. Amsterdam declared a full "Edison-light-week," and incandescent arches were cantilevered over the streets of Tokyo's Ginza district. An especially elaborate land-and-air link was strung between a studio in Berlin and loudspeakers in Ford's banquet room in Independence Hall, because Albert Einstein wanted to congratulate Edison viva voce, at the climax of the festivities.[232]

Dearborn's weather on Monday could hardly have been worse. Sheets of freezing rain beat down on the Edison and Ford families as they attended the president's nine o'clock arrival at the River Rouge transfer station. Hoover himself was soaked before he finished shaking hands, while his wife tenderly tried to shelter Edison under her umbrella.

Across the platform, an old-fashioned locomotive waited puffing, with coach, smoker, and baggage car attached. It was Ford's replication of the train Edison had ridden as a newsboy. He hustled his VIPs into the rear car, while press and White House personnel piled into the forward ones. Soon the train was jerking and coughing its way toward Greenfield Village. At a cruising speed of four miles an hour, that gave Edison—who had bucked up considerably after a day of rest—plenty of time to parody his old job. The car was furnished with a basket of fruit and candy for that purpose.[233]

"Candy, bananas, peaches, apples," he sang in a voice much hoarser than a boy's.

"I'll take a peach," Hoover said, producing a quarter.[234]

Edison's energy began to flag during the afternoon, when he accompanied the presidential party on a long, muddy tour of the village. At four P.M. in the laboratory, with Hoover, Ford, and a phalanx of photographers looking on, he assembled a replica of his original lightbulb of 1879. He worked with bushy-browed concentration, his hands still deft, aware that he would have to do it again in six hours' time, for an even more intimidating audience.

"Everything should be ready now," he said after finishing. "If only the vacuum is good."[235]

Jehl teetered up a stepladder, poured mercury into the Sprengel pump, and partially evacuated the bulb.[236]

"We used to seal 'em off too quick," Edison told Hoover. He gave the filament some battery current to burn off the occluded gases that remained. "Bring 'em up very high. We pumped out an enormous amount of air."

The president, an engineer himself, watched fascinated until the mercury driblets ceased to fall.

"Well, for all practical purposes that's enough," Edison said, peering closely at the outlet tube. "Seal it off."

Jehl heated, softened, and snipped the bulb's umbilical. When it had cooled, Edison reconnected the lead-in wires to the battery. A curl of light ignited in the glass. He sat back in his chair and beamed.

"Well, sir," Hoover said, "with that little invention, you've multiplied the light of the world a thousandfold."

By dusk Edison was again a frail old man. Stormy day became stormy night, compelling the Chicago skywriters to abandon their plans to glorify him aloft. Oddly, Greenfield Village and the facade of the Edison Institute were illumined with nothing but weak oil lamps. The dimness was intentional. After six months of mounting publicity, Edward Bernays was about to stage a coup de theatre that would eclipse anything the electric age had yet seen.[237]

At seven-thirty, live radio coverage of a white-tie-and-jewels reception for five hundred of Ford's most powerful friends—improbably including Marie Curie—began in Independence Hall. The lighting was almost as low as it was outside. It emanated from hand-dipped candles that threw shadows over the white walls and ceiling, while chandeliers and gold sconces stayed dark.

As they drank their cocktails under the Liberty Bell tower, they had the novel experience of hearing—along with millions of radio listeners around the world—the voice of an NBC commentator relayed from a hidden loudspeaker: *"Good evening, ladies and gentlemen of the radio audience. This is Graham McNamee speaking from Dearborn, Michigan, where Henry Ford and Edsel Ford are entertaining one of the most notable gatherings assembled in the annals of American history to honor Thomas Alva Edison."*[238]

McNamee was located not in the hall but six hundred yards away in the laboratory, along with the Fords, Edison, Mina, Hoover, and Jehl, who had set out another clutch of archaic lamp parts on a tiny table. Edison repeated his afternoon performance, minus the evacuation. He was deaf to McNamee's suspense-building hyperbole (*"Tonight is the climax of Light's Golden Jubilee, and what a climax. . . . the greatest tribute ever paid to any living man!"*) and could not understand why the ignition of his bulb had to wait until eight o'clock sharp, the official anniversary hour. It was not that anyone had been keeping careful time in 1879.

"Will it shine, or will it flicker and die as so many previous lamps have died?" McNamee mouthed into his microphone. *"Oh, you could hear a pin drop in this long room."*

"Light 'er up, Francis," Edison said.

Jehl respectfully declined the honor.[239]

Edison had to be helped to stand up. Again he connected the bulb to the battery, and again the filament incandesced. By prearrangement, Hoover pressed a button.

Back in the banquet room, chandeliers and sconces burst into radiance, "with the effect," one observer remarked, "of an eclipse running backward." The Liberty Bell pealed out the wonder of fifty years of electric light. Deerfield Village and Detroit lit up, along with many downtowns across the nation. A thousand rockets exploded in the rainy sky, while two Ford Tri-Motors took off from the company field. One released a coruscation of silverstar fireworks, and the other displayed, in glowing red signage under its wings, the name EDISON—written in the skies after all.[240]

AND NOW, LADIES AND GENTLEMEN

Ford had an automobile standing by at the laboratory to hurry his guests to Independence Hall. It was past two A.M. in Berlin, where Einstein was waiting to broadcast his congratulatory message. But Edison collapsed on the way. He was laid on a couch in an alcove off the vestibule, and Mina and the White House physician, Cdr. Joel T. Boone, rushed to attend him. After drinking a glass of hot milk, he revived enough to enter the banquet room, to a standing ovation.[241]

This reached a crescendo when Hoover, ignoring protocol, insisted that he take the seat of honor. Mina, seeing that her husband still looked pale, urged that the evening's speeches begin immediately.[242] The toastmaster, Owen D. Young of General Electric, agreed and spoke of "the vitality of spirit . . . aided by a little phosphorus" that had gotten young Al Edison tossed off a train at Smith's Creek depot nearly seventy years before. He compared it to the glow of radium, which touched off another ovation for Madame Curie. Walter Barstow, president of the Pioneers, informed the guests that just when Edison had illuminated the building they sat in, a memorial tower in New Jersey burst into light at the spot where he had defeated darkness in 1879. Barstow quoted the inscription on the tower's base: "The light once lit shall never dim, / But through all time shall honor him." Then to laughter, he added Edison's favorite saying, "All things come to him who hustles while he waits."

Young read out some congratulatory telegrams from Guglielmo Marconi, the Prince of Wales, President von Hindenburg of Germany, and Admiral Byrd, who was experiencing even worse weather in Antarctica. "And now, ladies and gentlemen . . . Mr. Thomas A. Edison."

Edison hitched himself up, to applause that was but a rustle in his right ear. His horror of oratory came not from shyness but from uncertainty about the volume of his own voice. Would those whom he could not hear be able to hear him?

What they did hear was a forceful if croaky light baritone, breaking when he pushed it too hard in the direction of a dangling microphone.[243]

"Mr President, ladies and gentlemen, I am told that tonight my voice will reach out to the four corners of the world. . . ."[244]

As far as he could see in the dazzling light reflected from gloss-painted walls and pilasters, it embraced a cathedral-like space lined with long tables and, beyond the transept, many more tables receding in perspective. Arranged along every front were some five hundred white-tied grandees and their women—Rockefellers and Morgenthaus and Rosenwalds and Kahns, Orville Wright and Lee DeForest, George Eastman and Will Rogers, university presidents and industrial tycoons. All of them were beholden to him for their telephones, dictaphones, stock tickers, and record players and movies, as well as the extra hours of work they could get out of their employees and servants, thanks to the billions of Edison bulbs now illuminating the world.

"I would be embarrassed at the honors that are being heaped on me on this unforgettable night," Edison continued, "were it not for the fact that in honoring me you are also honoring that vast army of thinkers and workers of the past"—his voice grew rough—"and those who will carry on, without whom my work would have gone for nothing."

Charles and Theodore and their wives, along with a large contingent of Mina's relatives, tensed with fear that Edison would break down.[245] But he drove himself to finish, his voice rising to a panicky yodel.

"This experience makes me realize as never before that Americans are—are s-s-sentimental, and this great event, Light's Gold Jubilee"—he began to weep again—"fills me with gratitude. I think—I thank our President and you all."

Edison turned to his right. "And Mr. Henry Ford, words are inadequate to express my feelings. I can only say that in the fullest and richest meaning of the term, he is my friend. Good night."

The last two words came out in a half-shout that took the last of his strength. He would not stay for the president's speech, and he had to be helped back to the anteroom by Dr. Boone. Lying there too weak to move, Edison heard nothing of Hoover's affectionately witty thanks to him for the gift of electric light.[246]

> *It enables us to postpone our spectacles a few years longer; it has made reading in bed infinitely more comfortable; by merely pushing a button, we have introduced the element of surprise in dealing with burglars. . . . It enables our cities and towns to clothe themselves in gaiety by night, no matter how sad their appearance may be by day. And by all its multiple uses it has lengthened the hours of our active lives, decreased out fears, replaced the dark with good cheer, increased our safety, decreased our toil, and enabled us to read the type in the telephone book. It has become the friend of man and child.*[247]

After this, Einstein's German tribute, broadcast through a storm of static, left few auditors the wiser, although some may have caught the words *Visionär, Ausgestalter und Organisator,* and at the end, his attempt at five words of English: "Good night, my American friend."[248]

FRAUGHT WITH GOLD

His apotheosis over, and his strength restored by two days of rest on the Ford estate, Edison returned to West Orange. "I am tired of all the glory, I want to get back to work."*[249] He arrived home in time to hear two pieces of catastrophic news. On Tuesday 29 October the stock market fell with such violence that ticker tape printers kept chattering far into the night, unable to keep pace with the volume of selling. And on that same day Arthur Walsh, a vice-president of Thomas A. Edison, Inc., announced that the Phonograph Division was ceasing all production, "in order to devote our great record plant to the production of radio. . . . This step is being taken regretfully because the phonograph for home entertainment was one of Mr. Edison's favorite inventions."[250]

---

* Edison's "glory" in October 1929 was amplified by the first honorary Academy Award, "in grateful recognition of your eminent service in the creation and development of the motion picture."

Neither denouement came as a surprise to Edison. During his birthday press conference earlier in the year, he had spelled out the consequences of overspeculation: "Ultimate panic. Loss of confidence." Now, and for the next forty-eight hours, many of the plutocrats who had sat listening to him the week before saw their wealth evaporating like acetone. None suffered more, in the long run, than Samuel Insull, who had leveraged his $2 billion empire of electric utilities to an extent that no amount of extra credit could save him from ruin, and eventual fugitive exile.[251]

There had been plenty of hints from Charles that the Phonograph Division was (like its founder) succumbing to sheer old age. The latest Edison electrical players and records were of superb quality, but it had proved impossible for them to succeed, given the company's well-earned reputation for stuffy design, dull repertory, and mediocre artists.[252]

Edison could only lash out at Charles in a final burst of impotent rage before returning to botanical research.[253] "I can't get my mind off rubber just now." At the beginning of December he announced his choice of goldenrod as the best guarantee of American rubber independence. After testing seventeen thousand plants, he was convinced it could produce a good tough polymer at sixteen cents a pound—less than the current spot price for foreign crude. But some years of development were needed to bring the weed to its maximum polyisoprenic richness.[254] Since his own years were obviously numbered, he would not stop for his usual Christmas at home.

On the sixteenth *Time* magazine reported:

Thomas A. Edison in a fringed muffler, Mrs. Edison, four servants, a dozen laboratory assistants and five carloads of laboratory gear and raw materials, all rolled southwards from New Jersey towards Fort Myers. . . . Inventor Edison, having celebrated the golden jubilee of his electric lightbulb, had signalized his annual winter hegira by an announcement that sounded fraught with gold.[255]

PART TWO

Defense

1910–1919

I N HIS SIXTY-THIRD year, Edison presided over an industrial complex so vast that only he knew what was going on in all its departments. "Say, I have been mixed up in a whole lot of things, haven't I?" he said, awed in spite of himself by the constellation of invention that swirled around his rolltop desk in the laboratory at West Orange.* From its farthest reaches nationwide—41 million Edison lightbulbs powered by six thousand municipal stations and one hundred thousand isolated plants—down to the pigeonhole in front of him, stuffed with notes of "new things" he meant to develop when he had time, the revolving mass had but one center of gravity. It was, however, expanding at a rate that threatened disintegration if his holding strength should fail.[1]

The six brick buildings that comprised his relocated laboratory of 1888 (itself an enormous enlargement of the old facility at Menlo Park) were now dwarfed by seven multistoried concrete structures, covering four city blocks. They and twenty-one smaller buildings scattered around the complex co-produced motion pictures, phonographs and records, primary and storage batteries, business machines, and chemicals. All were certified as fireproof—a vital attribute, considering the volatility of most of the materials that crammed them. The National Phonograph Company, as Edison's sound division was known, produced 130,000 records a day and six thousand phonographs a week, for an annual return of $7 million.† His movie factory shot out 8 million feet of nitrocelullose film stock a year. He employed

---

* Edison by the middle of 1910 had applied for 1,328 patents, or about one for every eleven days of his inventive career.

† Equivalent to $191 million in 2018.

more than 3,500 people, most of them highly skilled, few of them female,* all underpaid—chemists, cabinetmakers, talent scouts, diamond cutters, opticians, patent lawyers, screenwriters, lapidaries, machinists, and musicians, down to a little old Greek who did nothing all day in his lean-to except roast scraps of marble for lithium.[2]

Edison's commercial holdings extended far beyond the thirty acres of the West Orange complex. He owned, in addition to thousands of acres of mountain minelands upstate, a limestone quarry and the world's largest cement mill in the Delaware Valley, an equally immense chemical plant at Silver Lake, an electric car shop in Newark, a recording studio and showroom on Fifth Avenue in Manhattan, and a glass-roofed film facility in the Bronx that was larger than the Metropolitan Opera and shot two or three movies a week. He maintained agencies in London, Paris, and Berlin to handle the intricate marketing of his inventions under the patent laws of many countries, plus an

Edison film studio in the Bronx, circa 1910.

* Aside from movie actresses, the only women Edison employed were stenographers, packagers, and cooks.

export office at home that shipped tons of records and players weekly to places as remote as Madagascar, French Indochina, the Falkland Islands, and British East Africa.[3]

TOO NEAR THE SUN

Edison was, to outer percept and certainly in his own mind, a gifted businessman. Every brick and balance sheet that comprised the fabric and worth of Edison Industries derived from his inventive genius, dating back to the day he opened his first independent shop in Newark, forty years before. "I measure everything I do by the size of the silver dollar," he liked to boast—not choosing to remember the millions he had lost in a career remarkable for profligate spending and wasted opportunities.[4] Even when he restrained his natural impulsiveness and sought to behave like a canny Scot, he managed to lose again. Old associates still spoke bemusedly about the time Edison, arguing that he needed steady income, waved aside a huge British cash offer for telephone rights in favor of annual payments that he would have earned anyway, as interest on the lump sum.

One reason for his business failures was, paradoxically, the characteristic that had made him triumph so often over rival entrepreneurs: an impatient willingness, compulsion even, to take enormous risks. To this might be added such other quirks as his certainty that any idea, no matter how revolutionary, was realizable through sheer doggedness of experiment (witness the nine years it had taken him to perfect his alkaline storage battery), along with his habit of excitedly publicizing breakthroughs in advance, and his contempt for speculators, which did not stop him betting on himself. He was bored by what he called "the humbuggery of bookkeeping," while indulging an obsessive need to calculate costs to the last penny—although any accountant could see that budgeting was as alien to him as football.[5] Edison was both stingy with wages and overgenerous with bonuses when (on rare occasions) he felt that a colleague shared credit for an invention. He was personally honest and honorable, yet tolerant of whichever shady operator might help him beat another man to the patent office. If he had to choose between paying an overdue bill and emptying his bank account to buy a new piece of equipment, he did not see why he and his creditor should not face privation together.

Men who had to do business with him marveled at his inability to see money as anything of value unless it was invested in technology. Ralph H. Beach, president of the Federal Storage Battery Company, noticed him reach into his pocket one hot day for what he thought was a plug of tobacco. "In fact it was a wad of money bills that had evidently been there undisturbed for some time and possibly owing something to the sweat in his old alpaca pants." Edison gazed at the wad in obvious disappointment, and Beach suggested he could use some of it to buy a new hat.

"Yes I know, but I really haven't the time."[6]

If he had not traded away the securities of his greatest corporate creation, the Edison General Electric Company, to finance his greatest folly, iron mining, Edison might by now be as rich as Samuel Insull—the icily arithmetical assistant who kept him solvent when he tried, against his nature, to be an executive only. During those often perilous years, one of his English directors had said of him, "Like all creative and poetic minds he sees no difficulties where men of ordinary understanding require to make their ground good. This is one of the distinctive qualities of genius, their flight is so high and strong that they are apt to forget they may fly too near the sun and have their wings melted. This, I suppose . . . explains Mr. Edison's own pecuniary straits."[7]

The comment was made in private, and Edison would have scoffed to hear it. He flattered himself that he had, "beside the inventor's usual make-up . . . the sense of the business value of an invention." Yet as Insull's successor Alfred Tate observed, he was so arithmetically challenged he could not understand the figures in a balance sheet.[8] For all his laborious budgeting, he was cavalier about the worth of his own services. He either inflated them beyond reason or was naïvely amazed that any investor should offer him more than he expected. Nor was he a good judge of men, except in selecting (by attrition) laboratory staff who could stand his own pace of work. Optimistic and good-natured—as long as he was not crossed—he was quick to forgive associates who served him badly. Yet he had no sympathy for veteran employees who left him and fell on hard times, like Tate or Francis Jehl.[9] This had earned him a reputation for ruthlessness that was justified only in the sense that anyone out of his sight was thereafter out of mind.

Edward H. Johnson, the struggling milk carton merchant who, of all these forgotten men, had served him most faithfully and knew him best, cited haste as his most fatal business flaw. Edison was always in such a hurry to move from invention to invention that he would often leave a major one undeveloped, in order to experiment on a device as hard to sell as the tasimeter, which sought to measure interplanetary heat. Or he would lavish so many improvements onto something as marketable as his top-class Amberola record player that the Phonograph Works never had time to put the latest model into mass production. Over the years, by insisting on vertical control of all his companies and departments, rather than integrating them horizontally in approved corporate style, he had brought Edison Industries to the brink of financial collapse.[10]

Time had only reinforced the "opinion of many true friends," first expressed by Johnson in 1893, that "both the world and Mr. Edison would have been gainers if he had left the conduct of the purely business side of his affairs to associates of special commercial training and instincts."[11]

WHAT HUTCHISON WANTED

On 17 July 1910 a short, sleekly handsome entrepreneur of thirty-four escorted three naval officers onto the West Orange campus. Miller Reese Hutchison was no stranger at the gatehouse, having paid court to Edison for at least nine years in the hope of gaining business and other, more personal favors. By his own account, he had been a "worshiper" of the "Big Chief" since he was a boy in Alabama. After a privileged education in military schools and a Jesuit college, as well as polytechnic and medical institutes, Hutchison had flourished as an inventive electrical engineer, winning gold medals for a portable hearing aid that helped Queen Alexandra of Great Britain overcome her deafness.[12] His attempts to do the same for the most famous deaf man in America had failed when he realized that Edison *liked* being wrapped in a cocoon of near-silence. But that did not stop Hutchison from continuing to visit the laboratory, often with potential customers in tow, until he was so much of a persona grata that he was even allowed to conduct tours of the complex.[13]

Unlike the usual wheedlers who looked to Edison for jobs or endorsements, Hutchison was independent-minded and wealthy. He earned a fortune in annual royalties from one invention alone—the Klaxon horn, whose gargly, turkey-like call was one of the reasons the automobile was hated by peace lovers around the world.* His own luxury cars of choice were a Packard and a six-cylinder Pierce-Arrow, either of which put to shame Edison's little Bailey Electric Victoria phaeton.[14]

Exactly what Hutchison wanted by hanging around the laboratory nobody could yet figure, but the fact was that Edison, normally aloof from intimacy, had begun to enjoy his company and admire his social poise. They presented an amusing contrast when seen together—the younger man saturnine, elegant in dress and manners, smoking choice Havana cigars; the older white-haired and slovenly in suits that had often been slept in, chomping on a wad of the cheapest tobacco. Hutchison had an extraordinary voice, melodious and almost Levantine with its rolled *r*s ("thirty-fourr hundrred my-ils").[15] He addressed Edison as "Misterr Edi-sohn," and although always respectful in conversation, he was adept at scribbling the sort of smutty thigh-slappers that had the "O.M." rocking back and forth in his chair with laughter:

> *A young man with blackened eye was interrogated as to its cause. He replied, "I was kissing my girl good night and her elastic garter broke."*[16]

Hutchison's guests this day were young submarine commanders who wished to explore the possibility of developing the Edison storage battery ("Built Like a Watch, But Rugged As a Battleship") for underwater use.

Edison had experimented with defense technology before, working on a dirigible torpedo with W. Scott Sims in 1889, fantasizing aerial-dropped torpedoes and "dynamite guns" during the Venezuela crisis of 1895, and inventing an explosive illuminant during the Spanish-American War. But after that he had paid no more attention than the

---

* In 1910 Hutchison's Klaxon royalties alone totaled $41,921, equivalent to well over $1 million in 2018.

average newspaper reader to John Philip Holland's long struggle to sell the navy a revolutionary new weapon—the attack submarine. Only when President Theodore Roosevelt startled the world by vanishing into the depths of Long Island Sound for two hours in a Holland boat did Congress become seriously interested, and authorize the construction of seven bigger submarines in its defense appropriations for 1906 and 1907.[17]

That did not prevent Luddites in the Navy Department from continuing to resist improvements in submarine technology. "Innovation," Holland complained, "acts on these timid souls just as a sudden plunge into ice water." The Bureau of Steam Engineering was especially obstructive. It had mandated for years that submarines should be powered, illuminated, and controlled underwater by the force in its purview, instead of by electric motors connected to the Exide batteries Holland recommended: open-top, lead-acid cells that required lengthy recharging after the boat surfaced.[18]

To Hutchison, the bureau's objections were not unreasonable. A few weeks before, he had taken a dive with one of his current companions, Lt. Frederick V. McNair, Jr., in the submarine *Cuttlefish*. He was supposed to be demonstrating a marine speed indicator he had invented, but he became curious as to why McNair never dove at an angle exceeding fifteen degrees. The commander explained that a sharper inclination would cause the sulfuric acid in the vessel's battery compartments to slop over and "attack the steel plate of the main ballast tank." Should the tank then rupture under pressure, it would flood with seawater, the salt of which, split by the acid, would offer all on board a Hobson's choice between drowning or suffocating from chlorine gas.[19]

This had given Hutchison the chance to remind the navy that after a decade of ceaseless refinements, Thomas Edison had perfected an alkaline nickel-steel battery so benign that, even if keelhauled across Chesapeake Bay, it would throw off nothing sourer than a little iron chloride. What was more, it preserved steel through the use of noncorroding potash. Lighter and longer-lasting than lead units and almost completely reversible, it was designed for electric cars and trucks, but Hutchison felt confident that a giant version could be evolved for naval purposes.[20]

Hence the appearance now, at West Orange, of McNair and his fellow officers, traveling at their own expense. They spent two hours with Edison (Hutchison already functioning smoothly as interlocutor), briefing him on the liabilities of the lead-acid submarine battery and asking whether the capacity of his largest alkaline unit could be increased from 225 ampere hours to "several thousand." Going beyond questions as the meeting progressed, they told him "it was his duty" to help the navy solve one of its most dangerous problems.[21]

Edison agreed at least to try a range of experiments, and Hutchison wrote in his diary for the day: "The beginning of the Edison Storage Battery for submarine use and the beginning of my association with Thomas A. Edison."[22]

AN AUSPICIOUS DAY

Whether Edison saw this new challenge as "duty" or not, it had the attraction, always compelling with him, of difficulty.[23] The specifications his visitors had laid down were much more demanding than those for land batteries. They wanted an alkaline power pack strong enough to illuminate, operate, and maneuver a 105-foot, 273-ton vessel underwater for days on end. Less realistically, they hoped it would be deliverable at a price Congress could stand. This ignored the fact that the initial cost of nickel-steel cells was high, due to the extreme complexity of their internal design. The kind of monster unit McNair was talking about would cost at least $45,000—three times as much as a lead-acid equivalent.[24] If, as was likely, Edison took a long time to develop and test such a battery, its price could only soar, and Hutchison would have a hard time arguing that it would pay for itself in less maintenance and greater capacity.[25]

There was no question, however, that Edison stood to make a mint of money if he could produce a noncorrosive battery that would become standard on all U.S. submarines. Other navies around the world were sure to follow suit, and either order from him direct or buy foreign rights to his patents—a business he projected at $20 million a year.[26] But to win the vital first contract from Washington, he would need a supersalesman—charming, hyperenergetic, and expert in electrochemistry.

He did not have to look far. Miller Hutchison, who had the extra benefit of a military education, was confident of successfully lobbying everyone in the Navy Department, right up to Adm. George Dewey. On 25 August Edison authorized him to start doing so. Like a triggered dynamo, Hutchison spun into instant action. He was back within forty-eight hours to report that he had spent "an auspicious day" pitching the Edison battery, if not to Dewey, at least to Adm. H. Ingham Cone, chief of the Bureau of Steam Engineering. Cone had referred him to William Avery, his top electrical expert, and to R. H. Robinson, the naval constructor. Both men would have to pass on any matter of submarine redesign. Hutchison had not failed to treat Avery to a carriage drive in Rock Creek Park and dinner at the New Willard Hotel.[27]

Edison was sufficiently encouraged to say he would begin building an experimental big battery at once. But he could not give it much of his own time, because he was working on two major projects—movies that would talk and sing, and a supersmooth plastic to replace the hard wax on his current line of Amberol record cylinders.[28] Hutchison would have to supervise most of the prototype's development, and he should not expect an early success. Memories were still raw among Edison's chemical staff about the long agony of producing the A-4 automobile cell. It made them less than thrilled about the challenge of devising a unit ten times as powerful, soon enough to equip a new generation of submarines.

MERCY? KINDNESS? LOVE?

The advent of the second decade of the twentieth century, with the Western world increasingly, if unconsciously, preparing for a mechanized war, brought about such an angry escalation of the debate between science and religion that Theodore Roosevelt attempted to resolve it in an essay entitled "The Search for Truth in a Spirit of Reverence."[29] The philosopher William James virtually embodied the controversy, being both a psychologist and a quasi-mystic willing to believe, or at least speculate, that the soul was a detachable entity, capable of returning to earth after death of the body.

To Edison, this theory smacked less of Resurrection theology than of atomic physics. Insofar as he understood either, he had no doubt that Truth was scientific, and that reverence for it must therefore exclude faith. James's death that August prompted Edward Marshall, a feature writer for *The New York Times,* to ask Edison if he had ever discovered laboratory evidence of the soul.[30]

"Soul? What do you mean by soul? The brain?"

"Well for the sake of argument, call it the brain, or what is in the brain. Is there not something immortal in the human brain—the human mind?"

"Absolutely not."

They were sitting in Edison's great, dim laboratory library. He liked to keep it shuttered in summer, to block out the heat. But enough sunlight seeped in for Marshall to notice that the inventor's face, normally smooth and untroubled, broke into fine wrinkles when he pondered an abstract question. Sometimes he squeezed his eyes shut before answering.

"My phonograph cylinders are mere records of sounds which have been impressed upon them," Edison said. "Under given conditions, some of which we do not at all understand, any more than we understand some of the conditions of the brain, the phonographic cylinders give off these sounds again. . . . Yet no one thinks of claiming immortality for the cylinders or the phonograph. Then why claim it for the brain mechanism or the power that drives it? Because we don't know what this power is, shall we call it immortal?"

He insisted that the brain was a "mere machine." It could be willed to record an infinite number of things other than sound, but eventually always broke down and therefore could not be immortal.

"Is the will a part of the brain?"

Edison said he was not sure. "The will may be a form of electricity,*

---

* A century after Edison ventured this opinion, a team of American neurophysiologists endorsed it in their own language. They studied the movement of 1,019 neurons in twelve subjects exercising acts of free will and concluded, "There is substantial evidence implicating the parietal and medial frontal lobes in the representation of intention and in initiation of self-generated activity." Itzak Fried et al., "Internally Generated Preactivation of Single Neurons in Human Medial Frontal Cortex Predicts Volition," *Neuron* 69, no. 3 (February 2011).

or it may be a form of some other power of which we as yet know nothing. But whatever it is, it is material; on that we may depend."

He was willing to grant that certain aspects of "the thing we call life, or the soul," endured after death, in the chemical sense that all matter continues to exist through change. But change did not imply transference to another, imagined world. There was only the world of here and now, and it was plenty occult enough for any metaphysician. "Heaven"—the sentimental paradise where good souls were supposed to live forever—was nothing but "the ignorant, lazy man's refuge" from the mysteries that confronted him on earth. "There are plenty of savages, you know, who still call fire immortal."

Marshall asked how he, as a scientific materialist, would analyze the soul.[31]

Edison felt it could only be done at a microscopic level, by examining sentient units small enough to pass through glass. "Each part of us is made up of millions of cells," he said, and argued that the ability of skin to replicate an abraded fingerprint, for example, implied a physical if not spiritual uniqueness. His own individuality was nothing more than a "collection" of nervous or chemical or electrical impulses, just as New York was a collection of people "continually dying, moving away, and being replaced." He had about as much chance of an afterlife, from his cells holding together, as the city had of going to heaven.

Marshall tried repeatedly to get him to address the fashionable topic of psychic research, but Edison, his face wrinkling, would not be drawn. "I don't go into the psychic much," he said, dismissing its practitioners as "desirous of believing." With the naïve frankness that endeared him to reporters, he admitted to wishful thinking himself. Once, experimenting with a certain ore, he had selected what he thought were some random pieces to analyze.

"I assayed them very carefully, intelligently, and scientifically, and they showed 20 percent. I then took the same ore in quantities and crushed it, and assayed it, and it showed 17 percent. . . . I tried again and again, and each time the same result. I could not understand it. So I went again to the ore heap, shut my eyes, and grabbed, taking whatever pieces of ore I happened first to touch. . . . [They] assayed the same as the crushed ore. But if I took pieces while my eyes were open I always took bits which assayed high."*[32]

The fact that his eyes were capable, within milliseconds, of detecting a 3 percent differential in ore content did strike Edison as remarkable. He was more interested in showing Marshall a worn photograph he kept on his desk. It bore the stamp of a Russian portrait studio. "That's Mendeleev. See his autograph down at the bottom?"

Gazing at it as he talked, he explained that Mendeleev was the "great generalizer" who had discovered the periodic table of the elements. "Existing experimenters seem to be working, all of them, with details."[†] One day a man of comparable intellectual breadth would study the mysteries of the soul, and do so scientifically. "He will work through the material."

The phrase *through the material* obviously meant much to Edison, because he repeated it in a low voice. "That Russian is dead. Now where is his will? He was a very great man. His will was the greatest part of him. . . . What has become of that will?" He shook his head. "I don't know."

Marshall saw a chance to get him onto the subject, so far avoided, of the existence of God. "For that will to have entirely ceased when Mendeleev's body died would indicate a loose system in nature, would it not?"

"It would seem so," Edison said, "and yet nature's systems— nature's methods—are not loose. It's hard to figure out. Perhaps matter is getting to be more progressive. That may be it. But—God—the Almighty? No!"

The journalist had his scoop. Edison had expressed agnostic doubts in public before, but never atheistic ones. It was plain as he went on that, unlike Roosevelt and James, he wanted no compromise between faith and reason.

---

* Again, Edison anticipated the findings of modern neuroscientists that purportedly "random" selections are in fact deliberate choices made by the brain, often hundreds of milliseconds before it persuades itself to the contrary. See P. Haggard, "Human Volition: Towards a Neuroscience of Will," *National Review of Neuroscience* 9, no. 12 (December 2008).

† William James had voiced a similar complaint in his last book, *A Pluralistic Universe* (1909): "The over-technicality and consequent dreariness of the younger disciples at our American universities is appalling."

"Mercy? Kindness? Love? I don't see 'em. Nature is what we know. We do not know the gods of the religions. And nature is not kind, or merciful, or loving. If God made me—the fabled God of the three qualities of which I spoke: mercy, kindness love—he also made the fish I catch and eat. And where do his mercy, kindness and love for that fish come in?"

Edison continued to talk in this vein for some time. Marshall was struck by the forcefulness with which he expressed his convictions, his huge head tossing for emphasis and his face growing flushed. But he ended up grinning. "Nature seems to be a very undesirable member of society."[33]

"PROOF! PROOF!"

The New York Times made the most of Marshall's article, spreading it across the front page of its Sunday magazine section on 2 October under the headline " 'NO IMMORTALITY OF THE SOUL,' SAYS THOMAS A. EDISON." The reaction was extraordinary. Within two weeks, it generated at least three pamphlets, as well as a double avalanche of protest mail to the newspaper and to Edison himself. "Perhaps no utterance by any man of science," Marshall boasted two weeks later, "has created a sensation so extensive within the past decade."[34]

Fanning the flames, he interviewed one of the most outraged respondents, Dr. William Hanna Thompson, whose book *The Brain and Personality* Edison admired. Speaking as a man of faith rather than a clinical neurologist and former president of the New York Academy of Medicine, Thompson harrumphed that "people who do not believe in immortality are abnormal, if not pathological." He criticized Edison for making some "very unscientific" claims, while not refraining from one himself. ("The brain . . . offers a colored or distorted lens to the Personality that looks through it.")[35]

Preachers predictably assailed the blasphemer as an "intellectual anarchist" and ingrate who "seemed to have not one touch of disappointment" in denying himself the consolations of religion. "What metaphysical problems has Mr. Edison ever solved?" the Rev. Charles F. Aked, pastor of the Fifth Avenue Church in Manhattan, wanted to know. As an inventor, he had brilliantly dealt with technological ones. "But what has he ever done to entitle him to be heard as an authority on the human spirit and its relation to God?" The *Times* voiced similar sentiments in an editorial. It condemned the interview (which it nevertheless reprinted in six languages) as an example of how savants granted "preeminence in one or other domain of knowledge . . . often make amusing and even pathetic displays of overconfidence in their own judgment . . . in regard to matters lying outside the field of their special competence." The most eminent cleric to censure Edison was Cardinal Gibbons of Baltimore, who remarked that he had "maimed his own mind, just as Darwin did by a too one-sided exercise of its power."[36]

Edison was dumbfounded to have caused a theological scandal that had few freethinkers coming to his support. All he had done was voice, amid many expressions of uncertainty, a metaphysical opinion prompted by fifty years of empirical observations of nature. Nobody seemed to have noticed that he had discussed an array of purely scientific subjects with Marshall, including Hertzian waves, Brownian motion, attrition of memory in the Broca's fold of the brain, and the revelations of the ultramicroscope ("We may, eventually, be enabled to see the inner structure of matter"), as well as making an astonishing prediction: "The time will come when a man with a bad kidney . . . will be able to go into the open market and purchase a good kidney of some one else who has a good one . . . and have it inserted in the place of his imperfect one."

Replying to his critics in the pages of *The Columbian Magazine*, he said he could not help thinking the way he did.

> I honestly believe that creedists have built up a mighty structure of inaccuracy, based, curiously, on those fundamental truths which I, and every honest man, must not alone admit and earnestly acclaim. . . .
>
> I have not reached my conclusions through study of traditions; I have reached them through the study of hard fact. I cannot see that unproved theories or sentiment should be permitted to have influence in the building of conviction upon matters so important. Science proves its theories or it rejects them. I have never seen the slightest scientific proof of the religious theories of heaven and hell, of future life for individuals, or of a personal God. I earnestly believe I am right. . . . Proof! Proof! That is what I have always been after; that is what my mind requires before it can accept a theory as a fact. . . .

Moral teaching is the thing we need most in this world, and many of these men could be great moral teachers if they would but give their whole time to it, and to scientific search for the rock-bottom truth, instead of wasting it upon expounding theories of theology which are not in the first place firmly based. What we need to do is search for fundamentals, not reiteration of traditions born in days when men knew even less than we do now.

SOMETHING ELEMENTAL

The fire and brimstone Edison chose to bring down on his head could not have been better—or worse—timed to sabotage the publication, in early November, of a two-volume biography, *Edison: His Life and Inventions.* Its coauthors were Frank Lewis Dyer, general counsel for the Edison Laboratory, and Thomas Commerford Martin, former president of the American Institute of Electrical Engineers. Stoutly bound, gilded, and boxed, meticulously researched, respectful yet not sycophantic in tone, and featuring in facsimile the subject's signature of approval, it practically begged to be called "definitive." Volume 2 ended with an appendix of 156 pages, describing just nineteen of Edison's "twenty-five hundred or more" inventions.

After a month of silence *The New York Times,* which had harshly criticized Edison's religious views, conceded in an unsigned review that he otherwise deserved all the praise his biographers bestowed on him.[37]

> *There is something elemental about this man Edison, and the sense of it grows upon one in reading such a book. . . . More than once he has been called the greatest living American, and it is at least curious that a popular vote of one of the big daily newspapers and a poll of one of the electrical engineering journals gave him priority over all others considered. Somehow one thinks of Edison already as being as big and as typically American as his contemporary, Mark Twain, and as grouping with Lincoln and Franklin in largeness of mental mold and actual achievement along lines that are essentially of this nation.*

The reviewer noted that many pages, as revised by Edison, were essentially autobiography, "quoted in his own nervous and forceful language." At the same time it was technical enough to amount to a history of the last fifty years of electrical innovation, most of which Edison had dominated.

> *Yet his work does not stop short with electricity. It is interesting, even surprising, to learn that he is one of the largest makers of cement in America; to read of his gigantic efforts in the reduction of magnetic iron ore; to be shown that to him was committed the task of licking the early typewriting machine into practicable shape; to find that he invented the paraffin paper in which candy is wrapped; brought out the mimeographed copying press, and for eighteen to twenty hours a day applied his ingenuity against pretty nearly every real problem in the mechanical arts and sciences. At times the tone of the authors, as they dwell on one tour de force after the other, seems too eulogistic; but in the end it must be said that it would have been strange if they could have lived so close to such a versatile, indefatigable, resultful spirit without becoming enthusiastic.*

A PITCHFORK FOR CHRISTMAS

Edison's pious wife worried that the attention he was getting would encourage him to apply his "nervous and forceful language" to more and more subjects outside the range of technology. "It makes me sad to see him lose his old time simplicity," she wrote Charles. "He never used to assert himself but now upon all subjects he has something to advance. This miserable immortality idea is so upsetting."[38]

His heresies, though, were nothing new to Mina. She had realized as early as their courtship in 1885 that Edison was a cheerful infidel who delighted in shocking her pieties. Had he not been such a celebrity, even then, and so old-fashioned in begging for her hand, her equally devout father, a pillar of the First Methodist Episcopal Church in Akron, Ohio, might well have reserved her for somebody more comfortable in a pew.

After a quarter-century of going to church without him, she despaired of his salvation and concentrated on a new and more painful problem: what to do about her daughter Madeleine's infatuation with a Roman Catholic. They were about as opposite as any lovers could be—Madeleine a twenty-two-year-old flower of Bryn Mawr, bright,

witty, and impulsive; and John Eyre
Sloane, small, dour, bespectacled,
hard up, unsure of what to do with
himself at twenty-five except get up in
the dark every morning to go to six
o'clock mass.[39] His formidable mother,
Alice, shared the same conviction. She
would be more than a match for the
diffident Mina, if it ever came to an
argument over doctrine. But since
both women were determined Madeleine and John would never marry,
that prospect looked unlikely.

Madeleine shared something of her
father's irreverence (she teasingly
bought him a pitchfork for Christmas). Although not an unbeliever, she
failed to see why belief was so essential to John. His idea of a refreshing
weekend was to go on retreat and
ponder the remission of sins. This
bothered Edison less than the young

Madeleine Edison, circa 1911.

man's inertia. The "rising intelligence" he seemed to lack surged con-
trastingly in Madeleine, except that hers was bottled up by the social
conventions her mother imposed on her. "I want to be a free agent,"
she complained, desperate to excel at something other than household management and polite entertaining—which, along with regular
confinements, represented Mina's idea of fulfilled femininity. Madeleine felt a victim of male prejudice too. She loved to act in amateur
theatricals, but John disapproved. When, to her passionate excitement, a researcher in New York offered to pay her for help in processing the papers of E. H. Harriman, Edison curtly informed him that he
was affluent enough to support his own daughter.[40]

The unhappy lovers, held apart by prejudice and penury, occasionally breaking up yet always drawn back together, took refuge in a
private engagement that threatened to drag on for years. John kept

assuring Madeleine that all would be well, with divine assistance. "After all it is between God and you and me and you may be sure He will direct us when the time comes."[41]

THEY ARE PEACHES

Another Christmas present to please Edison was brought to Glenmont by Miller Hutchison on the evening of 21 December. It was an order from the Navy Department for the development and construction of a trial submarine battery, with no expenses spared. Hutchison reported that Admiral Cone had offered the use of USS *Cuttlefish* for tests once the device was ready. Clearly the high command was interested. Edison rewarded "Hutch" by introducing him to his family— a rare honor, since Mina seldom approved of mixing with the help.[42]

Early in the new year draft specifications were ready for the new S-type battery. It would consist of 102 alkaline cells, each standing five feet tall and containing nineteen positive and twenty negative tubes. When topped up with electrolyte, they would weigh 508 pounds apiece. This projected a power pack 25 percent lighter yet three times as capacious as the leaden mass that so inhibited the *Cuttlefish* underwater. The cells, moreover, were fully dischargeable, whereas acid ones could not stand a drop to one hundred volts without injury. They would last three times longer—ten years or more— and radiate enough electricity to drive a submarine 150 miles underwater at an easy five knots. Altogether the S-type battery would have an operating superiority of 92 percent, in Hutchison's excited calculation.[43]

Taking advantage of Charles's absence at college, he worked at becoming a substitute son to Edison, and succeeded sooner than he expected. He had special qualifications that endeared him to the Old Man and made it difficult for other sycophants at the plant to compete. One was his scientific understanding of deafness. Hutchison knew enough to feel the sense of exclusion behind Edison's lifelong pretense of being content not to hear most of the world's noises— a loneliness that had driven him, from puberty onward, to surround himself with the jostling camaraderie of men at work on machines. Hutchison also shared Edison's nocturnal energy, joking that he took vacations only "from 2 A.M. to 7 A.M."[44] Most congenially, both men had the positive imagination characteristic of inventors. They could see solid, finished engineering behind a pencil sketch and, with less rational clarity, the orders that were sure to result from any expression of buyer interest.

"If, within the next five years—the life maximum of lead cells in use in new boats—we get the battery business of the present submarine boats of the U.S. Navy," Hutchison wrote Edison, "we will sell 6,912 of the S-19 cells, equivalent [to] the gross business of $3,710,000.00." That was but a fraction of the billions of rubles, marks, lire, yen, and pounds that would flood West Orange if other navies of the world followed suit. At present, the United States had eighteen submarines, with ten more under construction. Germany deployed eight and Japan nine (both powers were rumored to be secretly and urgently building up the size of their flotillas), Russia thirty, France fifty-six, and Great Britain sixty-three. All relied on lead-acid batteries. Hutchison's next move was therefore to write, and have Edison sign, letters to the naval attachés at relevant embassies in Washington, inviting them to visit the laboratory and be briefed on S-type technology.[45]

He prided himself on his ability to write seductive copy, but the letters ("They are peaches") attracted just one representative of a major sea power, Cdr. Dmitri Vassilieff of the Imperial Russian Navy.[46] This was an unusual rebuff to Edison, whose celebrity was such that any chance to meet him was prized. It was his first experience of the extreme caution with which naval bureaucracies reacted to innovation.

Nor did he distinguish himself as a diplomat when Vassilieff showed up in full uniform. The commander was welcomed with an Edisonian harangue on Russia's ill treatment of Japan in the early years of the century. Afterward it was all Hutchison could do to persuade him to accept four smaller batteries for testing in Kronstadt. Rival attachés chose to wait and see how Edison's prototype performed.[47]

They were in no hurry, knowing his reputation for fanatical perfectionism. Undiscouraged, Hutchison applied for, and received from Edison, exclusive agency rights to market the S-type at home and abroad, for a 10 percent commission on each sale in lieu of salary. In the meantime he tried to stimulate interest in the battery by the old sales technique of making it sound hard to get. When a contractor to the Royal Navy, more interested in intelligence than business, inquired about its availability, he cannily replied that so many continental powers had approached him, "I cannot make any definite arrangements with you at present." And to Frank L. Dyer, he boasted that even Washington was going to have to restrain its impatience. "I told the submarine people we did not care to consider tying up with them at the present time, as Mr. Edison is averse to doing business on anything which he has not finished and thoroughly tested to his satisfaction. I therefore left the matter open, and their representatives departed disappointed, but hopeful."[48]

"INC."

Dyer was the most ambitious of Edison's senior aides. Having added the title of biographer to his other roles as company counsel and sales manager, he aspired to greater distinction. His hope had long been that *Edison: His Life and Inventions* would serve as a sort of pedestal onto which Edison the Colossus would step and turn to marble. Dyer could then reorganize the multicompany mess at West Orange (nobody knew quite what to call it) into Thomas A. Edison, Inc., and be rewarded with the new firm's presidency.[49]

Personal ambition aside, it was urgent that someone trained in corporate law prepare for a time when the laboratory would no longer be a fountainhead of Edisonian invention but merely the research arm of a great manufacturing concern. The National Phonograph Company, Edison's biggest dollar earner, was bled white from transfusions to unprofitable subsidiaries.[50] Two of these—Edison Storage Battery and Edison Portland Cement—seemed sure to pay off their enormous start-up costs eventually, because they each manufactured a superb product. So, for that matter, did National Phonograph: its cylinder-playing Amberolas were sonically superior to Victor's disk-playing Victrolas. But the evidence was unmistakable that consumers preferred the convenience of flat records. Victor's sales for the last year totaled $8.25 million to National Phonograph's $2.67 million. Even Edison agreed that he must adapt to disk technology or see the company go under. Over the last twenty years he had personally spent well over $4 million to hold his business empire together. Dyer made him understand that any further profligacy would bankrupt him. Only outside investment could help now, and the best way of ensuring that was to capitalize on his single greatest asset—his name.[51]

Reluctantly, he gave permission for National Phonograph to be reincorporated as Thomas A. Edison, Inc., and serve as the nucleus of a centrally structured organization run by one executive committee instead of dozens. He was made chairman of the board of directors, which then elected Dyer president, and Carl Wilson, chief of the Phonograph Works, general manager. Edison's compensation for the loss of his autocracy (and bets were off on how long he would stand for that) was the creation of an "Engineering and Experimental Department" allowing him continued control over all intellectual property issuing from the laboratory.[52]

The new company was registered on 28 February 1911. As a consolidation it was hardly complete. The battery and cement companies were kept off the books, for fear of frightening off investors. Still, a jumble of antiquated fiefdoms had been pushed part of the way toward the modern ideal of a professionally managed public corporation. "Inc.," as Dyer's creation soon became known, was capitalized at $12 million and employed about 3,600 people.[53]

YOU'D THINK HE'D HAVE AN APRON ON

Edison's metamorphosis into a corporation coincided with the silver anniversary of his marriage to Mina. They celebrated with a family party at Glenmont. The house was laden with gifts and flowers. Charles, now in his sophomore year at MIT, could not attend, so Madeleine put a photograph of him on the dinner table. "As a surprise," Mina squeezed as much of herself as possible into her wedding dress. Eleven-year-old Theodore at least was dazzled, and reportedly "fell in love" with her. After dinner Edison engaged everybody in a game of Parcheesi, which he had learned during his wandering days as a telegrapher.[54]

It was a treat for Mina to have her husband close for a whole evening, because he was in one of his periods of near-manic activity at the laboratory. Assigning Hutchison most of the responsibility for submarine battery development had by no means reduced his workload. He simply took on the weight of a project that he deemed much more urgent: restoration of the Edison Phonograph Division (as it was now known) to profitability. That meant coming up with a disk player and a compatible line of records that would not impinge on patents held by his competitors. "Such a drag on him," Mina complained to Charles. "Don't you think that Victor is gaining all the time?"[55]

She paid close attention to retail trends, and fretted about her husband's paradoxical refusal to accept that the cylinder was a doomed device. Soon Edison Records would be the only company, except for Pathé Frères in France, to keep producing it. His reasons for doing so were in no way sentimental. Having experimented with a telegraph-recording disk even before he invented the phonograph, he knew that the geometry of the cylinder made for more constant pitch. Its volute grooves spiraled from left to right without tightening or tapering, whereas disk grooves contracted toward the turntable spindle, slowing the speed at which the needle rode in its cut.[56]

When fresh copies of the Edison Amberol cylinder and the Victor Red Seal twelve-inch disk were played through the same horn, there was no question as to which sounded more natural, and which thinner and scratchier. But the very plasticity of the cylinder's hard-wax grooves allowed a sapphire stylus to carve away their definition, so its fidelity deteriorated, whereas the disk's gritty shellac, like most coarse things, endured.

Had it not been for an ill-reasoned injunction against him in 1905, Edison would have long before coated his cylinder blanks with one of the cellulose compounds he had pioneered in the early days of the phonograph era. Only now, having bought a competing patent, was he free to switch. But his star chemist, Walter Aylsworth—"one of the best experimenters I ever have known"—offered him something even tougher than celluloid. It was an infusible phenolic resin impregnated with a heterocyclic compound of ammonia and formaldehyde. The only trouble with this plastic was that its glasslike hardness, while admirably preserving the vertical incisions of the recording needle, caused the reproducer, or playback head, to ski-jump. Furthermore, each jump and subsequent touchdown shocked the sapphire point and fragmented the sound. Edison experimented successfully with a heavier reproducer, but the cost of continuous sound was more stylus wear.[57]

Searching for a varnish that would be kinder to sapphire, and good for both disks and cylinders, he patented his own "Composition for Sound-Records," a hard resin into which he melted crystals of halogenized naphthalene. The crystals felted together during cooling and solidification, giving him an end product of extraordinary tensility and strength. He laminated some tubes of German lignite wax stiffened with cotton flock, and found that both base and coating had the same coefficient of expansion—which meant none of the cracking that so often bedeviled plaster of paris cylinders during changes of heat or humidity. Delighted, he patented that process too. "My improved record," he boasted in his application, "is so durable that it may be dropped or even thrown on the floor with considerable force without encountering any objectionable injury."[58]

All the same, he had to acknowledge the superior durability of Aylsworth's plastic, which was nonfibrous and consequently smoother. If he could design a floating-weight reproducer it would not shrug off, it promised permanent fidelity. Since even Aylsworth had difficulty pronouncing the name of its condensed hardener, hexamethylenetetramine, they settled on the brand name Condensite. Edison allowed the chemist full patent credit and put it into bulk production at Glen Ridge, New Jersey.*[59]

Although he would cling to the cylinder as his preferred recording medium for another eighteen years, he was astonished at how good Condensite sounded when engrooved in disk form. But the disk had to be absolutely flat. If not, the vertical needle movement he insisted on (as opposed to the wall-banging, lateral swing his competitors preferred) tended to exaggerate any surface warp, with resultant distortion and pitch variations painful to his strangely sensitive remnant of hearing. This mandated an extra-heavy disk base, which might put off some customers.

Mina wished he would give up on the phonograph altogether, sell up, and retire. Seminole Lodge beckoned her, with its fruits and or-

* No sooner had he done so than a rival product, Bakelite, was developed by the New York chemist Leo Baekeland. For the next several years, Edison's Condensite Company of America had to fight a patent infringement suit from Baekeland, although its product was purer and harder. In 1917 Condensite was awarded priority, but by then Aylsworth was dead.

ange blossom, but Edison said he was too busy to go south. As it was, he seldom slept. "Papa is in the laboratory tonight working away at his disk," she wrote Charles on 6 March. "He cannot get the pure tones and it is worrying him greatly."[60] As always, Edison's solution to any problem was to pile experiment upon experiment (more than two thousand on the reproducer alone) until he dropped from fatigue. That same night Hutchison photographed the Old Man napping on a workbench in the chemistry building:

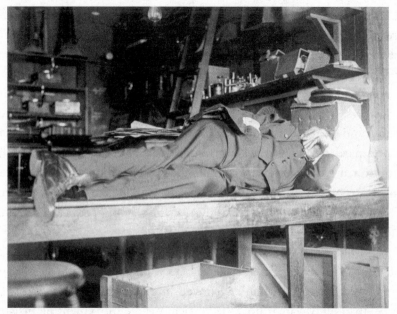

Edison asleep in his laboratory, 6 March 1911. (Photograph by Miller R. Hutchison)

If he did not come home at four-thirty A.M., only to bolt back to work after breakfast, he would stay away for days, until Mina went down and forced him to eat, bathe, and shave.

Edison was so driven, in both senses of the word, that when he had dry cleaning to drop off at the Armenian laundry on Valley Road, he would order his chauffeur to maintain speed and, *en passant,* hurl out his dirty suits. Rose Tarzian, the young immigrant inside the shop, got used to hearing the thump of the bundle on her screen door. Sometimes his vests would be virtually uncleanable, being burned by acid or spattered with wax. Beads of Condensite were of course unremovable. "You'd think he'd have an apron on, a leather apron!" she complained.[61]

After doing her best, Rose would return the suits to Glenmont, climbing the long slope of Llewellyn Park on foot. If Mina came to the door, she could count on a fifteen-cent tip. If the master of the house did, she got nothing.[62]

HE IS OPENING UP

Although Edison treated Hutchison and a *New York Times* reporter to a preview of a new "talking pictures" system he had devised, he said he was not satisfied with it yet. "I want to give grand opera. . . . I want to have Teddy addressing a meeting."* Characteristically, he waved aside its main problem, synchronism, sure that he would be able to fix it once he was finished with his recording project.[63]

He was, in any case, still involved in development of the submarine battery, if only because Hutchison sought the special intimacy with him that comes when men work side by side late at night at the same experimental table. They performed safety tests on the prototype, checking its water resistance and equipping it with an overcharge alarm. Its cells—suitcase-size six-hundred-pound steel jars—emitted large quantities of hydrogen at the beginning of their recharge cycle, so they subjected them to a series of internal detonations, to ensure the steel was thick enough to contain the explosive force.† "He is opening up to me more and more all the time," Hutchison wrote on 22 April. And a month later: "Long talk with TAE. Gradually making myself more valuable to him."[64]

Often as not, it was he who drove the Old Man home at dawn. Sometimes he would return at once to the laboratory, eschewing sleep altogether, in the hope that his dedication would be noticed. It was. Edison soon rewarded him with free office space in the storage battery building across Lakeside Avenue. That was enough encouragement for Hutchison to move his wife and four sons to a rental property in

---

* TR was at this time reemerging as a political force and possible candidate for a third term in the White House. Edison never made a talking picture of him but did issue four cylinder records of his campaign speeches in 1912.

† At one point Edison got Hutchison to detonate a cell inside a slightly inflated balloon, anticipating the modern automobile airbag.

West Orange, while keeping a beady eye on real estate opportunities in Llewellyn Park.[65] Discovering that Edison loved to be driven through the New Jersey countryside, he treated him and Mina to long jaunts in his Packard, and showed them where he lived. Mina began to be as suspicious of him as she was of Madeleine's rosary-rattling boyfriend. Was "Hutch," as he insisted on being called, also seeking a filial relationship with the world's greatest inventor?

As the weather warmed, she moved to detach her husband from his clutches, and her daughter from those of John Sloane. She declared that it was high time the Edisons—her Edisons, excluding of course Tom and William—took a grand tour of Europe. They should stay away all summer, exploring northern France, the Alps, the Danube valley, and Germany. Perhaps a reunion could be arranged with Marion Edison Öser, the half-sister Theodore had never seen, and Madeleine and Charles could hardly remember. Marion was married, happily by all reports, to a German army officer and lived in Mühlhausen.

Edison was not averse to a sabbatical. Having patented eighteen phonograph improvements in as many months and designed a prototype disk player (not to mention an apparatus to add sound and color to movies), he admitted to being exhausted.[66] He had not taken a real vacation since 1889, his winter retreats to Florida—when he took them—being little more than transfers from one laboratory to another. But he still would not leave until the disk player was ready to show at the National Association of Talking Machine Jobbers convention in July. After that he had yet another patent to file, for the reclamation of wash water from electroplate cathodes, and Hutchison needed help in the manufacture of nickel flake. . . .[67]

Mina, Madeleine, and Theodore sailed for France on 24 June, leaving Edison to follow with Charles at his leisure.

A GREEN SEVEN-SEAT, OPEN-TOP DAIMLER

"I want to get away and do a little worrying," Edison joked to reporters when he finally boarded the *Mauretania* at the beginning of August. He said he had been too busy to indulge that luxury at work, and now had at least two months to make up for it.[68]

On the first day out to sea, Charles came of age. He celebrated by smoking his first cigarette, a practice he would maintain as continuously as possible for the rest of his life. His father nudged him further into his future course with a gift of 505 shares in Thomas A. Edison, Inc., phonograph stock.* Charles by no means disliked the idea of taking the company over one day. But first, there were the fleshpots of Europe to look forward to. The astute Madeleine sensed that Charles had "Bohemian" tendencies. "He likes queer—out of the ordinary things and places."[69]

If by expressing a desire to "get away," Edison imagined he would escape his own celebrity, he soon discovered that it traveled with him.[70] He became a fixture in the first-class smoking room, where fellow passengers hung on his cigar-chomping monologues. One of them was Henry James. "The great bland simple deaf street-boy-faced Edison is on board and I have talked with him," James wrote in a letter describing the voyage. He added that he was touched by the kindness and sympathy with which Edison had asked after his favorite niece, a victim of depression. Charles was amused to see the author of *The Wings of the Dove* throwing paper darts on deck with his father, in an apparent investigation of the laws of aerodynamics. "The days of steam power are about to finish," Edison observed as it thrummed beneath him. "Flight will be the future transport phenomenon."[71]

News of his imminent arrival in London reached the highest levels of the British government. Arrangements were made to receive him in the House of Commons on the evening of 8 August, in spite of the fact that the Liberal government's Parliament Bill, the most controversial legislation since the Reform Act of 1832, was scheduled for debate that night. Hard as it was for Edison to comprehend, Britain was in a prerevolutionary state over the measure, which sought to deprive the unelected House of Lords of its power to control public spending.

His legal affairs representative in Britain, Sir George Croydon Marks, MP, met him after he checked into the Carlton Hotel and escorted him and Charles to Westminster. By order of the Speaker, they were accommodated in the Distinguished Strangers' Gallery of the Commons and looked down on a scene of extraordinary rhetorical

* This gift was worth $50,575 in 1911, equivalent to $1.4 million in 2018.

venom. The debate was dominated by Winston Churchill, who as home secretary of the governing Liberal party accused his old Tory colleague, Lord Hugh Cecil, of trying to provoke "riot and disorder" by resisting evolutionary change. Cecil declared that Prime Minister H. H. Asquith and his cabinet were "guilty of high treason" in seeking to overthrow a thousand years of aristocratic privilege. Arthur Balfour, leader of the opposition, defended a panicky amendment aimed at protecting the monarchy itself from violence at the hands of Liberal demagogues like David Lloyd George. Cries of "traitor" echoed across the aisle.[72]

Meanwhile Edison, unable to understand a word, pondered the deficiencies of the chamber's ventilation system. It was a hot night, and he asked if there were no means of cooling the room. Marks replied that generally, when temperatures became unbearable, iced water was sprayed onto the windows outside. Edison listened wide-eyed. "Do you tell me so? I could not have believed anything so stupid."[73]

Bored by the debate, which went on until after midnight, he sought relief on the terrace, where a steady procession of MPs paid homage to him. Lloyd George, a jovial little Welshman, asked if he could "invent something for getting bills quickly through Parliament." The Irish nationalist leader T. P. O'Connor received the same impression of naïveté that struck Henry James. "He is like a great schoolboy. . . . The simplicity of genius was never before so remarkably illustrated."*[74]

Next morning, as newspapers shouted news of the government's victory, a fire burned the Carlton Hotel to the ground. Since it was the traditional haunt of British bluebloods visiting town, Lord Cecil no doubt saw its immolation as symbolic of the vote. But by then Edison and his son had checked out, and were on their way to Folkestone and the ferry to Boulogne, where Mina, Madeleine, and Theodore awaited them in a green seven-seat, open-top Daimler.[75]

MR. VALENTINE

The car, rented for them by the ever-resourceful Hutchison, came with a chauffeur, as befitted a conveyance warranted to the British

---

* Edison was presented with a copy of the Parliament Bill signed by Asquith, Lloyd George, and other senior government members.

royal family.[76] Edison delighted in its size and power. One of his quirks was that he had never learned to drive, although he always sat forward, giving peremptory directions.

For the next six weeks he occupied this vantage point, enjoying panoramic views of northern France, the Loire valley, Burgundy, Switzerland, the Austrian Tyrol and Italian Dolomites, Hungary, Bohemia, Bavaria, Saxony, and Prussia. Every prospect was partially obscured by the American flag that Theodore—at thirteen a patriot contemptuous of all things foreign—insisted on flying from a pole on the front bumper. Edison protested in vain that the sight of Old Glory doubled or tripled the prices that innkeepers quoted them en route.

Sometime after leaving Paris, the travelers became aware that they were being followed by another car. It turned out to be carrying a representative from *The World* who, when confronted, protested that he was under orders to cover their every move in Europe. "You know we newspaper men have to do these things."[77]

Edison refused to have anything to do with him but nevertheless provided him with a scoop when the Daimler, approaching Interlaken, skidded into a ditch and had to be hauled out by horses. Thereafter the reporter could not be dislodged. The younger Edisons felt sorry for him, because he turned out to be underpaid, sickly, and endearingly incompetent.* They delighted in his name—Edward Abram Uffington Valentine—dubbing him "Feb. 14" for short. Edison yielded to their entreaties to grant him the occasional reluctant interview. When in the Austrian town of Bludenz he accidentally overdosed on strychnine† and nearly died, Mina nursed him with a calm competence that surprised her children. He was too weak to file his next report, so Charles wrote it for him.[78]

Impatient to get on, Edison showed little sympathy for the hapless scribe, nor for any of the dirt-poor peasants they saw as the Daimler continued east. (Charles remained behind, to follow with Mr. Valentine once the reporter recovered.) Their next destination was Buda-

* Valentine, Joseph Pulitzer's special foreign correspondent, published a novel in 1912 entitled *Hecla Sandwith*. It did not sell well.

† In 1911 small doses of this deadly poison were considered to enhance heart performance.

pest, which Madeleine and Mina both longed to see. If they imagined its remoteness would allow them to visit like any ordinary American family, they were progressively disillusioned. No matter how small the village or how large the city on either side of the Danube, from Klagenfurt to Vienna to Győr, crowds besieged them.[79]

In Budapest a wistful, bowler-hatted figure from Menlo Park days accosted them. It was Francis Jehl, whom Edison had sent abroad as an engineer twenty-nine years before and subsequently lost interest in. Jehl was now working for the Budapester Allgemeine Elektrizitäts Aktiengesellschaft, burdened with an invalid wife and resentful of the fact that he had never shared in the fabulous riches deriving from his pioneer involvement in incandescent lightbulb technology. As far as Edison was concerned, he hadn't either. But to Jehl, the great car, the royal suite in the Grand Hotel Hungaria, and the crowds in the street spoke gilt-edged volumes.

Alternately obsequious and querulous, he mustered the courage to tell his former boss that he, Edward Johnson, Francis Upton, and William Hammer felt unrewarded for their services. Johnson was now "a milkman," and Upton was "selling sand." Edison's response was to shrug and say they ought to have helped themselves.[80]

Jehl nevertheless proudly escorted him to Brünn on 13 September, in order to show off the theater he had fitted out with incandescent lights in 1882, on behalf of the Compagnie Continentale Edison. Next morning Edison and his party—amplified now by Marion and her German husband Oscar, traveling in Mr. Valentine's car—left for Prague. The Daimler got under way first, with a police escort, amid cheers and a shower of flowers. Jehl stood in the road with his hat off until it disappeared from sight.[81]

A SMALL BOY DASHING

Madeleine took an instant liking to her half-sister, a stout, vivacious forceful woman of thirty-eight. She also took to Oscar, an *echt deutsch* army officer who spoke little English. "He seems an awfully nice man, rather jolly or rather genial and good-natured and adores her."[82]

Over the years Marion had become more German than American. She even looked it, with her knotted blond hair and massive build. Fluent also in French, with an acute ear for opera, she had a sophistication that Madeleine admired, while lacking the means to live well. She was possessed of her mother's love of money, and like Tom and William, felt that Edison had never given her enough of it. She showed no envy of Madeleine's Bryn Mawr cachet, although she would one day remark that "my father's idea of an education was that I shouldn't get any."[83] In her heart, and on her lightly pockmarked face, Marion bore scars of Edison's neglect of her at age seventeen, when she lay ill with smallpox in France and he declined to visit her, or even write.* But her adulation of him was still so plain that he was flattered into a renewal of the love he had shown her after Mary Edison's death.

"My best invention?" he said, in response to a question from Oscar. "You own it."[84]

The couple's German came in useful—critically so—on Sunday 17 September, when the Daimler cruised into the Black Forest village of Lauf and smacked into a small boy dashing across the street. He was clubfooted and had probably tripped in an attempt to play chicken. An angry crowd collected at once. Edison and Mina jumped out of the car to try to prevent someone from lifting the boy, for fear he would choke on his welling blood. But then they saw he was already dead.[85]

It took some time for Oscar and Marion, traveling miles behind with Mr. Valentine, to arrive. The situation grew ominous until local police confirmed that the accident had been the boy's fault. Even so, Oscar's military bearing and Marion's interpretive skills helped quell the horror of the incident, enough for the Edisons to get some fitful sleep in an inn that night. An official inquest the next day cleared them of negligence, and Edison left four hundred marks behind to compensate the boy's indigent mother.[86]

The remaining ten days of the tour were hard to enjoy. Mina had been hoping that Madeleine would go to Italy for the rest of the year and, with luck, forget about John Sloane. But she had no heart to protest when the girl said she preferred to go home.[87]

Before kissing his other daughter goodbye in Dresden, Edison sensed her desperate desire for a car. He told her to buy a Mercedes-Benz and charge it to him. From Berlin, he sent the Daimler back to Paris, then took the rest of his family by train to Hamburg. On 28

* See Part Four.

September they embarked in the German liner *Amerika,* and the first person they saw on arrival in New York, snapping photographs as they came down the gangway, was Miller Hutchison.[88]

ANOTHER AND GREATER ARCH

Edison's oracular tendency, which had grown on him since his pronouncements on the afterlife, did not permit him to come home without letting American reporters know exactly what he thought about modern Europe. "What is my impression of the people on the other side? Well, I'll tell you. For the most part they are too thick—too wide."[89]

In a variety of interviews, he criticized the "lazy" English for consuming too much "beef and porter," dismissed the illuminations of Paris as "twilight" in comparison with those of Broadway, and complained about the feminine fashions he had seen there and in Prague. "Primary colors in a toilette are a sign of an undeveloped sense. . . . A woman's skirts should bow in curved lines from her hips." As for the urban scene east of the Rhine, "Something is wrong with the German aesthetic lobe. They ice their brains with too much beer. The result is beer architecture."[90]

This last comment outraged the scientists, writers, and industrialists who had welcomed him to Berlin, one with such respect "I felt inclined to kiss his hand." Sigmund Bergmann, who had known Edison for more than forty years, sent him an article entitled "Eine Bier-Phantasie Edisons" and asked him to deny his reported words, "so that I can pacify the people here, who are looking at this matter very tragically."[91] Edison cabled a quasi-corrective letter and tried to atone for his gaffe by praising Germany's phenomenal industrial growth, especially in the field of chemical manufacturing.[92]

He sounded more serious when he spoke about the belligerent nationalism that he had sensed in every country he visited. "They're all thinking too much about war—forts and guns everywhere and everyone on the lookout for spies." Even in Switzerland, where a man had been shot dead for picking strawberries on the wrong side of the Swiss-German border, there was fear of a collapse of international order. "I'm not a Malthusian," he told a reporter from the *Pittsburgh Telegraph.* "I don't believe in the agency of war in keeping down the population, though I think that if France had another tussle with another country, its wonderful intelligence would go far to meet supe-rior brute force." It was clear which "brute" power he had in mind. He worried about the extent to which war, and the glorification of it, permeated European history, saying that was why he had never been impressed by the Arc de Triomphe. "I always see beside it another and greater arch, thousands of feet high, made of the phosphate of the bones of victims sacrificed for Napoleon's personal glory."[93]

If every battle monument in Europe were inscribed with its true cost in blood and money, Edison said, there would be no new ones. However, there was now a deterrent that he believed would have the same moral effect: "fear of indiscriminate annihilation" brought about by the development of the flying machine. "A nitroglycerine bomb dropped from one of our modern airships will do more damage than whole days of fighting did in Napoleon's time." No sane political leader would ever contemplate such carnage. "In other words, inven-tion has got beyond the thirst for blood; the power of science that has been let loose must overwhelm aggressive diplomacy."[94]

A COUPLE OF QUARTS

On 18 October, nine days after the employees of Thomas A. Edison, Inc., welcomed their chairman back to work (Hutchison again snap-ping away), unofficial word came from Stockholm that he was to be awarded the $40,000 Nobel Prize for physics.[95] The news was, to say the least, consolation for his rejection, earlier that year, for member-ship in the National Academy of Sciences.*[96] But when the Nobel Foundation made its formal announcement, the prize went to Profes-sor Wilhelm Wien of Würzburg, a city rich in beer architecture. Edi-son maintained a dignified silence. If he had done so after his trip, instead of inflaming European sensibilities, he might have been be-medaled along with Marie Curie, who won the prize for chemistry.

Instead, he accepted from the president of the American Institute of Mining Engineers a gift made to his own Shylockian specification. It was a cubic foot ("Nor cut thou less nor more") of solid copper in-

---

* According to the physicist Robert Woodward, president of the Carnegie Institution for Science, Edison received only three votes "because of the profound prejudice among our academic colleagues against any kind of work not done in their characteristic ways."

scribed to him in appreciation of the boost his electrical inventions had given to the nation's copper industry since 1868. He mounted it on a pedestal in the laboratory library, preferring its tactile, 468-pound mass to the frippery of his other awards, of which he claimed to have "a couple of quarts" stashed somewhere.[97]

Boosted by his vacation, Edison confronted multiple business challenges that fall. "He certainly has come home with new energy," Mina wrote on 27 October. "It is so overpowering that it paralyzes me." He found that the 250 experimenters who depended on him for daily direction had been lax, during his absence, on several fronts—the most critical being their failure to adapt his prototype disk phonograph to the requirements of commercial production.[98] Although the model exhibited to jobbers both looked and sounded splendid, he had to be sure its complex technology did not price it out of the market. It would take him another year to achieve that confidence.

They had also stalled on the manufacture of one of his loonier inventions, concrete furniture—whose main virtue was that it tended to stay in place. Only the submarine alkaline battery project showed progress. The lead-acid lobby was already campaigning against it, a sure sign that Wall Street was seriously interested.

In appreciation, Edison appointed Hutchison as his personal representative at the plant and confided that as soon as Donald Bliss, his chief engineer, could be gotten rid of, he would have that powerful position too.[99]

A CIVILIAN OF CONSEQUENCE

Hutchison soon coined a variant to his title and had stationery printed that proclaimed him "Personal Representative of Thomas Edison in Naval Affairs." He arranged for his boss to be made honorary vice-president of the Navy League and on 2 November drove him, Mina, Madeleine, and Theodore to Staten Island to watch the Atlantic Fleet parade in New York Bay. It was the greatest display of American sea power yet seen, confirming news that the U.S. Navy now ranked second only to that of Great Britain.[100] In a frigid gale that failed to blow away the thudding of nearly four thousand cannon shots, twenty-four battleships foamed past the Statue of Liberty, followed by a five-mile chain of smaller ironclads. No longer white as they had been in Theodore Roosevelt's day, they presented a progression of gray, war-ready steel. Of particular interest to Edison and his party were eight submarines riding so low in the water that their saluting crews were at risk of slipping off each narrow whaleback.[101]

Later that month Hutchison escorted Edison to Washington for introductions to President Taft, Admiral Dewey, and officials at the navy yard. Edison was taciturn about their discussions of defense matters, except to predict that one day most of the engineering on U.S. warships would be electrical. A week later he welcomed two hundred officers and men from the Brooklyn Navy Yard to the laboratory for a lecture-demonstration on alkaline battery technology. Describing the S-type as the climax of his life's work in electrochemistry, he assured them it would enable a submarine crew to stay underwater for three months without breathing any fatal "acid gas." What was more, it was invulnerable to concussion. This was of particular interest to the sailors in his audience, some of whom might have suffered ear damage during the fleet exercise. They told him that violent sound waves, as from cannon fire, had neutralized many lead cells in the past.[102]

Slowly and subtly (and without awareness on his part, while he remained preoccupied with disk and talking picture development), Edison was being transformed into a civilian of consequence to the national defense. Hutchison wrote at the end of December: "I am ensconced here, right next to the greatest living inventor & apt to step into his shoes when he passes away. Brilliant future ahead of me & what others consider phenomenal advancement behind me. If every year of my life is as satisfactory to look back on, I'll be glad."[103]

A LITTLE GAMBLING

The first important visitor Hutchison escorted around the Edison plant in the new year of 1912 was a nouveau multimillionare from Dearborn, Michigan. Henry Ford, at forty-eight, had long been a genuflector at the shrine of Thomas Edison. He preserved as holy tokens some snapshots he had taken of him at a beach hotel in Brooklyn fifteen years before. In those days, Ford had been an aspiring gas-buggy designer in the employ of the Edison Illuminating Company of Detroit. Now, thanks to wildfire success of his Model T automobile, he was one of the richest men in America, keen as ever to become close to his idol.[104]

William J. Bee, Edison's resident expert on electric vehicles, was equally keen to have Ford divert some of his money into the Edison Storage Battery Company. He had sent him a portrait of his boss, flatteringly inscribed, along with an invitation to come to West Orange, as Edison "would be very much pleased to meet Mr. Ford."[105]

The pleasure was augmented when Ford allowed Bee to persuade him that the lightweight alkaline battery would make an ideal triggering device for automobile self-starters. He agreed to invest $1.2 million in whatever buildings and equipment Thomas A. Edison, Inc., would need to supply the Ford Motor Company with 450,000 type A cells a year, starting in 1913.[106] Overjoyed at the windfall, Edison sent Ford a letter in the curlicued calligraphy he reserved for momentous documents:

Friend Ford.
Billy Bee seems to be obsessed with the idea of having you do a little gambling with me on the future of the storage battery. Nothing would please me more than to have you join in. . . .

Up to the present time I have only increased the plant with profits made in my other things, and this has a limit. Of course I could go to Wall St and get more, but my experience over there is as sad as Chopin's Funeral March. I keep away.

Yours
Edison[107]

A MAN TO BE WATCHED

Edison turned sixty-five in February 1912, and decided that the alkaline storage battery, his most sophisticated invention since the movie camera, was "complete" enough to sell itself without further improvements. No matter how long Hutchison took to finish testing the submarine version (and the Navy Department's arthritic approval process was bound to take even longer) smaller A-cells were now pouring out of his factory at a rate that Ford's order promised to transform into an avalanche. Hutchison saw nothing but gold in its monetary moraine. Dazzled, he pitched for, and received, permission to act as advertising and sales agent for all Edison batteries. He was a fluent long-copy writer, and the media managers of the magazines who sold him space would probably find ways to express their gratitude.[108]

"I feel afraid of him," Mina wrote Charles. "He is so aggressive and has Papa so thoroughly under his thumb without Papa's realizing it that there is no telling to what lengths he may go— He seems to me a man to be watched." Her unspoken fear was that by the time her son graduated from MIT, "Hutch" would have amassed enough power to threaten Charles's future as heir to the leadership of Thomas A. Edison, Inc.[109]

Hutchison, whose appointment as chief engineer became formal that summer, was aware of Mina's fears, as well as that of Frank Dyer (ailing, overextended, and bullied by Edison at board meetings). He combated it by working longer days and nights at the plant than anyone else, the Old Man included, and by sending Charles lengthy reports of company activities, written with a disingenuous frankness that charmed the young man and reassured him that his future was secure.

I am very anxious to get something in shape for you to jump into when you get through College. . . . There isn't a job here in the Works that I would have on a salary basis. . . . I am so exceedingly fond of your father that I would work for ten years for nothing to help in any way if he happened to be in any such condition that he could not pay. As it is, I fully expect my commission end of the Government business* to amount to a good deal in the next few years, and meanwhile, I am doing all I can to promote the interest of the Battery Company and T.A.E. Inc.[110]

I am so exceedingly fond of your father. Mina worried that the fondness might be reciprocated. For most of his career Edison had been immune to flattery. He had always depended on acolytes to do

---

* By now, Hutchison was earning 20 percent on all battery contracts.

his will and treated them all, affectionately if distantly, as intimates. But his attachment to them had never been as quasi-filial as this one. There was something yielding about the way he accepted Hutchison's compliments, guffawed at his "coon" jokes, and allowed the younger man to publicize him as if he were a white-haired, benignly smiling cigar-store Indian. He posed for a couple of strange two-shots that Hutchison did not hesitate to circulate on company literature. One showed him apparently conversing with a submarine battery almost as big as himself. In the other, a withdrawn-looking Edison sat staring into space while Hutchison, a skilled Morse code sender, tapped out a message on his knee.

Edison receiving Morse signals from Hutchison, circa 1912.

THE INSOMNIA SQUAD

The photograph was deceptive, in that Edison, who had been attracted to the stage as a teenager, always enjoyed hamming for the camera. Helpless he was not, as staff in the Phonograph Works discovered when he embarked on a bout of disk development so protracted that his seven assisting engineers dubbed themselves "the Insomnia Squad."[111] It began behind locked doors around 9 September 1912 and continued with minimal sleep, soap, or shampoo for the next month and a half. Mina was not around to corral Edison home, due to consecutive absences in Maine on vacation with her children, and in Akron, where her mother lay dying. He took full advantage of his liberation, enjoying himself more, probably, than his bleary-eyed colleagues. When he boasted that they put in "more than twenty-one hours a day," he did not include a preparatory solo spell by himself, lasting 95 hours and 49 minutes by the laboratory time clock.[112]

Edison bore much responsibility for the desperate pace of the Squad's work, because in vowing to market the ne plus ultra of phonographs, he had with typical optimism assumed that it could be announced to the general public in October. Fourteen months after his call for the production of 3,500 disk machines, only 329 were packed—a frustrating situation, since the sales department had advance orders for nearly five thousand. The problem was not lack of supply—he had $800,000 worth of instruments stacked in the warehouse—but lack of records to issue with them. For that, his own obsession with sound quality was responsible.[113]

To the frustration of his executive committee, Edison rejected almost every test pressing he heard. Dust and other impurities endemic to the disk duplication process caused a slight surface noise that bothered nobody else. Cranking up the playback mechanism to its maximum amplification and cupping his right ear to the grille, he complained of loud "scratch." He would not approve any commercial pressing for release until it matched the clean sound of the masters he had cut in experiment.[114] That was the Insomnia Squad's challenge as Edison cajoled its members into action.*

After two or three days of progress, he posed with them for an ostentatiously "historic" photograph, as he had done once at Menlo

---

* The situation was especially critical because eager dealers had already ordered more than 175,000 unproduced disk records. Meanwhile the Edison Phonograph Works had run up an operating loss of $65,000 on top of a gross deficit for 1911 of $126,154.

Park when he was young himself, and some of his experimenters mere boys. It showed a group of not-yet-exhausted men stoking up on hamburgers, apple pie, and coffee at two in the morning.[115] They needed all the food they could get—"fuel for our physical energies," Edison called it—through mid-October, when he wrote Mina to say, "I have overcome with certainty the principal troubles."[116]

Edison and the Insomnia Squad at midnight "lunch," fall 1912.

On the twenty-sixth he patented three significant disk-molding improvements. One involved a controlled system of Condensite flow onto a rotating transfer plate of polished German silver, which tilted as it slowly spun, causing the varnish to bleed evenly across the plate surface before the rotation became horizontal. Thus Edison, who had not yet read Einstein, showed an instinctive sense of gyroscopic motion in relation to gravity. With equal ingenuity, he used centrifugal force to throw bubbles and dust granules in the varnish outward while the stock remained fluid. After it cooled and hardened, the rough periphery could be sliced away. The result, Edison claimed, was "a homogenous veneer free from imperfections," and the Patent

Office agreed, granting all three of his applications, along with fifty-eight others he had filed since the beginning of the decade.[117]

PRETTY CRUEL

That period happened to coincide with the political rise of progressivism, a largely white, middle-class, moralistic, and proregulatory insurgency drawing strength from the liberal wings of both major parties. In the election year of 1912 the movement rated a capital P with the founding of an official Progressive Party by bolters from the GOP. Its leader and formidable candidate for a third term in the White House was Theodore Roosevelt, running on the one hand (to use his favorite phrase) against the Republican president, William Howard Taft, and on the other against Woodrow Wilson, Democratic governor of New Jersey.

Edison had always been a loyal Republican, and with his ear so consistently jammed against phonograph grilles that fall, he might have been expected to pay little attention to the distant barking of ideological debate around the country. But he surprised the writer Will Irwin, while watching a trial of his A-6 battery on the Orange electric railway system,* by declaring for Roosevelt.

"I'm a Progressive, because I'm young at sixty-five," he said.[118] "And this is a young man's movement. There are a lot of people who die in the head before they are fifty. They're the ones who get shocked if you propose anything that wasn't going when they were boys."

Irwin was struck by the dreamy look in Edison's gentian-blue eyes as he watched his battery absorbing a recharge that would have melted any conventional lead-acid unit. He stood with hands stuffed in his pockets, talking half to himself, in the manner of a man not used to being interrupted.

It's the way the world goes—the young push ahead and do things, and the old stand back. I hope I'll always be with the young.

---

* Railcars designed by Ralph Beach and battery-powered by Edison had been in service in New York City and on the Pennsylvania Railroad since the summer of 1911. The batteries were underslung and propelled the car with twenty-four passengers well over one hundred miles per charge, often through heavy snow.

You see, getting down to the bottom of things, this is a pretty raw, crude civilization of ours—pretty wasteful, pretty cruel, which often comes to the same thing, doesn't it? . . . Our production, our factory laws, our charities, our relations between capital and labor, our distributions—all wrong, all out of gear. We've stumbled along for a while, trying to run a new civilization in old ways, and we've got to start to make the world over.

Edison spoke disapprovingly of the monarchical system of government in Germany, and that country's "great standing army," not subject to the will of its people. At least the United States was able to keep its polity in balance through regular elections and constitutional amendments. He said he was for Roosevelt's most radical proposal, the popular review of judicial decisions. The current Supreme Court was too powerful, and too conservative. "Precedent, all precedent!" he scoffed.

It occurred to Irwin that every one of Edison's thousand-odd inventions had been built on a precedent of some sort. Yet there was no irony in his claim to having been "a progressive always." His entire career had been a drive toward modernity.

"There's the matter of injured workmen," Edison said, citing the Court's opposition to employer liability laws. "A laborer loses his right hand in an accident. It's his capital. It's as though my plant should burn down without insurance. . . . I never heard a squarer and truer thing from Roosevelt than when he said the loss to workingmen by injury should be a tariff on the business, to be paid by the public in increased prices if necessary."[119]

Irwin was probably unaware that Edison's first example related to a particular amputee in his memory—poor John Dally, radiated to death by their work on X-rays—and Edison himself did not know that he would soon enough experience the pain of the second.

WE'VE HELD YOU DOWN

Edison also followed Roosevelt in embracing the right of women to vote—to the displeasure of his ultraconservative wife and, surprisingly, that of his younger daughter. Since Madeleine's sole attempt to get a paying job away from West Orange, she had been consumed only with the desire to marry John Sloane and have his children.* The female suffrage movement left her cold.[120]

Not so Lucile Erskine, a young independent journalist and summa cum laude graduate of Washington University, who boldly asked Edison in an interview what he thought of her sex.[121] His reply took her aback.

"It will be three thousand years— at the shortest 2,500—before women are the intellectual equals of men," he said.

She came back at him. "Haven't women any brains?"

"There's some there," Edison conceded. "A little, not much. But women haven't any cross fibers. That's our fault! We've held you down. But now you're beginning to evolute."

From the twinkle in his eyes, Miss Erskine realized she was being teased.

"There was a chance to hurl the name of Mme. Curie at him," she wrote afterward, "but the cruel lack of 'cross fibers' made one forget to put it in the right place."

CORPORATE REGRET

Woodrow Wilson had no sooner won the presidency in November than Frank L. Dyer lost his. Ever since the formation of Thomas A. Edison, Inc., there had been speculation among employees as to how long the Old Man could stand for something incorporated in his name to be run by somebody else. Edison blamed Dyer for the Phonograph Division's continued sales fallback behind Victor, choosing to forget that his own perfectionism (or obstructionism, as the long-suffering lawyer might call it) was a principal reason for the delay in getting its new, competitive products out.[122]

"The coming of the Disk Phonograph I hope will mark a period of great prosperity for you," Dyer wrote in his resignation letter, "and I think it better that my successor should take charge of the business at the start rather than later on." Both he and Edison knew well who that successor would be.

---

* A letter around this time from Madeleine to Mina makes clear that by then she and John were lovers. The document's woman-to-woman confidentiality implies that Mina was expected to keep the secret from Edison.

My present position is quite untenable. Many subordinates are reporting directly to you, and I have reason to believe that in a number of cases you have indicated to them that you have lost confidence in my ability or capacity. Rumors of this sort naturally spread very rapidly and destroy all possible authority. . . .

In my recent talk with you, you criticized me quite severely, but I do not think your criticisms were fair or just.[123]

Dyer pointed out that it had been he who pushed the disk project to begin with, and that the company's new line of Blue Amberol cylinders, only just coming onto the market, could have been released two years earlier, had Edison stopped fussing with its Condensite composition. He took credit for surging profits in the company's battery and dictating machine divisions and particularly in his own creation, the Motion Picture Patents Company—a hugely lucrative trust that dominated the movie distribution market. Despite the indignity of never having been allowed a corporate electric car, issued free to most department heads, he assured Edison that "I shall always entertain for you the strongest feelings of admiration and personal affection."[124]

William Meadowcroft announced Dyer's departure with the usual expression of corporate regret, followed by: "Mr. Edison takes the presidency in order that he may direct the policy of the Company in addition to the technical details which he has always had charge of."[125]

AND WHAT FLESH

At the end of the year Edison was working past midnight in his laboratory, with Miller Hutchison close by as usual. Distant steam whistles announced the arrival of 1913. The two men shook hands, and Hutchison wished his "Big Chief" well—as indeed he might, since Edison had just wiped off a record with potassium cyanide and was beginning to feel ill. But before taking to his cot in the library, he held a phonograph horn to his ear to hear the whistles better.[126]

Hutchison recognized the amplifier as part of Edison's new Kinetophone talking picture system, which he would soon have to introduce to reporters and exhibitors. He was not looking forward to the task.

It was bound to be confused with the unsuccessful audio peepshow device, also called a Kinetophone, invented by Edison and W. K. L. Dickson in 1894—not to mention their even earlier attempt to make the prototype Kinetograph camera responsive to sound. Hutchison doubted that this modern enhancement, a complex hookup of hitherto independent machines, would work as "perfectly" as Edison claimed when put in the hands of untrained or half-trained operators. So far his experience with the theater owners and projectionists he had signed up for the launch had not been sanguine. "If ever a fellow was up against a tough game," he wrote Charles Edison, "it is yours truly."[127]

Edison had been the first movie pioneer to equate pictures with spoken words and music: "I am experimenting upon an instrument which does for the Eye what the phonograph does for the Ear." But after the failure of the first Kinetophone, he had abandoned the idea.[128] For the rest of the nineteenth century in America, nickelodeon managers clicked coconut shells behind the screen to simulate the sound of trotting horses, hammered metal bars in time with on-screen blacksmiths, and blew bugles or popped Chinese crackers during battle scenes. Some employed hidden actors to utter aloud what their filmed counterparts were supposed to be saying. Rubber-lipped virtuosi who could imitate ship sirens or creaking floors or the whooshing of wind earned excellent money. Music, live or recorded, was a common background effect. Lyman Howe, the wandering "phonograph entertainer," cheekily used Edison machines to accompany movies from other studios.

Meanwhile in France, dozens of inventors pursued the chimera of *images parlantes* with a variety of systems that all, sooner or later, fell victim to the medium's peskiest problems, synchronization and amplification. The only way "live" sound recording (as opposed to a later dub) could be matched with cinematography was to position a phonograph as near as possible to the action, and have the cylinder roll in tandem with the camera. For as long as the phonograph was able to record before running out of wax—no more than two minutes—an illusion of synchronism could be enjoyed by cast and crew. But when cylinder and film were separately duplicated and installed in theaters of varying dimensions, it became almost impossible to maintain a convincing pas de deux. The devices had to be linked by an electric wire or geared shaft, generally run under the auditorium floor, and subject to such interferences as rat suicide or vibratory dislocation. Any skip or splice in the projecting reel might cause Sarah Bernhardt, melodramatically dying on-screen, to start talking like a man, or even worse, break into song. Audiences reacted with predictable outrage, and many an impresario went bankrupt on the huge costs of production and exhibition.

Even when so gifted a showman as Clément-Maurice Gratioulet premiered his Phono-Cinéma-Théâtre at the Paris Exposition in 1900, he had to rely on a projectionist skilled enough to hand-crank the film at varying speeds, while listening via telephone to the dialogue weakly sounding from a Lioretrographe player in the orchestra pit. The man must have had phenomenal hand-ear coordination, because at first Gratioulet prospered mightily, earning rave reviews for his presentations of such spectacles as the ballet *L'Enfant prodigue* and the duel scene from Rostand's *Cyrano*. "Here are beautiful sounds and beautiful gestures which are fixed for eternity," *Le Matin* declared. But the beauty the newspaper ascribed to the Lioretrographe was that of novelty, more than true acoustics. Amazement at being able to hear, as well as view, the exquisite Cléo de Mérode perform her *danse orientale* wore off when theatergoers realized they could see her do it in the flesh—and what flesh!—at the Folies-Bergère. Moreover she danced to the chimes of a live gamelan band, rather than indistinct noises from a tin horn.

Edison's most serious rival in the audiovisual field—and an unabashed infringer of his sound and movie patents—was Léon Gaumont, whose Chronophone apparatus featured several innovations, such as a transmission clutch, to improve synchronism. But Gaumont also tried to solve the amplification problem by squirting compressed air into the reproducer of his phonograph, which added more hiss than volume to its sound stream. And his device enabling the projectionist to adjust the RPM of the cylinder whenever the film jumped or lagged made for queasy changes of pitch.

Nevertheless, by the time Edison decided to resurrect his Kinetophone idea, the basis of all the French systems, Gaumont's *phonoscènes* were being successfully shown all over Europe. Some, with hand-tinted enhancements, made their way to North America. Their sound, however, remained thin and weak. Edison was confident that he could succeed where so many had failed. As president of two of the world's largest film and phonograph studios, he was in a unique position to combine the experimental resources of each. And as chairman of the powerful Motion Picture Patents Company, he could make sure no further Gaumonts arose to poach on what he regarded as his intellectual property.[129]

The new Kinetophone apparatus that he developed in an access-restricted, asbestos-padded tent athwart his laboratory* only superficially resembled its predecessor of 1894. There was still a wax cylinder for recording, betraying Edison's continued preference for that format over the disk. Only now it was a fat, foot-long drum that could hold six and a half minutes of dialogue or music, enabling Edison's talking pictures director, Oscar Apfel, to shoot the prison scene from Gounod's *Faust* in one take. The wax, moreover, was so pure and smooth under the recording needle it could have been frozen butter. It picked up the softest sounds—sighs, stealthy footsteps, creaks—from thirty to forty feet away, through a twelve-petal horn that expanded and tilted toward sound like a great lily seeking sunlight. The phonograph itself was immobile (it weighed seventy-four pounds) and unseen below the frame. This cramped lateral movements onstage, because Edison found that the horn's receptivity to voices faded at a compound rate when actors walked away.[130] As a result, the half-dozen features he prepared to demonstrate the Kinetophone's adaptability to various entertainment genres all had a centered, "tableau" look, in contrast to the fluid action of silent films.

A high-tension belt of unstretchable silk connected two wheels, one revolved by the phonograph axle, and the other driving a "synchronizer" that was in turn geared by means of a worm shaft to the camera. Thus the Kinetophone, unlike its French predecessors, recorded and shot at a speed controlled by the revolutions of the cylinder, rather than the rotations of the camera's shutter. When the film was printed and the cylinder duplicated in Condensite, neither could be

---

* Edison's idea in erecting this marquee was that its canvas drape would obviate the "echo effect" of solid walls.

edited, or the sounds and images would at once separate. "A varia-tion of one-fifth of a second is fatal," Edison admitted.[131]

The shorter and tighter the silk belt, the better the synchronism while filming.* But the reverse process of projection—with the pho-nograph in playback mode, hidden behind the screen and "talking" through a small gauze grille—almost always involved a lengthy exten-sion of the system, via pulleys, to the booth whence another unseen device (endlessly fascinating to children) sent forth its moving fingers of light. Every show, starting with the press preview Edison hosted at the laboratory on 3 January, required the services of a brace of opera-tors: one to crank the picture, and one to activate the phonograph on cue. This occurred when the opening credits (displayed in silence, to save cylinder space) faded from the screen and gave way to the image of an actor in full evening dress entering a luxuriously furnished room. He advanced, stationed himself between two potted plants, and opened his mouth.[132]

ALL THE RREALISM OF NATURE

"A few brrief years ago," the mouth said in a clear tenor voice, rolling its *r*s and articulating every syllable, "Mr. Thomas A. Edi-son prre-sented to the world his Kinetoscope.† Inventors the worrld over have endeavored to synchrronize the phonograph and motion picture. But it remained for Mr. Edi-son—"[133]

Was a mouth indeed saying these things, or was a diamond-point reproducer vibrating somewhere below the potted plants, camou-flaged with photography? To most people in the room, the illusion was total. Gasps of surprise and wonder could be heard on all sides as the actor continued reciting the words Hutchison had written for him.[134]

"—to combine his two grreat inventions into this one, which is now entertaining you, and is called the Kinetophone. The Edison Ki-

---

* Edison's directors clapped two halves of a coconut shell together at the start of each take.

† The reference is to Edison's motion picture camera, patented 31 August 1897. By 1913 the trademark terms *Kinetoscope, Kinetograph,* and *Kinetophone* were frequently confused in the public mind. For their changing meanings in the 1880s and '90s, see Parts Four and Five.

netophone is abso-lutely the first genuine talking picture ever prro-duced."

This was of course not true. Gratioulet's Phono-Cinéma-Théâtre had achieved a similar if less precise verisimilitude twelve years be-fore. But Hutchison, an early exponent of the art of movie hype, evi-dently thought the adjective *genuine* meant something.

"The actor," said the actor, moving freely about the stage, "per-forms exactly as he does upon the stage, moving freely about, and his everry word and every action are simultaneously recorded, with all the rrealism of nature."

He proceeded to demonstrate the Kinetophone's fidelity by smash-ing a china plate, blowing a horn and a whistle, and introducing some musicians, including a pretty girl who sang "The Last Rose of Sum-mer." In a crescendo of noise, he brought on a pair of barking dogs.

Edison, chomping on a big black cigar in his front row seat, chuck-led at the din and nodded at the actor's prediction that the world would be watching such performances "one hundred years from now." But he frowned when his signature flashed on the screen and he heard himself described as "that Wizard of sound and sight, Mr. Thomas A. Edison." Hutchison had not yet learned that the W-word irritated him.[135]

The show continued with six more demonstration shorts: the "Miserere" from *Il Trovatore,* a scene from Planquette's operetta *The Chimes of Normandy* with clinking coins and carillons, the quarrel of Brutus and Cassius in *Julius Caesar,* and three comic sketches that restored Edison's good humor. Afterward, however, he was cautious in accepting the congratulations of reporters. "No machine is per-fect," he said. "Man is not perfect." Nevertheless he could not dis-guise his pride in achieving a synthesis of all his experiments in phonography and cinematography. He said that he had "arrived at a place where 'the movies' are also to be known as the 'talkies.' "[136]

Within twenty-four hours the word *talkie* entered the vernacular. There was a rush by entrepreneurs, including Edison's conniving son William, to acquire Kinetophone exhibition rights.[137] The Chicago financier John R. Dos Passos offered a down payment of $1 million for a controlling interest in the venture. His envoy was staggered when Edison "just laughed" at the certified check, saying that he intended to "operate the machines and market them himself." The successful bidder, representing a combine of the nation's three largest vaudeville networks, accepted these conditions and named itself the American Talking Pictures Company. It contracted with Edison to manufacture three hundred systems and produce a steady supply of features to feed them. A national release date was set for 17 February, much to Hutchison's dread. The press preview had gone well because the room was small and the operators were well trained. But he did not see how he could ensure synchronism when the cord linkage expanded to the huge proportions of theaters like the Colonial in New York, let alone persuade unionized projectionists to learn a complex new technology. "This entire apparatus is the most unsatisfactory product we have ever turned out," he warned Edison. "I can see all sorts of trouble ahead."[138]

As far as Edison was concerned, that was Hutchison's problem. Never having cared much for movies as entertainment, he had an overriding interest in adapting the medium—with or without sound—for education.[139] Besides, he wanted to get back to the improvement of his disk records, which did not satisfy him and were still unavailable for general distribution.

Hutchison, having gotten the job of chief engineer through what he believed to be adroit manipulation of "the Old Man," was entitled to wonder who had manipulated whom. The fat commissions he looked to as Edison's storage battery sales agent had only just begun to accumulate, thanks to encouraging orders from train and delivery truck companies. But the navy was tying so much red tape around the installation of the S-type unit in a trial submarine as to raise questions about its willingness to switch from acid to alkaline cells. Meanwhile Edison had been quick to take advantage of Hutchison's status as an unsalaried employee, heaping responsibility on him for all the plant operations his title embraced.[140]

As for his secret hope that he would one day become president of Thomas A. Edison, Inc., it became increasingly obvious that Mina would not allow anyone other than Charles to succeed to that office. Relations between her and Hutchison exploded into open hostility in late January. "It makes me sick the way that man jinks," she wrote Charles. They came to a truce out of concern for Edison, who was vaguely aware of some people quarreling somewhere beyond the music room of the laboratory. Hutchison tried to reassure Charles in long, disingenuous letters that he was nothing but a faithful servant of his father. "Every one of these [Kinetophone] outfits that is to go into practical use is worth so much a week to the Old Man," he wrote. "I naturally am anxious to see as many of them in practical money-making as it is possible to get."[141]

Mina confirmed that Edison was facing one of his periodic cash flow problems, as a result of shipping thousands of expensive disk players while restricting production of the only records they could play. Acceptance of Dos Passos's offer would have dispelled the cloud of insolvency darkening over him, but independence mattered more to him than security. She could only wait for him to turn the cloud to sunshine, as he somehow always managed to do.[142]

Impatient as he was to closet himself again with the Insomnia Squad, Edison did what he could to help publicize the Kinetophone in the days leading up to its release. "Oh yes, I've plenty of time to see you," he said to a reporter asking to see the system in action. He led the way to his private screening room ("This is my experimental theater") and ordered the projectionist to put on the "Miserere" from *Trovatore*. The reporter boggled at the film's visual and sonic power. Out of the corner of his eye, he discerned that while he was watching it, Edison was watching him, with a strange, quizzical smile. "Truly, the man of practical science, noting the effect of his latest creation upon mankind!"[143]

At four P.M. on 17 February, Edison stood in the wings of New York's Colonial Theater to monitor the reactions of more than a thousand viewers to his portfolio of demonstration shorts. The program began in expectant silence, with the usual shutter-flutter emanating from the projection box. But when Hutchison's stentorian spokesman appeared on-screen and began to orate, there was a collective murmur of astonishment. The wonder grew when the pretty girl sang and Brutus and Cassius quarreled and Mephistopheles taunted Faust and a group of minstrels (two in blackface) launched into a medley of popular hits. The show climaxed with a chorus performance of "The Star Spangled Banner." When it ended, the audience sat spellbound for a long moment, then burst into applause and shouts of "We want Edison!" He remained out of sight while the calls, punctuated with rhythmic handclaps, grew louder. After five minutes Frank Tate, an American Talking Pictures executive, came on stage to say that the inventor was unavailable. That did not quell the bedlam, which lasted until Tate reemerged to say that Edison was already en route to another show at the Alhambra, in Harlem.[144]

It was lucky he chose to go there instead of downtown to the Union Square Theater, where he would have been humiliated by a ten-second slip in synchronism that had the audience hooting and jeering. During the "Edison Minstrels" short, the program announcer, wearing for some reason a powdered wig, sat down long before his amplified voice stopped speaking, while the singer he introduced launched into what *The New York Times* described as ten or twelve seconds of "fervent but soundless song."[145]

Theodore Edison claimed, with all the certainty of a fourteen-year-old, that union sabotage was responsible. In truth, the Kinetophone was much harder to operate in theaters than on the set. While the projectionist hand-cranked his machine, keeping one eye on the screen, he had to keep another on the synchronizer beside him, as well as listen through earphones to the sound of the distant phonograph.[146] Hutchison tried to make his job sound easy in an FYI letter to Charles:

> There is a little indicator on this device which shows the operator whether he is turning properly or not, and by operating this little indicator, he can shove his Kinetoscope [projector] ahead of the phonograph, or vice versa, as the case may be.
>
> The phonograph is, of course, located behind the screen. First the title is thrown on the screen from the Kinetoscope. Although the phonograph motor is running, the cylinder is not, and the reproducer is properly placed at the beginning of the record. After the title is shown, there is a blank space of one second, and just as soon as the blank space ceases to exist, and the picture comes on, the phonograph operator presses a button which throws in the clutch on the cylinder, and causes the phonograph to proceed to play or talk. . . . If the phonograph operator is a little slow in pressing this button, he will, of course, throw the outfit out of synchronism, and it is up to the operator of the Kinetoscope to hang back on the Kinetoscope until it is in step with the phonograph.[147]

Hutchison needed another half page to describe the workings of a supplementary telephone rig, which the projectionist, if he happened to have a third arm, could use to contact his invisible colleague. He complained of having to train twenty-one engineers to instruct operators in only eleven theaters, as well as dealing with fire inspectors and "unbusinesslike" impresarios. "I have never come across anything that has as many angles to it as this infernal talking picture proposition."[148]

But when the Kinetophone system worked well—which it mostly did at first—it succeeded so brilliantly as to promise a huge return to its backers. "EDISON'S TALKING PICTURES THE GREATEST SUCCESS IN YEARS," Edward F. Albee of B. F. Keith's Theaters telegraphed his regional managers. "THOUSANDS OF PEOPLE TURNED AWAY . . . STORMS OF APPLAUSE . . . WE WILL HURRY A MACHINE TO YOUR CITY AS SOON AS POSSIBLE." With the Orpheum and United Booking circuits joining in, shows spread to more than a hundred theaters. Hutchison had to dispatch his operational instructors farther and farther afield and ordered double-shift production of projectors to meet the demand. Foreign rights were sold to exhibitors in South America, Europe, and Asia. Edison seemed assured of at least $500,000 in royalties by the end of 1913.[149]

Audiences had difficulty believing that the sounds they heard were not emanating directly from the images moving before them. "It is all so natural as to appear almost uncanny," the *Philadelphia Item* reported. "I have heard a photograph bark," the syndicated columnist Arthur Benington wrote in *The World Magazine*. "I have heard a photograph squirt water from a siphon and splash in a bathtub." A music critic in Fort Worth, Texas, marveled at the synchronism of the *Faust* film. "The work was so perfect that the mechanized details were forgotten." Several reviews praised the beauty of the diamond-reproduced sound, and the fidelity that captured even a slight lisp in an actor's enunciation. "No, Silas, they can't fool me—there was a man back of that curtain," the usual little old lady was quoted as saying after a show in Pine Bluff, Arkansas.[150]

RATHER THAN WITH THEIR MINDS

Unusually, for a lifelong self-promoter, Edison never expressed much enthusiasm for the Kinetophone. He kept saying that it was a long way from perfection, that major problems had to be solved before the talking picture stood a chance of supplanting the silent. Among them were constricted stage action, limited feature length, underpowered amplification for large halls, and—most challenging of all—the recalcitrance or incompetence of operators whose numbers soon put them beyond Hutchison's instructional reach.

Edison tried to deal with the action problem by inventing an overhead miking system that extended sound intake to the limits of the stage. It consisted of an adjustable-height canopy impregnated with miniature receivers, electrically linked. "I collect the sound at a plurality of points . . . and transmit pulsations or impulses corresponding to the collected sound waves to a single recording device." He also allowed for similar reception, if needed, beneath grilles on the stage floor.[151] His patent application, dated 6 March 1913, was successful, but the system was apparently never installed at the Bronx studio.*

Perhaps because of his deafness, or the dismay he sometimes betrayed at being seen as a purveyor of mass entertainment, Edison was interested in talkie technology primarily as a means of elevating popular taste: he wanted "to make it possible for the poorest families in Squeedunk to see the same operas and plays that are produced in New York City for an admission price of five cents." He also recognized its enormous historical potential as a recorder of current events. Already he had a Kinetophone cameraman, James Ricalton, filming the war between Bulgaria and Turkey.[152]

It was not generally known that fifteen months before, on his visit to the White House, Edison had invited President Taft to become an audiovisual candidate for reelection. Taft had just completed a cross-country tour that lasted eight weeks and exposed him to more than three million people. Edison suggested he use the Kinetophone as a "campaign machine." He could record his stump speech, get the Republican National Committee to distribute it to theaters nationwide, and reach 60 million voters without missing a day at his desk. But Taft was the wrong person to lobby on this subject. There was nothing he loved more than getting out of Washington, so he passed on the

---

* Motion picture miking remained static until Walter Wellman devised the overhead boom on the set of *Beggars of Life* in 1928.

opportunity to pioneer an electronic medium that would one day define the democratic process.[153]

Edison showed much more passion—in his own words, "I was on fire"—when pushing his idea that film would be the educational medium of the future.*[154] His long-standing interest in the subject had been stimulated by the difficulty of finding words to answer some of the questions his son Theodore kept asking (just as young Al Edison had tormented a teacher in Milan, Ohio, sixty-six years before). Deafness, too, made him preternaturally aware of the value of lessons in things seen, not just described. The most he could hope for at present, given the militant protectionism of America's teachers, was that a scaled-down version of his projector, known as the Home Kinetoscope, would appeal to some progressive school boards as a classroom tool especially suited to lessons in geography. In time, its effectiveness (indeed its superiority to the oral method of many blackboard thumpers) should sell it to a much larger market. He could then increase the variety of subject reels to be produced by his studio. "We shall aim to teach not only geography but science, mechanics, chemistry, botany, entomology, and, in fact, all the regular branches of study."[155]

When a sample program of Edison instructionals was screened in sixteen New York schools, eleven of the audiences, consisting of senior staff, board members, city officials, and parents' associations, were highly enthusiastic, and six voted to buy a projector right away. The cost of the machine and the expense of renting films put off some other would-be purchasers. But a similar demonstration in Schenectady went badly. It took place at the annual general meeting of the New York State Principals Association, and the membership, composed largely of small-town pedagogues, rejected the Home Kinetoscope as a threat to their trusted "old ways" of teaching.[156]

This did not bode well for acceptance in states west of the Hudson, not to mention the *extremadura* of Texas, where an agent for the southwestern schoolbook publisher Silver, Burnett & Co. warned Edison that it would be many years before a regional school board

---

* Edison's educational movies project was a natural sequel to a series of semidocumentary, reformist shorts put out by his studio between 1910 and 1913. They covered such subjects as slumlords, tuberculosis, and child labor.

could entertain the notion of children studying "with their eyes and ears rather than with their minds."[157]

Edison refused to believe this. "Books," he blustered, "will soon be obsolete in the public schools."[158] This statement caused the biggest sensation since his denial of immortality in 1911, and a heavyweight delegation of teaching authorities came to West Orange at the height of summer to see if he could possibly be serious. They included the philosopher John Dewey, Leonard P. Ayres of the Russell Sage Foundation, and Arthur D. Dean of the New York Education Department. The visit was sponsored by the sociological magazine *Survey,* which reported it on 6 September in a symposium entitled "Edison Versus Euclid: Has He Invented a Moving Stairway to Learning?"

That Edison was serious was at once apparent. The delegation found that he had a production list of nearly a thousand educational "scenarios." Besides those already in the can, there were fifty or sixty ready to shoot, covering such subjects as astronomy, bacteriology, physics, forestry, fine art, and zoology. The technical excellence of the films Hutchison screened in demonstration amazed everybody, although reactions as to their effectiveness varied according to professional prejudices. Marietta Pierce Johnson, founder of a progressive school in Alabama, remarked that Edison had found a way to bring "joy" back to education. Rudolph Reeder, the superintendent of the New York Orphan Asylum, was impressed by the "unlimited possibilities" of observational instruction on film, while asserting that some subjects were still better taught with "words, words, words." Leonard Ayres marveled at an animated depiction of the Bessemer steel process and the beauty of time-compressed sequences showing crystal formation and the metamorphosis of a caterpillar. Edison, he thought, had devised "an educational tool of great value." However, the "very perfection of detail" that made cinematography so hypnotic made him worry that it would alienate students from one another. "When they sit silent in a darkened room, they are individual and exclusive. When they are making something material or abstract, because they need it in their business, they are active and alert. When they watch moving pictures . . . they are passive and inert."[159]

Predictably, John Dewey contributed the most thoughtful essay to the *Survey* symposium. "That Mr. Edison has a sound psychologic basis in relying upon the instinctive response of human beings to what moves and does something is unquestionable. . . . But I was also impressed by the fact that, after all, seeing things behave is a rather vicarious form of activity, and there is some danger of the better becoming an enemy of the best."[160]

CRAMPS

Dewey was not so cerebral that he did not boggle at the amount of money the Edison company must be investing in so ambitious a scheme, quite apart from its development of the Kinetophone and disk phonograph. Edison himself was so strapped at this time that he accepted a short-term personal loan of $50,000 from Hutchison. The latter had plenty of cash to spare. He had just sold the rights to his Klaxon invention for $142,500 and was happy to earn 5 percent interest on part of that windfall. But any employer with an ego less impregnable than Edison's would have felt embarrassed to be beholden to a subordinate.[161]

On 24 June the financial pressure on Edison eased, with a $100,000 second installment of Henry Ford's business loan, and royalties coming in from "Edison talkies," which he was now producing at the rate of five or six new titles a month. He announced that his next steps in the movie business would be "the production of multiple-reel screen dramas, colored pictures, and possibly stereoscopic films with the effect of actual depth," but he filed only one patent for the color process before showing symptoms of exhaustion and a return of his old enemy, gastrointestinal cramps.[162]

Mina insisted that he join her and the children for a summer vacation on Monhegan Island in Maine. She had been depressed for much of the year. What with Madeleine obstinately getting engaged to John Sloane, Charles falling for a girl in Boston, and the hated "Hutch" seeing more of her husband than she did, she complained that she was "crowded out" of the lives of her loved ones. "I can feel every minute the losing game, and it makes me feel unloving and hard."[163]

The most Edison would grant her of his company was a ten-day spell in late August. He prepared for it in typical fashion, working through the night on the eve of his departure and arriving at Monhegan more dead than alive after a three-day car journey. He remained ill throughout his stay there, suffering intense abdominal pain. Returning south with his family after Labor Day, he insisted on stopping off in Boston to meet Henry Ford. The motor magnate was there with another personal hero, the naturalist John Burroughs, and during a long morning of "chinning," as Madeleine termed it, a triadic friendship was born.[164]

Edison was pronounced "a very sick man" by a doctor who examined him in West Orange and diagnosed his ailment as either gallstones or an abscess on his gallbladder. Preparations were made for an operation, but applications of ice soothed his pain, and he was soon back in the laboratory, working up to twenty hours a day.[165]

"I am simply living up to the laws of my own being," he said to the writer John H. Greusel, who asked why he felt the need to deny himself food and sleep.

Greusel was unable to fathom what those laws might be or why they were so compulsive. "The strangest figure of our time," he concluded. "Aloof, enigmatic, unamenable to the rule of averages in human life."[166]

WOA, WOA, WOA

With the Christmas sales season approaching, it was urgent that Edison introduce his disk phonograph and complementary record catalog. When he did so at the beginning of December, the publicity campaign highlighted the jewel on which the whole technology rested.

Copywriters, bill posters, and sandwich board men made a mantra of the brand phrase EDISON DIAMOND DISC, so phonetically suggestive, with its repetitive dentals and sibilants, of the polished hardness of the stylus ("No Needles—No Trouble") and the clarity of the sound that poured from the hidden horn. The disks, unplayable on any other phonograph, were as extraordinary to look at as to hear: a quarter of an inch thick and inflexible as stove lids, with narrow grooves that packed in five and a half minutes of music, much more than the contents of a ten-inch Victor disk. No paper label obtruded on their glossy blackness, intensified by one of Edison's old laboratory standbys, lampblack. They had to be angled to the light before his portrait could be seen, impressed in halftone beside the spindle

Diamond Disc retail advertisement, 23 December 1913.[167]

hole, along with his name and signature and the record title, but—
bewilderingly—no performer credit. "I have very excellent reasons
for not putting the names of artists on our records," Edison informed
a jobber, without further explanation.[168]

The National Phonograph Company's old townhouse at 10 Fifth
Avenue in Manhattan was luxuriously refurbished as a four-floor
showroom for the new machines, issued in five sizes as models A80,
A150, A250, A300, and A450 ("Louis XVI Circassian Walnut, Metal
Parts Gold Plated"). When visitors realized that the numerals signified
dollars, it was all Edison's chief salesman, Percy Morgan, could do to
get them to listen to a sample Diamond Disc. Usually a minute or two
was enough to convince even skeptics that the Wizard of Menlo Park
had "done it again." Their reactions (which Morgan noted verbatim
and sent weekly to West Orange) almost unanimously expressed
amazement that recorded music could sound so full and sweet.[169]

This was also the general opinion of browsers and buyers at thirteen thousand stores around the country. Audio fanatics—already a
distinct species—agreed that the Diamond Disc phonograph's combination of floating-weight reproducer,* geared tracking, and records
of adamantine smoothness was superior to any other sound system
on the market, other than Edison's parallel line of Amberola players

---

* Edison claimed that his nonskip jewel holder was the result of 2,300 experiments.

Edison A-100 "Moderne" Diamond Disc phonograph, 1915.

and superb Blue Amberol cylinders.*[170] "I would have thought, had I not known differently, that the songs from the machine were really being sung by singers in the room," one of them wrote, giving Edison an idea for future publicity. A University of Chicago professor praised "the clear articulation, the plastic roundness of tone, and the fine balance of parts" of the A250 instrument, and although he already possessed a Victrola, he immediately treated himself to an upgrade.[171]

The willingness of such enthusiasts to spend half or a full month's salary on a player that accepted no other records bore out Frank Dyer's prophecy that the Diamond Disc would restore the fortunes of

---

* "Blue Amberols . . . when played with an Edison Diamond Reproducer . . . outperformed any other medium of reproduced music then available," the audio historian Roland Gelatt writes in *Fabulous Phonograph*. "The ears in Edison's recording studios were attuned with extraordinary sensitivity to the elements of good sound reproduction."

Thomas A. Edison, Inc.* Before long, the company indeed derived a large income from it. This was in spite of the fact that Edison, growing more autocratic by the month, did his perverse best to sabotage sales by imposing his own musical taste—or lack thereof—on everybody in the phonograph business, from performers in the studio to customers in stores.

He used the personal pronoun forty-seven times in an interview entitled "Edison's Dream of New Music," published in *Cosmopolitan* magazine. Acknowledging that he could neither read nor sing a note of music, he nevertheless declared that it was an art "in the same backward state today that electricity was forty years ago. I am going to develop it. . . . I shall also make the phonograph the greatest musical instrument in the world."[172]

Although Edison was not averse to Beethoven, or the occasional aria by a composer whose name ended in a vowel, his favorite repertory remained the moony melodies he and "the boys" used to caterwaul *legato e doloroso* in Menlo Park days, to the strum of Ludwig Böhm's zither—songs like "My Poor Heart Is Sad with Its Dreaming" and "I'll Take You Home Again Kathleen." He could not hear the latter tune often enough and recorded it numerous times. Its sweetness and simplicity were worth more to him than the unresolved harmonies of Debussy, which he likened to "interrupted conversations."[173]

That particular comment was acute, but professional musicians winced at some of Edison's other aperçus, which he voiced with the hortative smugness of Bernard Shaw. Mozart was "the least melodic of composers." He liked "the 7th Nocturne of Fields [*sic*]" because it had "no dissonance." After listening to 2,700 waltzes, he found that "they consisted of about 43 themes, worked over in various ways. . . . Of course, I do not include Chopin in this, as his waltzes are not conventional waltzes." There was apparently "no such thing as a definite musical term relating to time." As for the art in general, "I have already discovered that music is pitched too high."[174]

---

* Edison, aiming for wealthy connoisseurs, soon added even more elaborate disk players. By 1919, at the height of the phonograph craze, he was offering a luxury cabinet model at $6,000, or more than $87,000 in today's money.

His pickiness in the classical repertory extended to any Tin Pan Alley "hit" that he considered untuneful. He had no general prejudice against popular or vaudeville music,* even sanctioning, as the premier Diamond Disc release, a comic "coon" duet entitled "Moonlight in Jungleland," with chimpanzee chatter and birdsong obbligato. But he still insisted on approving every run of records that issued from his factory. As a result, the Blue Amberol and Diamond Disc catalogs grew at a slow pace quite unrelated to market demand. Jobbers became frustrated at the paucity of available titles and Edison's indifference to their repertory suggestions. Nor did his no-names policy convince them he was being anything else than perverse in withholding vital sales information.[175] Their protests grew so strident that he was called upon to explain it:

One of several reasons why I do not publish names of the singers is the "faking" going on in the musical world. There are many singers today with reputations upheld by advertising of the Italian & Jew[†] syndicates who never should be permitted to sing on any stage. They have no voices—just personality. The Composer & those artists who have beautiful voices [but lack] syndicated reputations, are ignored and the public made to believe that only Grand Opera artists can sing properly. The Victor Co. has carried this to the extreme. . . .

What I am trying to do is search the world for fine voices & instrumental soloists & to record & re-record their songs, etc. until they are musically perfect or as nearly so as possible & sell the records on their merits, giving the names after the public itself has given the verdict.[176]

The awkwardness of Edison's language suggested he did not altogether understand what he was saying. At any rate, the policy was soon reversed, and his artists got due credit—which was just as well, given his stinginess with recording fees. Rather than pay the enor-

---

* In 1917 Edison put out what is widely considered the first authentic jazz record, "That Funny 'Jas' [sic] Band from Dixieland." Arthur Collins and Byron G. Harlan vocal duet, Edison 5186.

† William Meadowcroft added the suffix -ish to this word when typing up Edison's statement for release.

mous sums demanded by stars of the caliber of Caruso and Paderewski, he looked for talent that was younger, hungrier, and willing to indulge a deaf man's belief that he knew more about music than they did.*

One who auditioned for him was Samuel Gardner, a twenty-year-old Russian-born violinist with great gifts but, as yet, no recognition. Instead of asking him to play, Edison, "very gruff, very kindly," asked him to comment on two violin records just received from Germany.

> He said, "They're very bad. These people who play have a shaky bow—woa, woa, woa." . . . I listened to one of them. The piece that was played was the "Ave Maria" of Schubert, arranged by Wilhelm. The first sounds I heard, I recognized a great artist immediately. . . . I heard a good strong vibration, very steady tone, and I wondered what he meant by the bad playing. That record was made by Albert Spalding.
>
> Then, he said, "I want you to listen to another one," the same piece by another player. Little different sound, but an artist. That was Carl Flesch. And this old man—I don't think he even knew the names of what he was listening to, he said, "These people have a shaky bow. They go woa, woa, woa." And I remember asking him, "How do you figure that out, Mr. Edison?" Well, he couldn't hear. . . . He gave me a microscope, a little glass, to look at the grooves. I looked and looked, but I didn't know what I was looking at. He said, "Don't you see how uneven those grooves are. It must be a straight line in the grooves."
>
> Wasn't much I could say.[177]

Gardner realized that no matter how Edison cupped his right ear to any music, he could not help receiving acoustic waves wrongly. The sea wall of his head had too narrow a sluice, breaking every high swell into foam. Because he was compelled to hear (or in this case, see) sound at the closest possible range, he could tolerate only the flattest undulations. What registered as full and rich to a normal ear, with the special overtones that made every instrumentalist's timbre and

---

* In 1911 the distinguished operetta composer Victor Herbert quit as Edison's music adviser, unable to stand any more professional insults.

every singer's voice unique, was torment to him, and he could not understand why nobody else flinched at the discord.[178]

"Mr. Edison," Gardner said, "that's not right. Your opinion isn't right."

Meadowcroft, who as ever stood at the boss's elbow, was horrified. "You musn't talk to Mr. Edison that way."[179]

Edison took no offense and asked the young man to record the "Ave Maria" without any left-hand vibrato. Gardner was desperate for the ten-dollar fee he would earn but could not bring himself to strip the bloom from Schubert's melody. "I'm just starting my career as a violinist," he pleaded. "I don't want to kill it right at the beginning."[180]

It occurred to him as he spoke that the cold, white, "spooky" tone Edison wanted might suit at least one piece: Chopin's Funeral March. He played it that way, hating the sound, and was rewarded with a check for ten dollars. At his insistence, the resultant record was issued without his name on it.

Gardner went on to have a long and honorable career as a performer, teacher, and Pulitzer Prize–winning composer. Asked in old age if he thought Edison's appreciation of music was hampered by poor hearing, he had a succinct reply. "His deafness had nothing to do with his musicality, because he hadn't any."[181]

WHO TOLD YOU YOU WERE A PIANO PLAYER?

Unlike most people with an aural problem, Edison went out of his way to publicize it as a professional asset. He willingly posed for a photograph to illustrate the *Cosmopolitan* article that showed him auditing a Diamond Disc with his right ear jammed right up to the speaker grille. "Beethoven, playing the sonatas that his deaf ears would not let him hear, formed no more pathetic picture than does Edison, with his gray head pressed against the machine that he made talk and sing," the caption read.[182]

It was just as well that the photographer did not know about the more extreme method Edison resorted to when he wanted to capture the last vestiges of a pianissimo emanating from a phonograph. "I hear through my teeth, and through my skull," he explained. "I bite my teeth into the wood, and then I get it good and strong." Many were the oak or rosewood Amberolas that he chomped in order to divert their reverberations into his brain. Because it was difficult for him to do so without slobbering, some cabinets lost their surface stain and looked as if they had been savaged by an enormous rodent. He even bit into the grand piano at Glenmont when one of the family was playing something he liked. A house guest that December, the educator Maria Montessori, was moved to tears by the sight of Edison attached to the frame, as though he were trying eat its sound.[183]

He insisted it was a "blessing" to be able to hear this way, because his cranial bone filtered out the haze of background noise—breathing, rustling, shoe creaks, heartbeats, subliminal vibration—that occluded the pure tones of music even in a muffled studio. "I have a wonderfully sensitive inner ear. I do not know that, in the beginning, it was any more sensitive than anyone else's, but for more than fifty years it has been wrapped in almost complete silence."[184]

What he called sensitivity was his inability at any distance to hear higher (or very low) musical frequencies. It threw the mechanical noises of sound production, such as the thump of a piano hammer, or the skitter of a violin bow playing spiccato, into abnormal relief.[185] Sound engineers were amazed that he could detect recording flaws they had missed in the studio. After subjecting an orchestral recording to a dental audition, Edison correctly traced a flaw in its sound to the top desk of the woodwind section. "The keys on that fellow's flute squeak." He used a felt-lined ear trumpet with a rubber diaphragm to measure the frequency of overtones by some method inscrutable to science. "I could strike any note on the piano anywhere and he could tell the exact vibrations," his music director, Ernest L. Stevens, testified. "I don't know how he ever did it. . . . It was remarkable, really."[186]

The same acuity, however, made Edison react pathologically to two effects essential to good tone production. One was the vibrato that so disturbed him in Gardner's playing. The other was tremolo, or rapid, single-note pulsations in the throat of a singer—an entirely natural phenomenon, albeit exaggerated by some show-off performers. To Edison, it was an aesthetic insult, "the worst defect a voice can have." He tried to stop it by making singers drink ice water before they stepped up to the horn, and on one occasion wondered aloud if taping a soprano's breasts flat might do the trick.[187]

When Sergei Rachmaninoff, arguably the world's greatest pianist, auditioned for a contract with Edison Records, Stevens neglected to warn him, "Don't play anything that's going to hurt the old gent's ears." After the first three thunderous notes of his Prelude in C-sharp minor, Edison interrupted to ask, "Who told you you were a piano player? You're a pounder." Rachmaninoff rose from the keyboard in silent outrage and reached for his hat. It was all Stevens could do to persuade Edison to let him record some further sessions, which included a crystalline performance of Liszt's *Second Hungarian Rhapsody*.[188]

Although Edison's aural dicta were more than most self-respecting musicians could bear for long, some—notably the lovely opera singer Anna Case—stuck with him because of his avuncular charm, the prestige of his name, and the unsurpassable quality of Diamond Discs. Their deep-lodged sonority and trueness on the turntable gave the illusion that the performers were somehow "present" inside the cabinet. Miss Case was the inspiration behind a dramatically effective advertising campaign that took advantage of this fidelity.

> *One day I walked into a shop, and they were playing one of my records. When I walked in the door, I started singing with the record and making my voice sound exactly like it. . . . They asked me to go on a concert tour with the machine. I gave a recital at Carnegie Hall, standing beside the machine, and copied the recorded sound. They didn't know when I was singing and when I wasn't. Of course, they could see my lips go, but by the tone quality, they couldn't tell the difference.*[189]

Other famous artists were hired to conduct "Edison Tone Tests" around the country, sometimes concealing themselves and the phonograph behind a curtain and challenging listeners to distinguish between live and recorded sound. The test results were equivocal enough to sell many millions of Diamond Discs through to the dawn of the electric recording era.*[190]

---

* In 2018 researchers at the University of California at Santa Barbara undertook to restore and digitize nine thousand Edison Diamond Disc recordings and make them publicly accessible under a grant from the National Endowment for the Humanities.

BUSINESSMAN JEKYLL

When Charles Edison reported for work at his father's plant in January 1914, he was twenty-three years old, a cheerful dropout from MIT, and had sown a considerable number of wild oats across the country, from Boston to Colorado to San Francisco. Although his seed-scattering days were by no means over, he was eager now to become a mature executive and learn all he needed to become second in command of Thomas A. Edison, Inc. If Miller Hutchison still nurtured a fantasy in that direction, Charles quickly dispelled it by visiting him at home one Sunday evening and grilling him until two A.M. on "all aspects of the business."[191]

Edison gave no sign of wanting to hand over power for some years yet. But neither did he try to impose his own management style ("An autocrat is the best kind of man to run an industry") on his son. Charles was both more willing and more able to hear the complaints of the Old Man's five thousand employees, whom he was distressed to find a demoralized lot. They had little corporate spirit, and were constantly on the lookout for jobs that paid better and abused them less. *We must never be paternalistic,* Charles told himself as he worked his way through department after department with the vague title of "Assistant to Mr. Edison."[192]

It was good for Mina to have him back home—not that she saw much of him at night. Like his father, Charles was usually out the door after dinner. But when they passed through the rock-walled gate of Llewellyn Park, the paths of father and son diverged. Edison swung left toward the laboratory, while Charles, mutating from businessman Jekyll into bohemian Hyde, headed for the railroad station and New York.

Mina clung ever tighter to Theodore, dreading the fast-approaching day when Madeleine would become Mrs. John Eyre Sloane and move in the same direction. Despite the efforts of both sets of parents, the young couple had overcome their own religious and emotional doubts and settled for a spring wedding. John had started an aeronautical manufacturing business in Long Island City, so they planned to rent an apartment in Manhattan. Conveniently for Charles, it would be in Greenwich Village.

ON THE BANKS OF THE CALOOSAHATCHEE

The Edisons went to Florida at the end of February for a final vacation together as an unbroken family. Madeleine was amazed to see a parade of Ford cars waiting to welcome them in Fort Myers, signaling the presence in town of her father's wealthiest friend. Edison had invited the Ford family and John Burroughs south for a long visit. He said it would be good for them to "get away from fictitious civilization."[193]

Madeleine liked the "awfully nice" Fords, but did not take to Burroughs. She found him aware of his own importance as one of America's most beloved writers.[194] Long-winded, simplistic, white of beard and low of brow, he carefully cultivated a folksy image not unlike Edison's, except that in his case it was unaccompanied by any hint of originality.

A cross-country automobile expedition to the Everglades in early March cemented the friendship of the three men, and presaged more such "vagabond" excursions in future. Hitherto, Ford had been the least popular of the trio, celebrated more for wealth than for charm. But he gave off a justified glow at the moment, having just announced a five-dollar daily wage for his workers in Detroit. This benefaction—far more than any other industrialist considered compatible with profits, and twice what Edison paid—had transformed the Ford Motor Company overnight into a mecca for skilled labor.[195]

Like most people rich or poor, Ford needed to be loved, but he was too attention-craving, too gauche in his enthusiasms (high-kick contests, bluegrass fiddling, health food) to hold on to public affection for long. Socially he was an incongruous combination of humor and humorlessness, intelligence and apparent idiocy. Rail thin and always immaculately dressed, eschewing the top-hat-and-cane uniform of other industrial magnates, he somehow lacked elegance. His bony awkwardness contrasted amusingly with Edison's relaxed ability to conform to the curve of any perch, whether it was a boulder or the shell of a rowboat. Ford could no more snooze in public than he could coax his jerky handwriting into calligraphy or match Edison as an easy, unhurried storyteller.

Yet there was much that drew them together, as former boy mechanics in Michigan—a shared disdain for Ivy leaguers, alcohol, and

haute cuisine; driving energy, graphic thinking, and delight in anything new. At this early stage of their relationship. Edison underestimated Ford's intelligence, just as he went too far, later on, in granting
him a poetic imagination. But when he noted that his friend possessed
"the practical ability of an Irish contractor foreman and a Jewish
broker," he used only one adjective that Ford would object to.[196]

After two weeks of birdwatching, fishing, and al fresco dinners on
the terrace of Seminole Lodge, Ford returned north impressed with
the beauty and tranquility of Edison's winter estate. He was more in
awe of his hero than ever, and ready, should the opportunity present
itself, to buy a similar property on the banks of the Caloosahatchee
River. "There is only one Fort Myers," Edison joked to a neighbor,
"and there are ninety million people who are going to find this out."
Mina and Clara Ford were cautious about the good-old-boy intimacy
developing between the two tycoons, and in no hurry to follow suit.[197]

Edison accepted Ford's adoration with the same affable equanimity
he displayed toward Hutchison and Meadowcroft and the dozens of
other moths, male and female, who had fluttered about his flame for
so many years. Although he was as liable as ever to irascible tan-
trums, they almost always related to business difficulties. Now that
his battery and phonograph divisions were booming (Diamond Discs
selling as many as fifty-seven thousand a day), he could open mail
from West Orange without misgivings.

He did not know that Charles had telegrammed Hutchison not to
send "any news, good or bad to Father unless absolutely necessary."
So for six weeks of warming weather, Edison was free to potter in his
bougainvillea-draped laboratory, chew cheap cigars (or a tobacco
plug when Mina was not looking), sleep twelve hours a day, and indulge his favorite recreation—automobile excursions over the rough-
est possible roads. It was hard to resist his childlike charm when he
was as relaxed as this. Madeleine begged her fiancé to come south and
hear her father rambling engrossingly over dinner on whatever
subject—parapsychology, physics, music, medicine—had his current
attention.[198]

Edison's tranquil mood was unbroken even when he heard, despite
Charlie's mail ban, that a fire had ravaged his movie studio in the
Bronx, causing $100,000 worth of damage. He merely expressed re-

lief that most of the hardware had been saved, saying he wanted to transfer production to West Orange anyway. A new talkie studio was being constructed on the second floor of the Kinetophone building, where he would improve talkie technology "to the limit . . . show the theatrical people that scientific people can beat them at their own game."[199]

He was indirectly acknowledging that Edison talking pictures, after their initial rampant success, had proved a commercial flop. There were just too few operators sufficiently trained to deal with the spread of distribution to thousands of theaters. When such a picture as *Mayor Gaynor and his Cabinet** was projected, amplified, and synchronized correctly, audiences still gasped at its truth to life. But more often the Edison advertising slogan "They Laugh—They Talk—They Sing" seemed to refer to talking and singing that bore no relation to screen action, and to laughter coarsened by jeers and catcalls. Complaints proliferated about actors taking a bow in mid-soliloquy and trumpets that blew ukulele music, as well as cramped stage movement and foreshortened plots. Hutchison felt that his misgivings about the Kinetophone had proved correct, but Edison was confident that its problems would be solved in time, as would those of the Home Kinetoscope and educational movies, which more and more school boards were rejecting. "There's no hurry," he kept saying. "There's no hurry."[200]

Madeleine seemed to feel the same about her wedding, which she was miserably inclined to put off because Mina and Mrs. Sloane were now squabbling about the religion of any future grandchildren. Eventually Mina capitulated to a private Catholic ceremony, providing it was not empurpled by the presence of a cardinal. Edison gave his daughter away at Glenmont on 17 June, and Mina wished the young couple well in a letter edged with black.[201]

NO HONOR OR GLORY

Eight days later Archduke Franz Ferdinand of Austria-Hungary was assassinated in Sarajevo.

---

* This series of talking-head interviews, now lost, was the first political documentary with sound.

As the chemistry of war percolated toward explosion in Europe and Russia, Edison like most Americans was concerned only with the pursuit of happiness in the world's freest, safest, and most technologically advanced polity. He had never confused the pursuit with attainment—"Happiness is only for the honest—that's a law that runs through matter as undetectable as gravitation"—but as long as the United States maintained its own peace, he saw no immediate threat to the national stability. To be sure, there had been a disturbing increase in violence on both sides of recent strikes organized by the socialistic Industrial Workers of the World. But they were mild compared to the prerevolutionary conflicts portending overseas, between emperors and peasants, autocrats and anarchists, colonialists and voteless majorities. Fortunately, three thousand miles of salt water separated Montauk Point from Land's End. Edison had seen all he wanted of the Old World on his recent tour of the nations now bristling at one another, and he was content to spend the rest of his life in the shadow of the Statue of Liberty, whose copper skirts he had "felt like kissing" when the *Amerika* glided past them on her way to the dock at Hoboken.[202]

Being an avid newspaper and magazine reader, Edison was well informed on political affairs without being particularly concerned with them. Except for his brief flirtation with Progressivism in 1912, he had never deviated from the orthodox, isolationist, pro-business Republicanism of his youth. Added to that, and congruent with his own domineering nature, was a strong belief in centralized power. "There's an open-mouthed philosophy of indolence today which finds a fine name in socialism. . . . I have more faith in governments based on oligarchy; the few govern the many through a law of evolution. The purest democracy shows that a few picked mentalities rise as instinctively to the ruling top as bubbles break on the surface of a stream. They are surcharged with the great initiative intelligence which contributes actively to the general good."[203]

If such a view made him a Social Darwinist, it did not extend to love of war. He read Friedrich von Bernhardi's protofascist *Germany and the Next War* with contempt, scribbling beside a passage in praise of bloodshed, "War kills off the best animals & leaves the degenerates to breed, a misapplication of Darwin's law." As for romanticizing battle as the breeding ground of heroes, "the thinking world can certainly find no honor or glory in it."[204]

SPEED HE HAD

Edison did not have to think much himself to infer, when the guns of August began to fire, that American industry would soon face a critical shortage of organic chemical imports from Europe. "Substitutes! Substitutes! We've got to find them. . . . It has been too easy for us to import our materials." He was himself the nation's largest consumer of German and British phenol, mixing a ton and a half of it every day into Condensite, the varnish that slicked Diamond Discs. It also happened to be a basic ingredient of high explosives, so foreign munitions factories would have a lock on it in the future.[205]

The chemist in him reasoned that phenol was a volatile derivative of coke. But few domestic coke ovens were designed to capture it. After inquiring in vain for an emergency supply of phenol from several chemical companies, Edison decided to synthesize the compound himself. Within three days he had invented a ten-step process of crystallization by sulfonation fusion. "It works beautifully," he told a friend, "and really it is indispensable." He then led another Insomnia Squad of forty draftsmen and chemists, working around the clock to design and construct a phenol facility at Silver Lake.[206]

Mary Childs Nerney once remarked that no one ever saw Edison rush at a thing: "Speed he had but not haste."[207] At all times, even in a crisis, he projected an air of catlike calm. Yet because his energy seldom slackened and he spent little time eating or sleeping, his achievements seemed sudden. The new plant opened on 8 September and became a cornucopia of the purest phenol he had ever used. So much of it poured out that he sold four or five surplus tons a day to envious competitors and expanded production to a second factory near Johnstown, Pennsylvania.[208]

His success enabled him to contract with coal tar companies to attach equipment to their ovens that sucked off rich gases for purification, liquefaction, and crystallization. One such extractor produced eighteen thousand gallons of benzene a day. He thus became a wholesaler of such valuable intermediates as antiseptic acetanilide, fragrant mirbane, toluene solvent, aniline salt, and—"Here's a jawbreaker," he used to say—paraphenylenediamine, the only known dye that turned gray furs black. Demand for it grew so great that he built a third plant to produce it exclusively. As a result, the discoverer of thermionic emission in vacuo found himself trading with furriers and fashion houses. Eventually Edison would have nine factories producing chemicals in short supply because of Britain's naval blockade of Germany.[*209]

As a pacifist, if not as a Democrat, Edison supported President Wilson's declaration of American neutrality in the "European" war. He was willing to profit by it—secretly selling phenol even to the German-based Bayer Corporation[†]—as long as he did not get into the armaments business. "Making things which kill men is against my fiber."[210] His conscience was untroubled by the fact that his S-type battery was designed to improve the performance of a torpedo-carrying vessel. Until the last day of summer, the submarine was still perceived around the world as a defense device, a protector of home ports.

But then on 22 September a German U-boat patrolling invisibly off the coast of Holland sank three British ships in less than ninety minutes. The news made clear that the war was going to be as different at sea as it already was on land, with the submarine and machine gun equally willing to abolish old notions of chivalry and fairness in warfare. It coincided with a secret report that made Josephus Daniels, Wilson's reform-minded secretary of the navy, receptive to everything Miller Hutchison had been saying about the dangers of lead batteries underwater. Submarine *E-2* had suffered an internal leak of sulfuric acid while taking a dive in the Atlantic. It was brought to the surface with difficulty, and all nineteen crew members suffered lung burns. Investigation showed that the acid had eaten through the walls of the ballast tanks and mixed with seawater, filling every compartment with the same chlorine gas that Germany would soon use against the French in Belgium.[211]

---

* He sold the last of them in November 1917, having profited greatly and restored much of his personal fortune.

† In the summer of 1915 the U.S. Secret Service investigated a briefcase mistakenly left on a Manhattan-bound train and found evidence of a $100,000 contract to buy and re-sell Edison phenol to German-American firms by means of a fraudulent "Chemical Exchange Association." The funds involved came from an espionage account at the German embassy. Edison was embarrassed when *The World* broke the story, although he had already committed the rest of his phenol surplus to the U.S. military.

Hutchison heard about the accident on one of his Washington visits. He leaped at the opportunity to invite Daniels to come north and see a pair of Edison S-type batteries being stress-tested in the Brooklyn Navy Yard. As an extra inducement, he suggested a preliminary visit to West Orange, where the secretary could get to know Edison over lunch, then drive with him to the Yard in one of Hutchison's limousines. Daniels not only accepted the invitation but, hearing that Edison had never been aboard a warship, arranged for a dreadnought and a submarine to be made available for his inspection on the appointed date, 10 October.[212]

The secretary was a portly, soft-spoken fifty-two-year-old North Carolinian, entirely unreconstructed in his vested suits and country bow tie. Wealthy and powerful as the longtime owner/editor of *The* (Raleigh) *News and Observer,* he had helped put Woodrow Wilson in the White House, and shared the president's patrician and racial prejudices. These were little in evidence now that he had moved to Washington and become an unctuous political operator. Only when his rigid Methodism was challenged, or when he allowed Hutchison to address him in slave dialect, did "Marse Josephus" reveal the bigotry that had made him a force in the Wilmington, North Carolina, race riot of 1898.[213]

Like most first-time visitors to the Edison plant, Daniels was awed by its size and complexity, and was even more humbled to meet the great inventor in his laboratory. He recorded his somewhat incoherent feelings on a Blue Amberol cylinder for the Phonograph Division archives:

> *The mecca of America is not in the national capital but at Edison's works. It is a great pleasure to see this wonder-working man at his task, and to find that although he is superman to all the world, he is very human. . . . In Europe today it is Edison who has made war more terrible, and therefore, let us hope, made it shorter, and that when this war ends, we will have no more wars.*[214]

It was not clear what Edison was supposed to have done either to worsen the war or to hasten its conclusion. Neither suggestion made much sense, unless Daniels, also a pacifist, meant that modern technology in general made the future frightening.

Edison assured a reporter, "I can't get interested in inventions for war," but from the moment he and the secretary arrived at the Navy Yard (to the sound of nineteen guns, and the sight of admirals saluting), he behaved to the contrary. Pacing the deck of the dreadnought *New York*, then descending into its supersecret control station, he marveled at the equilibrium of Elmer Sperry's gyrocompass. "It ought to have been discovered years ago—it's a cinch." He asked ordnance officers whether shells that smashed through armorplate were more lethal than those that exploded on impact. In the cramped torpedo compartment of submarine *G-4,* he boasted that he could easily devise a system of mechanical gills that would extract oxygen from seawater and allow the boat to remain immersed for months on end.[215]

The climax of the visit, as far as Hutchison was concerned, occurred when the commandant of the Yard showed Edison a steam-powered rig that was subjecting his S-type cells to a punishment no gyroscope could withstand. "Yes sir, we've rocked your batteries back and forth at all speeds and angles for the major part of two months and they haven't leaked yet."[216]

Edison was dismissive of the tossing, slamming machine. "Key it up, make it roll further and faster," he said. "The battery is all right."[217]

A BEAUTIFUL SIGHT

If Edison still nurtured hopes that his talking and teaching movies would succeed at a time when culture itself seemed to be in retreat, they went up in smoke at the plant on the night of 9 December 1914.[218] Just after sunset, at 5:25 P.M., spontaneous combustion took place in the film inspection building, a wooden, single-story structure crammed with nitrate stock. As the reels caught fire, they generated their own oxygen, transforming the little building into a tinderbox that soon touched off a nearby lumber shed, two alcohol tanks, and the five-floor "Wax House," where hundreds of cylinder blanks were stored, along with twenty tons of highly volatile phenol. That building became an inferno of such intensity that some of its concrete columns fused and flowed like candles.

Ladder companies from six surrounding towns fought to saturate the brick walls of the laboratory complex in the southwestern corner of the block. Their hose work was hampered by inadequate water pressure, even when a line from Edison's own artesian well was added to the main feed. A north-blowing wind came to their aid and fanned the blaze toward the carpenter shop and veneering department, both stacked with rare hardwoods. By six-thirty the fire was out of control. Undeterred by concrete or cinder block, it leaped east into the shipping, packing, assembly, and film print buildings, penetrating them through their dozens of wooden sash windows and crumpling tin doors like foil. Half an hour later the two main structures in the western yard, occupied by the Phonograph Division, were aflame too. The huge record building, holding nearly forty tons of Diamond Discs and Blue Amberol blanks, lit up with sequential evenness, window by window and floor by floor, as if a mobile flame thrower were advancing inside.

Among the crowd of twelve thousand townspeople who flocked to watch on the valley slope overlooking the plant was Edison. He was strangely calm, even cheerful after seeing that his laboratory was safe, screened from the wind by the long cement mass of the storage battery building across Lakeside Avenue. "Get Mother and her friends over here," he said to Charles. "They'll never see a fire like this again." *[219]

At seven-thirty a terrific explosion signaled that a benzene deposit had been breached. Multicolored flames shot into the night sky, illumining the landscape for half a mile around, while snowflakes fell indifferently.[220] The conflagration reached its height around nine o'clock, by which time it had engulfed thirteen buildings across more than half the complex. Destructive as it was to their contents, it collapsed only one top corner of a tall unit, number eleven, where finished Amberolas stood in crates, ready for shipping on the Erie Railroad. There the heat equaled that of a blast furnace. Slag dripped from buckling girders, and melted glass ran like water.

Edison surmised, as if he were monitoring an experiment, that some chemical vats on the fourth floor had burst. The resultant spill would have mixed nitric, hydrochloric, and sulfuric acids into an aqua regia solution, corrosive enough to crumble masonry.[221]

---

* At one point, a flushed and excited Edison was heard declaiming the last lines of Rudyard Kipling's poem "If."

The last open flames went out around midnight. "Mr. Edison, this is an awful catastrophe for you," an executive from the advertising department said in a shaking voice.

"Yes, Maxwell, a big fortune has gone up in flames tonight, but isn't it a beautiful sight?"[222]

The great fire of 9 December 1914.

When daylight came, he returned to his laboratory—wet-walled and sooty but intact—having been on his feet for more than twenty-four hours. He penciled a brief statement to give to reporters. "Am pretty well burned out—but tomorrow there will be some rapid mobilizing when I find out where I'm at." Then he stretched out on a bench, rolled his coat into a pillow, and went to sleep.[223]

WINTER SUNSHINE

"I see by the papers that the Edison factory has been largely destroyed by fire," Rep. Ernest Roberts (R., Mass.) said to Secretary Daniels that afternoon.

Daniels was testifying before the House Naval Committee on the need for an accelerated program of submarine construction, which he promised to push for if current tests on the S-type power pack were successful.

He said he had heard the same news. "The battery plant was not damaged, I am informed."

"The paper states that the factory was partially destroyed and 5,000 hands thrown out of employment."

"I do not know how that will be."[224]

Actually the damage was much less serious than the pyrotechnical display had portended. Edison's first estimate of his loss had been as high as $5 million; the true figure turned out to be $1.5 million. One workman had been killed in the first chemical explosion, but other casualties were few, thanks to the company's policy of regular fire drills. There had also been an efficient evacuation of vital documents, record masters, and portable precision instruments.* An astonishing 97 percent of the heavy machinery had withstood the heat and explosions, and the reinforced concrete walls and slab floors of the seven major buildings—all of them fireproof, except for their wooden elements—were largely intact.[225]

Far from laying off his employees, Edison hustled them into an emergency program of cleanup and retooling, while a construction company from New York worked triple shifts to make the standing buildings better than new. Within twenty days, six acres of floor space had been cleared. Square columns were rounded to hold more load, floor slabs were reinforced and slicked with his hardest portland cement. Partitions, which Edison disliked ("They make too much newspaper reading"), were reduced to a minimum, opening up vast spaces. When the vertical, flat, and cylindrical surfaces were painted white and winter sunshine streamed in through new, tilting, metal-frame windows, the result was as austerely elegant as anything later achieved by the Bauhaus school.[226]

Production of Blue Amberol cylinders resumed on the last day of December. Now more than ever, the Phonograph Division had to be

---

* Historic wax cylinders destroyed in the fire including the only recordings of such great nineteenth-century musicians as Hans von Bülow, not to mention Mark Twain telling jokes, do not appear to have survived.

the chief source of Edison's wealth. He needed all the profits it and his outlying chemical factories could rack up, since the fire insurance he carried paid out a mere $287,000 on a claim of $919,788. Far from being downcast, he radiated energy and excitement as he rose to the challenge of full recovery in the new year. "I am sixty-seven. . . . I've been through a lot of things like this. It protects a man from being afflicted with ennui."[227]

A GREAT RESEARCH LABORATORY

By the spring of 1915 Edison had created what amounted to a new plant, while his young efficiency expert, Stephen Mambert, made it the nucleus of a thoroughly modern corporation—Frank Dyer's dream of four years before. Mambert was a typical graduate of the progressive school of "management engineers," clerky, clean-shaven, and closely barbered, his neck movements constricted by a high, detachable white collar. Organization charts and budgeting—the geometry and calculus of business science—were his dry delight. Working companionably with Charles, who was still more at home in Greenwich Village than in West Orange, Mambert instituted Ford-like production and methods, demanded strict accounting of every purchase order down to the last paper clip, and put Thomas A. Edison, Inc., on the soundest financial footing it had ever enjoyed.[228]

Edison had his doubts about the company's burgeoning bureaucracy. "An 'efficiency' which submerges the individual," he remarked, "is an inefficiency." But it was a relief for him to hand over many executive chores and have more time to potter with new inventions—among them a portable, battery-powered searchlight that could throw a beam several miles. The idea for it had come to him during the fire. Willing as ever to jettison failed projects, he gladly took $50,000 from a Japanese entrepreneur for what was left of his talking picture business and looked around for some other large venture to engage him.[229]

There was no need to offer any help to his "personal representative in naval affairs." Since the beginning of hostilities in Europe, Hutchison had been selling storage batteries to the government as fast as the works on Lakeside Avenue could turn them out—most recently, seven thousand B-4 cells to operate wireless systems on warships.* Josephus Daniels had chosen USS *E-2*, crucible of last fall's chlorine-gas accident, as the first submarine to be equipped with Edison S-type cells, and he also approved their future installation in a larger vessel, the L-8, under construction in Maine. Edison priced the latter order at $90,000 and wrote the secretary: "Your telegram will cause the boys around here to lash me to the machinery to keep me from flying."[230]

Hutchison saw an opportunity to nudge the two men into a closer relationship, with a view to consolidating his own profitable position between them. He saw chains of zeroes accumulating in his bank account, like a submarine bubble trail, if he could only persuade Edison to abandon his pacifist sentiments and put his inventive genius at the service of the government. On 7 May a German U-boat saved him the trouble. It sank the Cunard liner *Lusitania* off the coast of Ireland, drowning 128 Americans and more than a thousand other civilian passengers. The tragedy caused even the most neutral-minded patriots to call for a program of "preparedness" to go to war if Germany ever struck again.

Among the first was Edison, who chose the Memorial Day weekend to make his own recommendations in a major article in *The New York Times*.[231] He advised against the creation of a large standing army and an overcommissioned navy but advocated enormous stockpiles of weaponry at strategic points along the nation's two seaboards: "All our war would be there." New battleships and submarines should be built with dispatch and held in drydock, thousands of military "aeroplanes" chocked for instant takeoff, and 2 million well-greased rifles kept in arsenals that could be got at by truck, instead of by trains, to speed distribution. Young American men, meanwhile, should be trained to spring to arms whenever their country called.

"I believe that in addition to this," Edison said, "the government should maintain a great research laboratory, jointly under military and naval and civilian control. In this could be developed the continu-

---

* Hutchison boasted a one-day tally of $415,000 in battery and phenol sales to Madeleine Edison Sloane on 20 March 1915. His commission was 20 percent, equal to $2.2 million in today's money. M. R. Hutchison to Madeleine Edison Sloane, 20 March 1915, PTAE.

ally increasing possibilities of great guns, the minutiae of new explosives, all the techniques of military and naval progression." When the time came—and sooner or later it would come—"we could take advantage of the knowledge gained through this research work and quickly manufacture in large quantities the very latest and most efficient instruments of warfare."[232]

Hutchison at once drafted a letter for Daniels to send Edison, begging him to help establish just such a "department of invention and development" for the navy, along with a board of eminent civilian scientists to supervise its operations. Daniels rewrote the document to include some of his own views and those of Franklin D. Roosevelt, his hawkish assistant secretary, and sent it off to West Orange on 7 July.[233]

> I feel that our chances of getting the public interested and back of this project will be enormously increased if we can have, at the start, some man whose inventive genius is recognized by the whole world. . . . You are recognized by all of us as the one man above all others who can turn dreams into realities and who has at his command, in addition to his own wonderful mind, the finest facilities for such work.
>
> What I want to ask you is if you would be willing as a service to your country, to act as an adviser to this board, to take such things as seem to you to be of value, but which we are not, at present, equipped to investigate, and to use your own magnificent facilities in such investigation if you feel it worth while.

Edison read the letter, then put in his out-basket with a scrawled superscript, *Hutch—note and return with comments.*[234]

YOU NOW RANK AS A COMMODORE

Many years later, when Edison was dead and Daniels was President Franklin Roosevelt's ambassador to Mexico, an aging and much diminished Hutchison recalled his excitement at seeing language he had drafted typed out on the Navy Department's heaviest stationery.[235] "There was my whole conception of the Board," he vaingloriously wrote Daniels.

I drummed it into Mr. Edison's head, until he took cognizance of the need and allowed me to use him as its sponsor. . . . I hopped on the Congressional to Washington, called on you, at your home, and said Mr. Edison would be glad to head such a Board if composed of men elected by the outstanding Scientific and Engineering Societies. . . . President Wilson appointed Mr. Edison and myself.

I will never forget the day we all signed the Oath and, jocularly, I asked Mr. Edison if he wanted to be measured for his uniform. "Uniform, h——!" he said. "But you now rank as a Commodore and really must wear a uniform," I replied. Turning to you, he said, "If I have got to wear a uniform, count me out. I want to be able to tell an Admiral to go to —— if he is in the wrong."

Hutchison was conflating, in retrospect, a pair of dates fifteen months apart. Nor did he mention the submarine disaster in between that could have sent him to jail. Daniels was either too tactful or too hazy himself to challenge the former chief engineer's memories, which were otherwise fairly accurate. Hutchison had even leaned on Edward Marshall, the *New York Times* writer, to publicize Edison's original call for what would eventually become the U.S. Naval Research Laboratory.

At the time—midsummer 1915—Edison's first priority was the composition of the proposed supervisory board. Daniels paid another visit to Glenmont on 15 July and approved his suggestion that it should be recruited from the Inventors Guild and ten major professional associations. These would be the American Aeronautical, Chemical, Electrochemical, and Mathematical Societies; the American Institutes of Electrical and Mining Engineers; the American Societies of Automotive, Civil, and Aeronautical Engineers; and the War Committee of Technical Societies. Each body would be asked to nominate two representatives, serving without remuneration as a patriotic duty.[236]

Edison notably excluded the National Academy of Sciences and the American Physical Society, on the grounds that neither was likely to nominate anybody "practical." He accepted the presidency of the board, and Daniels agreed to appoint Hutchison as his "personal assistant," on the tacit understanding that Edison's deafness would keep him away from most meetings. That made a round membership of twenty-four.[237]

Predictably, the scientific community reacted with outrage at being snubbed when the nominations were announced. Edison would soon enough learn to his cost that academic wrath was on a par with the *furor teutonicus* now ravaging Europe. But for the moment he could bask in the compliments he earned, from President Wilson on down, for putting his "genius" at the service of his country and publicizing the cause of preparedness. "The willingness of Edison to head the Board is a spectacular advertisement," wrote Waldemar Kaempffert, managing editor of *Scientific American*.[238]

The Naval Consulting Board, as it was officially named, posed for its first group photograph on the steps of the White House on Wednesday 6 October. Edison recognized only a few of the faces around him. Two were by no means friendly. Leo Baekeland, the Belgian-born inventor of Bakelite plastic, resented the success of Condensite and wrongly believed that Walter Aylsworth had infringed his patent. Frank J. Sprague's wolfish, frowning features had been forbidding enough thirty-seven years before, when Edison gave him his first break as a young inventor. If Sprague had ever cracked a smile since then, it was unrecorded in articles celebrating his brilliant achievements in the field of electric power. The gold medal of the American Institute of Electrical Engineers hung around his neck like a millstone, since it was engraved recto with the profile of Thomas Alva Edison. Sprague nurtured a grudge against his old boss for allowing the General Electric Company to erase his name from its history—forgetting that Edison had suffered the same corporate fate.[239]

A more affectionate face was that of Thomas Robins, Jr., inventor of the rubberized conveyor belt. He had developed it at the Edison mine in Ogdensburg, New Jersey. Forty-six years old, center-parted, Princetonian, and precise, Robins was a natural for the post of secretary to the board. Edison was also acquainted with William L. Saunders, a mining engineer, Elmer Sperry, inventor of the gyroscope, and the polymath Hudson Maxim, a bewhiskered eccentric equally devoted to poetry, penmanship, and explosives (as his prosthetic left hand attested). The rest of the board, consisting of sundry scientists, industrial executives, and engineers, were strangers.

When they went inside to meet the president, Wilson took the opportunity to announce his conversion to the cause of preparedness—

adding carefully that it should be "not for war, but for defense." He said that the army and navy would welcome "the cooperation of the best brains and knowledge of the country" to enhance national security.[240]

Anxious to dispel rumors to the contrary, Secretary Daniels organized a cruise down the Potomac that afternoon on the presidential yacht *Mayflower,* so that members of the board could get to know some admirals en route to the gun proving grounds at Indian Head. Edison could no more resist playing with the ship's communications equipment than a schoolboy.[241] The following morning, before the board met to organize itself, he was found at the aquarium in the lobby of the Post Office Department, so engrossed in the circulation of goldfish that a messenger hesitated to disturb him.[242]

At eleven o'clock he called his colleagues to order in the library of the Navy Department.[243] The board's first action was to elect Thomas Robins secretary. Edison then ceded the active role of chairman, for which his deafness obviously disqualified him, to Saunders.* Peter Cooper Hewitt, developer of the mercury-vapor lamp, became vice-chairman. It was agreed that the board should meet at least bimonthly at venues of its own choosing. Fifteen subcommittees were then appointed to advise on various aspects of naval and aerial defense. Edison did not sit on any of them. Instead he assumed the major responsibility of heading a committee of five that, on Frank Sprague's motion, would report as soon as possible on "the organization of a fully equipped and amply sustained laboratory for research and development . . . essential to the needs of the navy."

After lunch, he laid out his rough ideas for the facility. It should be built of indestructible concrete somewhere along the Atlantic seaboard, on "tidewater of sufficient depth to permit a dreadnought to come to the dock." There should be a large city nearby, "so supplies may be easily obtained," but—Edison's old desideratum for Menlo Park—not so near as to distract young researchers from their experimental work. The governing factors in its design were to be secrecy and security, with no visitation permitted. For maximum speed in developing inventions, it should also be a manufactory, equipped with a full range of shops,

---

* Edison thereafter became president of the Naval Consulting Board, but for the rest of his tenure, through 1921, he was loosely referred to as its chairman.

from a cast steel foundry and optical grinder to an explosives department, necessarily "separate from the main laboratory."

During a long afternoon's debate with the library door closed, Edison had to modify his original recommendation of complete civilian control as inimical to the navy. He agreed to let technologically qualified officers run the facility, as long as they did not impose "too much red tape." All its innovation, however, must come from outside the service, including freelance ideas that the board considered worthy of development.

Daniels interrupted the proceedings only once, with a reminder to the board that it had no legal status or funding yet. It should not call for a large increase in naval spending for fear of alienating pacifists on the House Appropriations Committee. This did not prevent a final resolution in favor of Edison's estimate that the laboratory would cost $5 million to acquire and at least half as much again to operate year by year.[244]

HIMSELF A MULTIMILLIONAIRE

"The soldier of the future will not be a sabre-bearing, bloodthirsty savage," the president of the Naval Consulting Board announced a week later. "He will be a machinist."[245]

Edison was holding a press conference in Chicago while his private Pullman car was hitched to the back of a train that would take him to the Panama-Pacific Exhibition in San Francisco and a reunion with Henry Ford. Clearly enjoying his new role as a prophet of preparedness, he said that the United States was "the greatest machine country in the world" and should be able, in time of war, to deploy mechanical agents of death twenty times more efficient than men on the battlefield.

"What do you think of the use of liquid fire and asphyxiating gases?" asked a representative of The New York Times.

"They are perfectly proper for use in defense, but not in offense. A man has a right to claw, scratch, bite, or kick in defending himself."[246]

Alarming as such sentiments were to Ford, a passionate pacifist, Edison's further declaration that he would keep the business of defense innovation out of the hands of government—"I am down on military establishments"—were even more so to his chief engineer.[247]

At the moment, "Doctor" Hutchison (as he now liked to be addressed, having gotten an honorary degree from his alma mater) urgently coveted the goodwill of the navy. His seat next to his boss at the recent board meeting—close enough to knee-tap Morse transcriptions of remarks Edison couldn't hear—represented another advance toward his dream of becoming equally famous. He would almost be Edison when deputizing for him at future meetings. This latest example of the Old Man's habit of feeding one-liners to reporters threatened to compromise the publication of a glossy booklet by Hutchison, entitled *The Submarine Boat Type of Edison Storage Battery*. He had sent copies to every vessel in the navy and was counting on the successful installation of S-type cells in the *E-2* submarine to make the battery universal and himself a multimillionaire.[248]

Piling promotion on promotion, Hutchison also planned to stage a media coup later in the month that would link West Orange with San Francisco and advertise Thomas A. Edison, Inc., as the most innovative company in history.

I HAD A GIRL ONCE

Edison had never been as sanguine about the S-type as Hutchison. Granted that its sealed, noncorrosive chemicals obviated chlorine-poisoning accidents, such as had caused the recent loss, off Honolulu, of submarine *F-4* with twenty-one aboard,* he knew that all storage batteries liberated hydrogen when electricity passed through them in reverse.[249] His own alkaline cells did so copiously at the earliest stage of recharge. This was no problem when they were installed in well-ventilated conveyances like automobiles and trains. But submarines, with their huge battery packs, had to be on the surface, with all fans going, to purge themselves of the odorless gas.

What worried Edison was that S-type cells continued to seep small amounts of hydrogen and oxygen after being recharged. In a letter to "friend Baekeland" written as his train crossed Iowa, he asked for advice on the problem. It was hardly necessary for one chemist to remind another that the formula $H + O$, at a certain level of concentration in a confined space, "reaches to an explosive mixture." He

---

* "The navies of the world," Edison announced after the *F-4* disaster, ". . . must expect catastrophes so long as they continue to use sulphuric acid in those vessels."

said that he had tried to absorb hydrogen in permanganate and also pumped it through unglazed porcelain—the latter an effective procedure, but impracticable underwater. "Won't you please think of other absorbers or methods & see what can be done."[*][250]

Henry Ford was in San Francisco when Edison and Mina arrived there three days later. He disapprovingly accompanied them on a cruise around the bay aboard USS *Oregon,* which had fired the first shot in the Battle of Santiago in '98. When Edison gave its big gun an affectionate pat, Ford said that as far as he was concerned, all warships were dodos, fit only for stuffing.[251]

The two men were making their first public appearance together as honorees at the Panama-Pacific Exhibition. Whatever glory Ford hoped to bask in paled in comparison to that accorded his friend, whom Californians had never seen before and who was to be honored on "Edison Day," 21 October, rather awkwardly flagged as the thirty-sixth anniversary of the electric lightbulb. But he clung close to his hero on the eve of the celebration, touring the exhibition grounds with him and cramming close whenever press cameras flashed.

At one point they stopped by the Western Union booth, where Edison sat down in front of an ancient telegraph sender.

"Where did you get this?" he asked the young woman in attendance, taking up the perforated tape and letting it spill through his fingers.

"It was made by you, Mr. Edison."

"Well, well! I had a girl once in New York that could send 119 words a minute with it."[252]

The exchange was more spontaneous than the one staged by Hutchison at the height of the festivities next day. It occurred during a lunch banquet in the California Building, and the audio connection he arranged between Edison's table and his own station in the library at West Orange went to ludicrous literal lengths, employing a multiple splice that linked sections of Samuel Morse's first signal wire, the first transatlantic cable, Alexander Graham Bell's first telephone cir-

---

[*] It took a month for Baekeland to come up with the idea of letting hydrogen bubble out of the submarine through a waterproof vent. Edison had to explain that what was needed was a way to vacate "without any gas leaving the boat to indicate its presence to the enemy."

cuit of 1875, a fuse from Edison's Holborn Viaduct power system of 1882, and other historic hookups.[253]

When Hutchison's cross-country "call" was piped into the lunch-room, it sounded loud and clear even to Edison, who was listening with the aid of a special amplifier. The chief engineer said he was reading his script under a "flood of mellow light" cast by an Edison incandescent lamp, powered by an Edison storage battery. Around him, he went on, were "several hundred of your friends," including Menlo Park veterans and all four of Edison's sons. Then he sprang a coy surprise: "This address is being made to you by your greatest favorite—the Edison Diamond Disc Phonograph. An Edison Granular Carbon Telephone Transmitter is transforming the sound waves into electric impulses which, after following the tortuous paths of copper between rivers and bays, over valleys, deserts plains, and mountains . . ."[254]

By the time Hutchison ran out of purple prose, his audiences east and west were more than ready to hear from Edison himself. Every-one knew that the Old Man never spoke over the telephone. But here he was, leaning into a mouthpiece and reading out yet more of Hutchi-son's advertising copy:

> *This is the first time I have ever carried on a conversation over the telephone. . . . A pretty big undertaking, but the engineers of the Bell system have made it easier to talk thirty-four hundred miles than it used to be to talk thirty-four miles. I heard the record of Hutch's talk very plainly. I should now like to hear a musical rec-ord. If you have one handy, I wish you would play that Anna Case record from Louise.*[255]

Later, when he adjourned to Festival Hall to receive a commemora-tive medal, he had to fight his way through a crowd so dense he lost both his wife and his hat.[256]

WHITE BLACKBERRIES

With a busy winter of naval-related work ahead of him, Edison took advantage of being on the West Coast to spend two weeks sightseeing by train and automobile. One excursion Henry Ford insisted on was

a visit to Luther Burbank's gardens in Santa Rosa. The little horticul-
turalist had a popular image as saintly as that of John Burroughs,
despite his desire to propagate a white super-race, in the same way he
bred white blackberries.[257]

Ford noticed, as Burbank led them around, that Edison was not
much interested in plants. But having recently met Harvey Firestone
in San Francisco, he raised the subject of rubber, predicting that it
would be the first vital import to be cut off if America entered the
war.[258]

"Could you devise a domestic equivalent?" Ford asked.

"I will," Edison said. "Some day."[259]

SHARE IT WITH NIKOLA

At sunset on 27 October Edison stopped by Universal City in the San
Fernando Valley to lay a plaque on the wall of Carl Laemmle's massive new film studio. The text lauded him as "the world's greatest
electrician" but said nothing about his two decades of dominance
over an industry he had himself founded. This was perhaps just as
well, because a federal court had just ordered the breakup of the "Edison Trust," as his Motion Picture Patents Company was generally
known. The judge found that a move by the trust to deny Laemmle
the right to thread independent films through a patented projector
had made it a conspiracy in restraint of trade in all aspects of the motion picture business.[260]

The decision was so much a reproof for Edison, and so much a victory for Laemmle, representing hundreds of maverick filmmakers,
that it was a wonder either man had consented to the ceremony. But
neither seemed to hold a grudge against the other. Edison could only
hope that the MPPC's right to license its products might be reasserted
by the Supreme Court. Otherwise his own studio, which depended on
trust royalties to stay profitable, would soon go dark. The future of
movies belonged to Hollywood, not to the Bronx or Fort Lee—and to
feature-length, star-studded, narrative scenarios, rather than the uncredited two-reelers he had specialized in.[261]

He returned home on 8 November to hear that he was once again
in line for the Nobel Prize. This time it was to be for physics, and
The New York Times reported that it would be awarded jointly to

Nikola Tesla. Edison declined comment, but Tesla was gracious, saying that "he thought Mr. Edison was worth a dozen Nobel Prizes." *262

In the event, they lost out to a father-and-son team of British crystallographers and were destined to die without receiving the honor.

A SHEAF OF PLANS

After three weeks of Western sun, Mina was mahogany brown but Edison remained pale, as if he had never left his laboratory. Mentally that was more or less true: his head was now full of defense technology requirements, from an invisible periscope (almost tauntingly wished upon him by the navy) to smokeless navigation lamps powered by the ebb and flow of waves. He wondered how the Brazilian firefly managed to contradict the second law of thermodynamics. "Its luminous organs are but specks, and the illumination generates no heat. I have studied that little bug for years, and tried it with the most delicate thermometers. . . . I'd give anything to know how he does it." 263

He was afraid that a closed-cylinder apparatus he had devised to burn off unventilated hydrogen in submarines might get hot enough to trigger an explosion. The navy had noticed seepage of the gas from the cells Hutchison had installed in USS *E-2* and was requesting a hydrogen detector that would warn its commander to surface and open all hatches if the diffusion level rose too high.264

Edison let his chief engineer deal with that problem while he drew up a sheaf of plans, including blueprints, for a naval research laboratory. He presented them to the Naval Consulting Board at its last meeting of the year, held at the Brooklyn Navy Yard and attended by three advisory admirals.

"This is the first speech Mr. Edison has ever made," Chairman Saunders joked, unaware that thirty-six years before, two thousand

---

* On 18 May 1917 the American Institute of Electrical Engineers awarded Tesla its Edison Medal. He reminisced on that occasion about working for Edison ("this wonderful man") as a newly arrived immigrant in 1884. (See Part Five.) For discussions of the internet myth that Edison and Tesla were bitter rivals, see Bernard Carlson, *Tesla: Inventor of the Electrical Age* (Princeton, NJ, 2013), 397ff., and the essay "Edison and Tesla" at http://edison.rutgers.edu/tesla.htm.

scientists had gathered in Saratoga Springs to hear a talk by the young inventor of the chalk-cylinder telephone.*

As speeches went, it consisted of the tersest possible comments as Edison laid sheet after sheet on the table, letting the drawings speak for themselves. They represented thirteen buildings and eight shops, and he was precise about the expensive equipment each would contain: "surgical apparatus . . . all universal tools . . . three five hundred KW turbo generator sets."

When he was through, the naval officers said he had convinced them such a facility would greatly speed up the development of new prototypes. Rear Adm. David W. Taylor, chief of the Bureau of Construction and Repair, thought Congress could be persuaded to invest $5 million in it. Rear Adm. Robert S. Griffin, chief of the Bureau of Steam Engineering, was not so sure. "If Mr. Edison will appear before the [House] Naval Committee with the plans and all the data that he has . . . it will make a very profound impression."[265]

No public exposure could have been less to Edison's liking, but he recognized his responsibility as board president. "I will go down to the Committee and explain that and fight for it if you want."[266]

A SPARK OF UNKNOWN ORIGIN

When 1915 came to an end, Hutchison wrote in his diary, "The year has been the happiest one of my life and I look forward to a most happy one for 1916."[267] In particular he anticipated passage by Congress of a bill, drafted by himself, providing for the installation of Edison S-type batteries on all American submarines.[268]

But just over a fortnight later, he fell from happiness to mortification. At 1:12 P.M. on 15 January a massive explosion rocked USS *E-2* at the Brooklyn Navy Yard, killing five men and injuring ten more. The submarine was in drydock at the time, undergoing modifications to its battery installation, and the whitish-gray smoke that shot out at the sound of the boom, along with a sailor still clinging to a section of the steel ladder, suggested a hydrogen blast.[269] Rescuers had to don oxygen helmets before entering the fume-filled interior. Along with the dead and wounded—some were burned beyond recognition, oth-

* See Part Six.

ers were pinned under mangled machinery—they found evidence that the boat's two hundred Edison cells had blown with such force that the engine-room bulkheads were curved back like cockleshells.[270]

The news reached West Orange late that afternoon. Hutchison rushed to inspect the wreck, while Edison, feeling far from festive, dressed in white tie for the annual dinner of the Ohio Society of New York. There was no question of him dropping out, since he was the guest of honor, and Secretary Daniels had come north to deliver a tribute to him. Both men looked stiff and grave as they endured what was supposed to be a jovial celebration. As usual Edison remained silent and, when accosted by a reporter afterward, said only, "I have no statement to make, except that the accident could have been due to any one of a hundred causes."[271]

Hutchison sent him a formal memo, blaming the disaster on lack of proper ventilation. It had been the coldest morning of the winter, with eleven degrees of frost at noon. "There were nine plumbers working in the boat, in addition to the crew present, and they doubtless wanted to keep warm." A preliminary survey by naval officials of the damage, which included a dismembered torso, indicated that the explosion had been caused by "a spark of unknown origin" igniting an unacceptably high concentration of battery gas. Daniels had no choice but to appoint a court of inquiry.[272]

It convened at the yard on Tuesday 18 January under the presidency of Capt. William H. Bullard. His youthful judge advocate was a lanky, tight-lipped lieutenant named Joseph O. Fisher. Another young officer, Lt. Chester Nimitz, served as counsel for the captain of the E-2. A retired naval engineer, Cdr. William H. McGrann, represented Hutchison and the Edison Storage Battery Company.

Hutchison did not endear himself either to Edison or to the court by talking to the press even before proceedings began. The New York Times headlined his self-defense in a front-page story that took care of what was left of Edison's reputation as a pacifist:

A FOREIGN NAVY USES EDISON BATTERY, TOO
M. R. HUTCHISON SAYS THREE SUBMARINES
EQUIPPED WITH THEM HAVE SUNK MANY SHIPS
HYDROGEN GAS NO DEFECT

AMOUNT THROWN OFF BY DEVICE
INFINITESIMAL, HE INSISTS—
E-2 EXPLOSION PURELY ACCIDENTAL[273]

He also blustered to *The Sun* that "the battery in the E-2 does not appear to have been injured in the least." As a result, "I see no reason to recommend to Mr. Edison any changes or alterations in the theory, construction, or method of installation of the Edison submarine type storage battery."[274]

Fisher at once brought out evidence that the "plumbers" Hutchison mentioned had been duct workers installing new, larger vents over the batteries. Lt. Charles M. Cooke, Jr., commander of the *E-2*, had expressed concerns about the behavior of his boat's power pack as early as September 1915. Some of the cells seemed to heat more than others, indicating an irregular rate of discharge. It was Cooke who had asked the Edison company, through Navy Department channels, for a hydrogen detector and an individual cell voltmeter. Nimitz showed that both requests were "held up by objections" from Hutchison, pending improvement in the submarine's ventilating system. The chief engineer had also advised that the ductwork be accompanied by a discharge of all cells to zero voltage, to even them up for sea service.[275]

Hutchison responded with yet another press statement, alleging that one of the *E-2*'s two ventilator fans had been idle on the morning of the accident.* Hence the commander of the vessel, not the manufacturer of the batteries, was responsible for allowing an explosive mix of hydrogen and oxygen to build up in the boat. When he appeared in the witness stand, Captain Bullard reproved him for public conduct prejudicial to justice: "Interested parties to this investigation will refrain in the future from quoting or giving articles to the public press." Red-faced, Hutchison said that reporters were entitled to know the full facts, since the battery seemed to be "on trial." This earned him another reprimand.[276]

He was lucky not to be on trial himself, for not warning Lieutenant Cooke, either verbally or in battery maintenance instructions, about

---

* An annotated diagram of the explosion in the Edison National Historic Park archive indicates that the main outboard battery exhaust was valved shut, and plates opened adjacent to the fans for increased inboard ventilation.

the hydrogen threat. Fisher kept harping on this dereliction, while failing to acknowledge that Cooke's worries about gas seepage had coincided with a study published by the navy itself on the vital importance of ventilating all storage batteries, Edison or Exide. There had been at least six lead-acid battery explosions in the Atlantic fleet during 1915—another fact ignored by the court. Hutchison suspected industrial lobbying behind the scenes.*[277]

He had to admit in further testimony that he "didn't know exactly" what was wrong with four quirky cells that the E-2's electrician, L. L. Miles, had complained of before the disaster. They lost their charge more rapidly than the others, which meant they began to recharge at zero, and liberated clouds of hydrogen while the rest of the plant was still powering down.[278] This caused a dispute between Hutchison and Fisher as to who was responsible for the gas buildup in the submarine— Hutchison as installer, or Cooke as caretaker of the cells.

Q Did you ever tell Lieutenant Cooke that the reverse cell of the Edison submarine battery, in closed circuit with other cells not reversed, generated gas to a greater extent than on normal discharge?
A Not that I know of. I did not consider it necessary any more than I would tell an engineer to keep water in his boiler.[279]

The judge advocate asked for Hutchison's haughty second sentence to be stricken. Bullard overruled him. "I think that does no harm."[280]

Later on, Fisher, still harping on premature reversal, saw an opportunity for some loftiness of his own. But Hutchison was better educated than he on the declension of Greek nouns.

Q This seems to be a phenomena that wasn't known to the officers of the Edison Storage Battery Company?
A I don't think that the phenomena—if you call it a phenomena—of gas given off by a reversed cell has been appreciated to a full extent by anyone.[281]

---

* Hutchison told Daniels that a detective in his employ observed "frequent consultations" during the course of the trial between a member of the court and representatives of Edison's principal competitor, the Electric Storage Battery Company.

It was a mistake for him to patronize an officer of the court. From then on, Fisher was determined to absolve the navy of all blame. He repeatedly referred to the *E-2*'s battery as "defective" and shouted at one point that Hutchison was a liar. This caused McGrann to object and demand an apology.

"I apologize for my bearing," Fisher said, "but my language stands."[282]

ALL THE MUCKERS WE NEED

By the time the court of inquiry was over and its secret report sent to Secretary Daniels, Edison had lost patience with prejudgmental headlines blaming Hutchison—and by extension, himself—for the deaths in Brooklyn Navy Yard. He issued a public appeal for industry support, accusing his competitors of "a colossal attempt to bring ruin to a product on which I have spent many millions of dollars and years of unceasing labor."[283]

He conceded that his chief engineer had been responsible for installing four nonsynchronous cells in the submarine, and negligent in not warning the Navy Department about the S-type's hydrogen potential until the day before the explosion.*[284] But Lieutenant Cooke should have kept the battery alive while the *E-2* lay in drydock and not allowed spark-prone metal work to proceed at the same time as the cells were being discharged inboard.

Daniels was sympathetic, and suppressed the report while appointing a technical board of examiners, including Nimitz, to decide once and for all if the Edison battery, properly handled, was superior to the Exide. This did little to ameliorate the damage that had been inflicted on Edison's reputation, just when his regular line of nickel-steel batteries was providing light and signal power to the nation's largest railroad systems, and traction to one-third of all electric trucks. In hindsight, he might have paid more attention to Hutchison's overdevelopment of the S-type. But for two years the monster battery had performed without fault on the navy's floating cranes in Boston and

---

* Hutchison had, however, been explicit about all kinds of battery gas evolution in his briefing booklet on the S-type, distributed throughout the U.S. naval command in the fall of 1915.

Honolulu—not to mention its secret success in foreign submarine operations. All that shining achievement was now bespattered with the blood that had to be hosed out of the *E-2*.[285]

"Yes, this is pretty bad," Edison consoled an anguished assistant. "However I can stand it."[286]

In the circumstances, he could have expected some aggressive questioning on 15 March, when he presented his plans for a national research laboratory to the House Naval Affairs Committee. But his charisma—in part the aura of world fame, in part the force of his jerky energy and blunt speech—cast a spell over the hearing room. There was also the always shocking effect of seeing how deaf he was, how dependent on Hutchison for shouted repetitions of everything said to him. It made him seem formidable and vulnerable at the same time. Perhaps in consequence, the *E-2* explosion was not mentioned.[287]

"The object of the laboratory is to perfect all the different details, or one unit of the war machinery, and do it quickly," Edison said in his opening statement.

> When I want to make a thing quickly, I put a hundred men on it instead of a few men, to carry it along for weeks and months; I put everybody in the shop on it. . . .
>
> In this laboratory I have all kinds of machinery; not manufacturing machinery, but all universal machinery, the same as they use in the great tool shops for making tools. . . . I can do almost anything in that shop. . . . I have laid it all out here; I have all the details of it, as far as I have gone, and the minimum amount for the land and the buildings and the machinery that we will want, I cannot figure it any lower than a million and a half, approximately; but I have left it so that you can increase it if you want it.[288]

Evidently Daniels, who sat listening, had persuaded the Naval Consulting Board to moderate its initial start-up estimate of $5 million. But he made clear that he foresaw "a very much larger laboratory" in future years, and showed a casual disregard for committee concerns about its upkeep in time of war.

REP. WILLIAM D. OLIVER (D., ALA.): How much would it cost?
EDISON: Well, we would work in three shifts of eight hours each—

never stop—and I should say over a million. You could string it along, if you wanted to.

REP. ERNEST ROBERTS (R., MASS.): Do you think you could get technical and scientific men enough to work three shifts?

EDISON: Yes, sir; I can get all the muckers we need—a lot of them.

HUTCHISON: Mr. Edison calls experimenters "muckers." He is president of the muckers' association in his own plant.

REP. ROBERTS: Suppose in a laboratory like this there was developed a satisfactory aeroplane, and the next day some inventor outside put on the market a superior aeroplane engine. How much do we gain by the enormous amount of money we have spent in the laboratory to perfect an engine?

EDISON: Well . . . Drop the other one and take the new one.[289]

At another point he had the whole floor laughing when Roberts asked him if patented parts might cause a problem in the manufacture of munitions. "I would not pay any attention to the patents," Edison replied. "Settle afterwards."[290]

Although he talked about the facility as if it were already built and managed by himself, he conceded with a shrug that it could be run by "the Navy Department, I suppose." But its creativity would depend on the services of civilian scientists and engineers working not for love but for money. "If the other fellow will pay $12,000 a year, you will have to pay 14,000, or you will not get them."[291]

"I move," said Representative Oliver, when he was through, "that we rise as an expression of our respect and appreciation."[292]

Spectators were treated to the extraordinary sight of twenty-one congressmen standing and applauding as Edison gathered up his plans and quit the testimony table.[293]

IF HE COULD ONLY FEEL RICH

After a vacation in Fort Myers sweetened by Madeleine's delivery of Thomas Edison Sloane, his first grandchild,*[294] Edison returned north to march in New York's great Preparedness Parade on 13 May.

---

* Madeleine's news coincided less agreeably with another of Beatrice Edison's avowed pregnancies. Her confinement was "expected" around the end of June, but thereafter she and Tom remained childless. Beatrice Edison to MME, 19 June 1916, CEF; Madeleine Edison to MME, ca. late August 1916, DSP.

Any hopes Daniels might have had that his celebrity recruit would support President Wilson's pacifist reelection campaign were dashed that same day, when Edison was quoted on the front page of *The New York Times* saying that Theodore Roosevelt was "absolutely the only man" to lead the country for the rest of the decade. "He has more real statesmanship . . . and a greater executive ability to handle the big international problems that will arise at the close of the war, than all the other proposed candidates put together." The paper also printed Roosevelt's emotional response. "My dear Mr. Edison: I am so profoundly touched by your letter concerning me, that I shall ask the Roosevelt Non-Partisan League to give the original to me. I wish to hand it over to my children."[295]

TR was not a serious candidate, although he had recovered from his attack of Progressivism and was willing to let his name be put forward at next month's Republican National Convention.[296] Ever since the sinking of the *Lusitania,* he had been the nation's most ardent advocate of intervention in the European war. All that was lacking to increase the impact of the endorsement was for him and Edison to stride together beneath the ninety-five-foot American flag strung across Fifth Avenue at 55th Street. But the colonel was detained at a Boy Scout function on Long Island. So Edison marched instead with his fellow board members and 125,000 other patriots, waving so many smaller flags that for eleven hours Fifth Avenue was a slowly flowing river of red, white, and blue *pointillé.**

He dominated the Preparedness Parade, stimulating a roar among bystanders as he strode along, waving and smiling, with the energy of a twenty-year-old. (Mina, wearing a large violet sun hat, tried to keep pace with him on the sidewalk, terrified that he would be attacked by pacifists.) But Edison's appearance of blitheness was deceptive. He was a careworn man at this time, sleeping uncharacteristically long hours and beset by money and other worries. "Poor dearie, if he could only feel rich once in his life," Mina wrote Theodore. Apart from the ruinous effect of the *E-2* disaster on Edison Storage Battery Company sales—just when the National Defense Act was about to create a huge

---

* The parade inspired Childe Hassam to paint his famous series of New York flag paintings.

demand for portable power—he was beset by labor unrest and a pollution lawsuit at one of his phenol plants. Although the breakup of the "Edison Trust" had not yet been confirmed by the Supreme Court, his movie business was in terminal decline. The House Naval Affairs Committee disappointed him by recommending a $2 million appropriation for his dream research laboratory—exactly as much as he had asked for, but far less than he hoped he might get, in response to his heavy hint, "you can increase it if you want." He drafted an angry letter to Sen. Benjamin Tillman, Democratic chairman of the parallel committee in the upper chamber, saying that if the $2 million was in any way reduced by Congress, "it would be better to drop the whole thing altogether."[297]

Another worry was what to do about Hutchison. Stephen Mambert and Charles were demanding that the chief engineer's profiteering at the expense of the Edison Storage Battery Company be restricted henceforth.[298] Looking askance at "Colonia," the mansion Hutch had acquired in Llewellyn Park (and staffed with three Japanese servants), they sought to impose new rules upon him that would restore the company's right to market its non-submarine batteries to the government. Edison felt obliged to agree, since Charles was now officially—as of 12 June 1916—chairman of the board of Thomas A. Edison, Inc.

THE HIGHEST KICK I HAVE EVER SEEN

After Theodore Roosevelt dropped out of consideration for the Republican presidential nomination in favor of Charles Evans Hughes, a fence-sitter of almost gymnastic equipoise, the way was clear for Edison to yield to pressure from both Tillman and Secretary Daniels to come out for Woodrow Wilson—to whom, after all, he owed some fealty in his position on the Naval Consulting Board.

He did so in a letter released by the Democratic National Committee for maximum impact on Labor Day weekend. "They say [Wilson] has blundered," he wrote. "Perhaps he has, but I notice that he usually blunders forward." The endorsement rated national headlines. It warranted a warm welcome from Secretary Daniels in Washington two weeks later, when Edison, Hutchison, and eighteen other members of the Naval Consulting Board officially became officers of the navy and swore to "defend the Constitution of the United States against all enemies foreign and domestic." *[299]

They were compelled to take the oath under the authority of the recent Navy Bill, which gave a broadside of instructions as to what Congress expected of them in return for its $2 million appropriation:

> *Laboratory and research work on the subject of gun erosion, torpedo motive power, the gyroscope, submarine guns, protection against submarines, torpedo and mine attack, improvement in submarine attachments, improvement and development in submarine engines, storage batteries and propulsion, aeroplanes and aircraft, improvement in radio installations, and such other necessary work for the benefit of the government service, including the construction, equipment and operation of a laboratory, [and] the employment of scientific civilian assistants as may become necessary, to be expended under the direction of the Secretary of the Navy.*[300]

Edison was less concerned by the contradictory length and vagueness of this list, which could be left to subcommittees, than by ominous signs that Washington bureaucrats disliked his idea of a naval research laboratory operating far from their purview. Already there had been a move on Capitol Hill, fortunately quashed in the Senate, to order its construction in the District of Columbia. All Edison's board colleagues joined him in protesting this preemption of their advisory privilege. "They attach great importance to having the location question decided upon after conference and investigation," Daniels advised Rep. Lemuel Padgett, chairman of the House Committee on Naval Affairs.[301]

At the general meeting of the board that morning, 19 September, Edison was elected head of a six-man committee tasked with reporting on some fifty possible sites for the laboratory. The other members were Sprague, Baekeland, Robins, Whitney, and Addicks. Secretary Daniels, attending as an honored guest, assured them, "I wish you

---

* From this date on, Daniels addressed Edison as "Commodore," unbothered by the latter's conviction that his own name was "Dannels."

gentlemen to understand that I have no views myself at all as to the place where it shall be located." But he privately allowed to Padgett, "It may be that Washington is the best place."[302]

On 6 October, Daniels, as a loyal Democrat anxious to reelect President Wilson, traveled to New York in a desperate effort to sock Edison and Henry Ford for campaign funds. American voters seemed to dislike both major party candidates equally, so Charles Evans Hughes's lack of charm was not proving the hindrance Democratic strategists had hoped for. There was a chance that "the Bearded Lady," as TR called him, might win. To emphasize the seriousness of the situation, Daniels brought the chairman of the Democratic Party, Vance McCormick, with him. The meeting, over lunch in the Biltmore Hotel, did not go well, as Daniels recounted in his memoir.

> I do not suppose anything so strange ever occurred at a luncheon in New York and elsewhere. . . . After the first course, Edison, pointing to a large chandelier, with many globes, in the middle of the room, said, "Henry, I'll bet anything you want that I can kick the globe off that chandelier." It hung high toward the ceiling. Ford said he would take the bet. Edison rose, pushed the table to one side of the room, took his stand in the center and with his eye fixed on the globe, made the highest kick I have ever seen a man make and smashed the globe into smithereens. He then said, "Henry, let's see what you can do." The automobile manufacturer took careful aim, but his foot missed the chandelier by a fraction of an inch. Edison had won and for the balance of the meal or until the ice-cream was served, he was crowing over Ford, "You are a younger man than I am, but I can out-kick you." He seemed prouder of that high kick than if he had invented a means of ending the U-boat warfare.[303]

When Daniels broached the subject of campaign finances, Edison made a convenience of his deafness. Ford was just as tightwadded, although he did consent to place a number of paid endorsement articles in national newspapers. These may have helped Wilson's subsequent victory, attained by a margin so slim that for fifteen days Hughes refused to concede.[304]

WE WERE IN NO WAY TO BLAME

Although Edison stayed away from almost all meetings of the Naval Consulting Board, he threw himself with passion into the tasks visited upon it by the Navy Bill. Making his first foray into ballistics, he invented a large-caliber, self-stabilizing shell that obviated the need for rifling and thereby reduced gun barrel erosion. He attended target practice exercises off the Virginia Capes and made sarcastic notes on the low standards of fire control* on even the newest battleships. The USS *New Jersey* had overly vibrating rangefinders, the *New York*'s weak searchlights were good only for letting "the Enemy know where you are," not one of the *Nebraska*'s 185 guns scored a hit, and the *Florida*'s "appalling" command communications network would infuriate a housewife calling her grocer. "Here we have the entire nervous system of the ship's battle organization," he wrote, "depending for its operation on a system that even in the piping times of peace will not stand the strain of ordinary drills."[305]

His main compulsion in the last weeks of 1916 was to persuade the other members of the committee on sites to recommend Sandy Hook, New Jersey, as an ideal location for the naval research laboratory. He stressed its easy connection by speedboat to Manhattan and Brooklyn, while not mentioning the equal proximity of West Orange. "Rough and quiet waters on the two sides, twenty feet or more at the old Railroad Dock where steamers for New York departed—now abandoned. The Government has a Railway running full length of the Hook. . . . There are 1300 acres and the Laboratory could easily get 100 to 150 acres and have the use of much more for special experiments." There was a fort and proving ground at the tip of the promontory, ideal for testing the big guns he intended to design and forge on the spot. Nearby on the mainland was a bluff with views of the whole of New York Bay. Edison thought this vantage point would be ideal for marine visibility tests, essential for the development of submarine detection technology.[306]

Baekeland, who favored building the laboratory in Annapolis, ob-

* Fire control is the aiming, balancing, and concentration of naval gunfire on moving targets.

served that Sandy Hook had "the very great defect" of being remote from the national capital. "If the Lab is to be a success," Edison countered, "it should be as far from Washington as possible." *[307] Other members of the sites committee shared Baekeland's preference for Annapolis, pointing out the advantage of having the Naval Academy next door as an intellectual resource. It was true that congressmen and navy bureau chiefs were more likely to pay visits to the small city on Chesapeake Bay than to a sliver of sand and saltings 225 miles farther north, but since the products of the laboratory would have to be approved, sooner or later, by those same officials, Baekeland saw an advantage in making it reasonably accessible to them.

The result, much to Edison's annoyance, was a report prepared for him to sign, declaring, "We are unanimous in favor of Annapolis." An overriding consideration, in view of "the great and regrettable reduction" in Congress's appropriation to the board, was the fact that the land available at the mouth of the Severn River was already owned by the government, and could presumably be gotten free. And instead of Edison's original idea of a troika command representing military, naval, and civilian interests, the report recommended a single navy officer in charge, responsible only to Daniels.[308]

After forty years of having his own way in planning laboratories, Edison was outraged at this attempt to coerce him. He refused to sign, on the ground that a facility so closely allied with the Naval Academy would become scientific rather than technological, and would manufacture theories instead of sophisticated new weapons. "I believe that I am right in re Sandy Hook & of a Rapid Constructing Laboratory," he wrote Hiram Maxim, "and I am going to stick to it. I shall never attach myself to a dead Government operated concern. If I can't get quick results & plenty of them then I will not play the game."[309]

Ignoring an appeal from Frank Sprague to accept a revision of the report that would include arguments for and against the Hook, he wrote a seventeen-point dissent of considerable force. It argued that uninhabited remoteness was essential to security, that the fragility of

* Edison had earlier prospected Fort Wadsworth and Governors Island in New York Bay, and even the Hudson Valley, but "on account of the ice, I did not go beyond Tarrytown."

such a wisp of sea-washed land almost duplicated marine conditions, that the dunes were perfect for "aeroplane" development, and that New York's wealth of specialist shops could be relied on to supply the most obscure raw materials at an hour or two's notice. He envisioned a secret factory of invention operating "on a war basis" twenty-four hours a day. "As to the management of the proposed Laboratory, I believe it should be civilian."[310]

Edison was humiliated when the full Naval Consulting Board rejected his minority report and accepted that of the majority. Sperry, acknowledging that the vote had been more political than practical, sent him an apologetic private letter. "This is written to assure you that you have been an inspiration to us all by your example; and your devotion to the cause . . . touches us all deeply. So, my dear Edison, do not for a moment be discouraged, because it may all come out for the best."[311]

Daniels was even more sympathetic. He knew that Edison had also been badly bruised by a leak of another report: the recommendation of the technical panel appointed to examine the E-2's power pack that "no Edison battery be installed on any of our submarines until further tests have shown that their disadvantages have been overcome." Its implication that nickel-steel technology was inherently inferior to that of lead-acid cells (which needed just as much ventilation when recharging, on top of their propensity to cause chlorine gas accidents at sea) had struck to the heart of the old inventor's pride. Hutchison warned Assistant Secretary Roosevelt's chief of staff, Louis Howe: "As you have not seen Mr. Edison when he is enraged, it would be difficult for me to describe to you the effect this article had on him. He has not gotten over it by any means."*[312]

Secretary Daniels was in an awkward position, feeling compelled to go along with the majority opinion while valuing Edison's more—as well as being indebted to him for his declaration of support for Woodrow Wilson. As gently as possible, Daniels asked him to consider switching his vote from Sandy Hook to Annapolis, for fear of a much more parochial alternative. "The feeling here at the department

---

* According to Hutchison, Edison's fury was such that he ordered the breakup of all the precision tools and dies that had gone into the manufacture of the S-type battery.

among the experts is not in favor of either place; they prefer the District of Columbia."[313]

Edison responded with bitter dignity. "I have it fixed in my mind, whether right or wrong, that the public would look to me to make the Laboratory a success, that I would have to do 90% of the work. Therefore if I cannot obtain proper conditions to make it a success I would not undertake it or be connected with it in the remotest degree."[314]

He mailed the letter on 23 December and spent the last days of 1916 in bed, having accidentally seared his throat with nitrous acid fumes. He also had a heavy cold that threatened to turn to pneumonia. It was a season of cold comfort, too, for his "personal representative" on the Naval Consulting Board. Hutchison wrote in his diary, "If I could go back one year and avoid the explosion of E-2, I'd give many thousands of dollars which it has cost us and especially me. We were in no way to blame, but the odium has gone all over."[315]

INFRA AND ULTRA

Edison's pulmonary prostration in January 1917 made it easier for his colleagues on the Naval Consulting Board to exclude him from a committee appointed to develop and design the research laboratory, at whatever location Congress eventually chose. The anger of many members against him for refusing to endorse Annapolis turned to sympathy, even fear, when newspapers reported that he was critically ill. "Should anything happen to wink out your life at this time," Hudson Maxim wrote, "it would be a human calamity of such magnitude as though another Atlantis were to go down under the sea."[316]

Maxim's imagery echoed an announcement by Germany that effective the beginning of February, *Unterseeboot* attacks on Allied ships would be extended to any American vessel suspected of carrying contraband. The Wilson administration still feigned a policy of neutrality in the war, but as any "hyphenated" German-American could tell, its real sympathies were Anglophile. Within seventy-two hours, Berlin made good on its threat by sinking the USS *Housatonic*, which was full of nothing more lethal than wheat. Wilson expelled the German ambassador from Washington and warned his government against any further "overt acts of war." The situation was fraught enough for

Secretary Daniels to suggest that Edison forget about board responsibilities and become an inventor again, developing secret defense ideas with the help of a twenty-five-man support team that the government would pay for.*[317]

Recovering from his pulmonary ailments, the Commodore plotted a series of experiments for the detection of submarines. He used his status as a senior naval official to persuade the Essex County Park Commission to lease him a secure location atop Eagle Rock, in the mountains overlooking West Orange. He had in mind an elegant, two-storied folly known as the Casino.[318] During the season it served as a restaurant, but now stood high, cold, and empty, its forty-mile vantage point perfect for his research purposes.† He wanted, for a start, to see if he could improve the vision of "splash observers," sailors perched at masthead to track the accuracy of fire and—if they focused on the right patch of ocean—the slight trail of foam that indicated a submarine approaching beneath them.

Pausing only to attend a fifteen-hundred-seat lunch at the plant in honor of his seventieth birthday, Edison established himself on the second floor of the Casino and returned to the study of optics, a science brought to his attention by Eadweard Muybridge in 1888. Sending along a pair of service volunteers to an ophthalmologist in New York, he wrote, "Please examine these two men and report if they will be O.K. for using Homatropin."[319] The drug—addictive and best kept under lock and key—dilated the pupils of the eye. He thought it might aid night vision, as well as the ability to pick out unfriendly shapes in dense fog. With medical permission, he administered it in solute drops, then put the men in a darkroom before a phosphorescent screen, tuned so low that for half an hour they might as well have been blind. Gradually they became aware of a patchy glimmering that resolved into readable letters.‡

---

* Edison was nevertheless confirmed as "president for life" by a vote of the Naval Consulting Board on 10 March 1917.

† The Casino, looking much the same as in Edison's day, is now a luxury restaurant in Eagle Rock Reservation.

‡ Edison had experimented with vestigial vision as early as December 1903, when he told a newspaper editor that the eye was "marvelously selective" in storing up low-intensity light.

This and other experiments along the infra and ultra ranges of the spectrum ("Want to keep all light out of the eye except Red & want different shades of red") taught him nothing beyond the obvious fact that the last thing a lookout needed, at sea and in sunshine, was a paralyzed optic nerve. The best he could do was build a massive hand-held visor that cut out ambient glare, not concerning himself about how it could be carried up a mast ladder. Going to the other extreme, he devised a low-slung glass bull's-eye for the observation of peri-scopes in silhouette, and he invented a submarine searchlight, after many measurements of the absorption of light in tubes of seawater. He tested the opacity of various chemical fumes as marine "smoke smudges," and the viscosity of oils that could be sprayed on the sea surface to blear periscope lenses. He mastered enough geometrical theory to bounce "dots" and "dashes" of light off tilted mirrors inside a ship, for emergency communications if its telephone wires were shot out. Perhaps his most ingenious radiant invention was a gyroscopic, disk-divided convoy lamp that flashed horizontally from vessel to ves-sel, uninflected by rolling, its sliced beams invisible at water level.[320]

WATER FALLING FROM WHITECAPS

By early spring Edison was in need of a laboratory with sea access, and quit Eagle Rock for Sandy Hook. Its panorama of New York Bay was wider and closer than the Casino's, and its secluded western shore, half holly forest, offered a sweep of quiet water for some bal-listic experiments he had in mind. The location was also ideally se-cure, being shared only by a Coast Guard station and Fort Hancock Army Base, which came with its own proving ground. He had only just built a shack for himself and his team on a pier south of the base when President Wilson asked Congress to declare war against Ger-many, in view of the continued "wanton and wholesale destruction of the lives of noncombatants" by U-boats.

Edison's first thought, even before Wilson signed the declaration on 6 April, was that his chemical and storage battery facilities were now vulnerable to domestic sabotage. He telegraphed Newton D. Baker, the secretary of war: "PERSISTENT RUMORS OR THREATS TO DAMAGE SOME OF MY MOST IMPORTANT PLANTS COMPEL ME TO ASK THAT YOU GIVE THEM IMMEDIATE MILITARY PROTECTION."

The request was referred to a distant department of the army, filed, and forgotten. However, the Office of Naval Intelligence considered his work to be important enough to supply him with a bodyguard.[321]

He threw himself into experiments with redoubled urgency. Conquering his aversion to men with degrees, he corralled four scientists from Princeton to advise him on the arcana of trajectory graphics, radio resonance, air and sea navigation, and gyroscopics. He was happiest, as ever, when inventing sonic devices—a waterproof microphone, an airplane direction finder, a depth charge that literally took "soundings"—or practicing chemistry,* as when he packed charcoal and soda lime into a gas mask that immunized masthead spotters from the narcosis of stack gas.[322]

One of his new recruits, Karl T. Compton, was a physicist well qualified to observe Edison as a body in perpetual motion.

> Barely taking time to say "how do you do," he took out his pencil and began to describe a problem which had been put up to him by the Naval Consulting Board—the problem of increasing the efficiency of the driving mechanism of a torpedo so that a larger amount of explosive could be stored in it without changing its range or size. He gave me a very brief history of the development of the present torpedo . . . and told me to come back and see him when I had a solution.
>
> In about three weeks I reported to him that I had found three fuels which seemed to offer possibilities. He disposed of these solutions in three sentences: "Fuel A can only be obtained in Germany. Fuel B has been tried but discarded because of the danger of explosions. Fuel C [containing alcohol] is no good because the sailors drink the damn stuff."[323]

COMPTON MARVELED AT the organic imagination Edison applied to the fabrication of low-resistance granules for his marine microphone. He bought hog bristles from a brush factory, electroplated them, then chopped the shining filaments with a microtome into tiny metal-rimmed platelets. Next he washed away the keratin in a solution he

---

* "I have always been more interested in chemistry than physics," Edison told a reporter in February 1917.

described as "the stuff men dissolved their murdered wives in,"* and packed the remnant rings into a diaphragm that he linked to a telephone receiver and amplified with a triode vacuum tube. After all this trouble, the granulated diaphragm did not satisfy, so he shaved down a disk of mica that worked much better. "He was uncommonly ingenious in figuring out ways of designing apparatus to do what he wanted it to do," Compton wrote, "and he was one of the most patient and persevering men who ever lived in carrying through his ideas to the last stage of comprehensive test."[324]

As finally rigged, Edison's underwater listening device let freighter crews hear the engine revolutions of a U-boat more than a mile away. It consisted of a cone dropped from the bowsprit ten or twenty feet ahead of the ship, where there was no sound of turbulence aft. The force of water flowing into the cone was counterbalanced by that of air compressing behind the diaphragm, allowing it to vibrate freely at any number of knots. He found that by swiveling the angle of reception, he could determine the path of an approaching torpedo. But the cone then became so sensitive that it registered distracting amounts of interference.[325]

Edison had to draw on all his acoustical knowledge to solve the problem. His explanation, in "Report No. 31" to Daniels, dated 30 April 1917, was not calculated to make the secretary look forward to Report No. 32:

> The greatest trouble is the sound of the water falling from whitecaps. But to day I have finally finished a lot of tests for cutting out the noise almost entirely by means of a mechanically operated resonance column.
>
> While listening for submarines, the water column moves up and down continuously, making a cycle in 6 minutes. Any movement of water by the movement of boat does not change the pitch, which is lucky. Any note from 2500 per second to 70 per second is picked out of the mass of sounds, making it conspicuous and by a brake on motor held there. If there are two more sounds and they repeat themselves and if no boat is in sight it is sure that they come from a submarine as no sound of high periodicity comes from the sea.[326]

* Potassium hydroxide, or caustic potash.

SHADOW SAILING

For the next year and a half,[327] Edison labored on land and at sea to perfect thirty-nine new devices, systems, strategies, and tactics of defense. Some of his ideas, such as a mast extension that lofted lookouts dizzily high in the sky, were landlubberly enough to amuse naval scientists.* But as long as a thing worked, he scoffed at their criticism: "My private opinion is that most of them lack imagination." He saw the war as a contest of technologies, not ideologies, and explored every notion that might help win it: a wireless telegraph message scrambler; a nocturnal telescope; cannon-fired steel mesh drapes to slow the momentum of enemy torpedoes; a turbine-headed shell that obviated the need for rifling; underwater coastal surveillance stations; a grease of Vaseline infused with zinc dust for rustproofing submarine guns; a silicate-of-soda fire extinguisher that glazed coal embers; and a water brake for quick turns of ship. In a particularly exuberant flight of fancy, he even proposed the dispatch of a fleet of self-steering skiffs to mine Belgium's Zeebrugge Harbor.†[328]

Almost more remarkable than this frenzy of invention—his last— was Edison's self-control in not boasting about it to reporters. He was as scrupulous about secrecy as he was about spending the government's money, constantly assuring Daniels that innovation could be economical.‡[329] Working twenty-hour days and financing some projects himself, he encouraged his academic recruits to work without pay, as a patriotic duty. The only perk he insisted on was a large yacht, and the navy's attempts to fob him off with inferior vessels in the early summer of 1917 made him threaten to "quit on experiments requiring a boat." Hutchison came to the rescue when he arranged

---

* When an officer of the Bureau of Ordnance objected to having to test some Edison projectiles, Daniels told him, "Commander, you may be right in this matter, but the public will think that Edison is right, so go ahead and test them." William L. Saunders to TE, mid-September 1917, PTAE.

† Edison's inventive flow was such that he begged Daniels in April, "Please do not send on other people's ideas. . . . I have more now than I could ever work out."

‡ When one series of experiments became costly, Daniels had to encourage Edison to "go ahead and spend as much money as will be necessary" to complete them.

with Assistant Secretary Roosevelt to lease a 210-foot "submarine chaser" to the Commodore indefinitely.[330]

USS *Sachem* had a crew of twenty and was captained by Lt. J. N. Patton, an experienced deep-sea navigator. Besides a large cabin earmarked for Edison, it had guest bunks for ten researchers, a conference room, and enough deck space to accommodate projectile launchers and observational gear.[331] Its only deficiency—not surprising in a warship—was that there were no amenities for a female passenger. The navy had a traditional prejudice against petticoats at sea, and Patton boggled when he heard that Edison wanted Mina to sail with him.[332]

This request had less to do with septuagenarian insecurity than with a deaf man's need for interpretive services. It was so unprecedented that Daniels stepped in to order the captain to comply. Mina was flattered to be acknowledged—for once—as someone Edison could not do without. But she looked forward with little enthusiasm to weeks, if not months, of being caged in a swaying stateroom, far from her garden, birds, and children. Theodore was back home from Montclair Academy, nineteen years old and ardent to sign up for military service, as John Sloane had just done. Mina, remembering the loss of another Theodore—her brother, killed in the Spanish-American War—was terrified that the boy might be in uniform when she saw him again. (She had no such worry about Charles, who had registered for the "managing director" class of draft exemption and claimed also to be going deaf.)[333]

President Wilson asked to see Edison and hear something of his plans before he went to sea. Impressed by their interview on 20 August, Wilson vowed to do anything in his power to help him. "I was an undergraduate when his first inventions first captured the imagination of the world," he told Daniels. "Ever since then I have retained the sense of magic which what he did then created in my mind."*[334]

Twenty-four hours later the Commodore and his wife were piped up the gangway of the *Sachem* at Hoboken. It set sail immediately for Sag Harbor, Long Island, where Edison had established a torpedo

---

* Edison, in contrast, remembered Wilson as a "conceited bookworm."

research station. He planned to use New London, Connecticut, as an alternate base for experimental cruises in the sound. After the first of these, Mina decided to make as much use as possible of onshore accommodations. "Papa and I sleep on a board bed with simply a mattress," she wrote Theodore, "and I can tell you it is hard."[335]

Captain Patton treated her with a strained courtesy that darkened to surliness as the weeks went by. Periodically she fled back to Orange for a few days of recuperative gossip with Madeleine, who was pregnant again and dreading John's imminent departure as a private attached to the Army Aviation Service in Washington.[336]

Meanwhile Edison saved Theodore from the draft by cannily advising the secretary of war, "I have come across many things that would be of value to the army," and would develop them if only he had a few extra engineers to help. Baker at once authorized him to hire "thirty men of the kind you have in mind." Theodore was the first qualifier, and Edison gave him an aircraft direction finder to try out at Hazelhurst Field in Mineola, New York. But Theodore's independent nature soon asserted itself, and he began work on a fearsome weapon of his own design, a self-propelled, unattached, toothed wheel loaded with TNT that was supposed to bite its way across the Western front, heedless of barbed wire, and explode in targeted trenches. It would keep him dangerously occupied for the duration of the war.*[337]

At first Edison enjoyed experimenting at sea. It was a novelty to be saluted as an officer of high rank, even though he stuck to his usual shabby suits and treated all hands on the *Sachem* with affable informality. But by Labor Day he was already bored by the claustrophobia of shipboard life and the slowness of port procedure, which was constantly subject to weather and communications delays. When Charles and Theodore paid a surprise visit to the yacht, Mina noticed how much he had been missing them, and how he clung to her after their departure.[338]

Another frustration for him was the apparent determination of the Navy Department, *pace* "friend Daniels," to block every one of his

---

* Theodore's long illustrated letter of 15 November 1917 explaining this device to his father shows that he was a born inventor.

Commodore Edison and the crew of USS Sachem, *1917.*

technological initiatives. He fumed over what he considered to be its anticivilian prejudice, although most of the rejections he received were respectful and exhaustively argued. In early October, he put a deputy in charge of ongoing experiments aboard the *Sachem* and took a temporary office in Washington to confront the bureau chiefs more directly.[339]

The suite Daniels found for him in the Navy Annex could not have been more imposing. It was the former sanctum of Adm. George Dewey, the hero of Manila Bay, who had died earlier in the year. But neither it nor Edison's own eminence had much effect on the line officers he tried to cajole. Hutchison was no longer available to help. His wings were so clipped by the management reforms Charles had instituted at West Orange, and by a new congressional ban on conflict-of-interest lobbying, that it was inevitable he must quit his job—but not so soon as to avoid responsibility for a sheaf of corporate lawsuits arising out of the *E-2* disaster.[340]

Edison expected to remain only a few weeks in the capital. His researches into marine optics had gotten him interested in camouflage and other trompe l'oeil phenomena, including the ultimate one of darkness. Before returning to his marine laboratory, he needed to make use of Washington's reference resources to find out where, and at what time of day, most U-boat sinkings took place in the war zone. The result was such an avalanche of charts and statistics that he used three aides to help collate them. In the process he began to see an offensive pattern around the British Isles that was not countered by defensive moves on the part of Allied shipping companies.[341]

"Just about what I expected," he said to Thomas Robins. "Those captains of the ships that do business from Norway to Greece don't know any navigation, they simply sail from one lighthouse to another, the Germans know that and sit there and wait for them."[342] He began a laborious effort to turn himself into a cartographer, on the assumption that Sir Eric Campbell Geddes, Britain's new first lord of the admiralty, would thank him for advice as to how the Allies could stanch the attrition of vital imports. In the process he discovered that statistical analysis was not much different from his deductive experimental method. By 21 November he had eight policy recommendations and forty-five "strategic maps" to send to Sir Eric. It did not occur to him that a member of His Majesty's Government might resent being patronized by an amateur American incapable of self-doubt. Anyone with "imagination," Edison wrote—again implying that few nautical men had it—could see that 94 percent of Allied losses occurred by day. Hence, "no cargo boat should enter or leave any English or French port except at night." This did not apply to convoys, which could more effectively be choreographed in at dawn and out at dusk, with such coordination that half the time, destroyer escorts would not be needed. A trinational routing office, working around the clock, would ensure that no lanes became overcrowded. Steamers from abroad should stop veering toward the lighthouses of the Irish coast, the Isles of Scilly, and the Severn Estuary, and steer a mid-channel course where U-boats were few. They should burn smokeless anthracite coal and drastically modify their silhouettes. Periscope observation, being low, depended on perpendiculars. "Cut off the masts, which are no longer of any use; cut down the smokestack to a minimum; close the gaps between the various deck constructions on the ships by canvas [flanks] to make an even contour. . . . In addition to this, all boats which cross the danger zone in an easterly or westerly direction should sail in line with the rays of the sun, or what I call 'shadow sailing.' "[343]

When he tried to apply his ideas stateside, "in the event that enemy submarines start operations along the American coast," he found that he would have to compile the charts himself, for lack of relevant records. The task kept him in Washington for another two months and caused him to brood more and more about governmental "red tape" and his thwarted plans for the naval research laboratory.[344]

Mina visited her husband at his hotel in mid-January 1918 and saw possible signs of depression. She itemized them for Theodore:

Suit a sight with spots
Shirt, dirty—when he changed, left studs & buttons in. . . .
Socks all holes
Glasses broken
Room, as if a cyclone had struck it
Suitcase not even unpacked[345]

A terse letter from Frank Sprague, complaining that the naval research laboratory was still unbuilt and unlocated ten months into the war, hardly improved Edison's mood. The irascible engineer pointed out that only one member of the Naval Consulting Board sites committee was responsible for blocking its construction at Annapolis. Because of the "intolerable" delay, sentiment in the Navy Department and on Capitol Hill had now swung toward the Bellevue Magazine site in Washington.[346]

"Frankly, we have no right to let this matter drag along further," Sprague wrote. He recommended the board vote unanimously in favor of Bellevue, or, what was even more embarrassing, ask to be relieved of expressing any preference whatever.[347]

Edison lost patience with the national capital and its endless political compromises. He ordered the *Sachem* to join him at Key West Naval Station in Florida, and he moved there with Mina at the beginning of February. William Meadowcroft sensed his acute frustration. "He does not wear his heart on his sleeve . . . only those who know him well realise how greatly he was discouraged."[348]

Explaining himself to Daniels rather than Sprague, Edison wrote on 4 March that he had always envisaged the laboratory as a place of rapid product development, not solely devoted to research.

> *Of course the board can do what in their judgment they think best, but they cannot expect me to agree to recommend what I firmly believe will be a failure. . . . I am so deaf that I have seldom attended meetings of this consulting board and am so entirely out of touch with it that it seems to be a species of deception for me to continue as its head, so I think I had better disconnect and work directly for the Navy, the board electing a young and aggressive man in my place.*[349]

A SORT OF GRUDGE

Daniels ignored this letter, perhaps on the grounds that a "president for life" would have to die to make his resignation effective. The alternative was for Edison to calm down on a tranquil, palm-shaded naval base as far south of Washington, D.C., as possible. He proceeded to do just that in Key West as the guest of the station commandant, Capt. Frederick A. Traut, a gentlemanly sophisticate who did much to restore his jaded opinion of naval officers. As an attaché in prewar Berlin, he had had extensive contact with Kaiser Wilhelm II and naturally knew much about German offensive technology. Traut insisted that the Edisons stay in his spacious villa on the base, and he further endeared himself to them by offering to find Theodore an uninhabited island on which to test his killer wheel.[350]

The arrival of the "boys of the *Sachem*" cheered Edison further. He at once began experiments with them, making use of the base's nautical, aerial, and wireless facilities and spending much time at sea, testing his quick-turn anchor. He even essayed a silvery sailor's beard. Mina sensed that she was again *femina non grata,* and removed in something of a huff to Fort Myers.[351]

Charles paid her a consolatory visit there, bringing with him Carolyn "Pony" Hawkins, his girlfriend of several years, a small serious woman furtive about her age and family circumstances.[352] One day the couple came in from the pier and announced they "wanted to get married right away." Mina was still struggling to recover from this when Edison telegraphed his approval to Charles from Key West: "IF YOU

HAVE DECIDED IT MUST BE THEN THE SOONER THE BETTER. CAN'T BE
ANY WORSE THAN LIFE IN THE FRONT LINE OF THE TRENCHES."[353]

The wedding, attended only by Mina, Charles's former nurse, and
a butler, took place on 27 March between a camphor tree and a cinnamon tree in the garden of Seminole Lodge.[354] Few spots in the world
were more fragrantly remote from the *Kaiserschlacht,* the German
spring offensive then reaching its peak in France. As Edison's mischie-
vous wire implied, Charles, by marrying at twenty-eight, had further
ensured he would not be called up anytime soon.

That did not stop another Edison *fils* from enlisting in the tank division of the army at the age of thirty-eight. Like many a lost soul in
ordinary life, William saw the war as his chance to prove himself as
something more than a playboy turned poultry farmer. He throve
under military training and soon won a sergeant's stripes. When the
time came for him to be sent overseas, Mina suppressed years of dislike and traveled to Gettysburg to wish him well.[355]

There was no question of sickly Tom doing anything as rashly brave
at forty-two. He had had no more luck growing mushrooms than trading away his famous name in dissolute youth. All he wanted was what
he had so long been denied: a job using his gifted hands somewhere in
Edison Industries.* The firm had recovered from the *E-2* disaster, and
under Charles's progressive leadership, its phonograph, battery, cement, and chemical factories were registering record profits. Fond of
Tom and increasingly self-confident as chairman of the board, Charles
decided to give his half-brother a break, once the war ended.†[356]

He also rid the company—at last, and at great expense—of the suave
services of Miller Reese Hutchison, canceling the last of his commission
privileges in exchange for a severance package of $112,589. Edison
made no protest. Hutchison resigned with worldly ambitions intact,
rented an office suite in the crown of New York's Woolworth sky-
scraper, and looked around for new business opportunities.‡[357]

---

* Between 1918 and 1934 Thomas A. Edison, Jr., was awarded ten U.S. mechanical
patents.

† Charles added the executive title of general manager to his chairmanship in January
1919.

‡ Hutchison prospered briefly, then became a victim of the postwar depression. By the
end of 1925 he was down to his last $275. He lived on until 1944, clinging to his title of

On 23 April, Edison wound down his experiments at Key West and transferred what was left of them to his garage at Glenmont, where he continued to putter and sketch with diminishing energy. The strain of working what he described to Josephus Daniels as "eighteen months steady for 17 hours per day" was at last taking its toll. So was a gathering suspicion that neither the navy nor the army would ever adopt any of his ideas. Madeleine noticed that the very sight of a uniform coming through the door at Glenmont drove her father into a sulk. "He seems to have a sort of grudge against the service." [*][358]

In August, Edison joined Henry Ford, Harvey Firestone, and John Burroughs on a camping trip along the Blue Ridge and Smoky Mountains. Burroughs was struck by his new tendency to sit apart and brood. "Occasionally around the camp fire we drew [him] out on chemical problems, and heard formula after formula come from his lips as if he were reading them from a book." [359]

When a submarine explosion almost identical to that of the *E-2* occurred in Brooklyn Navy Yard on 5 October, killing the captain and one officer, Edison noted that this time the hydrogen was given off by a lead-acid battery pack. This only reinforced his feeling that he and Hutchison had been discriminated against by the naval court of inquiry—even though a sympathetic Daniels was still suppressing its final report. [†][360]

By now the end of the war was imminent, but Edison continued doggedly to foist himself on the Navy Department, demanding a vessel larger than the *Sachem* to test the latest version of his underwater microphone. Daniels tried to make him understand that he had done more than enough to help defend his country. Edison's response was to say that the invention would be just as useful in peacetime as an anticollision device. [361]

"Doctor," and never ceasing to bask in the memory of having once moved among the great. "I spent the happiest days of my life with Edison. I knew him as did no other man."

* John Sloane did not improve the family atmosphere that summer by suggesting, to Edison's rage, that Charles and Theodore were evading war duty.

† The report was never released. Through his legal department, Edison aggressively fought the lawsuits arising out of the *E-2* disaster, with claims totaling more than half a million dollars. He settled them in 1919 for $66,000.

William got to the Western Front just two weeks before the Armistice on 11 November. He survived and was soon pestering his father for money to bring him home.[362]

Commodore Edison at Key West Naval Station, spring 1918.

THE THANKS OF THE NAVY

To Daniels's barely concealed irritation, Edison ignored the peace and kept behaving like a naval scientist through the summer of 1919. He failed to take a hint when Assistant Secretary Roosevelt withdrew the *Sachem* from service and provided him with a rusty substitute slated for demolition. When it too was withdrawn, in favor of a yacht lacking fifteen feet of bow, Edison asked if he was expected to end his marine experiments. On 10 September Roosevelt informed him that due to retrenchment throughout the service, it would not be possible to supply him with any further hulks: "I beg to extend to you now the thanks of the Navy for the efforts you have made."[363]

Edison replied with equal coldness, saying he would find it "quite satisfactory" to return to private life and "close my connection with the government." He did so without yielding to Daniels's plea that he bestow his valedictory blessing on a naval research laboratory to be built in the District of Columbia, "[on] government owned land and under the jurisdiction of the Navy Department." For the rest of his life he would chafe at the thought of this perversion of his original concept, presented to Congress with such high hopes three and a half years before, at the cravenness of the Naval Consulting Board in ultimately approving it, and most painful of all, at the failure of the armed services to accept a single one of the contributions he had vouchsafed them. "I made about forty-five inventions during the war and they pigeonholed every one of them."[364]

The only balm to his hurt in the last days of the decade came not from any official at land or sea but from an ambitious, Havana-smoking promoter of technological progress in the air. An immense Handley Page bomber flew low over Llewellyn Park and dropped a tribute to Edison *deus ex machina*:

THE GREATEST LAND TYPE AEROPLANE IN THE WORLD SENDS ITS GREETINGS TO THE GREATEST INVENTOR IN THE WORLD.

HUTCH.[365]

PART THREE

Chemistry

1900–1909

THE APPROACH OF Edison's fifty-third birthday, otherwise agreeably given over to freezing liquid carbonic acid at different soda strengths, was spoiled by a letter from his son William, a shifty Yale dropout who had promised, only a few months before, never to "darken the doors of your house again."[1]

Recently married to a young woman of "fast" reputation, whom Edison refused to receive, William wrote that they were living in New York and had "gained quite an axcess to society." All that was lacking to complete their happiness was an invitation to visit Glenmont. "Just for an evening that is all I ask. . . . I may have been a disobedient son but hardly a bad or worthless one."

In Edison's opinion, William merited all three adjectives. And his wife, Blanche, a Delaware doctor's daughter, was a spendthrift. No doubt the couple had worked through what little money William had inherited from the estate of Mary Edison, and now wanted "axcess" to some funds on his father's side. They assumed, as everyone did, that Edison was rich. In fact, his finances were seriously strained. Having squandered well over $2 million on an iron mine in the New Jersey highlands and sunk half a million more into a gold mine at Ortiz, New Mexico, he was in no mood to reinstate William on the list of his many dependents—along with Tom, another wastrel, with a wife even "faster" than Blanche.[2]

They were boys no longer, at twenty-two and twenty-four, respectively,* and cared little that their father had a second, much younger family to support. Their demands on him—William's alternately abusive and conniving, Tom's querulous and self-pitying—

---

* Marion, married in Germany, was about to turn twenty-seven.

were so continual that he relied on Mina to keep them off his back. Had she brought them up lovingly, rather than with a sort of dutiful affection, their vague memories of Mary might have been subsumed under a much more vivid experience of another "Mother." But try as she might, Mina could not conceal a natural preference for the flesh of her own flesh. Looking past her for a true sense of identity, they distantly perceived their father. He loomed as the only constant on their horizon, a mountain of familiar mass. Except that whenever they approached, it receded or faded. Was there, in fact, any father there?

Charles and Madeleine had no such confusion, and neither would Theodore when he grew older. Taking for granted the love of their parents for each other and for them, they accepted Edison's long absences from Glenmont as part of its domestic rhythm, just as they forgave him at home for being too deaf, or too abstracted, to pay them much heed. In a stream-of-consciousness letter written from MIT years later, Charles waxed nostalgic for the family circle that had never fully embraced his step-siblings:

> The sitting room, is it still as it was I wonder, the big window that looks out over the frosty lawn the litter of toys under it, the big table the fine old lounge. The moonlight just visible thru the high north windows as it filters thru the moving branches of the big maple tree. The canal coal burning brightly in the fireplace and near it the chair with the big glass lamp above it and in it the most widely respected loved and honored man in the world, reading, the pile of magazines on the floor, the leather covered books on the small table near the door, the chair beside it and the little foot stool and you, mother, in it. . . . And in the old south room the sister writing, writing, writing and the great dignified drawing room with its piano and the tall vase of american beauties or poppy flowers—nothing else—and the soft alabaster lamp in the drawing room. . . . The quiet dining room with its chilly exedra and the big uncomfortable den, the phonograph in Theodore's room and the white linen & bunch of red roses in my room and everywhere the touch & thought of one person and all the things I have dreamed of seeing again after all these years.[3]

Edison's chair and lamp in the sitting room at Glenmont, circa 1900s.

Madeleine shared Charles's adoration of Mina. They realized that her need to be assured of their love, regularly and often, was insatiable. And yet no amount of hyperbole could stave off her black depressions, which became more frequent as she grew older. She was the daughter of an inventor married to an inventor and could not shake off the neurosis that Edison cared more for his laboratory than her.

At thirty-four, she had long lost the teenage sexiness that had captivated him in the summer of '85 ("Got thinking about Mina and came near being run over by a street car"). Her firm contours had softened to plumpness, and her almost Indian "Maid of Chautauqua" glow was dulled by too much domesticity and too few winter vacations. She was by no means an overworked housewife, having a staff of eleven to keep the mansion clean, plush, and polished, her table loaded with food (but no wine—she disapproved of alcohol), and the estate and greenhouse immaculate.[4] Private schools and French governesses educated her children,* and a coachman and carriage were

* "We used to speak French quite as fluently as English," Charles recalled.

always on hand to drive her to ladies' luncheons and meetings of the local chapter of the Daughters of the American Revolution. Edison gave her a generous personal allowance, so she wore expensively dowdy clothes and could afford the best boxes at any opera or ball in New York.

For all these trappings and comforts of privilege, there was a grimness about Mina, buttressed by her staunch Methodism. She could not understand jokes, frowned on dancing and décolleté gowns, and deplored Edison's cheerful agnosticism. Every August she attended the Chautauqua Assembly, the dour adult-improvement festival her father had co-founded in upstate New York, to soothe her melancholy at spiritual concerts and lectures on such subjects as "The Problem of Suffering" and "The Teaching of Jesus Concerning the Industrial Order."[5]

It followed that she was horrified that boozy Tom had gone the way of so many millionaires' sons and married a blond showgirl, Marie Louise Toohey. Nor could she forgive William for telling Edison, "I could never love, [or] even like my stepmother. . . . I look upon her as the one who ruined our happiness."[6]

"NO ANS"

Edison's way of dealing with importuning mail—a substantial portion of the three thousand letters he received every year—was to scribble "No Ans" across the top of the first sheet. (Usually it was on the second that money was first mentioned.) He would leave it to his secretary, John Randolph, to decide if the supplicant at least merited a polite expression of regret. Randolph was good at trashing mail from religious maniacs, or desolate "widows" with masculine handwriting. But he felt less comfortable ignoring letters from Tom or William, even when Edison—a soft touch before they came of age—periodically threatened to cut their allowances for misbehaving. The secretary had emotional problems of his own, and could not help feeling sorry for them. Since his was, as it were, the only voice they heard in response to their appeals, and since he was the one who made out Edison's checks, they began to treat "Johnny" as an ally who might prevail on their father when they could not.

The question for Edison in the spring of 1900 was how to get both sons settled as far as possible from the fleshpots of New York. That became an urgent priority after William sent Mina a letter that came close to a threat of physical violence, and the yellow press described Tom and Marie ("late a splendid figure in pure pink meshings on the stage of the Casino") guzzling champagne *frappé de glace* at the Arion Ball in Madison Square Garden.[7]

Edison was less disturbed by that than by the way Tom, talking to newsmen, posed as his inventive heir apparent. "Reared in my father's own laboratory and educated by my father himself, I think I am capable of continuing the work which he will perhaps not live to finish." Randolph reported with annoying frequency that the young man was bouncing checks and selling his surname to all comers.[8]

Thomas Alva Edison, Jr., circa 1900.

On 9 May a flyer arrived at the laboratory announcing the appointment of Thomas A. Edison, Jr., as "consulting expert" to a new "International Bureau of Science and Invention," with offices in New York, London, and Paris. "The company's skilled technicians stood ready to "examine and look into any idea or ideas submitted to us (as per blank enclosed), giving their opinion of same, and if necessary making suggestions toward improvement." The bureau would help patent "any good invention" that resulted, in return for a two-thirds share of all subsequent profits. Its general manager, A. A. Frieden-stein, addressed a postscript to Edison saying Tom had assured him that "the above scheme had been endorsed by you. If so, would you kindly advise me to that effect, as I would not care to invest any money in any matter that was not strictly O.K."[9]

Having for years read similar missives touting the Edison Junior Improved Incandescent Lamp, the Thomas A. Edison Jr. & Wm. Holzer Steel & Iron Process Company, the Edison-Rogers Photo-scope Company, and the Thomas A. Edison Jr. Chemical Company (not to mention Dr. Edison's Obesity Pills and the Edison Electric Belt, which cured "all the ailments peculiar to women" by restoring strength to their "delicate organs"), Edison referred Friedenstein's pitch to a lawyer and turned to the construction of an invention more typical of himself, the longest rotary cement-burning kiln in the world.[10]

AN ENTIRELY NEW VOLTAIC COMBINATION

One day that May Edison stood on the west side of Manhattan, wait-ing for the Cortlandt Street ferry to Jersey City. Just two blocks away was Smith & McNell's restaurant, where once, famished, he had spent his last few coins on a plate of apple dumplings, a cup of coffee, and a cigar. It was the most delicious banquet in his memory, better than any he had subsequently been able to afford at Delmonico's. The dumplings were still available (at $2.95), and every now and again he would recommend them to a friend as "the finest you have ever had."[11]

All these years later, the streets he had lit jostled with horse-drawn traffic—overcrammed carts, cursing teamsters, and dogged drays whose manure and urine filled the air with such a miasma that a man needed the strongest cigar possible to counteract it. If New York was this jammed so early in the century, how long before it groaned to a standstill?[12] For two hours Edison jotted remedial ideas in his note-book.

Limited loads. Congestion. Resulting delay and expense there-
from. . . .
Solution:—Electrically driven trucks, covering one-half the street
area, having twice the speed, with two or three times the carrying
capacity. . . . Development necessary:—Running gear—easy. Motor
driver—easy. Control—simple. Battery—(?)[13]

Electric trucks and automobiles, as opposed to trolleyed streetcars and trains, depended on the lead-acid storage battery. It was thrillingly silent, but the payoff was a tire-flattening weight of lead plates, not to mention cells full of corrosive fluid sloshing between negative and positive poles, emitting an odor almost as acrid as horse piss. There were two alternatives, each with its own liabilities. Gasoline-powered vehicles were hard to start (their engines had to be hand-cranked into life, and could break a man's arm on the kickback) and laborious to drive, and called for crunching gear changes whenever they sped up or slowed down. In addition they were smoky and blaringly loud. Steam engine cars had to be water filled with annoying frequency, and in winter they took as long as forty-five minutes to warm up.* Being at least powerful, once they started puffing, they dominated the majority of the nation's eight thousand–vehicle "horseless carriage" market. But until one or another drive mode was made both practical and cheap, there was unlikely to be much lessening of the amount of manure on city roads.[14]

Edison's Cortlandt Street notebook indicated a willingness to bet that gearless, nonpolluting electric power would win out—if not for automobiles, at least for delivery trucks and cabs. What he had to do was to invent a reversible galvanic cell that was dramatically lighter and cheaper. It should compete with the high energy density of gasoline and be as clean as steam. It should generate current without surges or slumps, and enable many miles of traction before a recharge was necessary. It should tolerate short-circuits, rough riding over country roads, overcharges, and reductions to zero voltage. Admittedly these were huge imponderables, but his nature was to rise to

---

* In 1900 steam vehicles accounted for 40 percent of the U.S. automobile market, electrics 38 percent, and gas cars 22 percent.

such challenges. After ten years of carving up and crushing mountains, the scientist in him longed for a return to the atomic logic of electrochemistry.[15]

He refused to accept the shibboleth that lead, iron, and sulfuric acid were the only reagents that would ever generate enough current to move a car independently. It was "very beautiful in theory" but flawed in practice "because of the inherent destructive influence" of its liquid electrolyte.[16] At best, the massing of six or eight lead-lined, hard rubber cells* per vehicle caused a 15 percent loss of efficiency. Maintaining anything like that ratio for long required more skill and patience than most "automobilists" possessed—not to mention strength in lifting dud units out, a job that usually required two men. "If Nature had intended to use lead in batteries for powering vehicles," Edison declared, "she would not have made it so heavy."[17]

Without realizing that the question mark at the end of his Cortlandt Street notes portended the most agonizingly difficult project of his career, he began to look for an electrochemical yin-yang, a perfect counterbalance of positive and negative, attraction and repulsion, charge and recharge, and energy to mass, that surely existed somewhere in "Nature."[18] If not yet achieved, it was implicit in the primary battery invented a hundred years before by Alessandro Volta, the father of applied electricity: a cylindrical pile of acid-soaked cardboard disks, alternately separating thick medallions of silver and zinc, or copper and zinc. The damp layers reacted with the metal layers, generating a flow of power through the cell, from positive anode to negative cathode, the moment it was connected to an outside conductor. While copious, the flow was irreversible, draining away until the cell "died."

Gaston Planté's invention in 1859 of a secondary lead-acid battery that stored infusions of outside current amounted to a technological innovation almost as great as Volta's. As a boy chemist and teenage electrician, Edison had shocked and burned himself into an intimate understanding of both kinds of cells. But in adulthood, after nearly disfiguring his face with a splash of nitric acid, he had

---

* The words *battery* and *cell* are confused today to the point of interchangeability. In Edison's time, a battery was a group of cells.

wondered about the feasibility of a reversible traction battery filled with an alkaline electrolyte, perhaps doing away with plate electrodes altogether, so as not to waste its energy on the movement of dense metal.[19]

He took his first step toward this radical idea in 1889, when he manufactured an improvement to the Lalande-Chaperon cell, a primary battery that counterposed electrodes of positive zinc and negative iron in an aqueous solution of potassium hydroxide.* Its closed construction inhibited evaporation of the electrolyte and encouraged him to believe that a cell just as noncorrosive might be made rechargeable by an outside dynamo. To that end, for almost a year, he had been conducting regenerative experiments with copper oxide electrodes, dunking them into caustic solutions of varying strength. The results were unsatisfactory, because the copper either oxidized too much or would not reverse at all. He was to try fifty other combinations of metals and minerals, looking for "an entirely new voltaic combination," before the summer of 1900 was out.[20]

"LOVE IS A FOREIGN THING"

In their different ways of operating, Edison's two eldest sons ludicrously resembled the poles of a malfunctioning storage battery. Tom was the corrosible negative element, doomed to attract clinging ions like Mr. Friedlander, while William was the hard end, pulsing out a wild spray of electrons that sometimes threatened an explosion. Anything could touch him off—an imagined slight, a rumor, a landlord's demand for arrears—and just as quickly he could be moved to effusive declarations of love or good intent. He was as needy as his brother, but whereas Tom craved affection more than money, to William a check would always suffice.

Somehow in July Tom got the idea that his expectations as the son of an industrial tycoon were misplaced. Instantly he suspected a plot fomented by Mina to disinherit him.[21] His paranoia was plain in a letter received by Walter Mallory, the large, lugubrious engineer who ran the Edison Portland Cement Company.   ·

---

* Now known as the Edison-Lalande battery. By 1900 Edison was selling hundreds of thousands of these cells annually to railroad companies for use in signals.

Will you be so kind as to let me know as soon as possible whether my father has disinherited me or not. . . .

It is Mrs. Edison and a few of his friends? who have been instrumental in this matter. . . . Love is a foreign thing to my father but the world will know the true state of affairs pretty soon. Lies have been told him about me and my wife and he believes me [*sic*]—Let him do so but by God he will regret it and I will show him and the Miller gang that there is one son who is not a fop.[22]

So much for poor Tom, who draped his spindly body in elegant suits and wore high stiff collars to hide an attenuated neck. (William, in contrast, looked like a middleweight wanting to strip and fight.) The "Miller gang" were Mina's clannish relatives, who had always looked down on her stepfamily.

William went on to complain about the annoyance of having to earn a living. He was running an automobile agency in Washington, D.C., and not doing at all well:

I am compelled to seek a job as my meagre income is not suitable for my maintenance. My father if he was a true father would certainly look after my welfare. He never takes the trouble to find out whether I am dead or alive but I'll tell you what Mallory . . . I am tired of all this business and something dirty is going to happen before I meet my length in soil.[23]

Mallory forwarded the letter to his boss, who for some time had been allowing William $2,160 a year, more than the average salary of a college professor.* Edison ignored it, but when Blanche followed up with a hysterically scrawled six-page screed saying that the children of "The Greatest Man of the Century" should not have to live in such poverty, he permitted himself a rare show of anger.

I see no reason whatever why I should support my son, he has done me no honor, and has brought the blush of shame to my cheeks many a time. In fact he has at times hurt my feelings beyond mea-

---

* Edison allowed Tom a similar amount. At this time he was also paying mortages for both sons, but on what properties is unclear.

sure. For more than fifteen years I supported my family on less than
two thousand a year, & we lived well and after allowing you as
much as I do monthly to have you talk in the way you do shows an
utter lack of gratitude. Let your husband earn his money like I did.
I will continue to send the monthly installment until such a time but
in no case will I loan any more money or increase the monthly
amount.[24]

ALL MY DUCATS

In October Edison's first two patents covering "new and useful improvements in reversible galvanic cells or so-called 'storage batteries'" heralded his self-rediscovery as a chemist just when he had to
accept that he would never become another Andrew Carnegie. The
hard economics of mining and milling forced him to abandon his expensive new venture in New Mexico, where the gold sand was too
poor to process, and simultaneously close the iron-extraction plant he
and Mallory had built at Ogdensburg, New Jersey, with grandilo-
quent hopes, nine years before.[25]

Henceforth, apart from a scheme they had to adapt his long kiln
and leftover Ogdensburg machinery to manufacture portland cement,
Edison intended to spend as much time as possible with test tubes and
galvanometers: "I am putting all my ducats in the storage battery."[26]

The patent applications made clear that he did not expect early
success. He wrote that he had improved the performance of an alkaline cell by using the absolute neutrality of magnesium to prevent zinc
being "deposited in spongy form" on the negative electrode during
recharge. Instead, he got sizable clumps of it if the latter element was
copper oxide. But zinc itself was the problem. It was simply too soluble in an alkaline solution, the clumps tending to degrade after a
while, rapidly reducing discharge capacity, as "other experimenters
with batteries of this type" had already found.[27]

The last statement would return to haunt Edison. It suggested familiarity with the work of an obscure Swedish scientist, Ernst Waldemar Jungner, whose development of an alkaline automobile
"accumulator" was so similar and so simultaneous with his as to
arouse suspicions of mutual espionage—were they not separated by
two oceans and multiple barriers of language. Jungner, too, had in-

vented a variant of the Lalande battery some years back, and his first alkaline silver-cadmium cell had been patented in Germany on 26 August 1899, about two months after Edison began testing polarizations of zinc and copper in solutions of caustic potash. Around the same time Jungner had also applied for, but not yet received, an American patent for his basic battery.[28]

If Edison had any detailed knowledge of it at the time he executed his own first application, he would have to have read a recent article by Jungner in *Elektrochemische Zeitschrift,* forbiddingly entitled "Ein primär wie sekundär benutzbares galvanisches Element mit Elektrolyten von unveränderlichen Leitungsvermögen." This was not implausible, because he subscribed to a sister publication, *Elektrotechnische Zeitschrift,* and employed translators to help him keep up to date with foreign innovation.[29]

At any rate, the second of his 15 October patent applications showed a sophisticated understanding of cadmium-element electrochemistry, along with pride that he had conquered a problem that had defeated all "previous experimenters."[30] In language of the utmost precision, Edison described an invention more complex than any he had devised since his quadruplex telegraph of 1874. Its construction began with the rolling and annealing of two thin, rectangular nickel plates that were lugged to face each other, like infinity mirrors, once they had been respectively impregnated with elements of cadmium and copper. To that end he polished them with red heat and hydrogen before attaching the pockets—nickel too, and flat—to keep the cell as elegantly slim as possible.

The pockets and plates were perforated for later immersion in alkaline liquid. Next came the extremely intricate process of preparing two metallic powders to fill the pockets. One, for the positive plate, consisted of finely divided cadmium; the other, for the negative, was a similar division of copper oxide. Although so far the assembly of the cell had been a counterposing of opposites, the elements had to be manufactured differently. He obtained his cadmium by electrodeposition onto a platinum cathode, peeling off from it ribbons "exceedingly finely divided and filamentary in form, and of great purity." He washed them in water to remove any trace

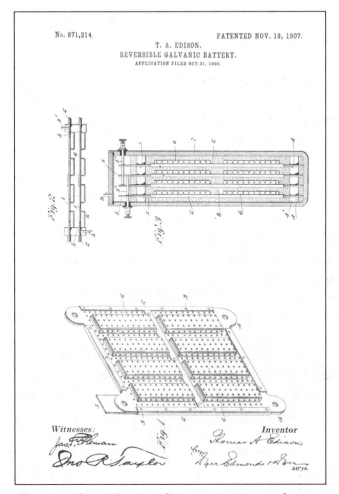

Illustration from Edison's cadmium-copper storage battery patent application, 15 October 1900.

of residual sulfate, then packed the filaments into the pockets—tightly enough to give them "coherence" yet not so tightly that the pocket lost porosity.

Delicate as this operation was, it did not match the difficulty of dividing the copper oxide. Here was where Jungner (or whomever else Edison accused of preceding him) had failed to create an effective depolarizer. All their efforts had been blocked by "the production of a small amount of copper salt, bluish in color, and which was soluble in the alkaline liquid." As the salt circulated and dissolved, it rapidly rotted both positive and negative elements, especially zinc. The containing cell had to be oversize to compensate, while its resistance increased in tandem.

"In consequence," Edison wrote, "reversible batteries using copper oxide as a depolarizer have never remained in commercial use and are now obsolete." His battery was unique in that he divided the oxide chemically, making it as smooth as the purest talc. "If . . . there is a single piece, no matter how minute, of dense copper, or even if the finely divided copper is compressed sufficiently to materially increase its density, a soluble hydroxid of copper will be formed." To avoid any such salt-causing impediments, Edison obtained his powder by reducing copper carbonate with hydrogen at the lowest possible temperature. This made it light, anhydrous, and insoluble, the last property being essential to the efficiency of storage battery electrodes. "When the copper has been thus secured in finely divided form, it is molded into thin blocks of the proper shape to fit snugly in the pockets of the plates. . . ."

By now, two-thirds of the way through his application, Edison was clearly reveling in the intricacy of what he was describing ("Grand science, chemistry. I like it best of all the sciences.") and in the terminology needed to protect every nuance from infringement.[31]

After the copper [filled] plates have been molded, they are subjected in a closed chamber to a temperature of not over five hundred degrees Fahrenheit for six or seven hours until the copper is converted into its black oxid (CuO). If higher temperatures are required, the density of the black oxid will be undesirably increased. After being thus oxidized, the copper oxide blocks are reduced electrolytically to metallic copper, and are then reoxidized on charging by the current until they are converted into the red oxid (CU_2O). In this form, the blocks are inserted in the perforated pockets 6 of the desired plates, which are then ready for use.

The finely divided copper originally obtained by reduction by hydrogen as explained, may be lightly packed in the perforated receptacles without being first oxidized by heat as described. I find, however, that when this is done, the efficiency is not so high as when the copper is first oxidized to the black oxid, because, unlike the cadmium, it is not filamentary in form, and its particles as originally produced do not apparently effect an intimate electrical contact with each other.

The last stages of fabrication were to brace and insulate the loaded plates in a nickel frame, connect them electrically, and slot the whole assembly—densely engineered, yet as easy to lift as an attaché case— into its nickel sleeve. It was then topped up with a solution of 10 percent sodic hydroxide and hermetically sealed, except for a one- way valve for the release of hydrogen bubbles on recharge. Two neatly protruding pole tips, positive and negative, completed the cell, which could be stacked with others, to stream as much power as desired.

In a final paragraph of description, Edison exulted in the duality of his design. During discharge, the cadmium became cadmous oxide and the cupric oxide became copper. During recharge, the metals and oxides converted back to their original state, and even the water in the electrolyte "respectively decomposed and regenerated, leaving the liquid in exactly the same condition and quantity after each dis- charge." There was so little evaporation that the cell hardly needed its refill cap. "In fact I find by practice that by interposing between the plates thin sheets of asbestos . . . which have been merely moistened with the alkaline liquid, nearly as good results can be secured as when the plates are actually immersed."[32]

It was a state-of-the-art battery that paid tribute, in its alternations of metal and damp fiber, to Volta's electric pile of a hundred years before.

WHEN SEEN AND FOLLOWED

Even as he signed his two new patents, Edison knew that he had done little to challenge the crude power of the lead-acid car battery. There were several things wrong with his cadmium-copper cell, starting with the prohibitive expense of both metals.* It was impractical, with an output of only .44 volts, and even if cheaper electrodes could be made to generate more energy per unit weight, the delicacy of its assembly boded ill for commercial production. Nor, despite Edison's claims, had

---

* The current price of cadmium was $1.20 per pound, as opposed to four cents a pound for lead.

he entirely solved the problem of blue-salt precipitation. By November he was back in his chemical laboratory, searching again for a perfect yin and yang of reversible galvanic power.[33]

It was a month otherwise enlivened by a letter from William, who seemed to have forgotten the paternal wrath he had recently incurred. Writing now in the guise of a concerned sibling, he reported that Tom's showgirl wife* had deserted him and was abusing her former connections:

> Marie Edison was seen going into the "Haymarket," one of New York's worst joints, with two strange men and a bad woman. She was making the rounds of the "Tenderloin" [District] and boasting to everybody that she was Edisons daughter in law and his favorite. . . . She seems to think she is playing all of us for "suckers." When seen and followed she was drunk and telling everything to these dirty people.[34]

William begged his father to do something to protect the family's "good name." Edison thought it wise to comply, arranging privately to pay Marie twenty-five dollars a week if she would stop identifying herself with him.[35]

Switching his attention back to the more congenial subject of cement making, he reconsidered the long kiln he had patented earlier in the year. It was intended to be the centerpiece of the great cement-making plant he was building at New Village, near Stewartsville,† New Jersey. Although the kiln was ready to ship, he could not resist the temptation to lengthen it during breaks from battery experiments. Walter Mallory found that Edison's new idea of long was whatever volume of tube would disgorge a thousand barrels of cement a day.

---

* Marie had left Tom bereft of cash and clothes while vacationing with him in early August on Lake George. She returned to New York with another man, while Tom, distraught, followed on another train. He could not afford to ride any farther south than Yonkers and was obliged to complete his journey on foot. On 16 August *The New York Times* announced Marie's return to the stage of the Casino Theater in a musical comedy, *The Liberty Belles*. Tom disappeared for six weeks, possibly on an alcoholic bender. Edison had to enlist William and the Pinkerton agency to search for him.

† Ruins of the plant, once the biggest in the world, may still be seen in New Village, New Jersey.

When the engineer warned him that output was 400 percent beyond the capacity of any burner in the industry, he came up with specifications for a kiln 150 feet long and nine feet in diameter, made up of fifteen cast-iron sections and rotating on fifteen bearings big enough to hold Nelson's Column.[36]

THE PRETTIEST PLACE IN FLORIDA

By the new year of 1901, with two patents pending and more than a hundred laboratory staff assisting him in further development of the alkaline storage battery, Edison was unable to keep his grand project secret from speculators. A Swedish corporation, Ackumulator Actiebolaget Jungner, had already been formed in Stockholm by his only competitor, but its name did not resonate on Wall Street as much as that of the Edison Storage Battery Company, capitalized on 1 February at $1.5 million.[37] Within eighteen days the New York trust attorney Louis Bomeisler offered Edison $3 million in cash for the right to market his new battery, even though it was still more of a theory than a product.

Edison hedged. Bomeisler assumed he was looking for other offers and wrote in some irritation, "I do not want to do a lot of work on a matter of this magnitude, and find when ready to close that I am bidding against the field. . . . [Hence] I suggested a figure which would be so high that you could not refuse it."[38]

It was in fact high enough to wipe out all Edison's mining debts and pay for his cement mill as well. But he kept politely putting Bomeisler off ("I do not want to dispute your arguments") until the lawyer, bewildered, realized he was a person who could not be bought.[39]

The same could not be said of Thomas A. Edison, Jr., whose eagerness to sell his own name caused a defrauded investor in the Edison-Holzer Steel & Iron Process Company to sue him for $400,000 in mid-February. Warned by a third party that Tom might end up in jail, Edison replied with weary déjà vu. "As I know nothing about this matter, I prefer not to have anything to do with it. As the young man is of age I am not responsible for anything that he does."[40]

With that, he decided it was time he took a break from Julius Thomsen's *Thermochemische Untersuchungen,* Gladstone and Tribe's

Chemistry of the Secondary Batteries of Planté and Fauré, and back issues of the *Journal of the American Chemical Society.* "I am going to Florida for a month to polish up my intellect," he joked to a friend.[41] For the first time in fourteen years he felt free to return to the estate in Fort Myers that he and Ezra Gilliland—more than a friend, once, before becoming more than an enemy—had bought together, back in the days when they called each other Damon and Pythias and Miss Mina Miller was Gilliland's gift to him beyond price.* Plump Damon was dying of heart disease now, and he too had long been a stranger to Fort Myers, so there was no chance of them bridling at each other across the twenty yards of garden that separated their twin houses.

In any case it was high time Edison checked up on a property he had allowed to deteriorate under a succession of vacationers and invalids since 1887.[42] Mina accompanied him, along with their three children, two relatives, and a maid. The visit was more chastening than nostalgic. Their caretaker had attempted to freshen the house with dabs of paint, but there were hardly enough beds for a party of eight, nor was there a cook to feed them. The Gilliland house now belonged to a multimillionaire—Ambrose McGregor, president of Standard Oil—and looked it, in contrast with its weed-fringed neighbor's. Nevertheless the surrounding park Edison had laid out with such symmetry in 1885 had lushly matured, and Mina with her gardener's eye could see much potential for bringing it back into horticultural balance.

For the next five weeks they made do, befriending the McGregors, eating out at the downtown hotel, and importing truckloads of soil for new plantings. Edison polished his intellect with a fishing rod, dragging a thirty-pound channel bass out of the crystal waters of the Caloosahatchee but failed in several seagoing attempts to land a tarpon. He told a reporter that he intended to make Fort Myers his regular winter home. "It is the prettiest place in Florida, and sooner or later visitors to the East Coast will find it out."[43]

---

* For the early relationship and ultimate estrangement of Edison and Gilliland, see Parts Five and Six.

WHICH?

One of the first things Edison did after returning north was to attend an electrical lecture-demonstration at Columbia University by Nikola Tesla. Although as fellow innovators in the field they had about as much in common as the rival power systems they personified—direct versus alternating current—their relations had always been distantly cordial.* Edison could be unforgiving of any former associate who tried to get rich on things he had taught them (as Ezra Gilliland could attest). But Tesla had brought his own genius to the Edison Machine Works in 1884, and taken nothing else with him six months later, when he left to form the Tesla Electric Light and Manufacturing Company.

Edison was late arriving at Havermeyer Hall, and the audience erupted with applause at his entry. Tesla was already at work displaying the light effects of his electric oscillator, but at the sound of cheers he looked up and, in the words of a reporter, "saw the greatest of all American inventors. . . . Mr. Tesla stopped his work and grasped Mr. Edison's hand, which he shook as he led him to a seat," to further cheers for them both.[44]

Another interested observer was Guglielmo Marconi. At twenty-seven, the Italian engineer was not much older than the students in the room. He came to West Orange on 16 April to look at Edison's old patent on "wireless telegraphy" and hear him ramble on for four hours about radiating sound waves that would one day encircle the globe and penetrate space.[45] This was hardly a revelation to Marconi, who had already beamed Morse signals across the English Channel. But the patent was of great interest to him and worthy of more businesslike discussion as his own experiments proceeded.

Shortly afterward a headline in *Western Electrician* speculated what the medium might best be called in the twentieth century— "SPARK, SPACE, WIRELESS, ETHERIC, HERTZIAN WAVE OR CABLE-LESS TELEGRAPHY—WHICH?"[46] Not even Marconi had yet suggested the word *radio*.

---

* Edison once sent Tesla a photograph of himself, eloquently inscribed "To Tesla from Edison."

A PACK RATHER THAN A TANK

On 21 May, the chief theoretician of Edison Industries, Dr. Arthur Kennelly, rose before the annual meeting of the American Institute of Electrical Engineers to announce that his boss had invented a new type of storage battery. After some nine thousand personal experiments, Edison had discovered a pair of metals whose electrochemical properties corresponded so closely as to permit the practical realization of the reversible galvanic cell.[47]

The paper Kennelly proceeded to present was technical, but its revelation that Edison had settled on a superoxide of nickel for his positive electrode caused what passed, in professional circles, for a sensation. Nickel was known to be nonconductive in its oxidized and reduced forms, and it was almost as expensive as cadmium if worked at all finely. Iron—his negative element—was a more predictable opposite, promising a large number of deep discharge cycles. Edison increased the conductivity of the positive electrode by mixing tiny flakes of graphite with nickel hydrate and tamping the resultant powders into the same pocketed plates he had used in his previous cell design. The graphite, pure crystallized carbon, took no part in the new battery's action except to provide microscopic conduits within each compound. They enabled a free flow of oxygen ions from pocket to pocket, in a solution of 25 percent potassium hydroxide—Edison's preferred alkaline electrolyte in all his battery experiments.[48]

Kennelly claimed, to the disbelief of some skeptics, that the nickel-iron cell had a storage capacity of fourteen watt-hours per pound, enough raw power to lift a load its own weight to a height of seven miles on a single charge. A lead-acid cell, in contrast, would rise only two to three miles before falling and making a significant dent in the earth's surface. Edison's battery was a pack rather than a tank, so solidly amalgamated that it could withstand all the shocks that automobiles were heir to, and (like its creator) superbly balanced. He admitted at the end of his presentation that the pockets had nevertheless caused Edison some trouble. Their expansion and contraction as they respectively lost oxygen, or recovered it, caused the "nickel" plates to swell slightly while the "iron" ones shrank, and vice versa during the next phase of the charge-recharge cycle. In either case, these internal pressure variations caused the cell's thin steel walls to—as it were—breathe in and out but, he insisted, "well within the elastic limits" of the metal.

Finally Kennelly half-answered the one question every Exide man in the audience wanted to ask: "As regards cost, Mr. Edison believes that after factory facilities now in the course of preparation have been completed, he will be able to furnish the cells at a price per kilowatt not greater than lead cells."

A NEW EPOCH IN THE CEMENT BUSINESS

A distance of sixty miles and seventeen stops on the Delaware, Lackawanna & Western Railroad separated Edison's two factory projects that spring. He became a commuter between them, supervising every detail of construction and pondering their likely output. One Saturday at 10:40 A.M. he arrived at New Village and took a tour of the cement mill from the quarry to the packing house. Seven of its eleven steel-and-concrete buildings were complete or nearly so, and the rest looked to be ready and fully equipped by midsummer.[49] Although what he saw was already the fifth-biggest cement plant in the country, he decided to increase its capacity from four hundred to a thousand barrels a day. He spent the afternoon brooding on-site, then took the five-thirty train home. Working from memory through the night and on into Sunday afternoon, he made a list of nearly six hundred necessary changes to the mill's design, including dimensions for new machinery and an order for two Carnegie steam shovels to open up more of the underlying cement rock vein.[*][50]

Only after the list had been copied and sent to the superintendent did he inform Harlan Page, a director of the Edison Portland Cement Company, that the firm would have to sell "say $400,000" of its preferred stock to pay for the modifications he required. "I am sure this Mill will establish a new epoch in the Cement business."[†][51]

Walter Mallory, part of whose job was to keep men like Page happy, knew from ten years' experience in the mountains that Edison

---

[*] Edison also designed the largest automatic lubrication system in the country to install at New Village. It distributed oil to ten thousand bearings across a distance of half a mile.

[†] By 1924 Edison-length kilns had become standard in the American portland cement industry.

designed more by instinct than by reason. "I cannot help coming to the conclusion," he remarked later, when the long kiln was turning and the plant was producing over eleven hundred barrels of cement a day, "that he has a faculty not possessed by the average mortal, of intuitively and correctly sizing up mechanical and commercial possibilities."[52]

WHAT HAPPENED ON TWENTY-THIRD STREET

The first suggestion in the American press that Edison was not alone in developing an alkaline car battery appeared in *The Marion* (Ohio) *Democrat* on 8 June 1901. "Mr. Jungner of Stockholm, a Swedish engineer, has invented a new accumulator,* which, in spite of its extraordinary light weight, is said to have a great capacity. . . . A vehicle equipped with this [device] made on a trial trip 95 miles, without it having been necessary to recharge. . . . The scheme seems to be similar to E's last invention."

The interest of Buckeye newspaper readers in Swedish auto traction was slight, but Edison evidently knew all about the accumulator by the beginning of July, when W. N. Stewart, an entrepreneur in London, offered him a chance to combine Jungner's European patents with his own. "The experiments of Prof. Jungner (who seems to be a most able chemist) cover a term of seven years, and are of great value. . . . I may say, also, that Prof. Jungner has a very high opinion of your work in this field, and that he makes no conditions of an embarrassing nature."[53]

Edison reacted dismissively, as he always did to direct competition. "I was surprised to learn that you had bought Jungner's patents. You will find that they have no value, because they are based on theory. An actual experiment will prove his patents bad in every particular."[54]

Instead, he sold his own past patents, plus any new ones he might win over the next five years, to the Edison Storage Battery Company for $1 million. He took only $100,000 in cash and trusted he would earn the rest in stock earnings. If he had known that he would be awarded sixty-two more electrochemical letters patent in that period, he might have valued his expertise more highly.[55]

---

* A term commonly used in Europe for the storage battery.

That was not the case with a movie patent he had been trying to profit from for years, which at this moment emerged from litigation and promised him fabulous royalties. On 15 July the U.S. Circuit Court in the Southern District of New York ruled in *Edison v. American Mutoscope and Biograph Company* that he was the original inventor of the Kinetograph movie camera and could therefore block others from capitalizing on any of its features without a license. The decision was subject to appeal but temporarily made Edison the most powerful film executive in the United States.[56]

Never having been interested in the creative side of moviemaking, he was content to leave his house director, the gifted Edwin S. Porter, in charge of production while he spent six weeks in Canada, seeking a source of nickel for his new battery. The metal was only slightly less expensive than cadmium or cobalt. He needed a private supply of it at cost if the chemical plant Arthur Kennelly had mentioned (rapidly rising at Silver Lake) was to be profitable. Assuming the battery survived a punishing series of vibration tests he had ordered, he planned to start producing it about a year and a half from now.

By early August he was cruising north through Lake St. Clair, the transitory body of water, half American and half Canadian, that divided Lakes Erie and Huron.[57] He had sailed these same waters as a child, on another voyage between opposites—from his birthplace in Milan, Ohio, to the big white house at Fort Gratiot, Michigan, where he had begun to be a man. There was no time now for him to disembark in Port Huron and revisit any boyhood haunts, because the ship was heading for the nickel-rich town of Sudbury, Ontario.[58]

He was swatting blackflies and prospecting with a magnetized needle for a seam to claim, on the day Porter set up a camera in New York and filmed a new Edison short, *What Happened on Twenty-third Street*. It caught the moment when a young woman, strolling the sidewalk in midsummer heat, stepped over a ventilating grille and felt her skirt billowing upward, to the voyeuristic pleasure of passersby.[59]

OMNIPOTENT POSSESSION

Edison returned home with a successful mine claim in his pocket—he had discovered dense deposits of nickel ore in the East Falconbridge area of Sudbury*—to find that he was once again a man with family problems. There had been a foiled kidnap threat against his younger children, and Madeleine's governess was so distraught over it she had committed suicide. Tom had managed to keep out of jail, but was bouncing checks and advertising something called the Wizard Ink Tablet ("We have testimonials from 1,000 banks") over the logotype of the Thomas A. Edison Jr. Chemical Company. William was quiet for the moment, but his lulls usually preceded storms.[60]

Business at least was good, despite the shock of President William McKinley's assassination on 6 September at the Pan American Exposition in Buffalo.[†] Newsreel coverage of his dying days and the assumption of power by Vice President Theodore Roosevelt had unfortunately rendered trivial some exquisite footage shot by Porter of the exhibition grounds illuminated at night—one slow pan resembling a spill of diamonds across black velvet.[61] Otherwise, Edison's film studio was doing well. So was National Phonograph, his recording firm, outselling every competitor despite the formation of an aggressive, disk-cutting newcomer, the Victor Talking Machine Company. The cement mill was complete, the chemical plant almost so, and the storage battery was getting some rhapsodic press comment. An illustrated article on its "wonders" in the *Rochester Democrat and Chronicle* showed a seated man holding up the slim metal box with one arm. "This latest achievement of Edison is probably destined to work as great changes in its way as did the electric light," the text commented. "The fact must be easily apparent to everybody that the ability to carry around in the palm of one's hand the power that can, so to speak, move mountains, would be almost an omnipotent possession."[62]

On 16 November Edison received a long letter from the inventor of the Phantoscope movie projector, which had become his own Vitascope in 1896 by right of patent purchase.[‡] Thomas Armat begged

---

* Edison's attempts to sink a workable shaft at Falconbridge in 1902 and 1903 were defeated by layers of quicksand. He eventually abandoned the mine.

† An aide to the stricken McKinley, remembering Edison's invention of the fluoroscope in 1896, begged his laboratory to send "an X-ray machine" to Buffalo, in the hope that radiation might somehow save the president's life. A crew was duly dispatched from West Orange, but McKinley's doctors ruled that he was too weak to receive radiation.

‡ See Part Four.

him to consider, in the light of his recent court victory over the American Mutoscope and Biograph Company, withdrawing from the case now under appeal. "Hopeful as you may feel over the result of that suit, you probably realize, as I do, that the decision, while probable, is anything but a sure thing." The case's cumulative effect so far had been to deny all parties to it the profits they would have earned, while "hundreds of ignorant or illiterate infringers of patents" had gotten rich at much less cost.[63]

Rather than continue to litigate, Armat wrote, the major players should agree to the formation of a consolidated motion picture company, or "trust," that would pool all their patents. "This combined action would establish a real monopoly, as no infringer would stand against a combination of all these strong elements." Edison's reward for cross-licensing his unmatched number of letters patent would be royalty and manufacturing privileges more than equal to their aggregate value.[64]

"Can I expect a prompt reply?" Armat asked, obviously aware that Edison was leery of any moves on his intellectual property.[65]

Edison referred the proposal to his studio head, William Gilmore. "Say that I cannot very well go into the matter by letter . . . also that I do not agree with many of his suggestions in regard to litigation." He was, however, interested enough in the trust idea to invite Armat to send an intermediary to discuss it with him. Eventually he said no, proposing instead a cross-licensing agreement to be negotiated after the appeals court ruled. Armat saw that Edison was gambling on a further victory that would make him so powerful as to be a monopoly unto himself.[66]

There was nothing for a weaker player to do but gamble on that gamble and await the court's decision. At least the idea of a national motion picture trust had been discussed for the first time, and it might be discussed again if Edison's dice throw turned up less than a six.

MONUMENTAL AUDACITY

Instead of flowers, centerpieces of tiny green lightbulbs ornamented the tables at a special dinner of the American Institute of Electrical Engineers in New York on 13 January 1902.[67] Larger, whiter bulbs at either end of the Astor Gallery spelled out the mysterious words POL-

DHU and ST. JOHNS, and yet more lamps, arranged in clumps of three, sporadically flashed the Morse signal "– – –." It signified the letter S, which the guest of honor, Guglielmo Marconi, had beamed across the Atlantic, from Cornwall to Canada, one month before.

The general glow bathed the faces of many other electrical titans: Alexander Graham Bell, Elihu Thomson, Frank J. Sprague, Carl Hering, William Stanley, and the institute's president, Charles Steinmetz. It also illumined Mina Edison, sitting alone at the high table.

"I believe I voice the sentiments of all," Steinmetz said in his opening remarks, "when I say that we are extremely sorry not to have with us the grand master of our industry, Mr. Edison." Instead, he welcomed Marconi as "another genius . . . who, taking up where Mr. Edison left off at the beginning of his career, has advanced beyond what others have done."

If that sounded like a reference to the wireless work of another notable absentee, Nikola Tesla, Steinmetz did not elaborate.* He turned the proceedings over to the toastmaster, Thomas Commerford Martin, who lost no time in reading a handwritten note from Edison: "I am sorry that I am prevented from attending your annual dinner tonight, especially as I should like to pay my respects to Marconi, the young man who had the monumental audacity to attempt, and succeed in, jumping an electrical wave clear across the Atlantic Ocean."

A COLD HEART LIKE MY FATHERS

That night Edison was lying ill in a New York hospital, suffering from an unusually harsh attack of the stomach pain that often troubled him. For the last four days he had subsisted on nothing but water. Mina's willingness to quit his bedside for the dinner indicated that he had begun to recover, but it would be three days more before he was allowed to drink some milk and eat a chop.[68]

William Edison could have chosen a better time to write to his stepmother complaining about his life as the "forlorn son of a great man." But tact had never been Will's principal virtue. Nor was he

---

* Tesla sent a message saying that he felt he "could not rise to the occasion" but congratulated Marconi on the steady advance of his "mind feelers." Six days later he applied for a wireless patent of his own.

subtle in appealing for sympathy from the one person who could re-
habilitate him in the family circle:

> *It is now over two years since I have gotten a line from you or home*
> *and it rather sticks in a fellows craw to be treated in this manner.*
> *Of course I believe in my heart that you would not treat me so*
> *coldly if it were not for my fathers wishes in the matter as I believe*
> *you have a good and not a cold heart like my fathers. . . . I don't*
> *blame you in the least as you have your own children and they oc-*
> *cupy all your time and devotion. We have sort of drifted away like*
> *a dead log down a slow flowing stream and its no easy matter to*
> *push that log back from its starting place and duced hard it is for a*
> *fellow to stand out in the starlight to find in what direction his*
> *home is when he has not a home that he can call his own. Often, I*
> *have half started for Orange but somehow the thought that came*
> *over me prevented such a step not knowing if any one would wel-*
> *come my outstretched hand or not.*[69]

Mina's experience was that whenever William stretched his hand
out, it was for money. But this time all he requested was a photograph
of the children: "To think I would pass them in the street and not
know them."[70]

She took Edison to recuperate in Fort Myers where, still frail in
mid-March, he heard that Tom and William had been arrested in a
fracas with police in Elizabeth City, North Carolina. Tom was charged
with being intoxicated on a public street, and his brother for having
struck the police officer detaining him. They had spent the night in
jail, Tom being released on payment of a fine of $7.50, and William
on finding—somewhere—$100 for a bond that committed him to ap-
pear in court later, on a charge of assault and battery.[71]

The same issue of *The Evening World* reporting this incident carried a news item equally if not more distressing to the young men's
"cold-hearted" father:

EDISON NOT INVENTOR OF MOVING PICTURES

The United States Circuit Court of Appeals handed down a deci-
sion this afternoon declaring that Thomas A. Edison was not the

inventor of "moving pictures" and that the various other machines
besides his are no infringements on his patents. By this decision the
Edison Company will have lost thousands of dollars in royalties.[72]

Edison's angry reaction was to reissue his basic camera patents, in narrowed form and once again sue every major studio in sight, including Mélies and Pathé in France. It was not a happy month for him, even though National Phonograph reported soaring sales of his new line of "gold molded" wax records, now being duplicated at the rate of ten thousand cylinders a day. He still needed to rest daily at noon after returning to West Orange in April.[73]

Better than bottles of medicine for his body and spirit as summer came on were some highly successful road tests of the storage battery. The first—sixty-two miles straight in a small Woods runabout—earned him a box of cigars from ESBC stockholders. He participated in some of them and became as addicted as Mr. Toad to the thrill of jouncing along country highways at dangerous speeds. "The sport of kings I call it—this automobiling at 70 miles an hour. Nothing on earth compares with it." Edison could not have attained that speed in an "electric"—more likely during a comparative run in a gasoline car—but mobility was the thing, and he would remain a road hog for the rest of his life. Oddly, for a man who needed to be in control, the act of driving itself did not suit him. After one or two tries that ended up in ditches, he settled for a seat up front next to the chauffeur, where he could see everything and let passengers behind enjoy his secondary cigar smoke.[74]

Mina was a nervous convert to his new hobby. "This afternoon we had a spin over to South Orange and back in one of the gasoline flying automobiles," she wrote her mother. "It was great sport but it made me feel like clinging to the sides every moment and felt myself drawn up to the highest tension for fear something might happen." When Edison bought two big White "steamers" for excursions upcountry, the family named them Discord and Disaster.[75]

THERE WILL BE A FIRE

Bulk orders were already coming in for the battery by midsummer. They were premature, because Edison was a fanatical tester. Until five of his prototype units had each withstood five thousand miles of rough riding in electric vehicles as large as a three-ton truck, he declined to go into production. Besides, he was still experimenting with combinations other than nickel-iron and showing a serious interest in cobalt.[76]

This did not stop him announcing in the July issue of *The North American Review* that he had achieved the "final perfection" of the alkaline storage battery. He wrote that the lead-acid cell could not be compared with it, being self-destructive as well as heavy. "A storage battery, to deserve the name, should be a perfectly reversible instrument, receiving and giving out power like a dynamo motor, without any deterioration of the mechanism of conversion." The alkaline cells he was currently testing weighed less than sixteen pounds apiece and showed "no signs of chemical deterioration, even in a battery which has been charged and discharged over 700 times."[77]

He allowed that an electromobile driven by his battery would be costly to buy, at $700 and up, but argued that its horsepower, unlike real horsepower, was cheap. A fifty-cent charge was all an Edison cell needed to propel a Baker two-seater eighty-five miles along a level road, and it did not have to be topped up with oats every day when it was not being used. Again in contrast to the horse, it could be relied on to work without regrettable sound effects. "The electric carriage will be practically silent and easily stopped in an emergency."[78]

Edwin S. Porter might have been expected at this juncture to come up with an automobile-featuring scenario, since he was under pressure to do something to rescue his boss's floundering movie business. But when he did come up with the idea of a film that told a dramatic story, instead of presenting a staged "turn"—like a boxing bout or a dance solo—he directed it around the more cinematic spectacle of a team of horses at full gallop. His *Life of an American Fireman*, which began shooting that fall, was so elaborate a production that the *Newark Evening News* felt obliged to warn its readers on 15 November, "There will be a fire on Rhode Island Avenue, East Orange, this afternoon."[79]

Porter liked to boast afterward that *Fireman* was "the first story film." That was not true, but it was nonetheless unprecedented in its use of temporal overlaps.[80] It anticipated Duchamp's *Nude Descending a Staircase* in breaking action up into a series of visual shards, each one angled differently yet integrated with those that came before or after. The paradoxical effect was of narrative speed and busyness of content, although the movie was (like the Edison kiln) more than twice as long as normal.

A drowsy fireman on watch duty sat half-dreaming of home (his wife, in a floating vignette, putting his daughter to bed) when an unheard alarm sounded. The fireman's sleeping colleagues awoke, jumped into their oilskins, and slid out of frame down a pole. In the stable below, the pole remained bare while horses, raring to go, were harnessed to their engines. The sliding firemen came on scene just in time to jump onto the departing equipages. Outside, it was—surprisingly—daytime, and the façade of the firehouse was quiet. Then team after team burst through the stable doors and galloped away. A crowd on a residential street corner watched as no fewer than ten engines sped toward and past the camera. On the outskirts of town, sparser onlookers saw the same procession approaching and slowing. A leftward pan tracked the hose wagon as it stopped in front of a burning house. Inside the bedroom upstairs, a little girl and her mother (also keeping irregular hours) were sleeping together while ominous puffs came through door and floor. Waking and choking, the woman rushed to the sash window and waved for help before falling in a faint. A fireman entered and axed the window open. The prongs of a ladder bumped against the sill as he hoisted the woman onto his shoulder, climbed out, and dropped from view. Moments later he climbed back in again and found Snookums still asleep in the smoke. As she too was carried to safety, two hose bearers came through the door and sprayed the bedroom with such force that its principal decoration, a framed plaque reading "THOMAS A. EDISON—TRADE MARK," nearly fell off its hook.

Down in the garden, the rescuing fireman was about to chop his way into the house. The woman appeared above him, waving through a sash window in a reverse of the image seen one and a half minutes earlier. The prongs of the ladder that saved her—and would save her once more—tilted upward as she fell back fainting. By now the ax-wielding fireman had gotten upstairs. No sooner had he broken open the window and brought her down, laying her tenderly on the wet grass, than she came to and with further frantic arm-waving told him that he had not completed his work. Hurrying back up the ladder, he returned to ground level with Snookums, and the feature came to an end with two nightgowned female figures embracing.

It would be thirteen years before a bit player on Edison's payroll, D. W. Griffith, rose to obliterate the memory of Edwin S. Porter as a film director.* In the meantime the older man pioneered a style that in future movie parlance might be described as "déjà vu all over again."

TOPSY

Considerably less entertaining, if horridly more watchable, was Porter's next feature, *Electrocuting an Elephant*. Filmed at Coney Island on 4 January 1903, it documented the last minutes of Topsy, a circus pachyderm of uncertain temper who had to be put away for killing three men in three months. The last had been a drunken trainer who thought it would be amusing to feed her a lighted cigarette butt. Nodding and swaying, she followed her handlers onto a pad electrified with six thousand volts of direct current and allowed them to strap her into place. For a few seconds she stood still, then white fumes billowed around her feet, and she toppled like a punctured airship. The camera held her in close-up as she lay on her side, until her left hind leg, stiffly extended, relaxed and sank.[†81]

THE NAME OF THOMAS A. EDISON

The desire of William Edison to protect his father's "good name," even while besmirching it himself, became a matter of commercial urgency that winter. Whether framed on the wall of a movie set or stamped on countless thousands of phonographs, dynamos, and other devices, the trademark

Edison's trademark signature, 1902.

* Griffith may be seen acting uncredited in the Edison feature *Rescued from an Eagle's Nest,* directed by Edwin S. Porter and J. Searle Dawley (1907).

† *Electrocuting an Elephant* has given rise to an internet myth that Topsy was deliberately killed by Edison in order to demonstrate the lethal danger of alternating current as

was an asset beyond price. Edison had cause to regret, just when he was preparing to sign off on the mass production of storage batteries and portland cement, that he had given the same name to his eldest son.

Joseph F. McCoy, who served him as an industrial and personal spy, reported that Tom had sold it to Charles F. Stilwell, throwing in the "Jr." for free. "Mr. Stilwell wants to put a Thomas A. Edison, Jr. Phonograph Company on the Market, he says there would be big money in it."[82]

Stilwell was Tom's maternal uncle, a former glassblower with plenty of hot air still left to spare. Edison regarded him with wary benevolence. He had made use of his family connection before, helping Tom organize the Thomas A. Edison, Jr. Improved Incandescent Lamp Company. McCoy wrote in his memorandum that their new venture trespassed even more rudely on Edison's personal territory. "He said that he refused last week $5,000 Five Thousand dollars he thinks he can get more."[83]

Apparently Stilwell had already tried to resell the Edison name to the Columbia Phonograph Company, a major competitor of National Phonograph, but failed because his intermediary (Tom?) "was drunk for more than (3) three weeks, and the Columbia people would not deal with them." If anything more was needed to seal Tom's paterfamilial fate, it was Stilwell's remark to McCoy "that Mr. Edison would only live a few years longer, at his death, he would bring suit against the National Phonograph Co. for using the name of Thomas A. Edison on their Phonograph and Record and supplies. He would make big money from that, as the Company would have to pay him, if they continued to use the name Thomas A. Edison."[84]

Edison *père* had to struggle between anger at Tom and compassion for Stilwell, who had recently gone blind and had a large family to support. Notwithstanding the pair's earlier collaboration in the lighting

---

opposed to his own preferred direct current. Her death was on the contrary ordered by Luna Park officials, who originally wanted to hang her. They were persuaded instead to adopt the triple method of poisoning, strangulation, and electrocution, with the approval of the Society for the Prevention of Cruelty to Animals. Edison had no part in the filming of the documentary, although it was issued under his trade name. For his role in the development of the electric chair, see Part Four.

industry, they were neither of them bulbs of especial brightness—
as evinced by Stilwell's naïve assumption that McCoy would not at
once alert Edison to their intent.

It was clear to Edison that Tom deserved a legal slap in the face that
would stop him from ever again participating in identity theft. At the
same time, having consistently refused to give him and William jobs
at the plant, he had to accept some responsibility for Tom's abject
condition. The young man was impoverished, depressed over the failure of his marriage, rooming in Newark with the Stilwells, and drinking heavily. He had taken to bed in one of his prolonged sieges of
paroxysmal head pain.[85]

But even now Edison felt unable to lay a symbolic hand on his son's
forehead: "Tom is either crazy or a very bad character." He allowed
Randolph to send him a terse note saying that they must come to an
immediate legal arrangement, with some guarantee of security on
both sides.[86]

In return he received a three-page, meticulously scripted outpour-
ing of bile, shocking to read from somebody as timid as Tom. It began
"Dear Sir," and continued:

> Since I left you some six years ago—my career has undoubtedly been a
> wild one—as everyone knows—but I am not at all sorry that I have
> had the experience—although I am sorry I have injured you in the
> manner I have—however this is done now and I will talk to you as man
> to man realizing that our hatred towards each other is very intense.
>
> I can honestly say that I never have had the slightest intention of
> doing you any injury—but your persistent refusal to take me back
> with you—I will admit has often caused me to give you little consid-
> eration in matters where I was personally benefitted. I know of no
> business deal that I have ever made—that I was not taken advantage
> of—having often been forced to enter into agreements to save myself
> from absolute poverty. . . .
>
> I never dared to ask your advice nor to consult you upon any matter
> whatsoever—for from the very first you gave me sufficient cause to
> consider you as my worst enemy and I still consider you today as such.[87]

The shapely undulations of Tom's pen seemed to calm him down
a little. He acknowledged that Edison had never done him any seri-

ous injury, "and you couldn't if you wanted to—for I have injured myself too much to have anyone else do it." He had rushed into his deal with Stilwell out of desperation, never having made one with his own father. There were, he confessed, some other name-selling contracts that Edison might find objectionable.* "My object in writing to you is to ascertain whether you are interested at all in their recovery—they are of course the only means by which I derive a living at present."[88]

Always, in letters of this kind, the begging note intruded. Having sounded it, Tom cast all dignity aside and verbally threw himself on his father's mercy. "I will sign any reasonable agreement with you—in which you can dictate your own terms—which will satisfy forever—an agreement which will deprive me of all future rights to the name of Edison."[89]

It was an abject surrender in a life marked by many. Edison had his legal department draw up twin contracts guaranteeing Tom and Stilwell respectively $1,890 and $1,350 per annum, in exchange for vows not to leech him again. Just to make sure they understood, he went to court anyway to prevent the Thomas A. Edison Jr. Chemical Company from selling any more Edison Magneto-Electric Vitalizers.[90]

SPONTANEOUS COMBUSTION

By mid-February, cement and car battery production had started at Edison's two huge new plants in New Jersey. Progress in each case was experimental and slow, but he predicted it would soon accelerate to a point where marketing could begin. He was uncowed by a threat to his pending copper-cadmium cell patent, filed by Ackumulator Actiebolaget Jungner, and announced by that company in English, presumably to get American attention: "The patent office of the United States has not agreed to Edisons claims, but has already made him several disagreeable questions. Some particulars in the case between him and Jungner will an interference jury decide."[91]

---

* Edison was bothered throughout his career by the attempts of imitators and swindlers to sell products under his name. This explains his furious reaction when his own sons misused it.

It was true that the Patent Office had agreed to hear Jungner's counterclaim to have preceded him in 1899 with a silver-cadmium cell patented in Britain, but Edison was sure that the examiner would find that early device inoperable. In any case, he was no longer interested in cadmium and had applied much more successfully for a patent on his nickel-iron combination.[92]

"At last I've finished work on my storage battery," he told a reporter who came upon him hunched over a yellow pad in his laboratory. "And now I'm going to take a rest." He threw a stub pencil down and dropped into an armchair. "I'm tired—very tired. I'm all worn out."

There was a twinkle in his eye as he said this, but he insisted that what he needed was an extremely long vacation, starting at once in Florida. He had a four-hundred-page notebook of ideas that he had never had time to develop. "I've made up my mind to drop industrial science for two whole years and rest myself by taking up pure science."[93]

Exhausted Edison undoubtedly was, and he did not mention the notebook again after he got to Fort Myers and started fishing. He took only a desultory interest in the sport, but it suited his deafness and love of being left alone. Little boats named *Madeleine* and *Charles* and *Theodore* bobbed alongside the dock that now extended far out into the Caloosahatchee, and a ninety-two-foot, double-deck steamer named the *Thomas A. Edison* was in service for excursions upriver. Mina had her own eponymous fishing boat, a twenty-five-foot naphtha launch. But she identified more with the house, extensively refurbished during the last two off-seasons, "and all so fresh and pretty." She decided that it should henceforth be known as Seminole Lodge.[94]

Edison enjoyed just a week of "rest" before news of a catastrophe at New Village reached him. There had been an explosion in the mill's coal blower that touched off a fire in the adjacent oil tanks. At least six men were dead, and scores injured, some burned so badly that their faces were crisped like bacon. Subsequent reports raised the death toll to ten, and ascribed the explosion to the spontaneous combustion of seventy tons of pulverized coal. Much of the plant was re-

Edison at Seminole Lodge, early 1900s.

ported destroyed, with damage—excluding lawsuits—estimated at several hundred thousand dollars.*[95]

It was the worst industrial disaster in Edison's career. He set to work on plans to increase safety at the plant, with no apparent thought of returning north to comfort widows and sufferers. As one of his aides remarked with mock envy, "Mr. Edison is fortunate among other men in having been born without feeling."[96]

Tom had a sense of that in June, when he signed a formal agreement to stop using his father's name commercially. He was welcome to do what he liked with "Jr." In exchange he was granted a weekly allowance of thirty-five dollars, every payment requiring a receipt. With typical naïveté he assumed he had been forgiven his peccadillos and the following month asked for a job at the laboratory. Edison was quick to disillusion him. "You must know that with your record of passing bad checks and use of liquor . . . that it would be impossible to connect you with any of the business projects of mine," he wrote. "It is strange that with your weekly income you can't go into some small business. . . . William seems to be doing well."[97]

* The Edison Portland Cement Company settled with the widows of the dead men at $500 apiece. Injured personnel had to go to court to recover any damages at all.

He did not mention that he had just approved a request from William to "borrough two thousand dollars" to buy an automobile garage in Washington, D.C. Edison's aloofness from his elder sons did not preclude him from treating them fairly when they attempted to succeed on their own. William was a good mechanic, and the time was propitious for him to get into the car business. He received his first installment of the loan on 17 July, the day after Henry Ford incorporated a new motor company in Detroit.[98]

"Now my dear father this is my last call on you," William wrote, with every appearance of sincerity. "I can promise that the William of several years ago is not the William of today."[99]

IT TAKES TIME TO DEVELOP AN INVENTION

When a representative of *Electrical Review* visited the laboratory that month, Edison hinted that he might become an auto engineer himself. He had just returned from a test of a twenty-four-horsepower gasoline tonneau and found it fast but unstable. "Look here, this is automobile data," he said, pulling out a red-leather-covered pocketbook stuffed with notes and graphs. "I am going to build a good machine." His would be all electric, geared for sandy traction, and "able to beat, or at any rate, keep up with, any gasoline machine on a long run."[100]

This led the reporter to ask the main question that had brought him to West Orange: why, after many announcements, was the alkaline storage battery still not on the market?

Edison became defensive. "We are making one set a day, and within a short time will be making two sets. We are not doing any advertising, because we have more orders than we can begin to fill. . . . The public doesn't seem to understand that it takes time to develop an invention." He said he had spent six years commercializing the electric lightbulb, eight on the telephone transmitter, and sixteen for the phonograph.[101]

The truth was that his battery, for all its theoretical simplicity, was more complex—and consequently harder to produce—than those previous devices. A tenacious problem was to prevent the iron electrode from being overwhelmed by the rising capacity of its nickel opposite, in order to ensure a smooth and constant voltage curve during the discharge cycle. As his chief chemist, Walter Aylsworth, put it, both elements had to be kept "in training." Compression of the graphite flakes in each at four tons a square inch was essential but almost impossible to maintain at a constant level, due to flexion of the cell walls. This affected the conductivity of the flakes and the performance of the unit.[102] Edison ordered the finest, strongest steel possible from Sweden, but he worried about its expense and fumed over shipping delays.

Another threat to the battery's economic prospects was the question of its originality. The Patent Office had dismissed Jungner's interference suit as expected, but did so by citing an obscure French alkaline-cell patent (Darrieus 233,083) that antedated Edison's by even more years. Now he heard that on 1 September Jungner had been granted a U.S. patent for some "new" storage battery refinements directly based on his own. Or so it seemed to Edison, who turned to the patent attorney Frank L. Dyer for help.[103]

He had hired Dyer full time a few months before to handle his accumulation of letters patent, which now numbered well over eight hundred and cost the company $100,000 a year in protective litigation.[104] Dyer also had a sophisticated understanding of movie rights. That qualified him to make the most of a surprise U.S. Court of Appeals decision granting the Edison studio full ownership of every film it had ever filed as a paper print at the Library of Congress.[105] Although technical, the ruling was of major consequence in the industry, and he was working to ensure that it restored the Edison studio's fortunes.*

Before the year was out, Dyer would serve as his boss's personal lawyer too. Tall, bespectacled, bookish, and precise, he was the perfect foil for an inventor impatient of restraint and bored by due process.[106] He was cool-tempered and adept at dealing with the strong feelings that arise when human relationships are codified. He venerated Edison without particularly liking him and felt sorry for Tom and William, seeing them as chips never to be reintegrated with the old block.

---

* Motion pictures on film were not subject to copyright protection until 1912. If, however, they were printed as a series of stills, same-size on light-sensitive paper, they qualified for protection under existing law. The Edison studio took the trouble to do this, contributing significantly to the preservation of movies that might otherwise have died in nitrate form. Its archive dominates the Paper Print Film Collection at the Library of Congress.

SHOT FOR SHOT

While Dyer prepared an aggressive case for the canceling of Jungner's new patent, Edwin S. Porter resumed production of Edison movies, among them a comic feature whose title sounded more innocent in the fall of 1903 than it would in a later age: *The Gay Shoe Clerk*.*[107] His major feature of the year was *The Great Train Robbery*. Effectively transporting theater viewers from their seats onto a train hijacked by murderous bandits, it was a pioneer "action film" and became the first blockbuster in American history. The scenes shot aboard an open-sided baggage car (hurtling along the same Lackawanna line Edison took on his trips to the cement mill) had the impact of authentic movement, as did an even more thrilling sequence photographed from above and behind the cinder-spraying locomotive. But nothing made audiences scream louder than the final brutal close-up, wherein the chief bandit cocked his revolver and expressionlessly fired straight out of the screen, turning shot into shot.

In perhaps unconscious acknowledgment that the age of the fixed camera had come to an end, Edison ordered the demolition of "Black Maria," the dark old box that had served as his movie studio ten years before.[108]

"THIS LATEST DISGRACE"

The year ended badly for Tom and William, the former checking into a sanitarium for vague medical reasons, and the latter incurring paternal fury after calling his garage in Washington the Edison Motor Company. "You are now doing me a vast injury," Edison wrote him. "You are being used for your name like Tom and as you seem to be a hopeless case I now notify you that hereafter you can go your own way & take care of yourself. . . . I am through."[109]

William, terrified that his loan was in jeopardy, apologized and hired a lawyer "to annul the company that I so foolishly allowed to come into existence." The business reconstituted itself as the Columbia Auto Company. Blanche wrote to say that she was now running

---

* Not to be outdone, the independent filmmaker Siegmund Lubin produced a feature called *Don't Get Gay with Your Manicurist* (1903).

it, with four mechanics working around the clock under "Billy's close attention." She clearly thought her husband was a commercial moron. All that was needed to make it a success was an extra infusion of capital. "We would like two hundred dollars to carry us through."[110]

Edison turned her down.

"You do not seem to think that I appreciate what you have done for me," William wrote him, "but on the contrary I do. . . . I would call your attention to the fact that I never had the business training that most fathers make their sons go through and it was not my fault as I repeatedly begged for a position at your works but in every instance was refused."[111]

The next post brought a note from Charles Stilwell to Randolph confirming that Tom was seriously ill and would undergo treatment, presumably for alcoholism, at St. James Hospital, Newark. He hoped that Randolph would keep "this latest disgrace to the name of Edison" from Tom's father.[112]

A PICNIC LIKE THIS

By January 1904 production of the Edison car battery had ramped up enough for it to be marketed at last. He designated it his E-type cell and made it available in three sizes and strengths. But his satisfaction in its initial brisk sales was clouded by Frank Dyer's failure to persuade Patent Office examiners that they had unjustly awarded Waldemar Jungner priority over himself as the inventor of a working alkaline reversible cell.[113]

Edison was so worried about the prospect of lawsuits, either from Jungner or Darrieus, that for the first time in his life he lobbied a major political figure. "Sir," he wrote President Roosevelt, "I have been before the Patent Office for thirty years and although I have felt sometimes that criticism on my part was warranted, I have been silent. Now I find that a great injustice has recently been done me, due to what I shall call incompetence or fraud"—he crossed out the last two words—"on the part of two of the examiners." He complained that the patent commissioner had taken "an arbitrary and practically antagonistic position" in declining his request for an official review of the case. "It seems to me that I am entitled to such an investigation, and if I can prove that an outrage"—he changed the word to *injustice*—"has been done, I shall be satisfied."[114]

Roosevelt lost no time in leaning on the commissioner, Frederick Allen. "Thomas A. Edison is a man who has done much for this country, and whatever can properly be shown him in the way of courtesy I should be glad to have you show." He demanded that the great inventor be granted at least "a full hearing."[115]

Encouragingly, Jungner's German patent was canceled in Berlin on 9 January, in a Supreme Court decision that found his battery to be "nonworking"—exactly the ground on which Edison was challenging it in the United States.[116] Commissioner Allen had no choice but to order a review of the case against the examiners.

As an advance birthday present for himself in February, Edison ordered two Lansden "electrics," a touring car and an express wagon.* He chose not to be intimidated by the growing popularity of gasoline automobiles (William and Blanche boasted from Washington that they had acquired the agency for the new Ford motor car), nor was he discouraged by complaints from early purchasers of his battery that it did not live up to its extravagant publicity. Far from generating more power per pound than a lead-acid unit, it averaged only a moderate 11.8 watt-hours for each cell. Its soldered seams developed microscopic pores that leaked electrolyte, and most annoyingly, every charge/discharge cycle reduced its capacity. Edison found that the latter problem could be solved by a time-consuming application of heat during reversal, but he recognized that it was a fault that must be eliminated for his battery to succeed.[117]

Having already spent $1.5 million developing it, he took a perverse pleasure in the need for further experiments. A long-suffering laboratory assistant remarked, "I could never get away from the impression that he really appeared happy when he ran up against a serious snag." The finished cells were beautiful things—shining slipcases packed with up to eighteen pounds of nickel hydrate, iron oxide, steel, and

---

* He also splurged $2,250 (more than $65,000 in today's money) on a thirty-six-foot yacht, the *Reliance*, in Fort Myers. In 1908 he bought the Lansden Electric Car Company outright, becoming in that sense a car manufacturer himself. Albion, *Florida Life of Edison*, 60; Millard, *Edison and Business*, 188–89.

caustic potash. There was something either masochistic or sadistic in Edison's insistence that randomly selected models should be dropped from upper levels of the laboratory onto the blacktop courtyard. ("Now try the third floor.") If they stopped working as a result, they were not solid enough, and the design had to be improved.[118]

Perfection, to him, was a state forever imminent and attainable, so he spent no time celebrating past achievements. When the American Institute of Electrical Engineers held one of its overly frequent black-tie dinners, this time to congratulate him on his invention of practical electric light, he endured it as usual in smiling silence. Yet the anniversary seemed to register on him, and in mid-May, when the New Jersey countryside was at its greenest, he accepted an invitation from the General Electric Outing Club to visit what was left of Menlo Park.[119]

At first he was depressed to see the house where Mary had died, tenanted now by Italian squatters, the decaying machine shop that had once thrummed with his first big dynamo, and the rusting hulk of his pioneer electric train, railless and sinking ever deeper into grass. Chickens roosted in the shed where Francis Jehl used to manufacture lampblack. At least the old laboratory building (so large-seeming when his father built it for him in '76) was in decent repair, serving as half firehouse and half theater for the farming community all around. There was still a depot of sorts, for visitors who could persuade the Pennsylvania Railroad to stop there, and a post office that continued to receive mail addressed to him. Otherwise the hamlet was headed for ghosthood.

Edison toured all of it except his former office building, occupied by a crotchety recluse. Afterward he cheered up and ate a late lunch of cold chicken and bread, sitting on a log beneath the giant trees. "I get tired of big banquets," he said. "But a picnic like this . . . I'm glad to be here."

It was a lovely afternoon, and he did not leave until sunset.

THAT WAS THE WORST

Presidential interest made Edison's challenge to the Jungner patent a red-hot item in Washington. Commissioner Allen prudently went on vacation and assigned his office's review to a deputy, assistant patent commissioner Edward B. Moore, who divided the challenge into three areas of review, ruled upon each, then asked five internal experts, each sworn to secrecy, to re-review them before presenting a final report to Alexander M. Campbell, assistant attorney general of the Department of the Interior, who approved it provisionally but requested the further endorsement of his boss, Secretary Ethan A. Hitchcock, who declined to be involved except to submit the swollen file to Roosevelt, who authorized Campbell to announce on 14 June that Edison had lost his case.[120]

The Patent Office was found to have acted correctly, with "absolutely no evidence of malfeasance," in awarding priority to Jungner's storage battery over Edison's. But in an obvious sop to his—and maybe the president's—feelings, the examiners were transferred to other departments.[121]

Edison was now faced with two choices. He could proceed with mass production of his battery and gamble on Jungner not having the money or the will to sue him for infringement. But if the gamble failed, he could easily be ruined. The other decision was whether, in view of proliferating complaints about the battery's unreliability, he should recall the units already sold and spend whatever time and money was necessary to fix his invention beyond challenge. Both alternatives were so painful that he postponed action on them through the fall.

In the meantime he had to deal with the news that Tom, alias "E. A. Thomas," had checked into a resort hotel on Greenwood Lake, New York, to escape vengeful business partners, while Clarence Dally, who had assisted him in countless X-ray experiments in the 1890s, had become a double amputee and was dying of radiation poisoning.[122]

Edison was lucky to have escaped a similar fate, having more recently succumbed to the temptation to experiment with the first known quantity of radium imported into the United States. William Hammer, one of his former lighting engineers, had worked with Pierre and Marie Curie in Paris and gotten their permission to bring nine tubes of radium bromide home, to see if it could be used to create new luminous substances.*[123] Edison bought one tube from him, under the impression there was little in common between radium and the glowing salts that had contaminated Dally. ("Don't talk to me about X-rays, I am afraid of them.") He found that over a hundred chemicals fluoresced when exposed to the mysterious element, and that it also inducted phosphorescence in a diamond ring. But when he sensed damage to his stomach and left eye, he gave up radiation research altogether. Dally died on 2 October.[124]

By the end of the month it was clear that the Edison storage battery was performing so erratically in the market that its total withdrawal was necessary. With thirty-seven thousand cells sold, this represented a huge loss to the company and another setback for Edison, after the failure of his mining ventures and his humiliation at the Patent Office. He ordered the recall under guarantee and threw himself and a team of eighteen assistants into a flurry of remedial experimentation. Working twenty-four-hour days in staggered shifts at two laboratories, they concentrated on the battery's inexplicably different rates of power attrition, cell by cell and auto by auto. Edison thought he had the problem solved after three months, but this hope proved illusory, and the team settled into an increasingly desperate quest for a chimera only he believed in—that of "complete reversibility" of electrodes. Fred Ott, one of his longest-suffering experimenters, had to perform so many gloveless adjustments to various counterbalances that hot potash seeped under his nails and made them bleed. When the pain grew unbearable, he would dip his fingertips in acetic acid to neutralize the alkali absorption. He could not sleep at night unless he lay with his burning hands propped above his head. "Of all the elusive, disappointing things we ever hunted for," a colleague remarked when it was all over five years later, "that was the worst."[125]

In January 1905, at the height of the initial phase of redevelopment, Edison was felled by an attack of chronic mastoiditis in his left ear. It necessitated a dangerous operation that reduced what little

---

* Pierre Curie asked Hammer, in return, for a sample of the tungstate of calcium that Edison was using in some private lighting experiments.

hearing he had left on that side to zero.* Examinations attendant to the procedure showed that the optic nerve leading to his left eye had thickened to a cordlike diameter. Whether this was the result of radiation was a question beyond current medical ken, but he had to live with partially blurred vision henceforth, along with greater deafness. This made him more and more difficult to converse with as he tried to keep interlocutors at bay with monologues or his particular brand of cornball humor—such as the one about the man with liver trouble who got rich after purchasing a spring in the San Joaquin Valley that cured fellow sufferers from all over the world. "Well about twenty years afterwards the man died, and at the coroner's inquest they had to take his liver out and kill it with a club."†[126]

Edison recovered quickly from his ear surgery, but because the knife had carved into the back of his temporal bone, his doctors ordered him to "do no brain work" before the spring. He ignored them and returned to storage battery experiments at once, declining even to recuperate in Florida. By February he had discovered that flake graphite, contrary to his earlier belief, was not stable when subjected to prolonged electrolysis. In the positive element, it frequently short-circuited, increasing resistance and so diminishing the capacity of the cell. This necessitated a search for some other insoluble material that would flake as well as graphite while maintaining a consistent contact with neighboring active particles. On 31 March he sought to patent a means of stuffing his electrode salt with minuscule, highly conductive "scales" of cobalt, alloyed with nickel to reduce oxidization to almost zero. Exquisite care was required to manufacture them. First he electrodeposited what he called "a mere blush" of zinc on a copper plate. The plate was then washed and transferred to an electrolytic bath, in which a film of cobalt and nickel (proportionately mixed by separate anodes) was laid upon it to a depth of only .0002 of an inch—hardly

---

* "It is certainly hard work to talk to him," Frank Dyer complained in 1906. Edison underwent a second life-threatening operation for mastoiditis on 23 February 1908, making his hearing even worse.

† A recording of Edison telling one of his stories in 1906 is available from Michigan State University's Vincent Voice Library at http://archive.lib.msu.edu/VVL/dbnumbers /DB500.mp3.

more substantial than the "blush" beneath. A third immersion, this time in a dilute acid, dissolved the zinc and caused tiny bubbles of hydrogen to rise off the copper and lift the alloy film. As it floated free, Edison wrote, it broke up into "small flakes or scales, which naturally assume a curved or curled shape—a phenomenon especially characteristic of cobalt." They were sized through microscreens before being annealed to a red heat in a hydrogen atmosphere, "which treatment effects a very perfect cleaning of the surfaces."[127]

No sooner had he perfected this exquisite process than the gross problem of sheet metal warpage reasserted itself. There was no point in tamping masses of scale-speckled nickel hydrate into pockets that would not keep them compressed. A month later Edison and Aylsworth filed a joint patent application for pockets that were tubular rather than cubical, still made of perforated steel, but closed top and bottom and prevented from bulging by virtue of their pipelike construction. When these "non-deformable" tubes were packed with active nickel hydrate and had absorbed their fill of the surrounding electrolyte, they should, according to the application, attain a "desired" internal pressure and hold it indefinitely.[128]

This supposition was accepted by examiners and moved toward eventual patent, but the idea of a constant balance between inner elasticity and outer rigidity proved to be illusory. Maddeningly, the cobalt scales kept shifting and short-circuiting inside the active mass, coating themselves with an oily insulant that none of Edison's chemists could analyze. Increasing the packing force from 6,000 to 20,000 pounds per square inch made the scales more conductive, but compressed the content of the pockets into something as hard as soapstone, with a consequent loss of porosity.[129]

And so the labor of "perfecting" a reversible alkaline galvanic battery dragged on toward summer and fall and who knew how many subsequent seasons. Even Edison wondered in moments of gloom if the thing would ever work. Whenever Yin bulged in the convex, it seemed to attenuate the concave tail of Yang.

MR. AND MRS. BURTON WILLARD

Tom's removal to the Valley House Hotel in Greenwood Lake, New York, had not succeeded in averting the wrath of his creditors. He remained there under a pseudonym, drinking heavily and battling ill health and depression, until he was rescued by Beatrice Willard, a young woman of inscrutable origin. At various times in her life she had been Matilda Heyzer, Beatrice La Montagne Heyzer, Miss Beatrice Matilda Heyzer, Mrs. Thomas Montgomery (widow of a ticket clerk at Madison Square Garden), and most recently Mrs. (or Miss) Beatrice Willard. She had almost as many birth dates as names, making her at last count either thirty-one, twenty-three, or ten years old.[130]

When Tom first wrote home about her, she was "Mrs. Willard," a fellow guest at the hotel who was nursing him back to health after his latest "nervous breakdown." After that it was difficult to deny the existence of a former Mr. Willard. Tom keen to return to New York in 1905 with Beatrice at his side, so he solved many problems by adopting the surname Willard himself. The couple intended to marry but could not until Tom's truant wife agreed to a divorce—an unlikely prospect for as long as Marie kept receiving weekly checks from Edison. Besides, she was a Catholic. It was accordingly as Mr. and Mrs. Burton Willard that Tom and Beatrice were now living in a leased house on Staten Island.[131]

Their sanctuary there was breached in September, when a doctor in New Brighton wrote to John Randolph to say that a certain "young man" needed institutionalization. "I would advise his removal next week to Dr. Laning's, Cornwall on Hudson." He was referring to a home known as the Cornwall Sanitarium for the Scientific Treatment and Permanent Cure of the Liquor, All Narcotic Drug Addictions and Nervous Diseases. It sounded comprehensively suited to Tom. "Not that I have a very great belief in that 'young man,'" the consultant wrote, "but I have seen some bad cases treated there."[132]

Randolph could not keep the news from Edison, who was going to be stiffed with the sanitarium bill. Dr. Laning advised that therapy for alcoholism alone would cost $100 a week, plus bed and board for the patient and his "wife."* He estimated that a month's stay would be enough to dry Tom out. Edison told Randolph to make whatever arrangements were necessary.[133]

When next he heard from his son, in late November, it was a cry of

---

* There is some evidence that Tom also had an opium problem at this time.

such pathetic gratitude as to distract him momentarily from battery development. "Words are certainly inadequate to express my appreciation for all your kindness tendered me during my recent illness," Tom wrote. "My entire system has forever and eternally rid itself of the poison that was hastily eating my life away—and a new form of manhood has enveloped me and transformed me to a character worthy of the name I bear."[134]

Now that the "stupor craze pain and worry" that had brought him close to suicide was over, he felt free to confess that "many times I hungered for commiseration but feared to appeal to you—as my bravery to bear censuring had long left me." Even now he felt that "the bridge that separates us" precluded him from hoping for a personal interview with the man he worshipped most in the world. However,

> With my new name my new life and new acquaintances—I am ready to start out in the world fully equipped to meet the demands of my business ability—This Father means that I must have an occupation. . . .
>
> I have long been an enthusiast in the interest of agriculture and I find that my greatest ambition and heart's desire is to possess a farm and start a mushroom business. . . . I am positively proficient in everything concerning the mushroom.[135]

That was news to Edison, though not the fact, broached by Tom on page four, that possession of any farm entailed the outlay of capital. "My idea is first to have you purchase me a farm—taking a six percent mortgage on same. . . ."[136]

Edison read the rest of the letter carefully and wrote across the top, "Randolph—Tell him to better rent a farm . . . if in time it was satisfactory it could be bought—Tell him to pick out a place & ascertain the rent & also the money required to operate one year & let me know & if satisfactory I will help him out financially."[137]

William Edison also felt the call of the wild that winter. After two years of working as manager, salesman, or grease monkey in a series of failed automobile businesses, he wrote his father from Waterview, Virginia, on stationery emblazoned "The Punch Bowl Island Game & Poultry Farm." He and Blanche were listed as proprietors. Just in time for Christmas, they offered "Mammoth Bronze Turkeys, English and Golden Pheasants, Homing pigeons, Belgian Hares, Imperial Pekin Ducks, Buff Cochins, Mexican Crested Quail, etc."[138]

It was plain to anyone familiar with Will's ever-changing letterheads that his new venture existed only on paper. He hinted that he would be looking at two other rural properties in the spring and hoped he would be able to afford a down payment on one of them.[139]

BREAKING STORY

Edison, who seemed to have a habit of poisoning himself around the turn of every year, inhaled so much hydrogen cyanide on 1 January 1906 that he had to hurry outside and clear his lungs with cold air. A lab assistant told Frank Dyer over lunch that "the old man persisted in handling such things as if they were milk . . . I suppose they will be the death of him yet."[140]

Dyer, aware that his boss's sixtieth year was approaching, had started keeping a diary, in anticipation of the day when he might contribute to—or even write—the official biography of Thomas Alva Edison. "That is just the sort of thing I would like to work up." He was also aware that William Gilmore, general manager of the prospering phonograph and film companies, had a thirst comparable to Tom's and was prone to unexplained prolonged absences from work. The day might come when Gilmore would not be welcomed back, in which case an ambitious, lawyerly executive more congenial to Edison would have a good chance to succeed him.[141]

Although Dyer had failed to prosecute his full case against Jungner's battery patent, that defeat looked Pyrrhic now that flake impregnation had so radically altered the technology. He encouraged Edison to talk to him on subjects other than law, and he was quick to record any biographical snippets that came out in their conversation:

Spoke to Edison today of my scheme to abolish speculation in stocks by requiring all purchases to be recorded in Washington, like deeds to real estate, under a Federal incorporation act. He talked continuously against it in a most lucid and interesting way, just as if he had opened his mind to a particular page bearing on the point and was reading it. . . .

I asked him where he had picked up so much information about Wall Street and he said that for five years he ran the stock tickers. Facts once absorbed by him are never forgotten.[142]

To John Randolph's relief, Dyer also became the main negotiant between Edison and his pesky sons, not to mention their even peskier wives. He was a lover of Victorian fiction and enjoyed dealing with the melodramatic crises they heaped upon him. On 8 February he met with Beatrice Willard and found her overeffusive yet otherwise touching in her desperate desire to marry Tom. Nine days later, in a denouement straight out of a dime novel, Marie Toohey died, aged twenty-six.[143]

Dyer had to handle three immediate consequences of the breaking story ("MRS. T. A. EDISON DEAD—CHORUS GIRL MARRIED INVENTOR'S SON AND LEFT HIM SOON AFTERWARD"). The first was the Toohey family's assumption that Edison would be glad to pay for the funeral. (He was not, but did.) Tom followed up by asking if his marriage to Beatrice could now take place. While awaiting Edison's decision on that score, Dyer visited the "Willards" in their Staten Island home and revised his opinion of Tom as a loser. The young man looked well and happy, excited at the prospect of a future in mushrooms. He played the piano with skill while Beatrice cooked a nice lunch. "It really was first class," Dyer wrote that night. "I hope his father will help him."[144]

Edison honored his promise to rent and equip a farm that Tom found in Burlington, New Jersey, but declined to receive him and Beatrice at Glenmont and would not hear of them marrying. They went ahead anyway on 9 July, after Tom informed Dyer that the ceremony was urgently necessary. Edison reacted to this news with surprising mildness. He said he guessed family life was the best thing for his son "and would probably keep him straight."[145]

Nothing further was heard of Beatrice's supposed pregnancy except a vague message from Tom in October, saying she was "getting along first rate." By then he had begun to quail at the task of upgrading his primitive homestead to the sophisticated requirements of a gentleman mushroom farmer. Winter was approaching—the time of most hard labor in mycological cultivation—and he was ailing again.

He was not sure he had the strength to tote, embed, flatten, and spray hundreds of wagonloads of horse manure compost, merely to begin the long, gassy process of fomenting a salable crop for next spring. In one of his many moments of despair, he suggested to Dyer that he might be better suited to a job at his father's cement plant. If not, he was thinking of becoming a professional photographer.[146]

On hearing this, Edison exploded with an anger such as Dyer had never seen before. He said he did not want his son near him or any of his factories. If necessary he would set him up in another business, but he was sure to fail at that, as he had at everything he ever tried. Tom was "no good, and a degenerate."

Dyer tried in vain to persuade him that Tom deserved a large degree of compassion. "It seems remarkable," he wrote in his diary, "that Edison should be so cold and vindictive."[147]

SMALL PALACES

As it happened, Edison was under considerable stress at the time. After more than ten thousand experiments on his car battery, he complained that he had not learned "one-millionth of one percent" of what there was to be known about alkaline electrochemistry. As fast as he and Aylsworth added improvements to the E-cells being manufactured at Glen Ridge, older units came back under guarantee for expensive repair or replacement. Ironically, the aspects of his business he had reassigned to others—moviemaking under Edwin S. Porter and record production under Walter Miller—were bringing in huge sums, thanks to the spread of nickelodeons nationwide and the phonograph's new status as an essential item of domestic furniture. It would be some years yet before the great mill at New Village, managed by Walter Mallory, recouped its start-up cost, but it was already disgorging six thousand barrels a week of the best portland cement and promised to be profitable sooner than the Storage Battery Company.[148]

Edison could at least take credit for the excellence of its product, having devised a self-measuring hopper system that fed parallel streams of limestone and cement-rock roughage into separately beamed scales, each set to tip according to weight limits specified by a chemist. The moment either scale tipped, a needle dipped into a cup of electrified mercury and shut off the hopper, precisely regulating the proportions of the resultant mix. This precision only partly explained the quality of his cement. Much more was due to his insistence on grinding slurry destined for the kiln to a degree 10 percent finer than the industrial average, so that at least 85 percent of it would sift through a two-hundred-mesh screen before calcination. After coming out the other end as granular "clinker," it was at first reground to a powder smooth enough to cream between wet fingers. But he discovered that such a consistency made for too-quick setting, so he adjusted his crushers to make a slightly grainier hydrate.* Concrete whitened with this paste developed formidable strength. "It is the coming construction for all great buildings," Edison boasted. "It won't bend, it won't break, and you couldn't burn it if you tried."[149]

Last May's catastrophic earthquake in San Francisco revived an idea he had had when the cement mill was first ready to roll. He saw low-cost, molded concrete houses replacing the fragile wooden boxes in which most Americans lived—houses that contractors would mix from cement (with a colloidal additive for grit suspension) and spill on the spot into prefabricated forms. A three-story house could be poured in six hours and set in less than a week. The forms could be detached, section by section, and used for as many duplicate dwellings as the neighborhood would tolerate. Not that they had to be identical. "There will probably be hundreds of designs," Edison told a representative from the magazine *Insurance Engineering*. "The architects will have a fine time, for they can pour statuary and all sorts of ornamentation while they are completing the walls. Thus, we will have small palaces renting for about ten dollars a month."[150]

"And the roofs will be made of cement also?"

"Yes, the whole thing—all poured cement construction."†[151]

He had to admit that the individual kits, consisting of nickel-plated cast iron parts, would be expensive, at around $25,000 apiece. But

---

* Edison's cement was too smooth for the comfort of his competitors. "It would have been better for you to have made a cement similar to our own," one of them grouched. Nevertheless the Edison texture became standard in the industry.

† Edison and his interviewer made the common mistake of using *cement* as a synonym for *concrete*. The former is merely an ingredient of the latter.

Edison and model cement house, circa 1906.

they would pay for themselves in frequency of use and universality of detail, molding mantelpieces, banisters, dormer windows, conduits for wiring, "and even bathtubs." Having made the investment, a contractor could pour a new house every four days. Each could be sold for $500 or $600, enabling millions of low-income Americans to become homeowners for the first time, with no need to worry about earthquakes, hurricanes, or fire. "I will see this innovation a commonplace fact," Edison promised, "even though I am in my sixtieth year."[152]

Builders around the country reacted with cautious skepticism. They had heard about concrete houses before, but never from a visionary so determined and capable. "This idea is one of the insanities of genius," said a concrete layer in New Castle, Pennsylvania. "Edison is crazy. He wouldn't be a great inventor if he were not. . . . A man who can solve the automobile question with cobalt batteries for electric machines can do almost anything else."[153]

AN ACCOMPLISHED FACT

As things turned out, Edison would be sixty-three before he poured his first few exhibition houses (including two in Llewellyn Park)* and introduced his radically redesigned car battery. By then cobalt scales were as much a figment of another season as last year's leaves, and he was already venturing into fresh fields. One was that of autobiography—or rather random reminiscences that he jotted down for a pair of insatiably curious Boswells. They were Frank Dyer and Thomas Commerford Martin, the editor of *Electrical World,* who had known him since 1877. Both men were blessed with clear, cool minds that showed in their prose. They took their task—a two-volume, authoritative study entitled *Edison: His Life and Inventions*†—with the utmost seriousness, but also with reverence, casting a hagiographical glow over nearly a thousand pages of otherwise scholarly text.

Edison's deafness prevented them from asking oral questions, so their "interviews" with him were scripted on both sides. His answers were fragmentary and nonsequential, for the most part jovial and unreflective, the memories of a man afflicted by no trauma worse than the deaths of his mother and first wife, and betraying little vanity beyond a fierce pride in his patents. He wrote most evocatively about his childhood in Ohio ("Lockwoods boy & I went swim he went down I waited then went home") and his early days in New York, before he set up shop as an independent inventor:

> *One day [in 1869] after I had exhibited and worked a successful device, whereby if a ticker should get out of unison in a broker's office and commenced to print wild figures, it could be brought to*

---

* These buildings, still in robust condition, may be seen as adjuncts to house tours of Glenmont, conducted by the National Park Service.

† Published in New York by Harper & Brothers in October, 1910.

unison from the central station and which saved the labor of many men and much trouble to the broker. [Marshall Lefferts] called me into his office and said, "Now, young man, I want to close up the matter of your inventions, how much do you think you should receive?" I had made up my mind that in taking into consideration the time and the killing pace I was working that I should be entitled to $5,000, but could get along with $3,000, but when the psychological moment arrived, I hadn't the nerve to name so large a sum, so I said, "Well, General, suppose you make me an offer." Then he said, "How would forty thousand dollars strike you." This caused me to come as near fainting as I ever got. I was afraid he would hear my heart beat. I managed to say that I thought it was fair.[154]

While the literary Frank Dyer worked with Martin and William Meadowcroft to polish the Old Man's grammar and tone down his breezy references to "J.C." and "jews," Dyer the corporate tactician discreetly maneuvered him into positions that benefited them both.* He seized on two 1907 court decisions recognizing rival cinematographic patents held by the Edison and Biograph companies to propose that the litigants move toward the establishment of a "trust" that would pool their rights with those of other major studios, to the exclusion of independent producers and piratical exhibitors. This led to the organization in December 1908 of Dyer's proudest creation, the Motion Picture Patents Company. Its participant moguls paid court to Edison at a celebratory banquet in the laboratory library. He was as genial as ever, and bored by the whole idea of movies as entertainment. After dinner, when the legal papers came out, he said, "You boys talk it over, while I take a nap." He retired to his cot behind one of the book stacks and woke up to find himself in executive control of 90 percent of the American film industry.[155]

He was much less amenable to Dyer's urgings that he permit the development of a disk phonograph technology to match that of the

---

* Dyer succeeded William Gilmore as business manager of the Edison plant on 23 July 1908.

upstart Victor Talking Machine Company. Inexplicably to Edison, many record buyers were opting for the novelty of flat hard records that sounded as if they had been sandpapered, rather than his cherished two-minute cylinders, marketed in heavy cardboard canisters that gave off a sweet savor of wax when opened—and which played melodiously through his phonographs with their big "morning glory" horns.[156] It was almost satisfying to him that a federal appeals court had voided his old patent on a celluloid recording medium, because as deaf as he was, he could not see that volume of tone was preferable to fidelity. National Phonograph's whopping sale of seven and a half million cylinders in 1907 suggested he was not alone in that prejudice. When American Graphophone, his principal competitor, launched a six-inch cylinder playable for four minutes, his response was to issue a new black wax record, the Amberol, that fitted the standard Edison mandrel but was microgrooved at two hundred threads per inch. It played as long and less harshly, but its "hills" were fragile and encouraged needle skip.* Dyer could only wait for Edison to admit that the days of soft sound were over. When the Amberol and its companion player, the Amberola, failed to arrest a sales slide of more than 50 percent by February 1909, Mina was able to report to Theodore, her most tech-minded child, that Papa was working on a new disk machine and having a terrible time with the reproducer: "I wish you were able to give him some points on it to help out with the difficulty."[157]

Whatever Edison's problem, it hardly compared with the nine years of intellectual, mechanical, and chemical labor that led up to his announcement on 26 June that "the storage battery is an accomplished fact." Reporters had heard him make the claim so often that the story rated few headlines and even fewer front pages. Only when the first 1.2 volt A-cells appeared in July—slender, light, and lustrous, so intricately engineered that they approached solid state—did awareness spread that Edison had pulled off a revolution in electrochemistry.[158]

---

* Although commonly described as wax or waxen, the Amberol medium was a soft-soap compound of stearic acid and sodium salt, hardened with ceresin and aluminum stearate.

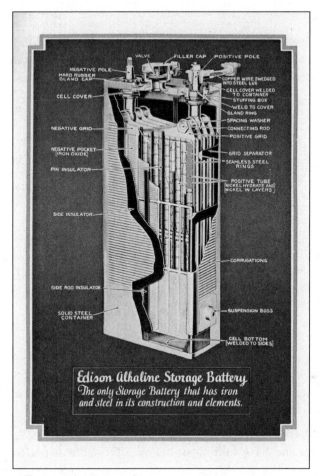

Edison's A-12 storage battery, 1909.

He had done it by returning to flake technology, using nickel leaf this time, plating it so thin that it floated on the air like gossamer. Two hundred and fifty sheets would have to be patted flat to match the thickness of a visiting card. They were drawn off rotating drums that he dunked, alternately and rapidly, into baths of copper and nickel electrolyte. Each deposition was washed and dried, the accumulated layers building up hardly more substantially than crossed shadows. When they laminated out at .3969 of a millimeter, the sheet was stripped and cut into tiny squares, which were then soaked in a solu-

Edison's A-12 storage battery, 1909.

tion that ate away the copper. This left 120 nickel flakes, each about
$\frac{1}{25},000$ inch thick.[159]

The subsequent electrode-loading process that had cost Edison so
many years and $1.5 million of his own money was performed with
the most delicate machines he had ever designed. One repeatedly
drove slender rods into the A-cell's perforated steel tubes, packing
them with seven hundred disks of powdered nickel hydrate and nickel
flake. The tubes were of spiral-wrapped construction now, reinforced
with seamless steel rings to withstand the insertion force of two thousand pounds per square inch. Their ends were mashed shut, much as
he had pinched the evacuation points of his lightbulbs, before they
were mounted in parallel on the positive grid. A similar but simpler
process tamped the flat negative plate pockets with exceedingly fine
iron oxide, unimpregnated with flake because its resistance was lower
than that of the tubes.[160]

Each cell consisted of five negative and four positive plates insulated with hard rubber and then rigidly sandwiched together, before
being canned in sheet steel and saturated with lithium-tinged potas-

sium hydroxide.* Edison boldly welded the can tops, to signal that they need not be opened for as long as the cell lasted—four years at least, by his calculation.† The only extrusions were twin tapered terminals, a lidded aperture for water replenishment, and a valve for the liberation of hydrogen bubbles.[161]

Not wanting to be humiliated again by a market recall, he subjected random cells of his new battery to brutal physical tests. Cells selected at random were thudded up and down on "a solid block, about two million times," then banged against "a brick or stone abutment five hundred times at speed of fifteen miles per hour at moment of impact." If they withstood this punishment at no loss of capacity, he could be sure they would pass the ultimate test of going into service in New York. "I expect soon to see every taxicab run by the new storage battery," Edison boasted. "Automobiles, too, and other pleasure vehicles."[162]

In some respects his dream came true. He soon had more orders than he could fill from cab, streetcar, and delivery companies, and the silent, odorless strength of his alkaline power pack so scared the manufacturers of lead-acid units that one of them issued a similar-looking battery called the "Ironclad," even though it contained no trace of anything ferrous. Over the years various sizes of the "Edison A" would power more than half of the nation's electric trucks, as well as railroad signal and time clock systems, miner's lamps, detonation devices, central station backups, and marine radios. An ambitious young engineer hanging about the West Orange campus would even build one big enough to drive a submarine. But Edison had lingered too long to combat the rocketing popularity of the "pleasure vehicle" powered by internal combustion. His battery went on sale a year after the Ford Motor Company of Detroit introduced its low-priced, high-mileage Model T

---

* Edison's addition of lithium hydrate to his electrolyte, significantly improving the A-cell's capacity and stability, anticipated by sixty years the development of the lithium-ion battery, now a staple in power science. Francis T. Bonner of the Manhattan Project described the innovation as "a real piece of magic," still not perfectly understood in the 1950s.

† He underestimated the longevity by more than ten times. Edison storage batteries were still working forty-five years later in various parts of the United States. In 2011 a researcher resurrected some eighty-five-year-old A-cells and found they still functioned perfectly.

car, hastening at a compound rate the day when the word *electromobile* would begin to sound quaint and drop from everyday speech.[163]

WE LIVE LIKE SQUATTERS

The end of the decade found Edison happy in the success of his cement and battery innovations and looking forward to the wealth they would surely bring. Apart from his now-serious deafness and chronic attacks of stomach pain, he was as energetic and scientifically curious as a postgraduate student. "Did you ever realize that practically all industrial chemistry is colloidal in its nature?" he asked a pair of old associates.* Having for forty years averaged one invention every eleven days, he was in no mood to take the advice of his cousin, old Lizzie Wadsworth: "My dear Alva . . . I think it is about time you rested that Brain of yours, you have given to the World enough & now take a well earned rest the rest of your life."[164]

He thought he might work productively for twenty years more before handing his businesses over to his younger sons. The elder ones remained splinters rather than thorns in his flesh, unfelt at times, hurting at others. Probably they would continue to do so as long as he lived (unless they carried out their occasional threats of suicide). "Burton Willard" was the sadder of the two, laboring with desperate energy on his mushroom farm when he was well, but having to beg for help when he was not. His latest series of brain seizures had prostrated him in a Philadelphia hospital for seven months.[165]

William and Blanche still engaged in moonlit flits from one "hell hole" to another, pursued by enraged landlords while splurging Edison's weekly checks on such essentials as champagne cruises in Chesapeake Bay. When Will's supplementary appeals for custom clothes and a Pierce-Arrow car were denied, he wrote his father so pettishly that Frank Dyer at last threw aside all lawyerly courtesy and replied, "I have no doubt that if we turned $100,000 over to you, it would be spent in idle foolishness in two months and that at the end of that time we would hear the same complaints and fault findings. . . . I have no sympathy with you when you act like a child."[166]

---

* Edison posed this question in April 1909, six years before the publication of an English translation of Wolfgang Ostwald's pioneering *Handbook of Colloid-Chemistry*.

Edison showed some interest in Will's design of a double-acting spark plug but could not help seeing it as a token of the new, unwelcome age of the gas-powered car. He sounded both worried and hopeful about the future when he lunched with an old friend, the artist and philosopher Elbert Hubbard. They were founder-members of the Jovian Society, a group of environmentalists promoting electricity as the clean energy of the future. Taking out a fresh cigar ("Just pass the matches, thank you!"), Edison launched into a polemic against other ignescent devices:

> Someday some fellow will invent a way of concentrating and storing sunshine to use instead of this old, absurd Prometheus scheme of fire. I'll do the trick myself if someone doesn't get at it. . . .
>
> This scheme of combustion in order to get power makes me sick to think of—it is so wasteful. . . . We should utilize natural forces and thus get all of our power. Sunshine is a form of energy, and the winds and the tides are manifestations of energy. Do we use them? Oh no; we burn up wood and coal, as renters burn up the front fence for fuel. We live like squatters, not as if we owned the property. There must surely come a time when heat and power will be stored in unlimited quantities in every community, all gathered by natural forces. Electricity ought to be as cheap as oxygen, for it cannot be destroyed.
>
> Now, I am not sure but that my new storage battery is the thing.[167]

PART FOUR

Magnetism

1890–1899

O N THE EVE of his forty-third birthday, Edison succumbed to an unusual attack of depression. He had only just recovered from the shock of seeing his proudest achievement, the Pearl Street generating station in New York, burned to the bricks, with seven of its eight great dynamos ruined. Thanks to quick emergency action by the local illuminating company, the system was back in service after only a few days. But the memory of seeing a powerhouse he had designed with such care standing blackened, drenched, and gap-windowed was still raw when he heard that Henry Villard, the multimillionaire president of Edison General Electric, wanted him to sacrifice a thousand shares of his stock in that firm, so that certain Vanderbilt interests—aided and abetted by the Drexel, Morgan banking house—could buy their way in.*[1]

This casual assumption that he would allow himself to be elbowed still further down the board table of a corporation bearing his name (and how much longer would the financiers maintain *that* trademark?) enraged Edison, at a time when he would have preferred to unload some of the less gilt-edged assets in his portfolio. Having agreed to accept stock certificates, instead of cash, from the treasurers of dozens of isolated central stations he had installed around the country in the 1880s, he held almost $4 million in currently worthless paper.[2] Having further agreed, perhaps foolishly, to Edison General's absorption of all the electrical manufacturing companies he used to control, in exchange for a cash payment that somehow also turned to paper, he found that his income had fallen from $250,000 to

---

* The Edison General Electric Company had been incorporated in 1889 as a merger of three Edison electric light manufacturing companies.

$85,000[3]—too little to keep the West Orange laboratory going, unless he delved into his own pocket.*

"Your request has worried me so much that it is the principal reason for breaking me down in spirits," he wrote Villard, saying that he needed a vacation in the North Carolina mountains to refresh him, body and soul.

> *I have been under a desperate strain for money for 22 years, and when I sold out [to Edison General], one of the greatest inducements was the sum of cash received, which I thought I could always have on hand, so as to free my mind from financial stress, and thus enable me to go ahead into the technical fields. To put it back into the business is something I have never contemplated. . . . I feel that it is about time to retire from the light business and devote myself to things more pleasant, where the strain and worry is not as great.[4]*

Villard dismissed this threat as the kind of cri de coeur to be expected on occasion from a temperamental genius. "No cause for worry," he replied. "With long rest I am sure you will return in better spirits."[5]

SPOTTED LIKE A LEOPARD

Actually Edison was remarkable for hardly ever losing his cheerful equanimity. Not since the death of his first wife, five and a half years before, had he shown such emotion. But the decade had certainly opened with a pileup of worries for him—in contrast to the 1880s, which had been one long crescendo of celebration and success. Now he was again a plaything of financiers (Villard suggested he borrow to make up for his lost income), contractually compelled to work up to nine hours a day, unpaid, on electrical problems that should have been handled by engineers in the field. That left another nine hours—eighteen being his habitual schedule—for him to devote to problems closer to home, such as a faltering start to the production of Edison

---

* Edison was, however, by no means a pauper in early 1890. His latest bank statement from Drexel, Morgan showed a balance of $465,440.25, or $12.5 million in 2019 dollars.

talking dolls, a sharp decline in phonograph and record sales, two looming court decisions that could well humiliate him, bad news from an experimental iron mine he had opened in Pennsylvania, and even worse news from Germany, where his eldest daughter had been struck by smallpox.[6]

Marion lay now in a Dresden hospital, her seventeen-year-old body "spotted like a leopard only more so," in the words of Elizabeth Earl, her governess and chaperone. She had been touring Europe for ten months, partly to finish her education but mostly to keep an ocean between herself and Mina. As far as Edison could see—which in domestic situations was a matter of millimeters—Tom and William had recovered from the loss of one mother and adapted to another without much difficulty. But Marion's memories of Mary went further back (the golden hair, the chocolates, the laughter). She could not forgive Mina for supplanting her. Mina had been unable to fill the void in the girl's heart, and Marion herself had decided to go abroad.[7]

Marion Edison as a teenager.

That venture itself cost Edison plenty, and now he was faced with thousands of dollars in medical and recuperating expenses, assuming Marion survived—there was some evidence that her case was hemorrhagic.*[8] Mina's budget for the rest of the family also showed inflationary tendencies. The boys had their names down for fall attendance

---

* "The abscess on her back [inflicted] permanent injury to the spine," Mrs. Earl wrote Mina on 10 March. "When they lanced it . . . she bled so profusely they feared for her life." Hemorrhagic smallpox is almost always fatal.

at St. Paul's in Concord, New Hampshire, a boarding school that was by no means cheap.* And she was pregnant with her second child.[9]

As long as Edison was flush, he willingly supported all the minor or impecunious family members who depended on him, from his eighty-five-year-old father to little Madeleine, not yet two. But his charity went no further than signing checks. He saw no reason to write sympathy letters in addition—not even to Marion, suppurating in a foreign sanitarium. She could count on all her bills being paid, no matter how long it took for her pitted face to heal. Surely that permitted him to ignore the suggestion of a past governess that what Marion needed most of all was "a loving letter from her *Father*."[10]

Affable to every stranger who waylaid him, generous with advice even to competitors, Edison was unaware of how often he hurt the feelings of intimates. He was at once gregarious and distant, willing to admit that "I live in a great, moving world of my own,"[11] like the flickering figures seen through the peephole of his Kinetoscope machine. Even when alerted to the pain, or loneliness, or shame, or other neuroses of people who were less successful than himself, he seemed puzzled that they did not cheer themselves up by embarking on some bold venture, as he was about to do.

MORE AND BIGGER ROCKS

Iron, more than fresh air, was what he sought in the mountains of North Carolina, on the basis of prospecting rumors. For years he had dreamed of becoming a mining mogul, in about the greatest conceivable contrast to his work as a laboratory engineer. The fantasy went back to his discovery in 1881 of a black beach at Quogue, Long Island—sheet upon sheet of powdered magnetite shifting and resettling in the sand every time the wind changed or the tide went out. Then and now, Atlantic Slope iron oxide looked like gold to him, although it was inferior both in quality and quantity to the great ferric oxide deposits around Lake Superior.[12]

He wore down many 4B pencils calculating how much low-grade Appalachian magnetite he would have to concentrate to undercut the

---

* The fee for two boarders at St. Paul's in 1891 was $1,200 per annum, or $32,280 in 2019 dollars.

price to local furnacemen of Michigan hematite, which was softer and richer but cost a fortune to haul east along a thousand miles of railroad. Edison's equations were conditional upon many variables, among them his ability to design excavating, crushing, grinding, and separating machines of such size and sophistication as to remain competitive even as other mills in the region closed and midwestern production increased.[13] Also he had to be sure that whatever deposits he found were enormous enough to guarantee he would never run out of ore.

After an unrewarding six-week search for iron formations along the Blue Ridge ("I go out to prospect a property here, I meet a negro, who refers me to another negro, and finally I find the mine—a dental drill vein"), he returned north convinced that his best bet lay in the wooded highlands of his own home state, near Ogdensburg.* He had recently acquired a sixteen-thousand-acre tract there on Sparta Mountain, including abandoned excavations left over from the days when the only iron was eastern iron. According to his estimate, the Ogden mine had a potential yield of 200 million tons of low-grade magnetite just in its central three thousand acres—enough to make him a billionaire if he exploited it the way he intended: "The ores of New Jersey are in the primal rocks and if these mineralized rocks can be worked commercially, there is more iron ore in the state of New Jersey than in any other area of equal size in the world." Just across the border lay the struggling but still active foundries of eastern Pennsylvania, and the anthracite mines that provided them with natural blast-furnace fuel. "The market is here in their midst," Edison wrote in a rationale for his scheme. He noted that local labor was cheaper, and the supply of skilled managers greater this side of the Alleghenies. "The only thing necessary is cheap ore. With abundance of that commodity . . . the center of iron production in the US would be brought back easterly many miles and Western [exporters] would not underbid the Eastern mills at their very doors."[14]

He needed no further inducements to resurrect Ogden under the

---

* Between 1888 and 1891 Edison conducted a mammoth dip-needle survey of the Appalachian ironlands from New York State to the Carolinas, buying up in the process title to "97% of all the concentrable ore" within practical range of eastern blast furnaces.

grand name of the New Jersey & Pennsylvania Concentrating Works. A phoenix rising from the ashes of his old Edison Ore-Milling Company, the plant would have all the features of a legally established corporation, with a small board of directors and a capital stock of $250,000. He was determined to keep it off the open market and was prepared to finance its future expansion himself, instead of depending, as in the past, on tightwads like J. P. Morgan for development money. For now he was content to be the major shareholder, with such loyal friends as Robert L. Cutting, Jr., Charles Batchelor, and Sammy Insull backing him up.[15]

A dinner in New York on 24 March, hosted by Henry Villard and attended by many senior figures in the lighting industry, confirmed his long-held suspicion that Villard was working on a merger of the Edison General Electric and Thomson-Houston Electric companies—a move that, if successful, would combine their respective strengths in illumination and motor technologies. Edison objected to the idea because Thomson-Houston would be able, through cross-licensing, to trade on his patents, and he was not sure he would receive fair compensation. Besides, Thomson-Houston had adopted an alternating current system, which was more efficient than his own and therefore detestable.* Whether or not the deal came off, he had had his fill of electrical invention after twenty years of connecting wires to other wires. It had left him no wiser as to what electricity *was*. Except when current pulsed under his fingers, or shocked or heated them, he had no sense of dealing with something substantial.[16]

As a boy reading R. G. Parker's *Natural Philosophy*, he had learned that science was divided between "ponderables" and "imponderables," that is, agents with or without mass.[17] So far he had dealt with the weightless phenomena of telegraphy, sonics, and light. Now he wanted to measure his strength against the material massiveness of the world. He longed in body as well as mind to crush more and bigger rocks than any man before him and to use magnetic force to drag the iron out of their dust.

An editorial in the trade journal *Iron Age* warned that such desire was addictive and potentially ruinous:

---

* See Part Five.

There is something very fascinating in the production from lean ore of a concentrate in which very few particles of foreign matter can be detected. No one who has approached this subject has escaped the glowing enthusiasm and the air of triumph of an inventor over such an achievement. . . . Yet there are very few of the promoters of such work who have an adequate conception of the costs and of the losses involved. . . .

We do not mean to convey the impression that magnetic concentration has not a brilliant field before it; but, generally speaking, hopes have been raised on the basis of underestimates of costs, which are sure to lead to disappointment.[18]

BLACK TOOTHPASTE

To Edison, the greatest of all mysteries, surpassing even that of electricity, was "the mystery of what passes between the north and south poles of the magnet."[19] That had not stopped him, in youth, from writing learnedly about the behavior of magnetism in self-adjusting telegraphic relays and wondering if it might be used to deflect the iron ray in the solar spectrum. He had gone on to invent a myriad of electromagnetic and pyromagnetic devices, including a rhomboid bridge that measured the integrity of various metals with extraordinary accuracy. His most imaginative use of the force had been to apply it to bent lamp filaments. They magically straightened when he passed a magnet close by: "One pole attracts while the other repels the charged carbon." Lately he had put on exhibit in West Orange a soft-iron magnet of great ore-separating potential. It was six feet long and two and a half feet wide, heavily wound with copper wire, and weighed well over three thousand pounds.[20]

Contrary to his reputation among pure scientists as being a technological experimenter only, Edison read deeply in the analytical literature and was familiar with all aspects of natural force theory, including the writings of Faraday and James Clerk Maxwell. His research into such phenomena as the effect of diamagnetism on magnetic stress, or the loss of conductivity in ferromagnetic materials, was limited only by his lack of mathematics. For that extra dimension of understanding, he had become dependent on Arthur Kennelly.[21]

Right now he could live with magnetic mystery, as long as what fell past the poles was a thin, broad stream of ore, with the blackish part of it wavering sideways while the shiny remainder dropped straight. The first separator he ever designed (out of the blue in 1880, at the height of his frenzy to perfect his incandescent lightbulb) had been so simple a device as to look almost silly: a wooden lamppost, a dangling hopper, a mounted electromagnet, a bifurcated bin. Yet it worked well on ferrous beach sand, using only gravity and magnetism, the most fundamental forces in physics.[22]

Edison did not pretend to have invented magnetic separation. He emphasized in a research paper coauthored with John Birkinbine and published in *Transactions of the American Institute of Mining Engineers* that various forms of it had been tried in the Adirondacks and countries as remote as Bohemia and New Zealand. But few of them matched the effectiveness of the seven separators Edison had patented since 1881, most recently one that combated the problem of too-dry ore clumping as it fell, so that grains of chert blocked the extraction of grains of iron. He fed crushed rock directly into a tank of water, then agitated it in a revolving, magnetized drum that freely dispersed all particles. The fines clung to the poles of the drum as it emerged from the water and could be scraped off afterward, like black toothpaste.[23]

For Ogden, he now sought to patent a sophisticated separator in partnership with his most talented engineer, the Scottish photographer William Kennedy Laurie Dickson. It monstrously resembled a secret camera they were working on, in its combination of drive wheels, spools, and rolling belt loaded with granular product:[24]

A snake feeder screwed ore, wet or dry, into a hopper that fed the fines one way, and the rough another, while four giant magnets pulled the iron powder onto the belt in a series of ripples that grew denser with every transit of the poles. Demagnetized as it proceeded, the powder fell from collecting pockets into a delivery container. Fans whirled away collateral dust and filtered it for "float iron." The overall system was unique in the way it repeatedly counterposed gravitational and magnetic forces to enrich the concentration of fines, resulting in a high degree of purity by the time they settled.[25]

In May, Edison ordered a series of preliminary tests at Ogden. He left them to be conducted by Dickson and returned to West Orange to plot a major expansion of the facility in the fall. If other eastern "iron men" were struggling to produce a thousand tons of ore a day, he intended to mine and mill five times as much.[26] In his haste to make plans, he probably never read a story that appeared in the *Pittsburgh Dispatch* on 11 May: "NEW IRON TERRITORY—INDICATIONS OF A RICH FIND IN AN UNEXPLORED FIELD OF MINNESOTA."

The range was called Mesabi, and its wealth of red hematite was reportedly so prodigious that three local entrepreneurs were already building a rail link from Duluth to export it to the world.

YOU OUGHT TO BE VERY THANKFUL

Marion Edison escaped the worst ravages of smallpox, thanks to the care of one of Dresden's best physicians. But her face was so pitted that she shrank from reentering European society until the crimson scars faded. Edison sent the doctor a magnificent set of silver and rented her a villa on the French Riviera, where she could recover in seclusion with Mrs. Earl. It would be months before Marion could be persuaded to leave the house without a veil. She continued to pine, mostly in vain, for some written words from her father, while berating herself for being a burden to him. "It makes me sick when I think of the money [I] cost Papa," she wrote Mina.[27]

For obvious reasons, the young woman could no longer be expected to attract an early marriage proposal and so relieve Edison of responsibility for her. Half homesick, half proud, wholly aware that another child was about to swell his second family, she clung to Mrs. Earl and accepted her fate as an invalid in exile.

Charles Edison arrived on 3 August. "You are a lucky woman Mama," Marion wrote when she heard the news, "and you ought to be very thankful you have one of the loveliest of men for a husband, a sweet little baby who will do you credit, money, beauty. For my part I don't know what else you want in this world to make it a Paradise."[28]

"SOCIETY OF HARMONIC CURVES"

Edison's "Ogden baby," as he jokingly referred to his other neonate acquisition, proved much less able than Charles to ingest processed helpings. Dickson ran some of the local rock through crushers rented from the Brennan company and found the resultant pulver far different from beach magnetite. It powdered in dry weather, abrading the oiled joints of machinery and penetrating the thickest of respirators. When wet, it sweated clay and clogged the rotating screens it was supposed to fall through. Ominously, Sparta Mountain's iron quotient turned out to be leaner than Edison had hoped, averaging only 16 percent. A previous generation of miners had carved away the four most workable seams, leaving it to him to figure how to excavate the rest.[29]

He was not discouraged when Dickson reported that the initial tests were "n.g. with a vengeance." Stimulated as usual by difficulty, he undertook to design new screening and drying systems that would meet all Ogden's challenges, and shut the plant, hoping to make it fully operational in the spring of 1891. That committed him and his nervous board to a carrying cost of $20,000 to $30,000 a month, just as news came of more and more hematite findings in Minnesota.[30]

In September an ambitious writer, George Parsons Lathrop, came to stay in West Orange, hoping that Edison would now have time to work with him on a project inspired by the phenomenal success of Edward Bellamy's utopian novel *Looking Backward*. Lathrop was the author of "Talks With Edison," a magazine article published earlier in the year that made much of his subject's affable approachability. He had been so impressed with the inspirational, quasi-poetic way Edison dreamed up inventions—"These ideas are occurring to me all the time"—that he suggested they collaborate on a science fiction novel, to be called *Progress*. To his surprise, Edison not only agreed but waved aside any question of a fee. It was enough that Lathrop would do the writing, while he, simply for fun, came up with futuristic notions to embellish the story. He even offered to illustrate it with his own drawings.[31]

Their collaboration was to be kept secret, under a first-serial-rights contract with the McClure newspaper syndicate. "There would be some money in it for you and me," Lathrop informed Alfred O. Tate, Edison's secretary and an intermediary in the arrangement. No doubt the novel could then be issued as a best-selling book. In that case he would offer Edison a share in the royalties, "tho' I fancy that it will not be much of a consideration in his eyes."[32]

At thirty-nine, Lathrop was no literary lightweight. He was married to Nathaniel Hawthorne's daughter and had published several volumes of fiction and poetry, as well as founding the American Copyright League and serving as an associate editor of *The Atlantic Monthly*. These achievements had not saved him, however, from the twin liabilities of a freelance writer's career, anxiety and alcohol.[33] Over the next nine months he was to discover, with frequent relapses into both, that Edison was possibly the busiest man in America.*

The delay-prone tempo of their "collaboration" was set in mid-October, when Edison sent Lathrop thirty-three pages of notes, scrawled so fast as to be barely readable in places. Some were surreal, others visionary, but most read like experimental prompts from Edison to himself, as if he had forgotten they were supposed to inform somebody else:

Lubrication at high temperature by the Bromine substitution
Mfr oxygen by passing over molten Titanium
Disassociation of all the Halogen group by incandescence[34]

Lathrop feigned delight at being vouchsafed such jottings—"I have copied them. They are immense!"—and returned the manuscript to Edison, red-penciled with many requests for elucidation. It was not clear to him what a "Society of Harmonic Curves" might be, or how vaporized mica might be turned into microfilm by electrical excitation. He also needed help on the technologies of cable telegraphy powered by "etheric force," screwless steamships, climate change, aerial navigation, hypnotizing machines, phonographic newspapers, Saharan canals, mother-of-pearl room panels, and colored music. Edison was unavailable for an interview but promised to record some explanations on wax cylinders. Lathrop waited for them in vain. He was not mollified to receive another batch of enigmatic memoranda and two sketches of an "air-ship." Edison finally saw him for a hurried discussion that left Lathrop no wiser than before.[35]

With his advance money dribbling away and McClure asking awkward questions, he tried to make literature out of the notes he had. It

---

* According to Alfred Tate, Edison was at this time supervising seventy-two projects.

was difficult to do so even when they were not technical. Edison's few efforts at science fiction could have been written by a schoolboy ("Person inside a non conducting chamber . . . passing limits of our atmospheric shield adjusted to attain speeds of 100,000 miles per second there being no friction in vacuous spaces") or by a tired man half asleep, drowsiness turning into dream ("Glow worm—not popular—striving for perfect steadiness, beautiful eyes.")[36]

Lathrop cobbled together some initial chapters and sent them to Edison for approval. Six weeks slipped by with no answer.[37] He could only plead with McClure for an extension of his contract and hope that whatever in the world was distracting Edison's attention would permit them soon to resume their imaginative journey into another.

PERSISTENCE OF VISION

That fall Edison, working secretly in the "precision room" of the laboratory, was developing a device far more fantastical, in its practical effect, than any novelistic machine. It was his Kinetograph motion picture camera, a radical redesign of the cylinder-based Kinetoscope he had conceived two years before.* W. K. L. Dickson (back in West Orange for the winter, while Ogden's new buildings arose) had persuaded him that a strip of translucent film, winding sideways across the viewer from spool to spool, could present hundreds of much larger, sharper images at a speed of ten frames per second.†[38]

The proof of this, around November, was *Monkeyshines,* a blurry sequence of movements by one of Edison's Greek employees in a

---

* See Part Five.

† Most film historians date Edison's development of the Kinetograph from the time Dickson rejoined him at the laboratory in October 1890. But as early as February that year, the *Orange Journal* reported: "For many months past Mr. Edison has been at work on a series of experiments in instantaneous photography which have at last been successfully concluded." In April *Western Electrician* described a mysterious Edisonian projection at the Lenox Lyceum in New York: "A magic lantern of almost unimaginable power casts upon the ceiling . . . such pictures as seem to be the actual performances of living beings!" That same month the *Minneapolis Times* stated that he was experimenting with a horizontal-feed spooled motion picture machine ("he calls it the Kinetograph") and also was planning to equip it with synchronized sound: "When it is completed . . . it will be possible not only to hear the voice of a person . . . but to see the person's face just as it was at the time the words were spoken, with every change of expression, the movement of the eyes, etc."

belted fustanella, energetically waving his full white sleeves. It lasted less than half a minute and was so diffuse at times as to resemble the pulsations of a jellyfish. Yet it was performance and photography combined, and history too, the first moving picture ever produced in the United States. Dickson and his assistant, William Heise, did most of the mechanical work.*[39]

The filmstrip that made *Monkeyshines* possible was cut from photosensitive cellulose nitrate plasticized by George Eastman of Rochester, New York. Thin, springy, and sweetly redolent of bananas, it came in rolls seventy millimeters wide. Dickson thought that unnecessarily broad. He sliced it in half before perforating one edge to fit the Kinetograph's sprocketed wheels. The result was a thirty-five-millimeter film that (to his pride as a forgotten old man) would become the standard stock of cinematography.†[40] *Monkeyshines* was followed by a better-focused sequel and a series of progressively improving "camera tests," but myriad drive and darkroom problems kept postponing the moment Edison could announce his invention publicly. The escapement mechanism chattered too slowly or too fast; patches of film emulsion frilled off the negative in development, leaving oleaginous images at the bottom of the trough; perforations snagged; torque yanked the pictures out of alignment.[41] "Persistence of vision"—a phrase Edison loved to use when explaining the eye's inability to separate a rapid succession of stills—began to look more like the orneriness of a new medium, as yet too raw to synthesize.[42]

PETTY SPITE AND LOVE OF REPUTATION

On 1 January 1891 Edison was annoyed to see himself advertised by *The Sun,* along with Robert Louis Stevenson and Rudyard Kipling, as the author of an important forthcoming work of fiction. The news spread as far as Germany, where it was described as an "electrical

---

* The complex chronology of the invention of cinema, involving simultaneous experiments and claims of precedence in France, Britain, and the United States, is a subject of unresolved debate by scholars in all three countries. Edison's relations with Étienne-Jules Marey and his pioneer work on the Kinetoscope in 1888 and 1889 will be discussed in Part Five.

† Dickson wrote that he got his filmstrip idea after a glance at Edison's "perforated paper automatic telegraph." Dickson, "Brief History."

novel" in two volumes that he would subsequently adapt for the stage. *The Hartford Courant* expressed mock horror: "Let him keep outside literature pure and simple where he belongs not."[43]

Edison blamed his collaborator for the story, calling it "an act of bad faith" and threatening to repudiate their partnership if any more embarrassing publicity ensued. Lathrop protested his innocence, pointing out that "the Sun's misstatement injures me, by ignoring my name." He was forgiven, but from then on Edison became even more inaccessible to him. When Lathrop appealed to Mina for the return of the sample chapters he had sent to Glenmont, they came back evidently unread.[44]

Except on the unique occasion when Edison shared a patent with W. K. L. Dickson on their new magnetic separator at Ogden, he disliked having his name coupled with that of any other creator— whether a minor talent like Lathrop or an inventor as brilliant as his onetime employee Frank J. Sprague. He had been careful in the 1880s not to involve Sprague in his experimental electric railway project, thus losing out on Sprague's later, seminal innovations in rail traction technology.[45] That false pride showed now on a visit to Buffalo, when he told a local reporter that Edison General Electric was "going to furnish the power for your street railway system."

"But is the company here not going to use the Sprague system?"

"Yes. But it is not known as the 'Sprague' system. It is the 'Edison.' We have absorbed and improved the Sprague."[46]

Edison General had indeed recently bought the Sprague Electric Railway and Motor Company, acquiring its invaluable patents and burgeoning goodwill (well over a hundred urban systems installed or contracted for since 1887) while providing it with the capital it needed to expand.[47] But Edison's royal-plural boast that he had "improved" its technology enough to substitute his name for Sprague's sounded like a slap at the latter for resigning after the acquisition. At thirty-three Sprague was hungry for renown and blamed both Edison General and Henry Villard for withholding it from him:

> Your company, instead of being managed in the best interests of its stockholders and to make the most of its property and connections, is conducted in the personal interest of Mr. Edison and his represen-

tatives and has become an active agent for my personal, professional and business injury, in which jealousy, petty spite and love of reputation, however attained, are the strongest motives. . . .

It contents itself with the promulgation of circulars known to every railway man in this country to be untrue, and has set out to do everything possible to wipe out the Sprague name and to give to Mr. Edison the reputation properly belonging to other men's work. . . . He finds most favor who is most abusive of all things Sprague, and he meets with a cool reception who does him the smallest reverence. The Edison fetish must be upheld, and the Sprague name abolished: that is the law. . . .

Not only Mr. Edison's subordinates and those who bask in the sunshine of their smiles, but Mr. Edison himself, forgetful of his dignity and jealous of any man who finds in the whole realm of electric science a corner no matter how small not occupied by himself, loses no opportunity to attack and to attempt to belittle me.[48]

Sprague, who went on to become an equally distinguished inventor of vertical traction systems, would complain for the rest of his life about the world's failure to recognize his genius, and its contrary insistence on keeping Edison at the summit of Parnassus.*[49] He never acknowledged that his achievements, great as they were, were confined to "the whole realm of electrical science," and that he would have been lost in such foreign fields as harmonics in music, illusionism in moving pictures, dispersion of rubber molecules in Soxhlet extractors, and magnetic separation in the highlands of northern New Jersey.

ROSY FOLIATIONS

Edison's impatience to start mining and refining at Ogden as soon as the ground unfroze—it had been a brutal winter—was stimulated, rather than slowed, by news in early February of a phenomenal gain in Edison General Electric stock. Speculators ascribed the surge to a rumor that Henry Villard, hard up for cash, was about to sell his majority shares to someone associated with "the Vanderbilts." No such

* In 1911 Sprague heard with modified rapture that the Institute for Electrical and Electronic Engineers had awarded him its highest honor, the Edison Gold Medal.

sale ensued, but it was not the first time Edison had heard the names of Villard and Vanderbilt bracketed in the same sentence. He was disagreeably reminded that the bulb he had invented, the dynamos he had built, the industry he had founded, had become corporatized far beyond his control.[50]

Mid-March found him heading north to gear up his new plant for production. A little mine train picked him up at Lake Hopatcong and began to ascend a back spur of Sparta Mountain. It passed the old Hurd mine, abandoned a century before, then rose further through thick forest still bare of leaves. No ferns or creeping vines yet obscured the errant boulders that lay everywhere on the slope, some looking ready to roll down and crush the train. When it reached twelve hundred feet, it puffed to a halt, leaving Edison to climb the last half-mile to Ogden on foot.

The slippery trail led to a quarry face known as Iron Hill, the most accessible part of his vast domain. It was a tabular sheet of gneiss four miles long,* running along the southeastern slope of the Beaver Lake anticline, parallel to the general corrugation of the Appalachian range. Wherever spring rains washed the rock clear of mud and snow, flecks of mica caught the light. Rosy foliations in the gneiss showed the dissemination of magnetite crystals, dense in a few places, disappointingly sprinkled in others. But who knew how deep and rich the seams might be, where they receded into the slope or dived almost vertically into bedrock? Edison guessed at least four hundred feet and a mile in each direction. He was prepared to carve away the whole mountain, if necessary.[51]

Seen from this vantage point between the ridge and a reservoir to the east, the New Jersey & Pennsylvania Concentrating Works had a certain flow-through continuity—albeit hampered by his decision to adapt as many of the old mine buildings as possible. He had spent nearly $54,000 on a new magnetic separation house, as well as adding some storage hangars and several miles of railroad track. The existing stone powerhouse now contained the four-cylinder, triple-expansion vertical engine he had treated himself to at the Paris Exposition in '89. A chain-and-bucket cableway was braced on the upper

---

* By 1898, this mass was already known to geologists as "Edison gneiss."

bench of Iron Hill, ready to convey hand-loaded chunks of ore (as yet he lacked a crane to lift larger boulders) four hundred yards down to the mill, where seventeen jaw crushers would bite them down to pebbles. From there they would proceed through a series of grinders and rotating screens that would reduce them further, from gravel to grains. Then the magnetic separators would divide them into iron "fines" for storing and shipping, or gritty "tailings" to a dump for separate sale as sand. Elsewhere in the congeries two millhouses loomed, one old and one new, a machine shop, and a black-towered pump sucking floodwater from an abandoned shaft in the heart of the complex.[52]

It was a raw and ugly scene in a landscape not yet broken into leaf. There were no accommodations for Edison's labor force of several hundred mostly Italian immigrants, who had to crowd into tenements in Ogdensburg, half an hour's trudge down the hill. For himself, there was at least the hospitality of a farmhouse to the east of the plant. He was invited to stay there whenever he came up from Glenmont, sixty miles away.[53]

TRAILING ZEROES

Harry Livor, the general manager at Ogden, tried to obey Edison's order to start milling at once. But the system was so intricate that it did not crank into a semblance of production until the beginning of April, and even then it was plagued by numerous mechanical and coordination problems. Raw ore from the quarry was often wet, filthy with clay or fibrous with torn roots, jamming the machinery and clogging screens. Or it gave off clouds of dust that mixed with grease on cableway wheels and abraded them, necessitating frequent shutdowns for replacement and repair. The jaw crushers were of frustratingly small capacity, and the six belt separators needed constant regulation. Undiscouraged, Edison went to Pennsylvania in search of foundry orders for his iron concentrate. He figured he could supply it at a richness rate of 66 percent (up from 25 percent in ore) for $5.28 per ton, earning a $2.62 profit for himself.[54]

John Fritz of the Bethlehem Iron Company was less persuaded by his figures than by the scope of his ambition. "Well, Edison, you are doing a good thing for the Eastern furnaces. . . . I am willing to help you. I mix a little sentiment with business, and I will give you an order for one hundred thousand tons."[55]

Or so Edison chose to remember the conversation, with his habit of trailing zeroes after any number that pleased him, like soap bubbles from a pipe.* Fritz's order was actually for one hundred tons a day, conditional upon the smelting performance of an advance consignment. The Pennsylvania Steel Company and North Branch Steel made similar commitments. Livor managed to deliver at the pace Fritz wanted but could offer only forty tons a day to the other customers, and the quality of his product declined. "Bethlehem complains iron running down phosphorus running up," Edison warned him. "Be careful or we will be ordered to stop shipping." Livor in turn grumbled that he was not getting enough marketing support: "There apparently seems to be no vigorous effort to dispose of our product. . . . Someone of some little knowledge of the business ought to be at the furnaces very quickly after the ore reaches them."

This was not the right tone to strike with Edison. Livor was soon dismissed, as was a "damned fool" of a mining expert who dared to predict that Ogden's quarry-and-concentrate method would never be profitable.[56]

"SUCH A HAPPY COMBINATION"

As Edison feared, Bethlehem Iron canceled its order after buying only a few thousand tons of his concentrate. It cited phosphorus levels, furnace blowback, and caking as the principal faults of Ogden fines.[57]

He decided that the only person who could get his grand scheme going was himself. That meant prolonged stays in the mountains and possible further redesign and reconstruction. He felt quite up to the task: "I feel that I am in my prime, and I suppose that I am a better man than I have ever been." But first he had a major patent infringement case to prosecute—*Edison Electric Light Co. v. United States Electric Lighting Co.*, in the Circuit Court of the Southern District of New York—that could be worth millions if Judge William J. Wallace

---

* See, e.g., his claim in 1906 that the Edison works at New Village produced "60,000,000" tons of portland cement a week, the correct figure being $600,000. In May 1891 he told a reporter that the amount of iron in the Ogden mine was "2,000,000,000,000,000 tons."

ruled in his favor. Then he had to prepare two prohibitively difficult patent applications, covering his still-secret Kinetograph technology. Dickson and Heise had improved the camera and its attendant player enough to begin to demonstrate them to carefully selected audiences. But there were enough competitive devices under development in France and Britain (Étienne-Jules Marey's exquisite "chronophotographs" of undulating sea horses and anemones had recently been featured in *Scientific American*) as to cast doubt on Edison's chances of winning any but the narrowest claims of exclusivity on his own.*[58]

It occurred to him that George Lathrop, still pining for his collaborative attention, would be the ideal scribe to publish an article on the Kinetograph that might subliminally influence Patent Office examiners in its favor. Lathrop jumped at the chance, and began to research a long piece for *Harper's Weekly*. Edison thereupon yielded to the temptation to start talking about it in advance, and scooped his chosen publicist.

At a meeting on 12 May with some commissioners of the great World's Fair planned for Chicago in 1893, he told them that he would exhibit something that would cause a revolution in home entertainment:

> Such a happy combination of photography and electricity that a man can sit in his own parlor and see depicted upon a curtain the forms of the players in opera on a distant stage, and hear the voices of the singers. When this system is perfected, which will be in time for the Fair, each little muscle in the singer's face will be seen to work; every color of his or her attire will be exactly reproduced, and the stride and positions will be as natural and varied as those of the live characters. To the sporting fraternity I will state that ere long this system can be applied to prize fights. The whole scene, with the noise of the blows, talk, etc., will be truthfully transferred.[59]

---

* Edison was familiar with Marey's pioneering work and could not fail to recognize its superiority to his own. He was also at least dimly aware of that of William Friese-Greene. The British inventor wrote to him on 18 March 1890, to say he was sending by separate post "a paper with description of Machine Camera for taking 10 a second." There is no trace of this paper in ENHP, but receipt of the letter was acknowledged.

Asked what the new invention would be called, Edison uttered its name publicly for the first time. "The Kinetograph. What does that mean? The first half of the word means motion, and the other half write. That is, the portrayal of motion."[60]

It was clear from his emphasis on sound effects, color, close-up camerawork, and projection that his imagination had moved far beyond the silent flickerings that Dickson and Heise had conjured up in a peephole box. He said nothing about the mechanics involved. "But that doesn't matter to Edison," *The Philadelphia Inquirer* remarked. "With him, to conceive is to execute. . . . He talks freely and seems to defy anybody to steal his designs even after he has given a clue to them."*[61]

Two weeks later he sat in court in Manhattan, tensely chewing a toothpick, as his patent lawyer, Richard N. Dyer,† summed up the Edison company's seven-year-old case against United States Electric.[62] Since the latter firm was now owned by George Westinghouse, Dyer's argument was a final offensive in the "current war" that had done so much to embitter Edison against his rival in the last decade.‡

Dyer argued for four hours that Edison's basic electric light patent of 1879 was original and unprecedented in its claim of a "receiver made entirely of glass," with conductors passing through into a carbon filament held in near-perfect vacuum. Consequently, an older and faulty patent held by Westinghouse, U.S. 204,144, amounted to invalid competition and did not entitle him to market lamps clearly modeled on Edison's own. Since United States Electric had been doing so since 1880, Westinghouse could owe Edison General Electric as much as $15 million in back royalties—not including another $2 million payable before Edison's patent ran out in 1897.[63]

The appellant complaint involved so many boring technical data, along with the deposit of seven volumes of evidence in front of the judge, that the public benches of the court soon emptied. Edison alone remained, in company with two or three newspaper reporters. Bored,

---

* One of the reporters who attended Edison's oracular presentation in Chicago was Frank L. Baum, a cub recently hired by the *Chicago Evening Post*. He was fascinated by Edison's top-heavy appearance. "Of medium height is the Wizard of Menlo Park . . . a massive head is his," wrote the future author of *The Wizard of Oz*.

† Brother and partner of Frank Dyer, the future president of Thomas A. Edison, Inc.

‡ See Part Five.

he willingly submitted to some sotto voce questioning about the Kinetograph from the representative of *The Sun* and invited him to
come and see it in West Orange.[64]

This gave both men an excuse to stay away from the rest of the
trial, which went on for several more days and ended with Judge Wallace promising a decision early in the summer. Meanwhile Edison,
back at the laboratory, not only demonstrated his "phenomenal machine" in action but drew a sketch of it for his guest, to George Lathrop's jealous distress:[65]

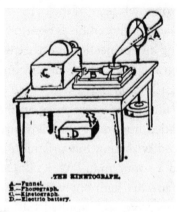

.THE KINETOGRAPH.

A.—Funnel.
B.—Phonograph.
C.—Kinetograph.
D.—Electric battery.

*Edison's sketch of his tabletop
Kinetograph, 28 May 1891.*

He explained that A was the sound amplifier, B the phonograph, C
the camera, and D a primary battery powering the whole synchronized system. It was clear that he still thought of the Kinetograph as
an audiovisual device, although for patent purposes he would have to
describe only C. Less clear from the sketch was whether that compo-
nent was designed only to shoot pictures or project them as well. He
insisted that the objects on the table were capable of recording whole
scenes of an opera in both sight and sound. "Marie Jansen comes out
and sings, and the band will play a charming waltzing minuet, and
then she dances around and the audience applauds."[66]

"How do you expect to do all that, Mr. Edison?" the reporter
asked.

Edison went into full imaginative mode. "I will get the company to
give a dress rehearsal for me. I place back of the orchestra on a table
a compound machine consisting of a phonograph and a Kinetograph,

with a capacity of thirty minutes' work. The orchestra plays, the curtain rises, and the opera begins. Both machines work simultaneously, one recording sound and the other taking photographs, recording motion at the rate of forty-six photographs per second."[67]

He said that in his opinion that shutter speed gave the most realistic illusion of continuous movement.* "Afterward the photographic strip is developed and replaced in the machine, a projecting lens is substituted for the photographic lens, and the reproducing part of the phonograph is adjusted. Then, by means of a calcium light, the effect is reproduced lifesize on a white curtain."[68]

Just such a sheet was hanging in his laboratory library. But Edison was coy about showing any projection more extensive than the one lit up within the Kinetograph itself. Running upstairs with the energy of a boy, he opened what looked like a plain pine box and displayed a ribbon of "gelatine" film three-quarters of an inch wide, perforated along one edge, threaded horizontally between two spindled, velvet-lined reels, and printed with tiny but pristine photographic images. Each frame portrayed a young man—W. K. L. Dickson—reaching by infinitesimal degrees for his hat. But when Edison closed the box, switched on its electric drive, and applied full power to the take-up spool, Dickson seen through an inch-wide peephole lens became a figure of miraculous mobility, uncovering, shaking his head, waving, and laughing. Only as the power scaled down did his movements become jerky and finally freeze.[69]

"I can put a roll of gelatine strip a mile long into it if I like," Edison boasted. He said that would accommodate 82,800 images a half-inch square and a half-inch apart and, at forty-six FPS, make for a moving picture of half an hour's duration.

The reporter noted that his math was faulty but did not have the temerity to correct a genius.[70]

The next day, 28 May, *The Sun* made the most of its exclusive story, running it as a front-page lead headlined "THE KINETOGRAPH— EDISON'S LATEST AND MOST SURPRISING DEVICE—PURE MOTION RECORDED AND REPRODUCED." Under the circumstances, there was

---

* Dickson appears to have convinced Edison of this. Marey shot at speeds ranging from thirty to fifty frames per second.

little more that Lathrop could report when his own piece came out in *Harper's* a few days later. He did, however, have a quote in which the inventor acknowledged his debt to such pioneers of "instantaneous photography" as Muybridge and Marey. "All I have done is to perfect what has been attempted before, but did not succeed. It's just that one step I have taken." Edison was referring neither to sound nor to projection experiments but to the precise coordination of his rotating shutter and leaps of frame. At forty-six advances per second—about as fast as the vibration of a hummingbird's wings—they found time to both expose and transpose the film, so a fresh square was ready for each new shaft of light.[71]

Lathrop was awed by the potential of the technology to distract human beings from one another and away from reality itself: "We seem to be nearing a time when every man may reach the old philosophical idea of a microcosm—a little world of one's own—by unrolling in his room a tape which will fill it with all the forms and motions of the habitable globe."[72]

BETTER THAN THAT

Edison was asleep at Ogden at noon on 14 July—he had worked right through the previous day and night—when Henry Hart, the mine superintendent, touched him.

"What is it?"

"I have good news for you."

"I know. The screening plates have come."

"Better than that."[73]

Hart handed over a telegram, and Edison sat on the edge of the bed reading it. Judge Wallace had upheld his electric light patent. After all the imitations, challenges, and outright infringements of the past eleven years—most annoyingly, those of George Westinghouse—his basic bulb of 1880 shone undefiled at last.

He could think of nothing to say but "Ain't it a daisy?" before joining Hart and his fellow miners for lunch.[74]

Westinghouse was certain to appeal, although the decision was grounded on such specific design details that his motive could only be to delay the date he would have to start paying royalties. Edison had no power in the meantime to demand the $15 million arrears United

States Electric technically owed him—the technicality being that his patent now belonged to Edison General Electric and would have to be litigated by the company's full board. A suit for so enormous a sum was bound to drag on far beyond 1897, and involve such commensurate costs as to bankrupt Villard in the process.*[75]

Nor, for the same reasons, could Edison expect to prosper much after the appeals court found in his favor. His experience with important patents was that seventeen years—the maximum protection period allowed by law—was scarcely long enough to defend them, let alone profit from their true worth. "What I have made has been because I have understood the inventions better, and have been able to manipulate the manufacturing of them better than the pirates."†[76]

He had not yet reached the point when, in extreme bitterness, he would complain of never having made a cent out of his patents in electric light and power. And now that the Kinetograph was publicized, he was quick to execute two patents covering it both as camera and player.‡ But he sympathized with any inventor who could not afford to fight for protection: "His certificate of patent is merely a certificate to the poorhouse."[77]

THE LARGE END OF THINGS

Edison had moved full time to Ogden, vowing to stay there half a year if necessary, because he felt he could not trust anyone else to manage the mine and the mills properly and solve the problems inherent in launching such a complex operation. A reporter found him there late one afternoon, just as the big engine in the powerhouse had ceased its

* *The Phonogram,* Edison's house magazine, predicted in 1892 that with all other lighting companies included as liable in the infringement decision, Edison General Electric was due as much as $50 million in back damages and $2 million a year in future royalties.

† Judge Wallace's decision was upheld by the U.S. Circuit Court of Appeals on 4 October 1892, more than twelve years after the issuance of the electric light patent. An identical opinion affirming the originality of Edison's invention overseas was handed down by an English court in *Edison and Swan United Electric Light Co. v. Woodhouse and Rawson* (1887).

‡ Edison made no attempt to patent the Kinetograph overseas. This was probably because he quailed at the cost and difficulty of claiming precedence over the rival inventions of Marey, Le Prince, Friese-Greene, and others. But he thereby lost millions and enabled such French competitors as Lumière and Pathé to make substantial inroads into the U.S. market.

throbbing. The sun was setting over Sparta Mountain, and cowbells tinkled in the valley. Yet a file of Italian laborers was heading up Iron Hill, where a pyramid of cream-colored rubble awaited transportation to the crushing plant.[78]

"We do not pause here day or night," Edison said, pointing to a row of arc lights near the quarry, ready to illuminate the evening shift. Taking obvious pride in the immensity of the scene—six miles in all directions, all owned by himself—he declared that there was enough ferrous rock in the mountain to be mined for at least a century.

A hill-shaking explosion came from the upper bench. "Well, there go five thousand tons more," he said, grinning.[79]

Like many another private person, Edison enjoyed confiding in strangers. "I like to begin at the large end of things. Life is too short to begin at the small end. The larger includes the smaller, the details grow out of the principle. . . . We are apt to be impressed by the boulder before us and not reason with the mountain above us, that the boulder rolled down from. Did you ever read Edgar Allan Poe's 'Domain of Arnheim'?"

He explained that it was the tale of a wealthy man* who loved beauty and sought, unwearyingly, to realize it on a monumental scale. "He sought to do rather than be known as achieving, and Poe says of him that in contempt of ambition he found the principle of earthly happiness."[80]

When his secretary, Alfred Tate, told George Lathrop that Edison was now working full-time at Ogden ("He has practically been retired from the world"), the bewildered writer vented his frustration in the angriest letter Edison had ever received. He reminded him that fourteen months had passed since they had first talked of doing a book together, "and you gave it your cordial assent, even suggesting the idea of cuts to be made from sketches of yours."[81] On the strength of that encouragement, "Mr. McClure has made me certain payments which I am not in a position to refund."

> I ought not to be left liable to be called upon to refund them, through delay on your part in completing the notes on which I depend. . . .

* Named, coincidentally enough, "Ellison."

I can understand how—preoccupied as you have been, and especially if you have somewhat lost interest in the plan of the book—my recurring to the subject may seem to you a sort of nuisance. But, on the other hand, I will ask you to try to realize what it is to me to be forced to hang around like a dog waiting for a bone—& not even getting the bone. . . .

It is only fair that you should give me a chance to consult with you about the book, in the same happy & genial spirit with which we began upon it. I have been willing to wait, to travel to the mines or anywhere with you, in order to carry the thing out. But Mr. McClure is now promising the story in his newspapers for October; & there is no time to wait any longer.

I am a man of my word; & you are a man of your word. I have praised you to the skies, right & left, as being a man not only of supereminent genius, but also faithful to his promises; whose word is even better than his bond—as you once told me it was. I wish to hold you to that belief, & be justified in it.[82]

Lathrop might as well have saved his ink. Edison had indeed lost interest in the novel. Obsessed with his mine and mills, he offered through Tate to compensate McClure. Lathrop considered it a debt of honor incurred by himself, and indignantly refused: "Nothing could induce me to accept pecuniary aid from Edison, although I appreciate his big-heartedness."

He thus faced years of privation, heavy drinking, and befuddlement while he tried to make plausible science fiction out of what he remembered Edison saying in their first few meetings. Eventually he would publish a pallid fantasy, "In the Deep of Time," that had two men exploring Mars on mechanized, antigravitational stilts. It attracted almost no attention, despite being advertised as "by George Parsons Lathrop in collaboration with Thomas A. Edison."* [83]

"CHASING AWAY FROM THE SUN"

One day that summer Edison was eating lunch under a tree at the apex of a large iron conformation when he noticed that the needle of

---

* Lathrop, emotionally damaged by debt and the loss of his only son, died of alcoholism in 1898, aged forty-six. His last literary project was a biography of Edison.

his pocket compass was trembling strangely. He had the momentary feeling that "signals sent through interstellar space might be responsible for the disturbance." Then he remembered that he was sitting in the center of a body of magnetite five or six miles deep. No matter how low grade, it was at least a million times more responsive to the electromagnetic flaring of sunspots than whatever deposits underlay the Kew Observatory in England, where solar radiation was measured daily.[84]

Intrigued by the notion that he might connect his own magnetic energy field to those on the sun, he strung a fifteen-wire copper power line on poles planted all around the iron bed, and ran it down to an ordinary Bell telephone receiver in the plant. He said it would enable him to listen in to sunspots, as well as observe them through his telescope. "Why, they are beautiful," he said to a reporter from, appropriately, *The Sun*. "The disturbances are tremendous. . . . Yes, sir, I can hear them with this telephone. . . . The next time there is any violent change in the sun's spots which disturbs the magnetic lines on earth I shall know it, and if 600,000 miles of hydrogen go chasing away from the sun I shall hear it."[*][85]

LOTS MORE RUBBER

Edison's peculiar delight in taking arms against a sea of troubles was never more evident than when Samuel Insull told him he was losing $6,000 a month on his Ogden venture. His reaction was to scrap much of the expensive machinery Livor had installed, order replacements of his own design, build a narrow-gauge railway along the foot of the western incline, and begin construction of an adjacent settlement, complete with post office, store, and saloon, to house his labor force of Italian and Hungarian immigrants. To nobody's surprise, the village was named Edison, New Jersey.[†][86]

A party of inspectors sent by *Engineering and Mining Journal* toured the plant early in the fall. Although some sections were idled

---

[*] Edison proposed in 1920 that a "scientifically-kept watch for interstellar signaling should be established in Michigan, where enormous masses of ore might be expected particularly to attract magnetic signals from space if any should be sent."

[†] Not to be confused with modern Edison, New Jersey, a town in Middlesex County that memorializes the original site of Menlo Park.

for refurbishment and Edison was coy about showing any of his new machines, they could see that he already excelled at quarrying and magnetic separation, if not yet in the difficult processes of crushing and refinement. They were particularly impressed with his cableway system, every suspended "skip" delivering four tons of rock to the crushers at only twelve cents a load. But they predicted that in view of the low iron content of local ore, Edison would still have to spend a fortune and deploy "the utmost resources of engineering skill" to compete with Mesabi ore at 64 percent iron. "With his surpassing genius [and] capacity for taking infinite pains, it cannot be doubted that he will ultimately achieve success."[87]

Another visitor to Ogden was Thomas Robins, Jr., a twenty-two-year-old rubber salesman looking for a job in engineering. He noticed some canvas conveyor belts being changed and asked Henry Hart how long they lasted.

"From six to eight weeks," the superintendent said.[88]

Robins examined a discarded belt. It was rubberized to protect it from the abrasive mass of tipped loads. But the laminate was so thin that he could penetrate it with his fingernail. Consequently the central strip, which bore the most weight, was eroded, and the edges were frayed where the belt had curled in its troughed bearings. He counted fifty conveyors in all, some of them longer than five hundred feet, and calculated they were costing the mill a fortune in replacements. What was needed was lots more rubber, so that resilience would replace resistance, and make for lighter belts lasting fifty times longer.[89]

It was an aperçu that would win Robins the grand prize at the Paris Exposition in 1900. In the meantime it endeared him to Edison, who let him perfect his invention on-site over the next several years, making Ogden the cradle of the world's first system of continuous mass-materials handling.[90]

SOLAR SURPRISE

Within six months of taking over management of the New Jersey & Pennsylvania Concentrating Works plant, Edison doubled the company's subscription capital from $500,000 to $1 million.[91] His fellow directors could see that he was prepared to double it again—and write checks of his own when they quailed—so sure was he that the day would come when orders flowed in, as fast as pure, phosphorus-free fines were trucked out.

His joyful sense of freedom to build and rebuild came in contrast to the impotence he felt as the owner of only 10 percent of Edison General Electric. In that company's stately headquarters in downtown Manhattan, Henry Villard reigned supreme, and the avid interest of Wall Street was increasingly felt at board meetings. Unrestrained by Edison, a habitual absentee, Villard was again trying to sell his majority holding and amalgamate Edison General with Thomson-Houston. In recent years the latter firm, run by a brilliant business tactician, Charles A. Coffin, had taken advantage of Edison's prejudice against alternating current systems and built itself up to the point that its paper worth, in early 1892, was $18.4 million, ahead of Edison General's $15 million. In fact, it was a smaller, less profitable concern, its products inferior and its business practices not far removed from larceny. Edison General, in contrast, ably served between four and five thousand customers and pulled in $1 million worth of business every month. It had fourteen acres of manufacturing plant to Thomson-Houston's eight. But Coffin saw that it had a weakness—a $3.5 million floating loan—that he could exploit, with the secret approval of his banker, J. P. Morgan.[92]

The news of their combined intent leaked out on Saturday 6 February, four days before Villard planned to announce it at the annual meeting of Edison General Electric trustees. It created an instant sensation. Ever since Judge Wallace had sanctioned the primacy of Edison's electric light patent, received wisdom held that his company would devour all its competitors. Instead, the shark was poised to swallow the whale. Villard put a brave face on it when he confirmed that "negotiations are in that direction . . . and are progressing rapidly."[93]

That weekend, in a macabre concatenation of electrical violence with the takeover of Edison General Electric, one of the largest geomagnetic storms ever recorded began to move across the surface of the sun, while in Ossining, New York, a convicted murderer, Charles McElvaine, prepared to be executed on Monday morning. Edison's connections with both events were more than metaphorical. He had wired his "cosmic telephone" at Ogden for just such a solar surprise and advised the authorities at Sing Sing prison that a sixteen-hundred-volt charge sent through McElvaine's wrists was likely to kill him faster than one through the head, blood being less resistant than bone.[94]

The execution was a reminder of a publicity campaign Edison would as soon forget, his battle in the late 1880s to brand alternating current as a lethal force ideally suited to capital punishment.* Reserving judgment on the merger until he heard from Samuel Insull, his personal representative at the negotiations, he sent Arthur Kennelly—currently testing the therapeutic effect of electromagnetism on the brains of a dog and a boy[†]—to monitor the execution, while he tracked the sunspot.[95]

At 11:32 A.M. on Monday McElvaine was strapped into Sing Sing's electric chair. It had been reconfigured so that his arms were forced down into two cans of salt water, wired in series to the prison's AC dynamo. "In the execution of Mr. Elvaine," the officiating physician told witnesses, "a new method, suggested by Mr. Thomas A. Edison, will be tried." An initial jolt lasting forty-nine seconds proved that Edison's theory of enhanced conductivity was wrong. McElvaine seemed still to be alive. One electrode was hurriedly applied to his skull. He died afterward, stiff in his straps and transpiring puffs of steam.[96]

Winter winds, meanwhile, kept blowing over Edison's poles on Sparta Mountain, frustrating his efforts to hear the crescendo of heliomagnetic signals impinging on observatories around the world. For the rest of the week he clung to his telescope, showing more interest in the cosmos than in Insull's efforts to protect him from the rapacity of Coffin and Morgan.[97]

"It was a beautiful sight, that aurora borealis last night, wasn't it?" he said on Saturday, exultant after the sunspot passed the solar meridian.[98] By then his fellow directors had officially approved the absorption of Edison General Electric by Thomson-Houston. The name

* See Part Five.

† Kennelly's experiments with "magneto-therapy," part of a major magnetism research program initiated by Edison at this time, anticipated by nearly eighty years the modern technology of magnetic resonance imaging.

of the resultant conglomerate had not yet been decided, and for the time being Villard was technically its president, but power had switched to more power, and soon the name of Edison General Electric would be shortened to just two depersonalized words.

PEOPLE WILL FORGET

Many years later Alfred Tate wrote that Edison blanched when he heard that he was the victim of a hostile takeover. "I never before had seen him change color. His complexion naturally was pale, a clear healthy paleness, but following my announcement it turned as white as his collar." Mina, too, ranted in old age about Insull selling her husband out and leaving him almost bankrupt, while laying the foundation of a vast fortune for himself.[99]

Memory tends to melodrama. Edison blustered at the time that he approved the merger.[100] Far from being impoverished by it, he believed Morgan's offer to exchange his 10 percent sharehold in Edison General Electric for a similar stake in the new company would "result financially to my advantage." And he was still the owner of several "large shops" not included in Thomson-Houston's purchase, the latest and most promising being his iron-concentrating works in New Jersey. He confessed to some disappointment at Insull's performance in the negotiations. However, "We are on the best of terms now. I expect he will come with me again when the consolidation has been completed."[101]

The first of his predictions, at least, turned out to be true. Morgan capitalized the combine at $50 million, making Edison richer than he had ever been in his life, with around $5 million in cash. Samuel Insull also did well, in spite of dashed hopes that he, and not Charles Coffin, would become its general manager. He was offered instead the post of second vice-president, two rungs below. No other Edison executive was so favored. This fueled angry speculations among his colleagues at West Orange and Schenectady that "Sammy" had sold them and the Old Man out.[102]

When, providentially, the directors of the Chicago Edison Company asked Insull to find them a new president, he suggested himself and was accepted. Edison let him go without protest. Unpopular as the little Englishman had always been, with his clicking, cash-register efficiency and Ozymandian sneer, he received a valedictory dinner at Delmonico's attended by Edison, Villard, and virtually every heavyweight in the electrical industry. Insull was still only thirty-two. Ahead lay all the glitter a lowborn lad could wish for—success beyond imagining, the beautiful actress wife, the thirty-one-thousand-square-foot mansion, the $20 million opera house—and in further prospect, the desiderata of pulp fiction: financial ruin, flight from the law, and death on a foreign railway platform, with only a silk handkerchief and the equivalent of eight cents in his pocket.* For the moment all Charles Batchelor, another guest at the dinner, could say was: "I think a very wise move for him."[103]

On 15 April the organization of "General Electric" was formally announced. Edison uttered no public protest about the exclusion of his name from its trademark. Nor could he consider himself snubbed, unless Elihu Thomson and Edwin Houston did too. He was appointed a director of the new behemoth but attended only one of its meetings.[104] The only hint he gave of deep hurt at being erased from the history of the industry he had founded came in conversation with his private secretary.

Tate, if you want to know anything about electricity go out to the galvanator room and ask Kennelly. He knows far more about it than I do. In fact I've come to the conclusion that I never did know anything about it. I'm going to do something now so different and so much bigger than anything I've ever done before, people will forget that my name was ever connected with anything electrical.[105]

PANTING OR SHORT BREATHS

In July Edison learned that his mining venture had so far cost him $850,000, including some $100,000 that could not be accounted for.

---

* Orson Welles once cited Insull, rather than William Randolph Hearst, as a role model for *Citizen Kane*—"a real man who built an opera house for the soprano of his choice." Unlike Kane, however, Insull lost his wealth when his $500 million electrical empire collapsed in the Great Depression. He was prosecuted by the federal government on antitrust charges and, although found innocent, never recovered from the attendant opprobrium.

A profit-killing amount of money was being lavished on labor that simply loaded and unloaded rock at either end of the conveyors. The jaw crushers took too long to do their work and often broke down, necessitating expensive repairs. The magnetic separators, plagued by screening problems, were concentrating only 47 percent iron—far less than the 66 or 70 percent he needed to match the richness of Great Lakes ore. He was still digesting this information when a stockhouse under construction at Ogden collapsed, killing five men and injuring twelve. Lawsuits alleging negligence were filed by bereaved families.[106]

A newspaper clipping he carried in his wallet read, "Thomas Edison is a happy and healthy man. He does not worry." As usual he countered the pull of bad news by pushing forward harder. Rather than continue to "improve" Ogden with ad hoc adjustments, he increased the capital of its parent company to $1.25 million, then shut the plant for a tear-down rebuild that would expand it enormously and make it a showpiece of automated design.[107] No sooner had a new separator house gone up than he decided it needed some screening towers, and should be constructed all over again.

"Is the Old Man all right today?" a fireman whispered to the chief rigger. "He told me to get it down to the foundations." Forty men were needed to do the job.[108]

Construction crews often found Edison working, eating, and even sleeping beside them. He loved hard labor and the luxurious tiredness it induced when he finally flopped onto a bed or the nearest heap of soft pea coal. In a letter to Mina that looked as if it had been scrawled upside down, he signed himself "Your Lover always the same (who sleeps with his boots on & smokes 23 cent cigars)."[109]

She could have used his company at Glenmont, because she had a discordant household to run. Marion, aged nineteen, was at last back from Europe. Her smallpox scars were sufficiently faded for her to face a reunion with Tom and William, themselves home for the summer from boarding school. It was an open question how long the boys would remain at St. Paul's, a school they both hated—and for that matter, what success Mina would have integrating Mary Edison's children with her own. For the time being she felt capable of managing both broods, with the help of a nanny for Madeleine and a nurse for little Charles. But to Marion, desperate to resume intimacy with her father, it was inevitable that sooner or later Mina would insist on privileged possession of him and the fruits of her own body.

The longer Edison stayed on Sparta Mountain, the more he lusted for that body—olive-skinned and stocky, not yet coarsened by the passing of youth. "Our dear little Mamma don't want to leave her nice home & come up to keep company with her lover—Why? no real love is the answer." He showed no awareness, as he teased, that she might feel the same way. In letter after letter he hailed "the 649th grandchild of Eve" with apostrophes of adoration: "Darling darling Billy Edison & 2 angels besides," "Darling Sweetest Loveliest Cutest Extra Billie Edison," "Sweetest on this ball of granite, verdure and H2O." His sign-offs were even more figurative: "With love Andesian in dimensions I am your Lover TAE," "With a kiss like the Swish of a 13 inch cannon projectile I remain as always your lover sure solid & unchangeable."[110]

He wrote about wanting to see her so much that he had resorted to searching around for a photograph of her and was frustrated by its inadequacy. Knowing that she enjoyed jokes about sex (behind a veil of Methodist decorum), he shared one that he heard at the plant. It was question: "How to recognize the Modern or so called Coming Woman." Answer: "By [her] panting or short breaths." In his next letter he said he had more stories of the kind to tell her when they met up. "I suppose you saw the point of the 'Coming Woman' joke, if not I will bring diagrams and explanatory notes."[111]

FAULT LINE DEVELOPING

In October Edison assigned Walter Mallory, an experienced iron and steel man fast becoming his closest associate, to supervise the transfiguration of Ogden. He reestablished himself in West Orange and worked with Dickson on an improved version of the Kinetograph, which he wanted to patent and exhibit at the Chicago World's Fair next spring. His announcement to that effect in *The Phonogram* magazine made clear, once again, that he conceived of the new machine as an audiovisual device: "The Edison Kinetograph is an instrument intended to produce motion and sound simultaneously, being a combination of a specially constructed camera and phonograph." With its thirty-five-millimeter film now feeding vertically rather than horizontally, a double row of perforations holding the frames steady as they were exposed forty-six to the second, and electrical connection to a recording device, it was the prototype motion picture camera of the coming century.[112]

While vast new structures arose on Sparta Mountain, a smaller and peculiarly ugly one was built by Dickson in a vacant lot behind the West Orange laboratory. Black-painted, pitch-roofed, and pinned together with great sheets of felt siding, it was windowless except for a small rectangle of red glass and an angled aperture open to the sky. It also lacked foundations, riding instead on a circular wooden path, so that at any time of day it could be aligned with the sun. It was the world's first movie studio, making use of natural light rather than the hissing, sparky flare of arc lamps or the soft glow of what were now "GE" lightbulbs. Dickson needed all the illumination he could get when shooting at forty-six FPS. He enclosed the rear of the stage in a fourteen-foot cone to give a dark background to foreground action and mounted the Kinetograph on rails, in order to dolly forward for limited zooms. There was a phonograph for sound-synchronism experiments, a central stove, and a darkroom in the rear. In inclement weather, the skylight could be closed with a flap of tar paper, black sealing in black. The shed became known as "Edison's Black Maria."*[113]

For a few months Edison and his elder daughter recaptured something of the closeness they had shared in the aftermath of Mary Edison's death. Marion rejoiced to have her adored "Papa" back at Glenmont and to find him no longer as cold to her as he had been when she first quit his hearth. But by the time Tom and William came home again for Christmas, Mina began to feel there were altogether too many of her predecessor's children in the house. She could not conceal her regret that Marion had declined the marriage proposal of a socialite she had met in Madrid.[114]

Embarrassingly for Edison, a gossip columnist publicized his domestic situation in *Town Topics* magazine:

---

* There is a full-size reconstruction of the Black Maria at Thomas Edison National Historical Park.

The Black Maria, circa 1893.

I am all the time running across charming newspaper accounts of the home life of one of the great inventors of the world, a genius that lives not very far from New York. The inventor is very happy in the possession of a young wife that [*sic*] is remarkable for her physical beauty and is devoted to him. They have two fine children of their own, and the inventor has several grown up [*sic*] children by his first wife. . . . Now, everyone knows how difficult it is to be a good stepmother, and therefore it is not at all strange that the inventor's wife is by no means the fond and generous type of the species that she attempts to have people believe. [She] is said to fancy that her treatment of her husband's children is quite all that it should be, but among her friends, I believe, it is held that her tolerance and gentleness are not remarkable.[115]

In January 1893 Tom, a sickly boy who clung to Marion, turned seventeen and refused to go back to St. Paul's. He said he wanted to work for his father. Edison saw a fault line developing in the family and decided to make Marion a present of their old home in Menlo Park. She was not yet of age, but he saw no reason to wait another year before handing it over to her. Marion was in some respects more mature than Mina, who had never known peripatetic insecurity, let alone faced death in a foreign country. He transferred the deed on the last day of the month, and Marion moved out of Glenmont a few weeks later. It remained to be seen how long she could stand living alone in a vandalized hamlet on the wrong side of Metuchen. As far as Edison was concerned, she was "now settled for life."[116]

After returning to Ogden, he tried with some irritation to soothe Mina's feeling that his heart was not all hers. "You are mean to doubt me as you did in your last letter . . . you are not a lover, only on occasions do you impress me as loving me, in any event it is not a strong deep love like mine, what little there is would easily be disturbed, someday Billy darling you will love me. . . . It is very cold here today, the wind is blowing very hard."[117]

NOT THE BEST TIME

In February the Philadelphia & Reading Railroad, a stressed spar of the nation's overextended transport industry, snapped and toppled into bankruptcy. Investors already concerned about a decline in Treasury gold reserves rushed to buy as much bullion as they could. Panic set in, just as organizers of the World's Fair in Chicago were preparing to celebrate American industrial might.

W. K. L. Dickson collapsed at the same time as the railroad. He was worn out by his multiple responsibilities as photographer, producer, performer, studio builder, and sound coordinator—in one experiment, playing violin while two young men self-consciously waltzed for the camera.*[118] On top of all these responsibilities, he had embarked on writing an authorized life of his employer, and had already published several advance chapters in *Cassier's Magazine*. Edison sympathetically treated him to a ten-week vacation at full pay in his house in Fort Myers. It did away with any lingering hope that they would be able, as promised, to exhibit a sonic version of the Kinetograph at the fair—or for that matter, even show the basic camera.[119]

He had already given up on an earlier dream, to be the official sup-

---

* A restoration of *The Dickson Experimental Sound Film* (1894 or 1895), the first sound film in movie history, has been jointly accomplished by the Library of Congress and the Rodgers and Hammerstein Archive of Recorded Sound in New York. It may be viewed at https://www.youtube.com/watch?v=Y6bowpBTR1s.

plier of electricity to the exposition, and fill its white palaces with the radiance of his greatest invention. Since he was no longer in the lighting business, it was a matter of practical indifference to him that George Westinghouse had won that honor by underbidding General Electric. More personally, it was satisfying to see Westinghouse scrabble not to infringe on his now-universal lamp patent, by supplying leaky, reconfigured bulbs that lasted about as long as candles.[120]

No sooner had President Cleveland opened the fair at the beginning of May than a second speculative juggernaut, the National Cordage Company, went into receivership. The stock market crashed. There was no doubt now that the economy was headed for a major depression. Hundreds of banks called in their loans, then failed themselves. Edison was somewhat protected by the variety of his investments in his own companies, but with all his children still depending on him—as well as Mina, with her love of fine food and good clothes—he had personal expenses of almost $3,000 a month.* It was not the best time to discover that he had been wrong in assuming he could build a new plant at Ogden as quickly as he had adapted the old. Instead of taking four months and costing $100,000, the project looked likely to drag on for another year and a half, at incalculable cost. And there were no more orders for the limited amount of concentrate he still had on hand.[121]

"The Ogden baby is sick," he said to Tate.[122]

He was unwell himself, suffering from an onset of diabetes that would trouble him for the rest of his life. Two insurance companies declined to cover him "on account of sugar," and he was able to register with a third only after stringent dieting. In other indications of stress, he borrowed $115,000 from Drexel, Morgan at the high rate of 6 percent, fired many employees, and complained about "professional sharks" continuing to infringe on his inventions. "I'm through with patents," he told a lawyer soliciting his business.[123] Over the next four years, he would send only five applications to the Patent Office—for him, the equivalent of a total boycott.[124]

---

* Budget figures prepared for Marion's information by John Randolph in the fall of 1894 indicate that Edison was currently spending $33,220 a year on household expenses, or just about $1 million in today's money.

INSIDE THE BOX

Dickson came back from Florida in time to help Edison mount the first public demonstration of the Kinetograph at the Brooklyn Institute on 9 May. The occasion was almost perversely uncommercial, being a lecture delivered by George M. Hopkins, chairman of the department of physics, to an audience of four hundred scientists. For once, Edison did not arrange newspaper coverage: nor did he attend the event himself. "These Zoetropic devices," he scoffed to Eadweard Muybridge, "are of too sentimental a character to get the public to invest in." He may have been embarrassed by his failure, two years after promising to show moving pictures with sound and color at the World's Fair, to come up with anything more impressive than the evening's tall, varnished box with a peephole at the top. As for pictures, all he had to offer Dr. Hopkins were some silent black-and-white experimental shorts.[125]

The professor chose a twenty-seven-second loop of three blacksmiths clustering around an anvil, sharing a beer, and forging a piece of white-hot iron. He was unable to project the action for communal viewing, but used a magic lantern to flash a few stills on the auditorium screen. The gradated differences between each frame were at least discernible. "Persistence of vision," he explained, "is depended upon to blend the successive images into one continuous ever-changing photographic picture." Using the lantern's radially slit disk shutter, he showed a spasmodic suggestion of movement. "In Mr. Edison's machine far more perfect results are secured," he said, explaining that its fundamental feature was an advancement system operating at hardly comprehensible speed. "This camera starts, moves, and stops the sensitive strip which receives the photographic image forty-six times a second."[126] He then invited his colleagues to file past the Kinetograph and bend over the peephole to watch *Blacksmith Scene* endlessly playing inside the box.*

Three hours went by before all were able to do so. Unless any of them had been overseas, and—by remote chance—seen private demonstrations of paper-roll motion pictures by Louis Le Prince, Étienne-

---

* *Blacksmith Scene* may be viewed online through the peephole of YouTube, at https://www.youtube.com/watch?v=FaFqr7nGsJM.

Jules Marey, and William Friese-Greene, this vision of a new medium was so strange as to defeat initial comprehension. Each scientist in turn applied his eye to the glass and was pulled from the bright auditorium into a flickering world where Lilliputian figures moved in chiaroscuro, their tiny hammer blows falling soundlessly.[127]

THE FIRST FEW CONCUSSIONS

Ogden, in contrast, was a crescendo of noise from August on, as Edison began to assemble and test the components of his new concentrating facility. First in order of process were the world's largest traveling crane, a 215-foot bridge rumbling on rails over the quarry and lifting overloaded skips pneumatically, with earsplitting hisses and snorts, and a six-ton electric elevator that thunderously spilled ore into the crusher building. Their combined cacophony, amplified by shrieking locomotive whistles, throbbing engines and dynamos, and the clatter of miles of conveyor belts, rose to hurtful levels when he invented and installed a pair of self-styled "Giant" crushing rolls in March 1894.[128]

These counterspinning, corrugated cylinders were six feet in diameter and weighed about thirty tons each. Along with four supplementary pairs of rolls, they were designed to reduce the most adamantine gneiss to powder. Only Edison, with his muffled hearing, could stand near them without wincing. "They have a surface velocity of nearly 40 miles per hour," he boasted, "and can strike a blow of 1,800,000 lbs."* The violence with which they did so came from his addition of a cabled friction clutch to their drive. It resolved the ancient conundrum of irresistible force meeting an immovable object by releasing the rolls to whirl free just before they bit into a boulder, so that momentum alone—seventy tons of chilled steel hitting a few tons of rock—did the fracturing.[129]

The first few concussions were enough to show that he had been unwise to mount his roll assembly on a wooden foundation. Drops from the hopper caused misalignments that either jammed the machinery, or threw boulders high into the air before they descended, spinning, and rode the rolls with the deceptive lightness of ping-pong

---

* Edison exaggerated the kinetic force of his crushers, which was more realistically the equivalent of seven tons, ample to shatter a five-ton rock.

balls. On such occasions the crew had to scatter to avoid flying fragments. Edison saw that nothing short of a bed of cast iron, and bab-
bitted bearings, could fortify the roll banks, both "giant" and
intermediate, well enough to stand a constant torrent of ore.[130]

The tests were a disaster, necessitating many more months of
crusher redesign that cost him another $200,000 and postponed—yet
again—any thought of getting the New Jersey & Pennsylvania Concentrating Works into regular production.[131]

THAT CROWD OUT THERE

Edison had no more luck in making a practicable combination of the
phonograph and Kinetograph when he returned to the laboratory. He
settled for the spring release of a coin-operated version of the peephole player Hopkins had unveiled in Brooklyn, along with a small library of "films"—his own word—to demonstrate the miracle of
photographed movement. Quashing hopes that the box might be
wired for sound, he renamed it the Kinetoscope* and announced that
the first reel made for it would feature the dancing biceps of Eugene
Sandow, "Strongest Man on Earth."[132]

For Sandow, alias Friedrich Wilhelm Müller, a German chain-
breaker who had wowed audiences at the World's Fair, the chance to
be associated with the most famous inventor of the age was more
than a splicing of superlatives. It meant that hundreds of thousands—
perhaps millions—of Americans would now be able to admire his
physique *in Aktion* and buy his various bodybuilding products. Nor
could the publicity hurt Edison, who needed the Kinetoscope to distract attention from the shutdown of his iron mill. He welcomed the
massive young man to West Orange and posed for a snapshot beside
him, taking care to stand a little higher, before escorting him to the
Black Maria. Once inside, Sandow stripped down to boxing boots
and a white undergarment that gave new dimensions of meaning to
the word *briefs*.[133]

The resultant forty-one-second "actuality," taken by Dickson and
Heise in a brilliant downfall of sunshine, beautifully caught the ripple
of his muscles as he clenched and writhed and twirled for the camera.

---

* *Kinetograph* henceforth meant only the talking camera.

But by chance, or more likely by design, the lighting emphasized some of his less mobile protuberances, with an attention to detail not to be matched in cinematography for seventy years.[134]

Eugene Sandow models for W. K. Dickson's camera, March 1894.

Nevertheless it was the likeness of Thomas Alva Edison, cast in bronzed plaster, that appeared on a pedestal in the forecourt of the first Kinetoscope parlor in New York on Saturday 14 April, two days before *Sandow* and two dozen other "moving pictures"* were due for exhibit. The ambitious lessor of the premises at 1155 Broadway was Alfred Tate, who like Insull before him had parlayed his job as Edison's private secretary into a variety of outside responsibilities. With his brother Bertram and a friend, Thomas R. Lombard, helping out, he spent the morning arranging ten Kinetoscopes for on-demand viewing. The oak cabinets were electrically linked in two rows and enclosed in a curving rail for patrons to lean against while moving from peephole to peephole. Framed pictures hung high on the walls, as if to emphasize the contrast between their stillness and the animated "shows" available below, at

---

* A word search of American newspapers in 1894 indicates that the phrase *moving pictures* was first used to describe the illusion of photographic movement when Edison announced his Kinetoscope on 10 March. Previously it referred either to still pictures that appealed to the emotions, or to mobile tableaux onstage. On 21 July the *American Encyclopedic Dictionary* announced that it was the first reference book to define "Kinetoscope" and [*sic*] "kinetograph." The word *cinema* would not enter the language until after the Lumière brothers patented their Cinématographe camera-projector in 1895. *Motion picture* appeared in 1896; *movie* around 1908. Edison, as has been seen, coined the word *talkie* in 1913.

twenty-five cents for five. The floor was glossed to reflect the varnished oak of the machines, and sprays of potted palm added a touch of salon-like elegance.[135]

By early afternoon all was ready for the opening on Monday. Tate and his companions retired to the back office to smoke and chat.

> We had planned to have an especially elaborate dinner that evening at Delmonico's, then flourishing on the southeast corner of Broadway and Twenty-sixth Street, to celebrate the initiation of the Kinetoscope enterprise. From where I sat I could see the display window and the groups who stopped to gaze at the bust of Edison. And then a brilliant idea occurred to me.
>
> "Look here," I said, pointing towards the window, "why shouldn't we make that crowd out there pay for our dinner tonight?"
>
> They both looked and observed the group before the window as it dissolved and renewed itself.
>
> "What's your scheme?" asked Lombard with a grin.
>
> "Bert," I said to my brother, "you take charge of the machines. I'll sell tickets and," turning to Lombard, "you stand at the door and act as a reception committee. We can run till six o'clock and by that time we ought to have dinner money."[136]

The trio never got to Delmonico's. There was such an inrush of patrons that Tate was unable to close the parlor until one o'clock on Sunday morning.[137] During the weeks that followed it became a magnet for oglers of both sexes. They admired Sandow's masculinity and the gyrations of female dancers and contortionists until Edison, embarrassed, ordered his bust to be removed.

Given the nonnarrative shortness of Kinetoscope films, the public's fascination with them derived, beyond prurience, from incredulity that movement, which by definition was a state of continuous change, could be both recorded and replayed. The fifty-foot loops magically kept boxers punching, barbers shaving, gymnasts somersaulting, Fred Ott sneezing, and Annie Oakley sharpshooting until celluloid fatigue set in—whereupon there was always a duplicate copy to wind onto the reels. Dickson and Heise were more interested in novelty than aesthetics, except when they filmed the Butterfly, Sun, and Serpentine dances of Annabelle Whitford. Her yellow hair and radiant, floaty costumes encouraged them to have a few strips hand-tinted, frame by frame. Privileged viewers were then able to watch Miss Whitford twirling amid undulations of colored gossamer that at one moment resembled wings, at another the petals of an enormous windblown flower.*[138]

Soon the Kinetoscope department of the Edison Manufacturing Company was selling $2,000 worth of players a week, plus Kineto-graph cameras and films, through three competing agencies. Purchase orders grew at a compound rate as new parlors opened up across the country. Over the next year Edison's income from his invention would exceed $250,000.† Yet he again refrained from patenting it overseas and again emphasized, in a handwritten statement published in the June issue of *The Century Magazine,* that he was not the only begetter of moving pictures. If the technology ever reached the point of pre-senting spectacles as grand as those of the Metropolitan Opera, he wrote, it would be due to "my own work and that of Dickson, Muy-bridge, Marié and others who will doubtless enter the field."[139]

Whatever authenticity Edison's graceful calligraphy (and misspell-ing of Marey's name) gave to this modest sharing of credit was com-promised by his declaration, a few lines earlier, "In the year 1887, the idea occurred to me that it was possible to devise an instrument that would do for the eye what the phonograph does for the ear."

If this alteration by one digit of the true chronology of his inven-tion was deliberate, rather than a simple slip of the pen, it gave birth to a lie that would make him and Dickson—who perpetuated it with fanatical insistence for the next forty years—morally suspect in the eyes of history.[140]

DAMN FOOL

The depression triggered by the previous year's panic reached its nadir in July. Exhibitors, nevertheless, were eager to invest in Kinetoscopes. Edison belatedly realized that entertainment was a public necessity,

---

* See, e.g., http://earlysilentfilm.blogspot.co.uk/2013/08/peerless-annabelle-symphony -in-yellow.html.

† Or more than $7.1 million in 2019 dollars.

especially in hard times. No longer could he pretend that his phonograph was a business instrument, best suited for stenographic purposes. He brooded over the success of Emile Berliner's rival disk-playing Gramophone and decided he could do better. Coin-operated phonographs were highly successful and a natural complement to peephole machines in amusement arcades. But first he had to wrest the commercial rights to his invention back from the hands of the ailing, failing entrepreneur Jesse Lippincott.[141]

He had sold those rights to Lippincott six years before, at the same time undertaking to manufacture phonographs for him exclusively, at the profitable rate of $250 apiece. The resultant North American Phonograph Company had struggled amid proliferating competition to keep paying him back. When Edison heard that it was $1 million in debt, he moved to push it into receivership.[142]

The suddenness and brutality with which he did so shocked Tate, who represented him on North American's board and who felt obligated to honor that company's many agreements with regional retailers—all of which he would have to abrogate if the bankruptcy suit went through. Rather than do that, he announced his resignation.

"What's the matter with you, Tate?" Edison said, turning on him in annoyance. "Why are you going to make a damn fool of yourself?"[143]

It was a split with yet another veteran of Menlo Park days, clearly less painful to him than to Tate, who for some time had noticed the growing willfulness of Edison's behavior.

> From the period of the fusion of the Edison General Electric and Thomson Houston Companies I observed a marked change in him in this respect. He seemed to repel discussion and his decisions became mandates issued from the depths of his own mind. If they were questioned he became impatient and merely reiterated them. . . .
>
> The iron bit into the flesh when I broke the link that bound me to a man I loved so sincerely.[144]

On 21 August North American Phonograph declared bankruptcy. Edison's bid of $125,000 for its assets was accepted, challenged by less agile rivals, and eventually confirmed by the receiver. He thus regained full rights to develop and market his favorite invention, creating for the purpose a new subsidiary, the National Phonograph Company: "I don't care to have anyone else have a lien on my brains." Tate drifted off to a life of wandering, indifferent achievement and was replaced in his managerial responsibilities by William E. Gilmore, a tougher executive better suited to the temper of the times.*[145]

ONE PARTICULARLY DISREPUTABLE ONE

That summer, notwithstanding the success of his film venture, Edison had to increase the stock of the New Jersey & Pennsylvania Concentrating Works to $1.75 million. He needed funds to reinstall the mill's crushers on a bed of cast iron, strengthen the traveling crane (which experts warned was too wide to be safe), and build something never seen near a metal mill before—a bricking house. This costly experiment was Edison's answer to complaints from smelters that Ogden fines had shown a dangerous tendency to "blow" in blast furnaces. He wanted to find a way to agglomerate the concentrate into Bessemer-quality briquettes, hard enough to stand heavy shoveling yet porous enough to absorb reducing gases at high heat.[146]

Every problem solved at Ogden seemed to generate a dozen more. The belts on the rolls began to slip at certain speeds, and buildings had to be reconfigured so many times that carpenters took little care over their work, cynically assuming that all of it would be changed sooner or later.[147]

Edison's way of dealing with every procedural obstacle was to throw himself at it, body as well as mind, until something gave way. He and Walter Mallory nearly suffocated when they crawled into an eighty-foot tower dryer to investigate a blockage above and were buried under an avalanche of ore.[148] This may have occurred just before Edison returned home for a rare family visit, imprinting on the memory of little Madeleine an image that would never fade:

One Saturday [Charles and I] were called in from play—scrubbed and combed & dressed to the nines to accompany Mother to meet

---

* Tate did not mention in his memoir that Edison, whose fits of anger were always short, gave him a farewell loan of $800, saying he could pay it back whenever he earned his "first stake" as a self-employed businessman. Twenty-six years later Tate took pleasure in sending him an interest-included check for $2,060.

him at the station . . . Mother—who was a very beautiful woman—looking exquisite in her flowered dress—ostrich feather hat, & lace parasol—the coachman—elegant in his livery—managing the high spirited team of bay horses . . . & the two of us—miserable but resigned in our starched ruffles because we realized that this was to be a great occasion: "Papa" was coming home!

Then the train arrived, puffing & blowing—black soft coal smoke—and from it emerged the most disreputable group of men I had ever seen—laughing & talking—they were dusty & dishevelled, their faces streaked with soot . . . and none of them looked as if they'd shaved for a week. I gazed at them in horror & then suddenly one particularly disreputable one detached himself and leaped into our carriage, kissed my mother most enthusiastically & we were off—my Father had arrived.[149]

If Marion had been there, she might have recalled a much younger but equally filthy Edison besmirching her own mother's fine linen. But as expected, she had been unable to stand her rustication in Menlo Park and was back in Germany—whence she applied, now, for his permission to wed Oberstleutnant Karl Hermann Oscar Öser of the Royal Saxon Army.

"I at last love some one better than myself," she wrote with her usual engaging frankness. "I hope dear Father you will make a flying trip to Europe to see me married. I have a good reason for wishing you very much to do this for me. Because I don't throw my money away people think I am an imposter and not your daughter."[150]

Like her rapidly maturing brothers, Marion was afraid that when the time came for her father to divide his kingdom, he would not do so equitably. Mina was a more powerful influence on him than all of them combined, and she was bound to fight like Goneril to acquire the largest slice possible, in favor of her own privileged brats. That at least had been Marion's obvious belief when she left—angering Edison so much he had refused to see her off, and had kept Mina from doing so too.[151]

He was soothed by the simple sincerity of Oscar's request for his daughter's hand, but not enough to cross the Atlantic to give her away. Assured by intermediaries that the lieutenant was a decent man who loved her, he gave his approval. It would take many months of *gemütlich* residence in Neusalza-Spremberg before Marion apologized for "the way I acted before I left America." She blamed her old traveling companion, Mrs. Earl, for making her doubt his goodwill to her, Tom, and William. "She it was who told me that you had settled all your money on Mina so that we would get none of it."[152]

THE GREATEST GENIUS OF THIS OR ANY OTHER AGE

In September the Thomas Y. Crowell Company announced the forthcoming publication of *The Life and Inventions of Thomas Alva Edison,* an imperial quarto volume of nearly four hundred pages with 250 illustrations, co-written by William Kennedy Laurie Dickson and his sister Antonia. It was an expanded compilation of the biographical articles they had been publishing about Edison in *Cassier's Magazine* and was billed as "the first complete and authentic story of his life," reflecting years of collaboration between the authors and their subject.

Edison received an advance copy and gave it a qualified testimonial: "Although I have not had time to read it through carefully, after a casual glance I must say that it is extremely well gotten up."[153]

His glance may not have extended to the final line, which described him as "the greatest genius of this or any other age." He was used to superlative salutations and quite aware of his public stature, but the Dicksons elsewhere gave him enough praise to embarrass an egomaniac. This was unfortunate, since the book contained much biographical information derived from Edison himself. *The New York Times* reviewed it favorably. "No one can help admiring the man who is revealed in these pages. Starting with nothing, he has acquired almost everything that men prize. The boy who sold papers on the Grand Trunk Railway forty years ago is today known and honored in every country in the world. . . . The popular notion is that Edison will discover everything if he shall live long enough."[154]

CATASTROPHIC FOLLY

Edison spent lavishly building the new Ogden, liquidating all his General Electric stock in the process.[155] At first the reopened plant seemed set to become its designer's dream: a fully automated *fons et origo* of purified magnetite, cheaply delivered in unlimited quantities to revived foundries on either flank of the Adirondacks.

In mid-October he put its machinery into experimental motion, aware that a malfunction at any point along the line—twenty-two sequences of pulverization, separation, and refinement—could jar the whole into immobility. The first disaster occurred in December, when one of the ore elevators split and fell. It necessitated a total rebuild of all three, plus complex adjustments to the crushing machines they served. The new bricking facility produced, after a number of false starts, some cakes that had encouragingly high levels of magnetite, but they were too few and too crumbly, bound to shatter en route to the foundry. In damp weather they absorbed water like sponges. Edison was obliged to shut Ogden down for yet another winter. He ordered the construction of a larger, more sophisticated bakery and set about developing a resinous binder, not anticipating that the "briquetting problem" would torment him for the next several years.[156]

Twice in the early months of 1895 he called on his fellow shareholders for cash infusions. Alarmed that the mill was costing $1,200 a day just to maintain, they declined to increase their stakes. Gloom gathered among Edison's engineers, all of whom regarded the giant rolls as a catastrophic folly.[157] He alone remained convinced that when their kinetic action was accelerated to the point that they outperformed the explosiveness of dynamite, the plant would usher in a new age of automated magnetic mining.

MORE THAN ENOUGH GLORY

Emboldened—and personally enriched—by the rush of peepshow exhibitors around the world to buy Edison machines and show Edison films, W. K. L. Dickson chose this time to publish *History of the Kinetograph, Kinetoscope, and Kineto-Phonograph*, a monograph that reflected much glory upon himself as the great man's closest aide. Or so it seemed to Edison, hypersensitive as ever to any presumption of intimacy. He thought that his tribute in *The Century Magazine* to the photographic innovations "of Dickson, Muybridge, Marié [Marey] and others" conferred more than enough glory to go around.[158]

Now he saw that same tribute reproduced, along with a full-page portrait of himself, as the opening spread of a volume that otherwise paid him only passing attention. The text that followed was evidently written by Dickson's sister Antonia, who cultivated a high literary style. ("With its great flapping sail-like roof and ebon complexion, [the Black Maria] has a weird and semi-nautical appearance, like the unwieldy hulk of a medieval pirate-craft or the air-ship of some swart Afrite.") Dickson, identified as the book's designer, managed to attach his extremely legible signature to most of the illustrations, including two bizarre self-portraits. One showed him posing à la Napoleon with hand tucked inside coat, while the other was a trick photograph of his severed head on a platter.[159]

His principal provocation, however, was to append "by request" an article from the *American Annual of Photography* that described him as "a clever young electrical engineer" who was "co-inventor with Edison of magnetic ore separators." This resulted in a rare outburst of Edisonian rage, dictated for the record to a stenographer:

> *I object to the little book gotten out by Dickson. The part about Dickson being a co-inventor in the magnetic separator etc., is incorrect, as there is no co-invention in the Ogden business with Dickson or anybody else. . . . Mr Dickson will get full credit for what he has done without trying to ram it down peoples throats. . . . I am not especially stuck on having my own photograph in the book, it looks too much like conceitedness and self glorification on my part and the public never takes kindly to a man who is always working his personality forwards. It's the thing they want to know about and not the man for whom they do not care a D—.[160]*

Having thus convinced himself, if nobody else, of his personal humility, Edison cooled down. But Dickson would never get the "full credit" he deserved as a pioneer of American cinematography.

AN EXPERIMENTER OF THE HIGHEST TYPE

Edison was about to reopen Ogden in mid-March when news came that Nikola Tesla's laboratory in Manhattan had been destroyed by fire. Although the Serbian inventor was a wealthy man, on the strength of his brilliant innovations in alternating current electricity and wireless power transmission, he had neglected to insure the property. He was seen walking through the ruins, a storklike figure, picking up a piece of brass, blowing the soot off it, then tossing it aside in tears.[161]

"I am in too much grief to talk," he told reporters. "What can I say? The work of half my lifetime, very nearly; all my mechanical instruments and scientific apparatus. . . . Everything is gone. I must begin over again."[162]

Edison reached out in sympathy to his stricken colleague. He knew what it was to begin over—and over and over. "I have received a letter from Mr. Edison offering me the use of his workshop in which to continue my experiments," Tesla announced to reporters. "He has shown me the greatest kindness and consideration. I do not think, however, that I will accept the offer."[163]

Ever the loner, he said he would look for temporary quarters in the city and try to resume work there. Many observers thought it was more probable he would lose his mind. Just weeks before, he had confessed that his current experiments were "so beautiful, so fascinating, so important," that he had virtually given up on food and sleep. This was hard to believe, since Tesla was a regular, solitary diner at Delmonico's and ate pathological quantities of meat.*[164] Nor, with his frail constitution, could he keep the same kind of hours as Edison did without damage to himself. He admitted as much: "I expect I shall go on until I break down altogether."[165]

In all respects except that of creativity, the two inventors were opposites. Tesla at thirty-nine was a melancholy celibate. Edison at forty-nine still had a healthy libido and had carried off two teenage brides, impregnating both of them repeatedly.† If he was egotistical, his vanity concerned only work, while Tesla's megalomania had no bounds. *The New York Times* went too far in reporting that "personally they are warm friends," but they admired each other despite their professional differences. Edison restrained his contempt for alternating current enough to praise Tesla's "amazing" success in transport-

---

* According to one source, Tesla counted his jaw movements while chewing, and always used eighteen napkins.

† In addition to bearing three children, Mina Edison had at least one and possibly three miscarriages.

ing hydroelectric from Niagara Falls, while Tesla let it be known that he had "the utmost faith in the genius of Mr. Edison."[166]

THE VANISHED PRECURSOR

Just when Ogden was thundering back to life, news came that Kineto-scope business in the United States had gone into a sudden slump. All three of Edison's principal exhibitors—the Latham Company, Maguire & Baucus, and Raff & Gammon—reported peak sales in January, followed by precipitous falloffs of 72, 92, and 95 percent, respectively. Evidently the novelty of a device that showed moving pictures through a peephole had worn off. The clenching of Eugene Sandow's buttocks did not encourage repeat viewings, except by a furtive minority of patrons.[167]

Frank Gammon begged Edison to transform the Kinetoscope into a projector that would entertain large seated audiences, not just one standing viewer at a time. He got nowhere. Edison had lost interest in screened images after failing, four years before, to throw any that were larger than ten inches wide.*[168] The difficulty with projection was that it called for intermittent movement—forty-six film-tearing stops and starts per second—as each frame passed between a light and a lens. Otherwise, it would not reproduce in detail twenty feet away—much less at a hundred. Kinetoscope loops ran smoothly just below the eyepiece, with tolerable clarity.

Edison the manufacturer in any case preferred to sell multiple ma-

---

* Edison never explained why he abandoned the projection method (calcium-lit images, slotted rotary shutter, magnifying lens, and screen) that he tantalized the *Sun* reporter with in 1891. It is described by the Dicksons in chapter 22 of *Life and Inventions*. They refer specifically to "exhibition evenings" in the "projecting-room" of Edison's photographic department, its walls "hung with black" to prevent reflection from "the circle of light" [sic] emanating from the screen at the other end, and "the projector" similarly draped so as to expose only "a single peephole for the accommodation of the lens," connected to an electric motor running with "a weird accompanying monotone." They even report that some images were "projected stereoscopically" with "a pleasing rotundity." These evenings can have occurred no later than September 1894, when their book went to press, and presumably no earlier than October 1893, when *Cassier's Magazine* published a shorter version of this chapter that made no mention of projection at all. If the images thrown on the screen were as "life-like" as the Dicksons claim, then Edison was the father of movie projection. The enlargements, however, were apparently "not . . . much more than ten times the original size [of a 35-mm. frame]," too small for commercial viewing.

chines to parlors rather than single machines to theaters. He paid little attention when Dickson, choosing his words carefully, said that the Latham Company was building a projector for the specific purpose of screening Edison pictures. He chose not to mention that he had designed it himself and was spending many evenings in New York with the Latham brothers, secretly discussing the prospect of joining them in the organization of a full-scale production company once the machine was perfected.[169] Such a studio would of course compete with Edison's, making Dickson a pending, if not yet actual traitor to his boss.

After twelve years of service, he was in terror of being found out and fired before he could be sure of security with the Lathams. They were little known and underfunded, entrepreneurial dwarfs in comparison to the giant he had done so much to mythify. Tempting as it was to accept a $125,000 start-up stock offer from them, he earned a good salary and substantial royalties as chief of the photographic department at West Orange. Until recently he had also run a profitable sideline selling portraits of Edison, shots of the laboratory, and paper print film strips, all copyrighted under his own name. That extra income, however, had been cut into by William Gilmore, who as general manager had forced him to transfer most of the copyrights to the Edison Manufacturing Company. Because of this Dickson hated Gilmore, and felt supplanted by him as the Old Man's favorite aide.[170]

In fact, that fluid title currently belonged to Walter Mallory. Edison had always winked at Dickson's photo trafficking, since it served his public image. But he agreed with Gilmore that it must stop. His sudden attack on Dickson for aggrandizing him (after years of pretending not to notice) implied that Gilmore had advised him to distance himself from an associate whose days were numbered.[171]

As indeed they were, once the general manager got wind of Dickson's negotiations with the Lathams. On 2 April Gilmore accused him, in Edison's presence, of corporate treachery. Dickson blustered that he had only been spying on the competition and demanded that Edison choose between him and Gilmore. His wish was gratified, but not the way he hoped.[172]

Later it turned out that he had also given creative advice to American Mutoscope, a film company ambitious to advance beyond Edison

Manufacturing in both peepshow and screen-machine technology.*
All Edison would say publicly was "We are not the best of friends."
Dickson became Mutoscope's globe-trotting cameraman—a career
step-down from the prestigious position he had enjoyed under Edison. In later life he became, like Edward Johnson, Francis Jehl, Alfred
Tate, and many other alumni of Menlo Park, a pathetic claimant for
the notice of historians and biographers. He lied about his coinvention of the Kinetograph and Kinetoscope with such fanatical
insistence—advancing every initial date by one year, in order to claim
precedence for Edison over Marey and Friese-Greene—as to obscure
the fact that he had done most of the work himself, and deserved
principal credit.[173] Not until almost a century after he quit Edison's
employ did a fragment of pristine cinematography, filmed in the garden of a Yorkshire house on 14 October 1888, prove that Louis Le
Prince was the vanished precursor of them all.[174]

A BUMBLE BEE IN FLOWER TIME

Edison's Kinetoscope business continued its precipitous slide through
the spring. In a vain attempt to compensate, he introduced the
"Kineto-Phonograph" he and Dickson had rigged up the year before,
renaming it the Kinetophone. It was a combination cylinder and reel
player equipped with two sets of rubberized earbud tubes, so that
couples could watch peephole films together and hear background
music blasting away as they did so. The instrument made no attempt
at more subtle synchronizations and sold only forty-five units.[175]

Leaving Gilmore to deal about the news that the Latham brothers
had simultaneously and successfully demonstrated their "Eidelo-
scope" projector in public, Edison returned with relief to iron mining.[176]

"The mill as it stands today is the largest crushing plant in the
world," he boasted, in a draft report to investors. "It has double the
capacity of the great crushing works of the Calumet & Hecla copper
mines of Lake Superior." When brought to full concentrating power,

---

* It ultimately became the American Mutoscope and Biograph Company, under which
name Edison sued it in a marathon patent infringement case that was eventually resolved
in his favor.

Ogden should be able to produce "from 1400 to 1600 tons of Besse-mer briquettes daily." It was modern in both machinery and method, automated "to the limit," so that one day it might run under a single supervisor. "This venture has all the elements of permanent suc-cess."[177]

Nothing so ugly, certainly, had ever befouled the Appalachian sky-line. At latest count—because he kept adding extensions—the works consisted of thirty-nine major structures at the crest of Edison Road above Ogdensburg. It was dominated by the cathedral-size magnetic separator building and webbed together with so many bridges, cranes, conveyors, steam pipes, power lines, and busy little railways that it looked like a compressed red-painted city. North, south, and west of its littered fringe, cliffs of gray rock were being blasted into slow re-treat, while the surrounding forest (dangerously creviced in places by the shafts of ancient mines) retreated too, leaving behind a litter of felled or dying trees. A perpetual dust boiling out of the mill whitened every upturned surface and the clothes and hair of workers and ex-ecutives alike. Those who cared about what they breathed wore sponge-filled rubber snouts. Seen from a distance of ten or twelve feet in the pale gloom, they could be mistaken for pigs walking upright.[178] Only when rain rinsed the filthy landscape did Ogden temporarily become a place where a man could feel clean.

That was not a sensation that Edison—to whom the plant was paradise—seemed to care about. Slovenly as he was at the laboratory, here he emulated the shabbiest of his employees. But the big head under the brown cap (slashed open and laced at the back for extra room)[179] and the clean-shaven jowls (either chomping on a stogie or bulging with plug) flagged him everywhere as "the Old Man," a be-nign autocrat always willing to stop, swap yarns, and spit.

"Today has been hotter than the seventh section of Hades reserved for Methodist ministers," he wrote Mina on 9 August. She was mak-ing her annual pious retreat to Chautauqua. "The dust in the air was frightful. . . . I feel lost in not going home to see my darling dustless Billy. What am I to do without a bath, some smartweed seeds have commenced to sprout out of the seams of my coat. . . . Think of it Billy darling your lover turned into a flower garden."[180]

Twelve days later he had more serious problems to report. The bricking plant was plagued by a breakage rate of over 50 percent, and there were frequent accidents farther up the line, causing labor unrest and expensive safety changes. Edison had to sell another batch of General Electric shares just to keep production going through the summer.[181]

> Everything seems to go wrong and I fear we shall have to close the works for want of money. . . .
>
> I have had 6 hours sleep in 4 days & am trying to pull it through. Especially the Bricker for when that goes the whole problem is solved & we actually know what we can do. I shall run till Saturday night & if I have good luck will probably go ahead if not I shall probably shut down until I see my way clear for money. While raising money I can have time to get a rest and go over the whole thing carefully so that when we start up again we will be OK—Mallory is the most dejected man you ever saw. The master mechanic and Mr Conley are completely discouraged while your lover is as bright & cheerful as a bumble bee in flower time.[182]

It occurred to Edison that he could drastically cut costs by reducing the number of men he employed, now that the major phase of plant construction was over. He boggled at a group photograph Dickson had taken in 1894, showing at least four hundred workers massed in the mill yard and clustered like ants on rooftops and steam lines. Then, they had looked like a contented lot—even the pitmen and muck makers and coal passers who earned no more than $1.30 a day. But now they were threatening to walk off their jobs, due to Edison's refusal to pay extra wages for overtime work.[183]

He saw an opportunity to speed them on their way when he heard that a strike meeting would be held at the end of the workday on Thursday 22 August. Five minutes before quitting time, Edison put up a large sign outside the assay house reading WORK IS SUSPENDED AT THE MILL EMPLOYEES WILL BE PAID IN FULL SATURDAY THE 24TH.[184]

The result was an angry mass exodus over the weekend and a resumption of mill operations on Monday with a residual force so small that Mallory was surprised to see how well much of the line operated without human assistance. His spirits lifting, he persuaded himself

The Ogden mine workforce, circa 1895.

that soon, thanks to Edison's unstoppable drive, "we will be able to turn out product at a very considerable profit."[185]

Decades later, when the red city had disappeared and the forest reclaimed the mine, Mallory wrote:

> You cannot live with a man without learning a great deal about him; that he wears two or three suits of underwear instead of many sweaters and coats when it is cold; that he steps out of his clothes at night, leaving them on the floor so they will be easy to step into again next morning; that he loves pie; that he is inordinately fond of smoking cigars. Little things, all of them, but they set off the big things. There were many big things. All of us associated with Edison knew, from the first, that we had to deal with an extraordinary man.[186]

"A MORE ORIGINAL GENIUS"

From now until the end of the decade, Edison spent the bulk of his time at Ogden, working an average of sixteen to eighteen hours daily and returning home to Glenmont only on Sundays. He stopped and started production so often ("New problems to be solved come up every day") that his predictions of imminent fabulous success began to sound fabulous indeed. The mill crushed more money than magnetite, and to keep himself, if not it, flush, he had to accept a $15,000 retainer from General Electric to develop a squirted-cellulose lamp filament.[187]

This nuisance did not last long, but another distraction temporarily, and irresistibly, diverted him in the first week of 1896. News came from London that the German physicist Wilhelm Röntgen had discovered a mysterious green "X" ray that emanated from an electrified glass vacuum tube and caused a barium platinocyanide screen nine feet away to fluoresce, even if a sheet of the thickest cardboard was placed in between.[188] The ray had a similar and even eerier effect when the body it penetrated was that of a human being, making solid flesh resolve itself into a mist in which bones stood out with a sharpness half erotic, half frightening. "*Ich habe meinen Tod gesehen,*" Röntgen's wife said after he irradiated her hand and wedding ring. "I have seen my own death."[189]

Edison was instantly consumed with desire to explore, and possibly exploit, this electromagnetic phenomenon. Ten hours after hearing about it, he began to construct a special darkroom in West Orange. "How would you like to come over and experiment on Rotgons [*sic*] new radiations," he wrote Arthur Kennelly. "I have glassblowers and Pumps running and all Photographic apparatus. We could do a lot before others get their second wind."[190]

He soon produced X-rays of his own, and was shooting and printing radiograms by the first week of February. Journalists plagued him for images to print. William Randolph Hearst begged "as an especial favor" a picture of the human brain. Edison failed to achieve this ne plus ultra of invasive photography with the standard Crookes tube Röntgen had used. Instead he designed a range of variant bulbs, with platinum wires and thinner glass to enhance emission. Imaging interested him less than the invisible, undeflectable streaming of light that was not light: "I want to see if the Roentgen rays are really perpendicular to the cathode plate, or if they curve between the cathode and the anode in a manner analagous to magnetic rays."[191]

Absorbed in his work, he paid scant attention to a report from

Paris that the Lumière brothers had perfected an intermittent-action projector, the Cinématographe, and exhibited moving pictures to a paying audience. When William Gilmore suggested that he move at once to acquire the rights to a rival American projector, the Phantoscope,[192] he agreed with the equivalent of a shrug and plunged back into radiation experiments.*

One evening that month a representative of *Metropolitan Magazine* found Edison in his darkroom, regulating the balance of electricity and airlessness in a long bulb. He seemed "oblivious to everything in the world but the gradation of light within the tube." Coaxed out of his trance, he said that he was now trying to determine whether the X-ray was "ethereal" or "allied to coarser matter." The vagueness of these terms indicated that his researches were still unscientific. He was enjoying, as boyishly as the four young laboratory workers assisting him, the thrill of exploring a new technology that so far seemed benign. They worked far into the night, photographing the effects of various degrees of radiation on opaque substances. Every strip received a twenty-minute exposure. Not until two A.M. did he leave the photographs to dry and invite the reporter to join him and "the boys" (one of whom turned out to be Thomas Edison, Jr.) for dinner. He could not stop talking, as he ate, about the practical implications of Röntgen's magic ray.[193]

Somebody managed to change the subject from fluorescence to incandescence and asked if he would undergo a test to see if the human retina stored light. Edison agreed. He sat for two minutes with his eyes shut, then opened them to the glare of a lamp positioned inches away, on top of a camera. He withstood the glare for another two minutes, after which all lights were switched off and the camera simultaneously clicked. It captured a momentary double gleam at the back of Edison's vision that, when printed, made him look less human than feline, a great cat in the dark.[†194]

Edison was by no means the only, or even the first experimenter

---

* These experiments were briefly interrupted by the death on 26 February of Sam Edison, age 91. Edison traveled to Port Huron for the funeral.

† Edison was to repeat this experiment in 1917, when researching the optics of night vision for the U.S. Navy.

with X-rays in the United States. Röntgen's announcement had galvanized many of the country's finest electrical engineers, including Elihu Thomson, William F. Magie, and Nikola Tesla. Aware that they were all, like himself, venturing into a strange new world, he ceased trying to keep ahead of the competition and on 18 March published the results of his research so far in *Electrical Review*. Among them were two findings that illustrated the Carrollian contrariness of X-ray behavior: first, that the lower the vacuum and the dimmer the fluorescence within the tube, the greater the radiation without; second, that the sharpest "shadowgraphs" were registered by the shortest bulbs and got sharper with distance.[195]

On the same day and in the same periodical, Tesla, who had taken almost a year to recover emotionally from the loss of his laboratory, described his own radiation experiments. He boasted that thanks to his recent invention of an oscillating steam generator, he had succeeded in throwing X-rays forty feet or more. And he, too, had tried to photograph a brain—in this case, his own. He had bombarded it at close quarters for more than half an hour, but found only that the treatment made him sleepy. Edison sent him an encouraging note: "I hope you are progressing and will give us something that will beat Roentgen."[196]

Notwithstanding their mutual—if guarded—goodwill, an article in that month's issue of *Scribner's Magazine* did its best to set them up as David and Goliath. The author, C. Benjamin Andrews, president of Brown University, opined that Tesla was "a more original genius than Edison" because he had eliminated the wiring inside lightbulbs and sent bolts of high-tension current through his own body. "He surrounds himself with a halo of electric light and calls purple streams from the soil. His aim is to hook man's machinery directly to nature's." Edison let this hyperbole speak for itself. But the "genius" comparison rated headlines in many newspapers across the country and set him and Tesla up as rivals for glory at the National Electrical Exhibition, scheduled to open in New York on 4 May.[197]

Coincidentally, they were both developing fluorescent lamps that they hoped to be able to show in time. Tesla claimed his would have 250 candlepower and a light efficiency of 10 percent; Edison aspired to 12 or 15 percent efficiency and said that by coating the inside of his tube with a secret material, he had harnessed some of the electromagnetic energy of X-rays: "I have turned them into a pure white light of high frangibility."[198]

He kept the coating (calcium tungstate) secret only because he had used it in another device that was sure to be the sensation of the exhibition if it, too, could be made ready for display. Edison the chemist had discovered, after testing eighteen hundred fluorescent salts in 150 tubes, that the phosphor $CaWO_4$ came brilliantly alive when its fused crystals were excited in a vacuum. He applied it to the screen of a portable visor, flared in shape and fitting snugly to the face, that took an instant X-ray picture—even a moving picture, if desired—of whatever lay before it on a radiant box. "You can see all the bones of the body and the heart beat plainly," he told Mina.* This capability of what he called his "fluoroscope" was of obvious benefit to medical personnel who, in emergency situations such as a shooting, would not want to wait two hours for a radiograph to be exposed and developed. He therefore declined to patent it and sent an early model to one of his fellow experimenters, Michael Pupin of Columbia University.[199]

"It is a beautiful instrument," Pupin wrote in surprised gratitude, saying that he had demonstrated "its miraculous power" in three public lectures, to much applause. He doubted, however, that fluoroscopy "would entirely supersede the photographic method of diagnosis in surgical work," where record keeping was vital. To that end, he was already testing the idea of contact prints that could be taken directly from Edison's "very excellent" screen. "Your success will be received with great delight by all scientific men."[200]

Edison was not used to reading such compliments, especially on stationery headed "University Faculty of Pure Science." Touched, he replied that he was working on some other tubes that "I think will surprise you and aid you in your scientific investigation, which is out of my line."[201]

---

* A young American dancer, Loie Fuller, visited Edison in his darkroom at this time, and had an epiphany of herself performing in costumes permeated with his radiant salts. They experimented together with initial success, but the fluorescence kept fading. In her later career, Fuller won fame for light-based choreography, while also becoming an amateur expert on radiology.

There was no point in being secretive anymore about his calcium-tungstate bulb, because the fluoroscope quickly preempted it as a news sensation in the weeks preceding the electrical exhibition. Advertisements guaranteeing that "Edison will be there. And Tesla will be there" only increased the demand for tickets, as did an announcement that any attendee with the courage to hold a hand under the magic machine could get a bone examination for free.[202]

The distraction saved Tesla from having to answer too many embarrassing questions about why his own lamp was not on display. Edison had never thought much of him as a lighting engineer and knew he was having difficulty with it. But he begged the editor of *Western Electrician* not to publish a "foolish" letter comparing their respective molecular-impact systems unfavorably to the new glow discharge tube of Daniel McFarlan Moore. "I don't care what is said, but Tesla is of a nervous temperament and it will greatly grieve him and interfere with his work. . . . It must not be forgotten by Mr. Moore that Tesla is an experimenter of the highest type and may produce in time all that he says he can."[203]

His request was of course unheeded, and press speculation grew that Edison and Tesla were rivals. Unnoticed in the general publicity was a brief report in *The New York Times* that both men complained that long exposure to X-rays hurt their eyes.[204]

PERFECTLY BAKED AND HARD AS GRANITE

As the technology of moving pictures advanced, patent and copyright offices were deluged with so many Greek and Latin brand names, most of them ending in *-scope,* that even entertainment lawyers had difficulty remembering the difference between the tachyscope, eidoloscope, mutoscope, bioscope, parascope, veriscope, magniscope, and kalatechnoscope, not to mention the cinematograph, centograph, projectograph, and kineopticon. Edison was fortunate in being so famous that he had merely to attach his name to the Kinetoscope and Kinetograph to imply that they were somehow superior to the rest. Gilmore urged him to bestow similar cachet upon the phantoscope, even though it was the invention of Thomas Armat, a young engineer who had licensed it to him in exchange for exhibition royalties.[205] He agreed to market the projector as "Edison's Vitascope" and "Edison's

Latest Triumph" and, in doing so, became party to a deal that did much to harm his reputation for proud individuality.*

However, it also did much to restore his wealth, because the Vitascope was a culture-changing success. Audiences gasped at the onrush of great waves in *Rough Sea at Dover* and at the prolonged intimacy of *The May Irwin Kiss*—shocking when blown up more than life size and the Edison studio's biggest hit that year. "Can genius go farther?" the *Los Angeles Times* marveled. "We have been made to hear the voices of our distant friends, and now we are able to see them move and act."[206] It remained only for Edison to present himself to the camera, which he did, teasingly, by letting a young newspaper artist, J. Stuart Blackton, sketch him while he remained offscreen.

The Edisonian forelock and black brows were given gestural attention, with slashing sweeps of charcoal.[†‡207]

On 11 August Henry Ford, chief engineer of the Edison Illuminating Company in Detroit, had a chance to worship the same features "live" during an industry convention at Manhattan Beach, Long Island. He surreptitiously photographed Edison snoozing straw-hatted on the porch of the Oriental Hotel but did not dare to approach him until around midnight, when Edison sat drinking beer with a number of Menlo Park veterans.[208] Ford worked up the courage to tell him that he had designed and driven "a little gas car." Although Edison was himself beginning to think of the storage battery as the ideal power source for horseless carriages, he reacted encouragingly: "Keep on with your engine. If you can get what you want, I can see a great future."[209]

Less than a month later, at the state fair in Providence, Rhode Island, two "electrics" averaging fifteen miles per hour beat five gas-powered Duryeas in the first track race for automobiles ever held in America.[210]

By then Edison was back at Ogden, where at last the briquetting

---

* Edison's license lasted only a year. On 30 November 1896, he brought out his own highly praised projector, and in 1897 reassigned the Phantoscope patent to Armat.

† https://www.youtube.com/watch?v=lW3uIm82hpY

‡ Blackton took a bow at the end of this ninety-five-second short and went on to become a major movie producer and the father of film animation.

plant was registering big improvements. "Made 13,000 bricks without a miss," he wrote his wife in triumph. "They came out perfectly baked and hard as granite. Now everything is known all will work and we are getting things to completion."[211]

Mina had heard such effusions so often, and had pined for his company so long, that it was hard for her to simulate excitement. She was more inclined to complain—or in his language, "growl"—about their many separations. She had recently turned thirty-one, and like Mary Edison before her, she was becoming stout. Edison tried to make her feel less neglected by sending extravagant endearments. "Darling Billy (Constitutional Growler). . . . I just simply love you to pieces. . . . with kisses so thick that 40,000,000 X ray lamps couldn't penetrate. . . ."[212]

Meanwhile on the Mesabi Range in Minnesota, steam shovels were pushing aside thin topsoil and lifting out pyramids of high-grade hematite, thirteen tons at a time. The price of that ore was steadily falling, just as shipments were rising—from 621,047 gross tons in 1893 to a frightening 2,884,372 tons this year. Edison's only response was to scale up the size of his assault on Sparta Mountain. He let newspapers know that he, too, was ordering steam shovels and was ready to go into full production at five thousand tons of ore a day, if a sample order of his briquettes tested satisfactorily at the Crane Iron Works in Catasauqua, Pennsylvania.[213] In that case, other eastern furnacemen were bound to switch to satiny Ogden agglomerate and rejoice that they need never again choke over trainloads of red rubble from Duluth.

As if to remind the world he was still an inventor, he rounded off his laboratory work for 1896 with a flurry of new product announcements: a superior projector to replace that of Thomas Armat, a spring-driven home phonograph, an improved wax for cylinders, an "autographic telegraph" that sent dotted script and sketches, and a lightning-fast current breaker that, combined with Tesla's coil oscillator, greatly increased the electromagnetic force of an X-ray machine.[214]

A demonstration of this hookup at the Kentucky School of Medicine on 23 November showed how radiation could be used to aid surgeons in the extraction of bullet fragments from human flesh— a frequent chore in that part of the country. It also tantalizingly suggested what scientific miracles might be achieved if the two "geniuses"

could be persuaded to form a team. But given their opposing personalities, that was as unlikely as an alliance between Oscar Wilde and
the Marquess of Queensbury. Already Edison had joked about the
nonappearance of Tesla's fluorescent lamp—"If Tesla has a light why
don't he show it?"—while Tesla criticized him for X-raying the eyes
of blind people to see if they registered any internal light patterns: "Is
it not cruel to raise such hopes when there is little ground to them?"[215]

On one thing they were agreed: that by toying with X-rays too
much, they had done some mysterious damage to themselves. Until
more was understood about the pathology of radiation, they preferred to return to safer research.[216]

"The fact is, there is really a terra incognita bound up in crystals
and salts," Edison told a visitor to his laboratory. "Just come out here
in the workshop and see how my assistants have suffered from the
bombardment of these rays." He led the way into another room and
got Clarence Dally to hold out his arms and hands. They were swollen out of all proportion, as if they had been pounded with clubs.[217]

Edison seemed more interested than sympathetic. He wondered
aloud if focused rays might not be effective in killing tuberculosis ba-
cilli or clarifying cataracts. "I can blindfold you, and yet cause you to
see objects by means of the X-ray. . . . I know there are those who say
that such a thing is impossible, but you cannot laugh a fact out of
court." *[218]

THE LAST NOTCH

On New Year's Day 1897 Edison was in Catasauqua to observe the
start of a long series of tests on Ogden briquettes. He was accompanied by his inseparable aide Frederick Ott, who a few years earlier
had become history's first film star by pretending to sneeze for the
Kinetograph camera.

"This is Freddie," he said to Leonard Peckitt, president of the
Crane Iron Company. Peckitt was an Englishman and thought he said
"Friday," in reference to another famous factotum.

---

* According to the appropriately named *Niagara Falls Cataract,* 20 November 1896,
Edison pressed his aching eyes shut after a long spell of work on X-rays and found that
he could still see his hands.

Later, when they were alone, he asked, "What does Friday do?" and Edison, deaf, replied, "Nothing."

"If he doesn't do anything, why do you want him?"

"Because he never falls asleep. That's what I pay him for, to keep awake. Whenever I want him, he is there. The other damned fools are always asleep when I want something."[219]

Peckitt put Edison up at home while Ott stayed at the local hotel. At mealtimes he was fascinated by his guest's obliviousness to whatever was served.

> He never asked for anything, never would express a preference, never helped himself. He ate and drank what was before him. If you put nothing on his plate, he did not miss it. He then told endless stories.
>
> If you put a glass of wine before the Inventor, he drank it. It made no difference whether the brand was sherry or champagne, he said nothing, but the glass was always emptied without comments. He [just] went on talking and cleaned his plate without any trouble.[220]

Edison was gratified by the initial yields of his briquettes in the blast furnace but kept pushing for more and more revolutions of the blower to produce more heat. When Peckitt demurred, for fear of an explosion, he scoffed and said he had deliberately wrecked a $25,000 crusher at Ogden in order to see how much load it could stand. "Now I can design and build one that will do as well as she did before the last notch was added."[221]

There was intense interest throughout the Pennsylvania iron industry in the ongoing tests, which usually began at two A.M. Trade reporters hung out in Peckitt's office to hear Edison tell stories while the furnace was prepared for casting. One night he interrupted himself to roar out, "Hi, there, what are you doing? What the devil now? Someone kick him."

Ott had fallen asleep.[222]

Edison stayed in Catasauqua for a week, long enough to see that the tests were going to be positive, pending a final report from the works. Shortly before returning to West Orange, he sent a telegram to one of Peckitt's rival smelters, S. B. Anderson of Andover, New Jersey: "Come to breakfast. Have 11,000 tons for you." When Peckitt asked what it meant, he explained that Paterson had mocked his briquetting method by saying, "I will eat all you make."[223]

MILE OF MAGNETS

The test results were more than positive, they were extraordinary. Peckitt reported that the briquettes had caused a 33 percent increase in furnace smelting, and he was confident of reaching 50 percent if they were supplied to him in large quantities. They reduced ore to a precipitate that "showed unusual strength, and was, in fact, the strongest and toughest foundry iron we have ever made." Considered technically, it "could not be better, as the purity of the briquettes enabled us to make an iron very low in phosphorus and sulphur."[224]

Even more pleasing was an order from the Crane Company for as much ore as Edison was willing to deliver. Along with news that his film and phonograph business was booming, and that William McKinley's election to the presidency had brought about an end to the long depression, it was the nicest possible present for his fiftieth birthday, at Ogden in February. Mina sent up a congratulatory cake with model miners and little electric lights.[225]

Edison needed no further encouragement to authorize yet another expansion of the plant, intending a switch to full commercial production in the spring. He commissioned two Vulcan steam shovels (one of them, at ninety-three tons, the biggest ever built), and studded his giant rolls with steel "slugger" knobs and more than doubled their rotary speed. He also invented a device that made dust a lubricant rather than a coagulant and lengthened the line to what the *Harrisburg Daily Independent* breathlessly described as a "mile of magnets," totaling 480 separation processes.[226]

The trouble with his goal of refining five hundred tons of concentrate a day was that the bricking facility could handle only half that output. Unless he installed another fifteen fabricators and eight new furnaces at a cost of $50,000, both stockhouses would soon be swamped with fines. Walter Mallory had to send a begging letter to investors, telling them that Edison had already spent over $1 million of his own money at Ogden ($200,000 on the giant rolls alone) and could use some help from "friends" now that his product was in demand.[227]

This appeal coincided with intelligence that the price of Mesabi Bessemer ore, which had dropped as low as $3.25 a ton in 1896, was now falling so fast that it might soon approach the two-dollar mark. Edison insisted that he could ship at seventy-eight cents a ton, but he was not famous for arithmetic.[228] The president of the American Institute of Mining Engineers said at its annual convention that a more likely price for Ogden ore was $4.08 per ton. In that case, "it is not probable that the Edison Works can be run continuously at a profit." The most that could be said for the remarkable venture on Sparta Mountain was that it was "a monument of perseverance in original research which certainly deserves our admiration."[229]

Under the circumstances, Edison's backers again declined to give him any more money, so he again had to shoulder the cost of a vital improvement.[230]

At least he could save on one minor expense while doing so—the wage he had been paying his eldest son as a general mechanic at Ogden. Tom had just turned twenty-one and come into a $17,309.91 legacy left him by Mary. In a letter half aggrieved, half supplicatory, he wrote to say he wanted to strike out on his own. "I feel that I have never pleased you in anything I have ever done. . . . I don't believe that I will ever be able to talk to you the way I would like to—because you are so far my superior in every way that when I am in your presence—I am perfectly helpless."[231]

Edison ignored Tom's request for a special assignment that would give him a chance to show that he, too, was an inventor. In a series of letters to Mina, the young man demonstrated only that he was a world-class whiner. "Why is it I am unhappy? why is it I feel alone? . . . why am I so backward? . . . I love but I am not loved."[232]

He went west and south for a few months but inevitably, like a small moon in irregular orbit, yielded to the pull of its star. By May he was back at the plant, doing laborer's work. His father gave no indication of noticing that he had been away. "I can say," Tom wrote to Mina, "he has not even looked at me."[233]

Edison hardly had time to look at a clock. He was busy doubling the size of the bricker plant and building a larger powerhouse, patenting Ogden's screening system, designing new machinery (in one case, forty-eight versions of a single device), taking delivery of the Vulcan steam shovels, and running out of money fast. In August, after yet another start-stop, he had to sell his stock in the Edison Electric Illuminating Company, just to maintain the mill during closure. "I am full of vinegar yet, although I have had to suffer from the neglect of absent minded Providence in this scheme."[234]

It was plain even to him at summer's end that when Ogden next opened, it must stay open and prove itself to be the inexhaustible cornucopia of iron that he had so long promised. Otherwise it would be forever known as "Edison's folly," a multimillion-dollar mockery of his past achievements, good for little more than the production of sand.

By late September the new bricker machines were ready to receive. He put the entire line into operation and gave reporters unrestricted access to it for the first time.[235] The result was three major articles in *The Iron Age, Scientific American,* and *McClure's Magazine,* exquisitely illustrated and sharing a common tone of reverence. "Mr. Edison appears in the new light of a brilliant construction engineer grappling with technical and commercial problems of the highest order," the trade journalist wrote. "He pursues methods in ore dressing at which those who are trained may well stand aghast. But considering the special features of the problems to be solved, his methods will be accepted as economically wise and expedient."[236]

What awed the visitors, apart from the fact that they were looking at the biggest iron-concentrating works in the world, was the way Edison had synchronized all its operations, integrating mechanical power with the natural forces of gravity, momentum, and magnetism. High on Iron Hill, the steam shovels loaded slabs of black-grained gneiss weighing as much as six tons into skips that ran down a tilted track toward the mill, one every forty-five seconds, simultaneously returning "empties" to the quarry.

The Traveling Crane lifted them to the top of the Crusher House, whence they thudded ten feet down into the whirring cleft between the giant rolls—an abyss no man could look into without fear. Shattered in less than three seconds, they dropped through to a set of intermediate rolls and got chewed into stones. Elevator no. 1 took them for further decimation by the first and second thirty-six-inch rolls, whereupon they lost their plural identity and became a speeding mass of rubble. Except in the hottest weather, this reduction usually showed sign of dampness. It passed through the twenty-four-inch rolls to elevator no. 2, slid onto Thomas Robins's rubberized first conveyor belt, and was roasted in dryer no. 1. Elevator no. 3 and the second and third conveyors, then carried it, smoking, to the three-high rolls, which beat it into gravel before it fell through fourteen mesh screens, in a zigzag trajectory that allowed only the finest pulver to reach the separator building. There, from a great height, it fell again, a thin gray curtain of dust that paled as three progressively stronger twelve-inch magnets deprived it of its iron specks. Even now the black draw-off was not rich enough for Edison's purpose. He subjected it to more heat in dryer no. 2, reconcentration by the fifty mesh screens and eight-inch magnets, cleaning and dephosphorization in the dusting chamber, and a final refinement by the four-inch magnets before it was conveyed to an immense stockhouse, ready for caking and baking.

The process whereby Edison transformed his fines into hard briquettes that were, paradoxically, both porous and waterproof could have been devised only by an inventor equally versed in chemistry and physics. He mixed the iron powder with a warm binding material whose formula was a trade secret. Then he transferred it as dough to a row of die-block machines that took it, cut it, and compressed it three times (under squirts of oil to prevent adhesion), the last plunger slamming down with a force of sixty thousand pounds. This all happened at a disgorge rate, per machine, of sixty briquettes a minute— each a squat black cylinder three inches in diameter and nineteen ounces in weight. They were superheated for well over an hour before shipping.

Theodore Waters, the writer from *McClure's*, marveled at Ogden's total automation. "The never-ending and never-resting stream of material constantly circulates through the various buildings . . . and not once in its course is it arrested or jogged onward by human agency." He was also impressed by the economical way the mill recycled its waste products. On the last day of September he watched a conveyor unrolling from "the magnet-house" and pouring what looked like a cascade of gold onto a hill-size dune that shimmered strangely in the sun. It was a mix of quartz, feldspar, and lime phosphate tailings, and the shimmer came from its sharpness—unlike the dull obtundity of beach sand. Builders and the manufacturers of abrasives prized it, so to that extent it was gold, of a sort, in Edison's pocket. He even sold the dust from his dephosphorizing chamber to paint companies, who thickened their pigments with what had once been the rock of Sparta Mountain.[237]

HOT CAKES

Edison needed every cent he could earn as Ogden cranked into full production. The artist William Dodge Stevens sketched him for *McClure's,* while Waters went down the line, and caught a knot of intense worry between his eyes.

*Edison sketched by William Dodge Stevens,
Ogden, 30 September 1897.*

He had just mortgaged the Phonograph Works at West Orange for $300,000. That same day, he humiliated himself by borrowing $11,175 from his eldest son's inheritance.[238] Tom did not take kindly to the transaction, even though Edison paid him 6 percent on the loan. Edison already owed him $4,500 on a "bond" agreement of doubtful validity. The two withdrawals pretty well swallowed up all that Mary had left Tom.[239] But he was as usual helpless in his father's hands. All he could do was continue to send self-obsessed letters to Mina from Asbury Park, New Jersey, where he had settled after leaving Ogden for the second time, to nobody's regret.

Hitherto, Tom had written in a rounded script creepily imitative of Edison's own. But now it was spiky, the hand of a different person, and an unstable one. It tilted ever more to the right, as if he were losing equilibrium. On 27 November he wrote Edison to say that he was ill, due to "the treatment I have received at the hands of my family," and might not survive. "I sincerely hope—that if this does prove very serious to me—you all will feel better satisfied. . . . I am for a little while longer your affectionate son." The signature "Tom" reverted to upright characters, but was inked so small as to be almost illegible.[240]

In another letter, to Mina, he sounded cheerful and full of ambition. He had invented "one of the finest incandescent lamps in the world" and was hustling it with great success in New York. It was called "the Edison Junior Improved" and would "sell like 'hot cakes'—in fact I never can fill my orders—for it is simply remarkable. . . . I intend to have ten thousand agents on commission. . . . I will control the market of the world or bust." He went on excitedly for eight pages, ending with: "I wonder what father will think when he hears about this. He very probably won't believe it."[241]

Tom guessed right. Edison was aware that the lamp was derived from the fluorescent tube they had worked on together the previous year.[242] He also heard that some unscrupulous "backers" were hoping to cash in on the fame of Tom's surname. On 5 December *The Sunday World* featured a huge drawing of the young man inside a lightbulb. Across its base was bannered a new version of Tom's signature, writ large now and so much an imitation of his father's that Edison hastened to file for trademark protection. Even more infuriatingly, the article below declared that "a new personal power is risen in the world of invention" and quoted Tom as saying he would soon build a lamp factory "in Menlo Park, N.J."[243]

It was clear that Tom's paternal fixation was degenerating into a fantasy of reincarnation. Unless he was checked soon, he might well claim to have invented the phonograph. Mina wrote him a few kindly words of caution and was rewarded with an effusion that could have been penned by Little Nell: "That letter that has never left its sacred place—nearest my heart—binds me nearer and nearer to you darling Mother."[244]

WHAT THEY MOST WANTED

At the annual board meeting of the New Jersey & Pennsylvania Concentrating Works on 12 January 1898, Edison boasted that Ogden was now so automated that he had been able to reduce the workforce from 400 to 78. This economical-sounding figure was deceptive, because the mill was again not running. He had closed it to deal with a drying problem and could not say when production might resume. But at least the problem—mud and ice clogging the ore—was caused by success: the steam shovels had proved more effective than dynamite in gobbling up chunks of mountain, "and now the rolls will take anything that can be put into the hopper." That meant that the lower parts of the line had to be adjusted to deal with an *embarras de richesses,* not least the stockhouse, which was already crammed with unbricked fines.[245]

Edison talked on for a while in his usual optimistic way, before asserting (with Walter Cutting's eyes upon him) that the plant was underfinanced: "I am in negotiations at the present time with a syndicate to furnish operating money until the Company has its own funds; but as to the money needed to liquidate the Company's present indebtedness and that for test expenses, insurance, and leases, nothing has been done except by myself."[246]

He said he was still prepared to bankroll the plant, "as far as I am able," in confidence that it would soon become profitable. Mallory spoke next and cast doubt on the "soon" with a recital of cold numbers: Ogden, capitalized at $2.25 million, had so far cost $2,091,924 to build, equip, and test, and sold only $158,591 worth of iron and sand.[247]

Under the circumstances, it strained belief that Edison (who was privately trying to borrow $15,000 from his father-in-law) had just told reporters that he intended to build a $1.5 million gold mine southeast of Santa Fe, New Mexico.*[248] But such was the positive spell he cast that his fellow directors reelected him president and accepted his offer to lend the company $51,500 over the next six months. As Mallory remarked to Theodore Waters, "What has been said of his personal magnetism has not been overstated."[249]

On 9 February Edison wrote Mina that he had cleared the stockhouse and was about to start milling again. He was working sixteen-hour days and "making good progress on 3 or 4 inventions to raise money." The same could not be said for his eldest son, who had gone to Florida, reportedly on doctor's orders. "Tom wrote a horrible letter to William saying he was deserted by his family, that he was lying at the point of death, that we would probably never see him again."[250]

Mina had received a similar cri de coeur from Tom himself. She was the principal recipient of Tom's letters, which came at the rate of two or three a week and formed an ongoing record of manic depression. "I enquired of my agent at Fort Myers," Edison went on, "and got word that Tom & friend were having a fine time and had just come in from a Deer hunt."[251]

Being a depressive herself, and by nature compassionate, Mina sympathized with Tom to a degree. He had indeed been dangerously ill, with what sounded like inflammatory rheumatism. She dreaded that he might come "home" to live, as he occasionally threatened to do.

William was another claimant on her responsibilities as stepmother. He was in his freshman year at Yale and hated it there. Mina did what she could to give both what they most wanted—expressions of love and payment of bills—but she was expecting another child of her own in July and had little time to spare for two malcontents who should have long since grown up.[252] Edison had even less. As far as he was concerned, they were out in the world he had entered at age twelve, and could drown if they chose not to swim.

---

* Edison was serious. He bought a two-year lease on fifty-four thousand acres of low-grade gold sands at Dolores, in the Ortiz Mountains, and by the summer of 1898 he had put a preliminary dry placer plant into operation extracting gold by a magnetic drum process. The mill grew to Ogden-like proportions, cost him half a million dollars, and was a complete failure. "I lost the usual amount," he joked.

William Edison, circa 1898.

YOUNG DAYS

Patriotic young Americans that spring smelled war coming between the United States and Spain over the cause of Cuba Libre, a movement to win freedom for the last major European colony in the New World. One of the first Yale men to vow to fight if President McKinley issued a call to arms was Mina's brother Theodore Miller. A twenty-three-year-old postgraduate law student in New York, he was well acquainted with both William and Tom, and admired neither. At least William (who dropped out of college in early March) wanted to join up too. Tom manifestly was unfit to serve. He talked vaguely of "going away soon," but at the same time he posed as the president of a new $100,000 lighting company, with two maternal uncles as his backers. "He is a queer boy," Theodore wrote his father. "I am very sorry he is allowed to do these things and have talked to Mina but she says Mr. Edison says he can do nothing."[253]

Congress declared war on 25 April, and within six weeks Theodore and William were enlisted as privates, the former in the First U.S. Volunteer Cavalry, or "Rough Riders," and the latter in the First New York Regiment of Engineering Volunteers.[254]

For Edison, toiling obsessedly on Sparta Mountain, the Spanish-American War proceeded as little more than a twelve-week geopolitical disturbance in the Antilles, its rumbles inaudible amid the satisfying roar of his mill.[255] He took enough notice of it to offer the navy a night illuminant consisting of calcium carbide and calcium phosphite packed into shells that would explode on contact with water and flare long enough for enemy ships to be detected four or five miles away. But as he would discover later in life, the government was not much interested in civilian defense ideas.[256]

Incomprehensible as it might have been to Mina, awaiting her baby in the green peace of Llewellyn Park, Edison so loved being at Ogden that coming home on Sunday was something of a chore for him. He even lost interest in his laboratory. It languished for lack of assignments from the Old Man, although the rest of the West Orange campus thrived with burgeoning production of phonographs, moving pictures, cameras, and projectors.

By now Edison had resigned himself to the fact that because of the Mesabi phenomenon, iron prices were never going to rise more than a few cents above the historic low they had registered in May. (Bessemer was selling as low as $2.25, and non-Bessemer even lower at $1.75, almost sixty cents less than what he had hoped to get for his iron when he began mining at the beginning of the decade.)[257] But such was his pride in the magnitude of his achievement so far that he began to think of bigness as an economic advantage in itself. He would prevail by building more Ogdens, each four times the size of this one.

Besides which, he was happy at the plant, happier even than he had been starting out at Menlo Park in '76 with Charles Batchelor and "the boys." Batchelor was mostly absent now, semiretired and wealthy on his share of their mutual inventions over the years, but Edison did not miss him. He had a different set of "boys" to josh around with now, and in their rough masculine way they enjoyed mountain life as much as he did. On rare days off they played baseball, boxed, or bet

their wages on rattlesnake and cock fights in a pit dedicated to bloodshed. Edison allowed the sale of beer at the company store, rather
than encourage the smuggling-in of banned hard liquor. Racial violence occasionally broke out in the Summerville settlement, where
laborers lived in a squalid clutch of frame houses and outside priv-
ies.[258] It derived from Old World hostilities of no interest to the
"Americans" more comfortably quartered in Cuckoo Flats, or the
hotel Edison had built for visitors and senior management. He could
often be seen daydreaming on the porch, or wandering up to the
quarry in his enormous straw hat and duster to watch the steam shovels at work, a sight that endlessly fascinated him.

"I never felt better in my life," he reminisced years later: "Hard
work, nothing to divert my thought, clear air, simple food . . . very
pleasant." In old age Dan Smith, his big mine rigger, looked back with
similar nostalgia on these "young days . . . the happiest time of my
life."[259]

Neither man, however, was on active duty in Cuba. On 1 July, in a
coming together of names, Theodore Miller fought beside his commander, Col. Theodore Roosevelt, at the battle of San Juan and was
fatally wounded. Ten days later an anguished Mina gave birth to a
son. He was baptized Theodore Miller Edison.[260]

William survived through the cease-fire on 12 August and thereafter lay sick and bored in Puerto Rico, begging his father to help him
get an early discharge. As if this were not enough family distraction,
at a time when Edison urgently needed to give all his attention to
problems at the mill, Tom also complained of ill health, along with his
usual financial straits. He was then reported to be summering in a
camp in the Adirondacks with a showgirl, scandalously unchaper-
oned, by the name of Marie Toohey. Tom denied press rumors that
they were engaged, and when she reappeared in his apartment in October, he insisted that she was only "nursing" him.[261]

THAT HOLE IN THE GROUND

Personal, professional, and meteorological crises converged on Edison in the last days of 1898 with a suddenness that drove him closer
to panic than ever before in his career. A blizzard whitened Sparta
Mountain and was followed by a long period of intense cold. Many

laborers walked off their jobs, never to return. Hitherto, during the depression years, they had been insecure enough to tolerate the low wages and primitive accommodations Edison begrudged them. But now there were better prospects elsewhere. Mallory warned that unless decent housing was built at the plant, Ogden for all its automation would never be a paying proposition. In fact, it was already at the point of financial collapse.[262]

On 2 December Edison wrote in desperation to Mina, who was visiting with her family in Akron, "I must have the $12,000 of the General Electric bonds in exchange for Phonograph Works bonds without these the works would have to be shut down." He needed the cash from this transaction on the fifteenth and a further $5,000 in Northern Pacific bonds before Christmas: "You had best make a flying trip home." Three days later he cabled her, "I think you better return by the tenth . . . very important i am feeling quite ill." In an attendant letter he admitted, "I am very worried about things."[263]

She hurried to help him, but before she could get back to Glenmont, weather and worker attrition obliged him to close the plant for what looked ominously like the last time.[264] Meanwhile the proliferation of yellow press articles about the erratic behavior of "Thomas Edison, Jr." annoyed Edison so much that he sent his son a message, via William, threatening legal action unless he stopped abusing the family name. "The old man says he is through with you," William wrote, enjoying his mission. "Also he says that you are in debt and furthermore that you have married this actress."[265]

Whether in fact Tom had spliced himself to Marie was not clear, but he responded with outrage on 17 December, addressing his father as "Dear Sir." He asked why it had been necessary to reprimand him by proxy. "However I understand this is one of your characteristics." In a shrewd blow, he pointed out that he was not the only Edison in debt. "If you knew how to handle your own achievements—what have you today—ask the financial world—they know. . . . People are through putting money into your inventions—and—as a consequence they are through with the name of Edison for good—otherwise I would be a rich man."[266]

This was uncomfortably near the truth of Edison's current situation, and Tom's separate reply to William showed that he knew its full dimensions. "He couldn't raise a dollar on anything—he put two millions out of his own pocket in the Mill—simply because he couldn't get it from any one else."[267]

Edison gambled what was left of his faith in magnetic mining on anticipated borrowings from the National Phonograph Company, although William Gilmore resented having to play Peter to his Paul. For the rest of the month—"What, is it Christmas already?"—he sought relief from worry in reading. But the books and periodicals he studied had less to do with iron concentrate than with the golden pyramid of sand piled up outside the separation house. Before the year was over, he was ready to apply his tailings separation technique to the production of portland cement.[268]

He would continue to insist, until the end of the century, that his great experiment at Ogden would succeed. But events tinged with a sense of finality, or of significant change, kept cautioning him, throughout 1899, that it would not—that alternative avenues of research and development were open for him to explore. For the better part of a decade he had embraced the problems that Ogden taxed him with, delighting in his ability to solve them one by one. But when, early that year, he compiled a list indicating there were 183 more to be tackled, he could no longer ignore the odds against him.[269]

On 17 February Mina's revered father Lewis Miller died—the second great grief to hit her in seven months, hastening her passage toward middle age. Edison himself, having just turned fifty-two, was white-haired now. Three days later he heard that Tom and Marie had married, in a Roman Catholic ceremony that at least confirmed the seriousness of their relationship—as did news that the girl had given up her stage career in order to be a full-time wife. "She will not go back without my consent," Tom was quoted as saying, trying to sound like a man in charge for once.[270]

When spring came, Ogden remained shut amid a nationwide surge in iron ore demand that Lake Superior mines were only too pleased to satisfy.[271] Mallory told backers that $100,000 was needed to get the works started again. If necessary Mr. Edison would contribute yet more cash in exchange for stock, but first, decent housing had to be built for the labor force.[272]

By then Edison was back to working full-time at the laboratory, educating himself in all aspects of portland cement production. In one sleepless twenty-four-hour stretch he designed what would become the largest cement mill in the country, right down to minute details of plumbing, lubrication, and ventilation.[273]

On 15 April he organized the Edison Portland Cement Company, capitalized it at $11 million, and began looking for a suitable site in western New Jersey. The following month he attended the annual electrical exhibition in New York, where all the talk was of electric automobiles. He denied that he was building such a car for himself, but told Mallory that he had an idea in mind for a light, efficient, durable storage battery that was "absolutely not to work with lead and sulphuric acid." He began experiments on it at once, and soon had more than a hundred technicians detailed to the project.[274]

Tom told reporters in July that he had "severed all connection" with his father and would pursue an independent career as an inventor: "I think he is too wise a man to bother over the inevitable." William, no longer in uniform and furious not to be offered a job at West Orange, made a similar bid for independence just before he turned twenty-one on 26 October. "What little money I receive in a few days I will invest in picture machines and a small factory," he wrote Edison. "If I fail it will be my loss." Within three weeks he, too, would marry and embark on a lifetime of proving that to be true.[275]

In November construction began at Ogden on a few rows of worker houses arrayed along the ridge of Sparta Mountain. Edison and Mallory both talked bravely of reopening the mill when the houses were occupied and leaves were back on the chestnut trees. But when Ogden's balance sheet came in at the end of the year, it emphasized the vanity of human wishes. Since 1890 the New Jersey & Pennsylvania Concentrating Works had cost $2,600,942 to build and run, and sold a mere $180,688 worth of product. Edison was due $334,611 plus interest for cash advanced, and there was no money to reimburse him. He already owned most of the rest of the company's stock in unredeemable paper. There was just one unfulfilled contract on its books: a two-year-old order for five hundred tons of briquettes for Bethlehem Iron, conditional on low phosphorus content. If he chose to honor it, at a time when the nation's stockhouses were awash in cheap ore, he would have to do so at a competitive price that was sure to drive him even further into debt.[276]

The memories of aging men, recorded after the plant became a clutch of ghost buildings, varied as to who made Edison realize it was facing bankruptcy. Charles Batchelor recalled bringing him a press report of John D. Rockefeller's takeover of the Mesabi field, portending a vast increase in production, and shipments all over the world via the Great Lakes and St. Lawrence waterway. Improbably, though, Batchelor said the article made Edison burst into laughter before exclaiming, "Well, we might as well blow the whistle and close up shop."[277]

Thomas Robins claimed to have been a witness to the incident— "The Old Man and I were sitting astride a plank which rested between two horses"—but he remembered no laughter, only Edison's kindly concern that if the mill shut, "this boy" would be out of a job. In 1899, however, Robins was long gone from Ogden, and so was Batchelor.[278]

The account that rang most true was that of Walter Mallory, who for most of Edison's time on the mountain had worked with him day and night. He said that when he first got the bad news, he could not bring himself to pass it on. Edison sensed something was wrong, and the two men avoided each other for three days, postponing a moment they both dreaded. Eventually they met in a bleak bedroom at the Lake Hopatcong Hotel. Mallory locked the door. Speaking as loudly as possible to give himself courage, he reported what he had heard.[279]

Edison sat listening on the edge of the bed, nervously tugging at his right eyebrow. His initial reaction was predictable: "Yes, it's a problem, it's a problem. But worry won't solve it. Brainwork will."

Mallory ventured to disagree, knowing that no amount of thinking could alter the laws of supply and demand.

Edison waved him silent and brooded for a long time, still tugging at his eyebrow. Then in a calm voice he said, "We will stop work here immediately."[280]

There was little to stop, apart from the construction along the ridge. The mill itself had been inert for a year. Edison marked some machinery for transfer to a limestone field near Stewartsville, which he had chosen as the site of his new cement plant. But before that happened, he was determined to fulfill his contract with Bethlehem Iron, at the latest low prices. Ogden, his too-heavy phoenix, must try to rise one more time before sinking back for good. *[281]

The Ogden mill under snow, late 1890s.

No matter how little Bethlehem paid, he would not accept bankruptcy. His sense of honor, incomprehensible to Tom and William, demanded that he settle all the mill's debts. He displayed no embarrassment over its failure and for the rest of his life would look back on his mining days with nostalgia. Only once, on a return visit, was he heard to say, "I put three million dollars down that hole in the ground, and never heard it hit bottom."[282]

---

* The plant did reopen for a few months in 1900 but failed to satisfy Bethlehem Iron's low-phosphorus requirement and was forced to dispose of its remaining briquettes below cost. It closed finally for dismantlement at the end of that year.

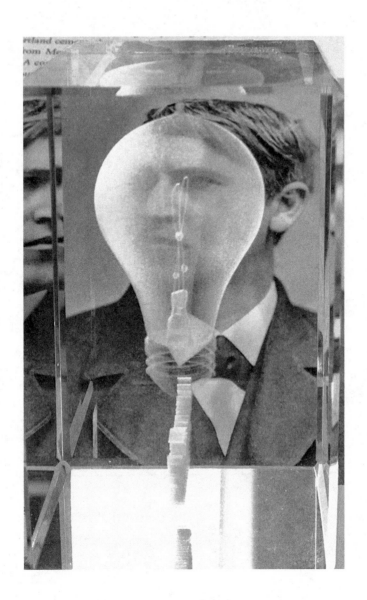

PART FIVE

Light

1880–1889

OVERLEAF: *Thomas Edison refracted.*

I N HIS THIRTY-THIRD year Edison embarked on what he afterward called "the greatest adventure of my life . . . akin to venturing on an uncharted sea."[1] His challenge was to take the small incandescent thing he had just perfected—history's first reliable electric bulb—and turn it into a vast urban illumination system, every part of which would have to be invented, manufactured, and installed by himself.

Since the extraordinary display of linked lamps that he had staged at Menlo Park on New Year's Eve 1879, he had moved in popular esteem from being the "wizard" of recorded sound to the "genius" of electric light. He scoffed at the latter label, which had been overused since the deaths of Beethoven and Goethe. "You know well enough I am nothing of the sort," he remarked to an old associate, Walter Phillips, "unless we accept Disraeli's theory that genius is prolonged patience. I'm patient enough, to be sure."[2]

The trouble with being so endowed was that Edison always made himself available to visitors, few of whom had the delicacy to wonder whether they were not intruding upon him. Overnight, and for as long as the Menlo Park light show lasted, his rural laboratory had become a fashionable destination. He had achieved a spectacular public relations coup in opening its doors to the public.[3] But the financiers committed to support his looming "subdivision of the light"—potentially the most revolutionary invention since the telegraph—were aware that every hour he spent sharing cigars and jovial indiscretions with strangers was an hour to the advantage of rival electrical engineers striving to overtake him both in the Patent Office and the marketplace. The key elements of his high-resistance, coiled-carbon, fused-glass, evacuated lamp were known, and it was critical that he protected them at once, or Edisonian lightbulbs would soon sprout like snowdrops on both sides of the Atlantic.

"If I had my way," an attorney in charge of his patent assignments wrote, "I would not allow half the publicity that has been given by Mr. Edison." Already Menlo Park was having to deal with the consequences of it. The depot, formerly so little used that Pennsylvania Railroad trains sped by without stopping, became mobbed every late afternoon with hundreds of curiosity-seekers. Most wanted simply to gawk at the great inventor. A large number of lighting professionals, including George Westinghouse, Charles Brush, Edward Weston, Elihu Thomson, William Sawyer, and Hiram Maxim, came bent on industrial espionage.[4]

THE LIGHT OF THE FUTURE

Allowing for artistic license, an illustration in the New Year issue of *Harper's Weekly* represented what visitors saw before they walked up the plank road that led to the research complex at the top of the slope. The immediate prospect was dominated by Edison's house set against the sunset, its front yard already darkening. Marion and Tom could be seen playing on a swing, while three adults, a man and two women, strolled close to the picket fence on Thornall Avenue.[5] They were drawn very small, but the man—bandy-legged and waving his hat at some passing horsemen—did not look like the owner of the house. It was most likely Charles Stilwell, an assistant glassblower at the laboratory, who roomed there. In that case, the two female figures were Charles's sisters Alice Stilwell, twenty-seven, and Mary Stilwell Edison, twenty-four, distinguishable by her plumpness.

The hat, the wave, the fence, and the riders, along with a flagpole in front and windmill out back, were part of the iconography of a million Currier & Ives prints, except that some windows of the house glowed with a new kind of light—as did a white post on the opposite sidewalk that seemed to be crowned with a halo tilted out of perspective. The artist clearly did not know what to make of it. It was the first streetlamp in the world to shed incandescent light, and the first of fifty-eight fishbowled bulbs arrayed elsewhere up the slope.

At the moment, they shone down on nothing more than a few hundred square yards of buildings and scrawny fields rising to a wooded

Edison's house in Menlo Park, January 1880.

crest.[6] There were hardly enough houses for Menlo Park to call itself
a hamlet. Only three of them, strung together by droops of bare copper wire, gave off a light similar to Edison's. It was soft, with a pale
orange tinge, and steady, different from the lambency of oil lamps or
the whiter glare of gas.[7] Nor did the various pools of radiance con-
join. Each pole cast little more than a dissolving nimbus. The nearest
thing to brightness came from everyone's destination—the upper floor
of the laboratory, a long, two-story shed in a field of its own, flanked
by a brick office/library, glassblowing house, and machine shop.

To early visitors from New York, the "Village of Light" described
in newspapers was at first a disappointment, its display hardly to be
compared with the dazzle of arc lamps along Broadway. But when
darkness came and the countryside around Menlo Park receded into
blackness—except for a few distant, lantern-lit farm windows—the
miracle that Edison had wrought became more apparent. Here, illuminated in the midst of nowhere by an invisible generation plant,
were globes of glass that gave off no fumes and sooted no ceilings,
that did not ignite anything, that could be held in the hand as they
burned, discharging nothing but a pleasant warmth, that could be
dunked into a bank of snow, or even drowned in water, without going

out. They were silent, flicker-free, each as bright as sixteen candles,* and reportedly flattering to ladies of a certain age. Incredibly, they did not break when dropped from a height of six feet onto the wooden laboratory floor. They could be clustered like flowers or scattered like stars, yet a single switch turned them all on or off. Nor did they mind having their series interrupted. One bulb could be unscrewed from a chain of twenty, and the other nineteen would continue to shine.[8]

Hard for nontechnical visitors to understand was that the tiny, unlookably bright horseshoe inside each unit had originally been paper. Edison's craftsmanlike assistant inventor, Charles Batchelor, had perfected a way of cutting U-shapes out of bristol board and carbonizing them at white heat until they shrank into stiff, shiny-black "filaments"— a word new to electricity, coined by Edison himself.[9] Clamped to platinum wires and incandesced in a vacuum of 1 million atmospheres, they burned without consuming themselves, some for hundreds of hours. How long that was on average, nobody at the laboratory was ready to say.

Souvenir hunters able to reach nine feet high stole every streetlamp they could unscrew, not knowing or caring that Edison needed to keep statistical tabs on each one. Shorter vandals took advantage of the crush in the laboratory to pocket tools and test tubes and carve squares of wood out of the workbenches. "Gumshoe" security men had to be hired to intercept thefts or breakages of precious equipment. A representative of the gas industry was caught trying to short out the room's entire electrical circuit with a jump cable threaded through the sleeves of his coat.[10]

Edison often passed unnoticed in the crowd. Few had yet seen a photograph of him, since newspapers were able to print only engraved portraits that failed to register his animated gaze. Nor did his rough workman's clothes distinguish him from anyone else in the laboratory crew, unless the white silk scarf, knotted under one ear in lieu of a collar, caught attention. He lounged around with a cigar in his mouth, a hank of hair falling over his huge pale forehead, looking like one of the hoboes who periodically wandered up from the depot.

The impression of vagrancy was not altogether deceptive. Over the

---

* The modern equivalent of sixteen candlepower would be around ninety-five watts.

last several months Edison had become so obsessed with illumination technology as to become almost a resident of the laboratory. Mary, despairing of seeing her husband at home, would send dinners up the boardwalk that grew cold in his office. "His assistants say that he would forget to eat his meals or go to bed if he were not reminded of those things," *The Sun* reported.

> The other day, while returning from New York, he alighted from the train at Menlo Park, forgetting that he had his little daughter aboard. As the train was about to start on its way to Philadelphia, the conductor recognized the child. "Are you not Mr. Edison's little girl?" said he. "Yes, sir!" she answered. The conductor led her to the platform. Some distance ahead her father was seen hastening to his laboratory, entirely forgetful of his negligence.[11]

Absent-minded as Edison was (even during night receptions he kept on his enormous floppy hat), he focused on every question from the crowd, cupping his right ear and responding with old-fashioned courtesy. He was repeatedly asked how there could be no air inside each bulb—or as one visitor put it, "How do you extract the vacuum?" In the plainest possible words, leavened with jokes, he explained how Ludwig Böhm, his Bavarian glassblower, puffed each one into shape before plugging it with a separate semimolten base that intruded the filament assembly and two platinum "lead-in" wires. Again and again he demonstrated the operation of mercury pumps, blowpipes, and necking shears. Only when he dropped the Latin term *in vacuo,* or cited the identical coefficients of contraction and expansion obtaining between platinum and glass, did he betray the fact that he was, after all, a professional inventor.*[12]

"He is partially deaf and very modest," the correspondent of *The*

---

* Robert Friedel observes that the crowds attaching themselves to Edison at this time represented "a new relationship between advanced technology and the common man. Edison's electric light was as mystifying and awe-inspiring as any invention of the age. . . . The wizardry of scientific technology was now a source, not of distrust, but rather, of hope. This attitude toward the powers of science and technology was one of the nineteenth century's most important legacies, and no single instance exemplifies it better than the enthusiasm with which the crowds ushered in the new decade at Menlo Park." Friedel and Israel, *Edison's Electric Light,* 89–90.

Times cabled back to Britain. "Yet when he finds that his visitor really sympathizes with him, and is not a 'professional expert' whose object is only to criticize, he warms up into one of the most entertaining men I have ever met. . . . He has not the slightest trace of that self-assertion which is often the accompaniment of success."[13]

What Edison did have was a lively awareness of the value of public relations. He seemed to enjoy chatting to the proletariat as much as to scientists and financiers. "When T. A. hain't got his thinking cap on," a local farmer remarked, "he's just as jolly as a schoolboy." Visitation reached three thousand a day after the Pennsylvania Railroad laid on extra trains east and west. Even Edison saw that the laboratory should revert to privacy, lest it literally split its sides. His father had carpentered it well, but by the second week of January the walls needed buttressing with telegraph poles.[14]

Thereafter only bona fide scientists or officials connected with the Edison Electric Light Company were allowed to disturb him as he assembled the resources—human, mechanical, intellectual, and financial—required to consummate his urban power plan. His old friend and counsel Grosvenor P. Lowrey was one of the few who understood the immensity of the task ahead, and in moments of dread, worsened by rheumatism, he doubted that he could do it. But the flawless performance of the lamps during their recent exhibition comforted him. "I am writing this by the light of the future at Edison's table," Lowrey scribbled to a friend one evening when the laboratory was quiet. "The same light which was burning when I began, is sound still . . . having burned three hundred hours before. Should it sustain another test, the economy and durability of the lamp is demonstrated."[15]

THE GREAT DISCOVERER

Lowrey's optimism was echoed overseas by no less a voice of authority than *The Times* of London. It praised Edison's achievement in a column and a half of dense print that could have been written by a publicist for the Electric Light Company.

Mr. Edison is determined to maintain his place as the great discoverer of the age. After a silence of some months he has again come before the world as the inventor of a system of electric lighting which he claims to have made complete at every point. . . . The new light, as Mr. Edison's machines will furnish it, admits of being employed for every purpose of public and of private use. It is as manageable as gas has been. It can be raised to an intense brilliancy beyond any that gas can reach, or it can be turned down to a thread. There is no difficulty in laying it on. A thin wire in connexion with the generating machine is all that is necessary for this. It is light without heat. However carelessly it may be used, there can be no danger of fire from it. . . . It gives almost exactly the tone of ordinary daylight. With all these advantages it is, moreover, the very cheapest light that has ever been produced.[16]

Members of the British electric engineering establishment rose as one to protest the idea that such things could have been achieved by an unschooled American huckster who did not have the decency to wear a beard. The electrochemist Joseph Swan was first to declare his own preeminence in the field. "Fifteen years ago," he announced in *Nature* magazine, "I used charred paper and card in the construction of an electric lamp on the incandescent principle." Swan was honest enough to admit, "I did not then succeed in obtaining the durability which I was in search of."* He did not explain why he had never filed a precautionary caveat, or provisional specification,† to protect his light, while continuing to experiment with it on and off, beyond the ken of the English trade press.[17]

That lack of notice was put to right by periodicals such as *The Electrician* and *Chemical News,* not to mention *Saturday Review,* which on 10 January published an unsigned polemic entitled "The Great Edison Scare." Some of its blows were shrewd:

What a happy man Mr. Edison must be! Three times within the short space of eighteen months he has had the glory of finally and triumphantly solving a problem of worldwide interest. It is true that each time the problem has been the same, and that it comes up

---

* Later Swan backdated his invention another five years, to 1860, and after his death, his children added a further five, to 1855.

† A *caveat* in nineteenth-century American parlance, or *provisional specification* in British, protected the main elements of an invention in advance of a more formal, detailed finding.

again after each solution, fresh, smiling, and unsolved, ready to receive its next death-blow. . . . His friends may look forward to a
long and equally happy future, crowned at periodical intervals by
similar dazzling and final triumphs; for, if he continues to observe
the same strict economy of practical results which has hitherto
characterized his efforts in electric lighting, there is no reason why
he should not for the next twenty years completely solve the problem of the electric light twice a year without in any way interfering
with its interest or novelty.[18]

It would be ten months before Swan produced a filamentary lamp
of his own, and it proved to be virtually identical to those on display
at Menlo Park. This did not stop John Tyndall, the revered professor
of natural philosophy at the Royal Institution, from endorsing Swan's
retroactive claim to have developed such a lamp in the 1860s. Hence
Edison's was "cursed by a total absence of originality." But both inventors were pursuing a chimera, in Tyndall's opinion. Eighty years of
research had proved that "the most economical form of electric light
is, and in all probability always will be, the arc lamp."[19]

It was true that Edison and Swan were just the latest in a historic line
of electric-illumination pioneers going back to Sir Humphrey Davy,
who in the early years of the century had coaxed both arc and incandescent light out of a mass linkage of voltaic cells. Since then scores of
inventors had tried to turn either of these incompatible radiances—the
one harsh, flaring, and consumptive, the other weak and maddeningly
variable—into a light that could be relied on and be competitive with
gas. As early as 1840 William Grove had made a spiral of platinum
wire incandesce for a moment or two inside a glass tumbler.[20] The international arc lamp fraternity, notably Pavel Yablochkov of Russia,
had been successful enough in illuminating some great public places
with their sputtering "candles," but that kind of light would never be
accepted by ordinary people at home and at work around the world.

The question opening up now, of whether Swan had indeed preceded Edison in finally achieving a viable incandescent lamp, was not
going to be settled by Tyndall's say-so, or for that matter by the only
lamp Swan could so far hold up in evidence—a carbon rod enclosed in
a flask that looked like a urological appliance. He had demonstrated it

in early 1879, several months after Edison's premature achievement of a platinum-spiral lamp, but long before the first little carbonized "horseshoes" began to shine in globes around Menlo Park.

By all reports, Swan's rod gave off plenty of light for a minute or two, and also plenty of soot, which indicated either imperfect vacuum or too much exposed carbon. Yet there was an ominous indication in this month's United Kingdom caveat listings that Swan understood filamentary technology as well as Edison. He gave notice that he intended to seek patent protection for a process of improving bulb evacuation by the use of heat to expel occluded gases out of the incandescent element, whether "rod, filament, or lamina."[21]

Before the end of January, British official opinion had solidified against concession of any priority to Edison.* The editors of *Nature* dismissed his lamp as "a hopeless failure, wrong in design, wrong in principle, useful only in showing how singularly devoid of sound scientific knowledge a clever practical man might be."[22]

"In short my Boy you are not loved over here by these fellows," Edward Johnson wrote him from London. "You have committed the grave error of having succeeded."[23]

The same went for Paris, except that the precedents cited there by Théodose du Moncel, the ranking French authority on illumination technology, were mostly the work of his compatriots. He informed readers of *Le Temps* that Edison was nothing more than "a very ingenious and fecund inventor" who could not claim to be "au courant with the subtleties of electrical science." It was impossible to believe that his new lamp, "this horseshoe of carbon, so fine spun, so delicate," did not degrade when incandescing. Du Moncel advised European commentators to withhold their praise until Edison confirmed that he had lit fifty or sixty bulbs from a single generator. For himself, he would continue to be wary of "the pompous announcements that come our way from the New World."[24]

---

* The consensus endures in modern Britain. See, e.g., the timeline under "Joseph Swan" in *Grace's Guide to British Industrial History,* https://www.gracesguide.co.uk. It states that Swan "obtained a UK patent covering a partial vacuum, carbon filament incandescent lamp" in 1860, whereas he did not even apply for such protection until 27 November 1880 (UK 4933 of 1880), more than nine months after Edison's anticipatory British patent (UK 4576 of 1880) was approved.

Edison was philosophical enough about foreign anti-Americanism to shrug such comments off. His patience was more strained by the abuse of compatriots who claimed to know more about his invention than he did. In a widely published letter, Henry Morton, president of the Stevens Institute of Technology, called his light "a conspicuous failure" and sure to remain so, because as "everyone acquainted with the subject" realized, stringing together a large number of bulbs "involves an immense loss of efficiency." Tests of the "identical" carbons of such electrical pioneers as Siemens, Weston, Brush, and Maxim had proved over and over that incandescence was a short-lived phenomenon.[25]

Edison read Morton's letter beneath the glow of eighty-four lamps suspended from the ceiling of his laboratory. "He should investigate first and animadvert afterward," he said to a reporter watching him.[26]

ALL THE OPERATIONS CONNECTED THEREWITH

On 27 and 28 January Edison alternately received and executed the two most historic patents of his career: U.S. 223,898, "Electric-Lamp," and 369,280, "System of Electrical Distribution." The former guaranteed seventeen years of protection to his basic bulb—or should have, if eleven of those years were not to be spent fighting jealous lawsuits.*[27] The preamble to the latter application showed that the main components of his plan to illuminate lower Manhattan were already integrated in his head, if nowhere else:

> To all whom it may concern:
> Be it known that I, THOMAS A. EDISON, of Menlo Park, in the State of New Jersey, United States of America, have made certain new and useful Improvements in Furnishing Light and Power from Electricity. . . .
> The object of this invention is to so arrange a system for the generation, supply and consumption of either light, or power, or both of electricity, that all the operations connected therewith requiring special care, attention, or knowledge of the art, shall be

---

* Edison's parallel British patent, UK 4576 of 1880, was approved ten days later. See Part Six for the final approval of U.S. 223,898, arguably the most important of Edison's 424 lighting and power patents.

performed for many consumers at central stations, leaving the consumer only the work of turning off or on the supply as may be desired.[28]

The four thousand–plus words and set of complex diagrams that followed lent weight to Lowrey's concern that no human being would be able to design and build such a system before the Edison Electric Light Company's capital (currently standing at $300,000) ran out. Edison on the contrary regarded his scheme as so much a fait accompli that he labeled the streets in one drawing "Cortlandt," "Broadway," and "Maiden Lane." Another drawing, of four identical power "districts" arranged in a grid, each with its own "central station" and interconnecting "conductors," not only anticipated the look of electronic circuitry of a century thence but made clear that he dreamed of lighting whole cities.[29]

"I will here state," Edison wrote, "that all the devices for translation of electricity into work are arranged on the mutiple arc system, each device being in its own derived circuit, the effect being in substance to give each a circuit from the generating source independent of the circuit of all the other devices."[30]

The last word denoted every machine and immobile link that interposed between raw coal stored in the central station and light, or power, pouring out the other end. Young Francis Upton, in a magazine article timed to coincide with his boss's application, extended the arc of translation even further, from sunshine to artificial sunshine, but Edison concentrated on the specifics of getting the job done.[31]

He began with the prime motors, steam engines whose belts and shafts caused a group of generators to whir up a mass of electromagnetic energy, or "field of force," that could be duplicated to any number the market called for. Having raised his favorite subject of electromagnetism, Edison treated himself to a somewhat rambling disquisition on the "extremely long" magnetic core of the bipolar dynamo he had patented the year before, to power his lightbulb experiments. He would have to rewrite this section of the application extensively to meet Patent Office objections, but when he did, he would make the important claim that "currents of the desired high electromagnetic force can be generated in armatures of low resistance, and the waste of energy in the form of heat in such armatures will be reduced to a minimum."[32]

Next came the copper cable conductors, or "mains," along which the current flowed at a pressure controlled by regulators that sensed the fluctuating demands of customers turning their lights on and off. This maintained the high resistance, at farthest remove, of Edison's uniquely efficient lightbulbs. He explained that a common differential of low resistance was what made the lamps of other inventors uneconomical. His use of multiple arcs meant that he could wire in any number of circuits without appreciably weakening the output of the generators. To ensure uniformity of pressure, he envisaged photometric test lights at the central station, along with galvanometers at any desired point, so that any drop or surge would be indicated by a change of light or deflection of a needle.[33]

"For distributing the current thus generated and regulated," Edison went on, "I prefer to use conductors within insulated pipes or tubing made water tight and buried beneath the earth, provision being made at suitable intervals for house, or side connections." He had seen the crazy cross-hatch of telephone and telegraph wires that shadowed some streets in downtown New York, and he did not intend to tangle with it.*[34] If running conduits under the city's sidewalks was going to cost the Electric Light Company vast amounts of money and labor—not to mention permission from city officials, and all the plain brown envelopes that entailed, then the investment must be budgeted for. The gas industry had installed its own piping decades before and gone on to profit enormously.†

Insulating pneumatic pipes, however, was a less fraught task than ensuring that no water or rodent teeth reached the copper in electrical conduits. Reliable seals were most important at nodes where branch

---

* "No bird could fly through their network," Edmund C. Stedman wrote of downtown New York's tangle of wires in the 1880s. "A man could almost walk upon them . . . they darkened the street and the windows below their level."

† Edison had it in mind to make as much use as he could of the existing gas fixtures in every building he electrified, even down to adapting the mantles of chandeliers. It would save him money and at the same time enrage his competition—both agreeable prospects.

wires ran up lampposts, or horizontally down side streets in subsidiary mains, branching out yet farther into every house or business establishment willing to subscribe to this newfangled system. That would require the emplacement at entry points of tamper-proof meters. Then the buildings themselves would have to be wired with derived circuits that fed light or power to as many switchable bulbs or live outlets as the customer wanted to install.[35]

Edison of course already saw his patented globe as the crowning flower of this gigantic electric tree. But he felt free to continue developing it while he gave notice there were components he had yet to invent, such as safety fuses and centrifugal governors to control the running of motors as demand rose or fell. In the meantime he claimed to have listed thirteen wholly original contributions to electromotive science, and he sought protection for them in the plainest of words: "A system arranged as thus provides for all the conditions precedent to an economical and reliable utilization of electricity as a lighting or motive power agent."[36]

For the rest of the year he isolated himself at Menlo Park, turning his laboratory and its adjacent lots, sidewalks, and fields into a roughly one-third-size model of his projected "First" lighting district in New York City.[37] He intended to generate enough power on the spot to illumine eight hundred lamps. Pennsylvania Railroad trains reverted to their customary policy of not stopping at the depot unless by advance reservation, but this did not lessen the public's fascination with Menlo Park. At night, passengers traveling to or from New York crowded to look for it when the call came, "There's Edison's light!" Out of the darkness ahead a few bright pinpricks emerged, swelled, and whizzed by in a momentary splatter that soon gave way to darkness again.[38]

INERT BODIES

If Edison had been remarkable through his twenties for industriousness and executive will, he now became freakish in both respects—to his employees, an Übermensch; to his financial backers, an uncontrollable fantasist, half-genius, half-fool; to rivals, a publicity whore of no especial originality; to his wife and children, increasingly a stranger;

to Patent Office examiners, a tireless nuisance, filing sixty applications in 1880 alone.[*][39]

The scope of his project dwarfed anything in the history of electrical engineering to date and was, besides, so new in most of its parts that he could not think of embarking on it without the recruitment of an expanded and intellectually upgraded team of helpers. With the exception of old James Mackenzie, who had taught him telegraphy as a boy, and "Pop" Edison, who came and went with the unpredictability of a septuagenarian Huckleberry Finn, they were all young. Their number rose from sixty-four in the spring to about seventy-five in the fall.[40] Edison had his pick of job applicants and paid them little or nothing to start, on the grounds that those with talent would soon earn their worth, and those lacking it, or requiring a normal amount of sleep, would drop off by natural selection. Consequently he rarely had to fire a man.

Charles Batchelor, his impassive, black-bearded deputy of the past nine years, remained indispensable—faithful, meticulous, dull, a cool English breeze whenever Edison blew too hot. Francis Upton combined the manners of Phillips Academy and Princeton with a mastery of mathematics and scientific theory.[†] Edison nicknamed him "Culture," and John Lawson, an argumentative assayer who insisted that basic oxides required special heat treatment, as "Basic" Lawson. Martin Force, the laboratory handyman, inevitably became "Fartin' Morse." The cerebral Charles L. Clarke, who had a master's degree in science from Bowdoin College, was hired at twelve dollars a week for electrical systems analysis, but proved to be more valuable as a draftsman, his sketches as precise as steel engravings. William J. Hammer at twenty-two was already a gifted electrical engineer, close-cropped, military-neat, supercilious to anyone junior to himself. These included

---

[*] From 1880 through 1883 Edison would file for 321 patents—more than at any other time in his career and more rapidly than any inventor before or since. This total does not include seventy-eight other applications that he alleged were stolen from him and sold by a corrupt patent attorney, Zenas Wilber.

[†] Upton was the victim of a typical Edisonian tease. Asked to calculate the volume of a pear-shaped bulb, he spent several brain-cracking hours numerically integrating its curves in three dimensions. Before he finished his *quod erat demonstrandum,* Edison reappeared and asked if it would not be simpler to fill the bulb with mercury and weigh the contents. Dyer and Martin, *Edison,* 277.

Members of the Menlo Park laboratory team, 22 February 1880. From left, Ludwig Boehm, Charles Clarke, Charles Batchelor, William Carman, Samuel Mott, George Dean, Edison (in skullcap), Charles Hughes, George Hill, George Carman, Francis Jehl, John Lawson, Charles Flammer, Charles Mott, James MacKenzie. (Library of Congress.)

the teenage office boys "Johnny" Randolph and George Hill, as well as neighborhood urchins who hung around the lab hoping to steal cigars or explosive chemicals. Stockton "Griff" Griffin acted as Edison's private secretary, a job Randolph would one day inherit. Francis Jehl, nineteen, was passionately interested in electricity, but had such bovine strength that the general manager, William Carman, made him responsible for keeping the vacuum pumps topped up with mercury, which was much heavier than lead. Wilson Howell was a bespectacled youth eager to do odd jobs without pay, as were several other aspiring lab workers hopeful that Edison would eventually take them on.[41]

A large Germanic quotient affected Menlo Park's habits of delegated procedure and fanatical record-keeping. Johann ("Honest John") Kruesi, the master machinist, was *Schweizerdeutsch*; Ludwig Böhm, Edison's leather-lunged glassblower, his assistant blower William Holzer, and the chemists Otto Moses and Dr. Alfred Haid were all German-born; John and Frederick Ott and Francis Jehl had grown up speaking German at home; and even Upton, Yankee to his finger-tips, had spent a postgraduate year at Berlin University studying under Hermann von Helmholtz.[42]

Despite their common work ethic, everyone had to adapt to Edison's decidedly un-Teutonic attitude toward the clock. His day was punctuated only by breakfast at seven and midnight "lunch," and he was capable of forgetting about each. When tiredness overwhelmed him, usually around four A.M., he would curl up beneath the stairs like a tramp and sleep on a pile of old newspapers. As a result it was not uncommon to see inert bodies at various points in the building and at various times of day or night, while experimental activity went on busily around them.[43]

MYSTERIOUS BLUE HALO

Allowing for national and other prejudices, Edison's doubters were correct in saying that he had not yet developed a perfect lightbulb. The paper-derived filament was brittle and hard to install, breaking sometimes even in Batchelor's nimble hands. It glowed beautifully when seated on its carbon clamps, but the glass crown beneath had a tendency to crack around the lead-in wires, causing a loss of vacuum and consequent oxidization of the carbon.[44]

These were thermal problems that Edison was confident of solving. He was mystified, however, by the tendency of his lamps to darken inside after a week or two of life. It was as if an invisible soot—"carbon vapor," he called it—were being given off by the filament that became apparent only as the molecules thickened on the glass close by, clouding it at first, then blackening it. Yet soot was the product of flame, and there could be no fire in his airless bulb, only incandescence. Nor was the blackening uniform. From a certain angle it seemed to show a negative shadow of the filament, most noticeable on the positive side of the horseshoe. William Hammer related it to a blue fluorescence that appeared around the clamps and was weirdly responsive to magnetism, draggable from one pole to the other. Edison thought the blueness was gas given off by the clamps, but when he substituted copper ones, the same "halo" wavered about them. He painted a filament with zirconic oxide, and the blueness deepened to violet. Fascinated, he inserted a wire between the poles and ran it out to a terminal via a galvanometer. The needle at once showed that there was an arc of subsidiary current linking them. This did not, however, explain the "carrying by electrification of the carbon from one side of the carbon horseshoe," a phenomenon that was reversed when the current was reversed. Perhaps heat loosened their cohesion within the baked body of the filament itself, and caused them to migrate to the cooler surface nearby. "The amount of such carrying," Edison wrote, "depends upon the resistance of the filaments, the degree of incandescence, the electromotive force between the clamping-electrodes, and the state of the vacuum."[45]

He did not understand, even though his language repeatedly edged toward it,* that he was on the verge of discovering electronics.[46] The theory of the electron, or charged subatomic particle, would not be propounded by J. J. Thomson for another seventeen years.† What now became known, half mockingly, as the "Edison Effect" was thermal electron emission.‡ It was a nuisance as far as lighting was concerned, yet novel enough for him to ask the Princeton astronomer Charles A. Young to make "an examination of the mysterious blue halo by spectroscope." The results were inconclusive. Edison continued to experiment with wireless molecular transfer for three years and eventually patented the phenomenon, with a view to "the utilization of this discovery for indicating or regulating electromotive force." But he never realized its world-changing potential as radio.§[47]

---

* Edison also used the phrase *molecular bombardment.*

† In 1972 the molecular biologist Gunther Stent famously raised the question of "prematurity" in science—discoveries or theories too much in advance of contemporary knowledge to be explored seriously until years later.

‡ As now exemplified in the vacuum tube diode.

§ Edison's U.S. Patent 307,031 was the first ever granted in America for an electronic device. It features a drawing and description of a two-element vacuum lamp that in essential details anticipates the diode "invented" by John Ambrose Fleming in 1904. Fleming had previously used the Edison Effect to rectify radio waves. Two decades later Edison was annoyed to read Fleming's claim in his autobiography *Fifty Years of Electricity* to have been the first to realize "that a carbon filament incandescent lamp with a plate sealed into the bulb could be used to rectify high-frequency alternating currents." Edison scribbled in an angry marginal note, "Absolutely untrue & he knows it is untrue."

A SIMILAR CONSTELLATION

By March, Edison had 220 lamps burning night and day around Menlo Park. He invited Charles Young and another Princeton physicist, Cyrus F. Brackett, to visit the laboratory and make an independent assessment of his generation system. The result was a report, published in the June issue of *American Journal of Science,* so astonishing that the academic community as a whole refused to believe it. Brackett and Young found that Edison's bipolar dynamo had a total efficiency rating (electrical output proportionate to mechanical input) of 89.9 percent. Even if it consumed four points of that output internally, it still made a mockery of the theoretical maximum potential/industrial average of 70 percent.[48]

Two further physicists, George F. Barker of the University of Pennsylvania and Henry Rowland of Johns Hopkins, reported almost as favorably in the same periodical on the thermal efficiency of the Edison bulb. "Provided the lamp can be made either cheap enough or durable enough, there is no reasonable doubt of the practical success of the light." Again, this praise was widely dismissed, instilling contempt in Edison's bosom for the pure-science fraternity that time would increase.[49]

He could have lit hundreds more lamps after adding a two-ton dynamo to the smaller units already in the machine shop, were it not for the hours it took to manufacture every bulb by hand. The glass had to be blown, the filament baked and mounted and wired, the air pumped out under blowtorch heat, the evacuation point sealed and cooled, then the whole tested, not always with success. Some perfect-looking specimens just would not light, or did so dimly, or flared and burned out due to reventilation. Microscopic cracks appeared not only at base level but at the "two o'clock spot" on the round of the globe, for a reason nobody could figure. On average, however, the bulbs shone for 686 hours, with a quality consistent enough that Edison was emboldened to accept a commercial order that required the opposite of street lighting.[50]

It came from the railroad tycoon Henry Villard, who was building a steamship, the *Columbia,* for service on the Pacific coast. Villard had attended Edison's "Village of Light" exhibition and wanted to float a similar constellation out to sea—specifically, to sail around the Horn on his ship's maiden voyage to San Francisco in early May. Although this fantasy disrupted Edison's plans for a lighting district in New York, it was irresistible from many angles, not least that of publicity. The shortness of the deadline helped weld the Menlo Park team together as a productive unit, while the *Columbia*'s compact hull—332 feet from bow to stern, with a beam of 38½ feet—enabled him to integrate at close quarters all the elements that would one day comprise his city district. Villard called for 120 lights in "all-glass chambers," one in each first-class stateroom and chandeliers in every saloon.*[51] He provided enough rear hull space for four 110-volt dynamos, three of them belted to a countershaft driven by vertical engines and connected in parallel to the light circuits. A switchboard in the engine room sent power throughout the ship via stranded cables insulated with soft rubber tubing. The strands (cotton-covered and painted red or white to indicate polarity) radiated in seven independent feeder circuits, each of which subdivided again to feed lamps distributed among the upper and lower decks. For extra safety Edison invented an array of tripping devices, with fusible wires in each circuit and single-pole breakers fixed to the saloon lamps in tiny glass tubes, so that in the event of a power surge, no drops of molten lead alloy would fall on anybody's dinner jacket. Also new to his illumination technology were the keyed sockets and brackets that held each lamp, and the ceiling fixtures that allowed chandeliers to sway gently at sea. Switches were locked inside rosewood boxes sunk into the ship's paneling, accessible only by stewards.[52]

Despite these protective devices, Villard's shipbuilder was so afraid of electrical fire that he refused to have anything to do with the system. Edison and Batchelor therefore gained the useful experience of supervising the installation themselves. By the time Upton arrived with the bulbs, borne in an immense basket and individually wrapped like fresh eggs, they had in effect created their first "isolated" lighting plant.[53]

The *Columbia* itself became the world's first all-electric ship. It lay fully lit at its pier in the East River on the evening of 27 April, when

---

* Edison was not asked to provide lights for navigation, that function being performed by the much more powerful arc lamps of Hiram Maxim.

Edison escorted his wife aboard for a celebratory reception. A prom-
enade band on deck serenaded several hundred of Villard's elegant
friends as they danced below in the ballroom and went forward for
supper, bathed all the time in soft incandescent light. The occasion was
a rare treat for Mary, who did not have much opportunity to show off
her fine wardrobe at Menlo Park. And it was yet another public relations coup for Edison. His fixtures attracted more admiration than any
other of the ship's lavish appointments. The *Columbia* set sail for the
Horn ten days later, streaming its long lines of glowing portholes past
New Jersey and Delaware until the horizon blotted them out.[54]

A MOST BEAUTIFUL ACCIDENT

Without doubt, the most bored worker at Menlo Park that spring was
"a red-haired, freckle-faced Irish boy with a face like a hop-toad"
who was seen sitting all day out of doors, dipping cordwood rail ties
into a barrel of boiling asphalt. He was rendering them nonconductive for an experimental electric railway Edison was building north of
the laboratory. It was another transportation project financed by
Henry Villard, and was gladly undertaken by Edison as an opportunity to study the laws of motor mechanics and load balancing—both
of which were important aspects of his urban lighting plan.[55]

The track (upon which two other urchins nearly fried themselves
when they stood on opposite rails and shook hands) ran uphill across
open country to a wooded ridge for about a third of a mile, then curved
west for another third before looping back toward the laboratory.[56] If
its trajectory was slyly patterned on that of a gigantic filament, the
resemblance was lost on Villard, who had been born Heinrich Hilgard
in Speyer, Bavaria, and was not known for his sense of humor.

Edison excited the railroad circuit with two of the same generators
he had installed on the *Columbia*. He assigned one of his top engineers,
Charles T. Hughes, to turn a third into a locomotive by bolting it flat
onto an iron truck just big enough for two men to sit behind. The driver
controlled the traction of two massive fore wheels with a long friction-
gear lever that came in handy as a vaulting pole at times of imminent
derailment. One wheel drew power from its rail and transmitted it via
a brass hub and brush to the spinning bobbin, or armature, of the
motor, while the current rushed on and out the other wheel. Edison's

engine made its first trial run on 18 May and proved powerful enough to pull two cars carrying twelve to fourteen passengers, or the equivalent weight of freight.[57] As such, it was a partial realization of a vision he had had four years before in the Midwest, of driverless electric trains loaded with corn crisscrossing the plains on wheels that would "grasp the track like an iron hand," deriving their power from wind dynamos.

With refinements such as an electric headlight, signal bell, and fringe-topped observation car, the train became a popular tourist draw, although on hot days the odor of armature wafting back from the engine, mixed with that of the tar-soaked sleepers, could offend delicate nostrils. Mary Edison waited for a cool evening before she took some of her friends on what was probably history's first electrically illuminated railway excursion.[58]

By early June the "Edison Express" was attaining speeds of forty miles an hour, enough to whiten what was left of Grosvenor Lowrey's hair. "We ran off the track," he reported to his fiancée after a day on the railway that threatened to be his last.

> I protested at the speed on the sharp curves (designed to show the power of the engine) but E. said they had done it often & finally when the last trip was taken I said I did not like it, but would go as long as Edison did. The train jumped the track on a short curve, throwing Crucy [Kruesi] who was driving the engine, with his face prone in the dirt and another man in a comical somersault through some underbrush. Edison was off in a minute, jumping and laughing & declaring it a most beautiful accident. Crucy got up, his face bleeding & a good deal shaken; & I shall never forget the expression of voice & face in which he said (with some foreign accent) "Oh yes, pairfeckly safe!"[59]

Edison applied for a patent on various aspects of his railway, but claimed no overall priority on the system.* He emphasized to reporters that Werner von Siemens had invented and operated an electric train in Berlin the year before. When news broke in July that an American en-

---

* Edison's 1880 locomotive and cars can be seen at the Henry Ford Museum in Dearborn, Michigan.

The Edison electric train, Charles Batchelor driving.

gineer, Stephen Dudley Field, had been awarded letters patent for a locomotive looking remarkably like his, he reacted with jovial unconcern.
Field's claim rested solely on the novelty of a trailing arm that took
current from a conductor running between or to one side of the tracks.
"It is a curious thing how vague the ideas of the general public are on
the question of patents. . . . A man . . . draws an entire machine with
his 'improvement' in it, and people think he has invented it all."[60]

The good humor and objectivity of remarks like these, exemplify-
ing Edison's absolute refusal to be discouraged in any endeavor (even
when the Patent Office declared his application an "interference"
with Field's), came as a tonic to pessimists like Lowrey, who worried
that he was playing with ships and trains when he should be devoting
all his energies to the light. Portly, bug-eyed, fiftyish, and bruisingly
widowed, the little lawyer had known and loved Edison since 1869.[61]
He had always felt responsible for protecting his client from the push
and pull of too many ideas fighting for precedence at any given time.
Now the time was especially critical. As corporate counsel for the
Edison Electric Light Company, Lowrey knew that its directors were
concerned by the accelerating rate of Edison's laboratory expenses, in

contrast to what appeared to be halting progress toward his announcement of a central station in New York.

The appearance was true. Week by week Edison was confronted by
a proliferation of development problems that would have caused any
project manager less positive to see failure looming ahead, like the
still-unsolved blackening of his bulbs. One day Lowrey, confessedly
"ultramarine" with depression over the Electric Light Company's finances, came out to Menlo Park to be cheered by his client, and was
not disappointed: "An hour with Edison has restored [my] spirits. . . .
Perhaps I'd better marry him, since he cures me."[62]

GOD ALMIGHTY'S WORKSHOP

Mindful of his own remark on the tendency of "people" to intuit the
whole from the particular, Edison waved aside a growing number of
press suggestions that his municipal lighting scheme for New York
was a chimera. He ascribed them to lobbying by the gas industry. "I
am superseding a system of artificial lighting in which is invested
about $850 million," he said to a representative of *The Boston Globe*.
"This cannot be done in a day."[63]

The reporter, more objective than Lowrey, scrutinized him as he
talked, and got a distinct impression of monkishness.

> He resembled a young man who had spent several years of proba-
> tion in the novitiate of a Roman Catholic religious society. He had
> a tired appearance; his face was almost expressionless and his gen-
> eral ensemble made a suggestion of close confinement within doors
> and unceasing application and thought. . . . His eye is brilliant,
> emitting a sort of electric light that bespeaks keen penetration and
> rapidity of perception. It illuminates his whole face, which is other-
> wise passive. . . . His sandy hair is streaked with gray.[64]

That summer, while Francis Upton calculated the market mathematics of electrifying lower Manhattan, and Kruesi dug up Menlo
Park's red clay to bury an experimental conduit system, Edison and
Batchelor absorbed themselves in filamentary experiments. The lamp
factory was due to begin operations in the fall, with a projected annual output of half a million units, and they had to have the basic

bulb standardized by then. There had been enough failures among lamps tested in the laboratory to recall what du Moncel had said about the atrophy of incandescent elements. Edison distrusted the mealy texture of his bristol board carbons: "Paper is man made and not good for filaments." No matter how hard and shiny the little loops baked, they could not be relied on for equable distribution of heat under electrification.[65]

For week after week the two men cut, planed, and carbonized filaments from every fibrous substance they could get—hickory, holly, maple, and rosewood splints; sassafras pith; monkey bast; ginger root; pomegranate peel; fragrant strips of eucalyptus and cinnamon bark; milkweed; palm fronds; spruce; tarred cotton; baywood; cedar; flax; coconut coir; jute boiled in maple syrup; manila hemp twined and papered and soaked in olive oil. Edison rejected more than six thousand specimens of varying integrity, as they all warped or split: "Somewhere in God Almighty's workshop there is a vegetable growth with geometrically powerful fibers suitable to our use." *[66]

In the dog days, as heat beat down on straw hats and rattan parasols, the idea of bamboo suggested itself to him. Nothing in nature grew straighter and stronger than this pipelike grass, so easy to slice from the culm and to bend, with its silicous epidermis taking the strain of internal compression. It had the additional virtue, ideal for his purpose, of being highly resistant to the voltaic force. When he carbonized a few loops sliced off the outside edge of a fan, they registered 188 ohms cold, and one glowed as bright as 44 candles in vacuo. That particular specimen, being cheap Calcutta bamboo, blued at the clamps and went out after an hour or so. Splints from the Far East proved to be of much finer grain, and carbonized so well that they could take a white heat that melted the platinum clamps they stood on. Böhm blew a new pear-shaped bulb to accommodate their typical bend. In a decisive experiment on 2 August, some Japanese samples lasted nearly three and a half hours at the dazzling incandescence of 71 candles—well over four times as much light as was needed for commercial purposes. Another, reduced to

---

* He even tried such eccentric materials as myrrh gum, macaroni, asphalt, fishing line, cork, and banknote paper.

the comfortable glow of sixteen candles on a current of 110 volts, burned for an astonishing 1,589 hours. On the evening it registered that record, William Hammer ran up the laboratory steps bulb in hand, like an eager knight brandishing the Holy Grail, to share the news with Edison, Batchelor, and Upton. An impromptu conga line developed behind the four men as they danced in serpentine fashion around the workbenches, then downstairs and out into the night, singing and cheering.[67]

From that day on, the words *bamboo* and *filament* were synonymous in the shop talk of Menlo Park.

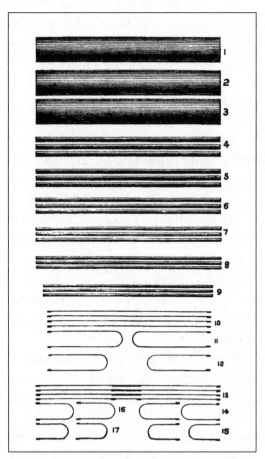

Stages of splitting and shearing a splint of madake bamboo into filaments ready for carbonization.

LONG RIPPLES

John Kruesi was the most gifted mechanic in Edison's employ, Swiss-trained in geometrics and physics, equally adept at precision machining (he had built the prototype phonograph) and the hard labor of laying out the world's first underground electrical distribution system. His long arms and slope-shouldered stoop seemed to incline him naturally toward any manual task that lay within reach. He was so objective in addressing technological problems that he had to be kept away from investors. Edison tried without success to make him understand the difference between truth and "deferred truth."[68]

Nevertheless, Kruesi had intuition enough to render his boss's sketchiest diagrams into logically functioning models. The most inspired of these was a feeder-and-main principle of distribution that at one stroke solved a problem the mathematicians had been struggling with all year: how to conduct electricity through block after city block without using enormous amounts of copper. At an estimated eight hundred thousand pounds for just nine blocks, costing in excess of $700,000, the metal could have made it impossible for Edison to undercut the price of gaslight as much as he needed to, if customers were to allow him to snake wire into their premises.[69]

Kruesi's invention, just as vital as that of the lamp itself, replaced the "tree" system he had originally planned. That had essentially been a trunk of copper emerging from the central station and thinning into branches and stems that then "translated," to use his own word, into leaves of platinum and carbon. The massiveness of the trunk was necessary to convey as much electrical sap as possible to the top of the tree. Even so, Upton had warned of a 30 percent drop in power there, because of resistance along the way.

"The object of this invention is to obviate such danger and to maintain practically throughout the entire system an equal pressure," Edison wrote, in the first of two patent applications for his feeder-and-main concept. He drew a blocky square labeled "CS" for "central station" and surrounded it with four square borderlines that expanded symmetrically, as did the lots, blocks, and districts of a typical American city. Each side of the CS square radiated a pair of lines that fed north, south, east, and west, into the resultant grid.

Long ripples showed where each feeder ran to its destination main. They graphically, if unintentionally, conveyed the flow of current around the whole, in an exquisite distribution of balanced forces. Its practical effect at 110 volts* was to reduce copper cost by seven-eighths, and almost completely absorb the energy loss to be expected of lamps farthest from the central station, with no visible dimming of candlepower anywhere.†[70]

When Sir William Thomson, Britain's most eminent electrical scientist, was asked why no one else had dreamed up a system so simple, yet so efficient, he replied, "The only answer I can think of is that no one else is Edison."[71]

In physical reality, the system was more complex than it looked on paper. Edison's second application—one of seventy-seven that he executed for the distribution system alone—featured zoom-in diagrams and explanatory paragraphs not likely to win him speedy approval from the Patent Office: "In Fig. 6 is shown direct or main feeding circuits 1 2 and 5 6 with lamp-circuits 3 4 and 9 10 with branch feeders 7 8, 15 16, and 21 22 leading into side streets, supplying lamp-circuits 17 18, 19 20, 23 24, and 25 26, the branch feeders being derived circuits from the main feeders." But his overall claim of providing a consumption circuit that spread drops in voltage so widely that the candlepower charge of individual lamps remained, to the naked eye, imperceptible, was so strong that letters patent were almost immediately granted him in Canada, Italy, Belgium, France, Austria, Australia, New Zealand, Spain, and India. His American patent would not arrive until after the "lamp-circuits" he designed were glowing five thousandfold around the First District in Manhattan.[72]

Kruesi and his gang of six diggers had no sooner finished interring Menlo Park's subterranean experimental conduits—five miles of wired, four-by-four pine scantling boxes, each sixteen feet long—than two weeks of rain liquefied the clay that covered them. Red dribbles

---

* Edison's choice of this potential, over the much lower voltages favored by his competitors, eventually became standard in the United States.

† Edison also invented a "three-wire" connecting system that reduced the copper content of his conductors by a further two-thirds.

leaked into some of the boxes and short-circuited the conductors, even though each pair lay in grooves well sludged with coal tar and capped with extra wood. The entire grid had to be exhumed while Kruesi applied himself to the unstudied subject of insulation. He wrapped various lengths of copper cable with white rubbercloth, muslins, and marline hemp, all soaked with hot coal tar or cold paraffin or linseed oil, or smeared with resinous gums, or stewed in black pitch, pine tar, cottonseed oil, and various other proofings, but none were sufficiently water-repellent. Edison gave young Wilson Howell free rein of the laboratory library and chemical room to boil a series of compounds, some so noxious that even Otto Moses, used to pungent odors, was driven to seek fresh air. Eventually a blend of "refined Trinidad asphaltum boiled in oxidized linseed oil with paraffin and a little beeswax" was chosen, and fifteen men and boys deployed to apply it to the cables. They elevated the bare wire on sawhorses and straddled it in groups of three, each pair of hands tightly winding a spiral of sticky muslin ribbon that advanced, inch by inch and layer by layer, toward an end that never seemed to get any nearer. When, however, it eventually did, the triple-wound cable was found to resist the leakage of both current within and water without.[73]

The first fully insulated line was reburied and reconnected in time for Election Day, 2 November. It was looped to the laboratory and ran for a mile northeast, parallel to Thornall Avenue and the railroad. When evening came on, Edison, a staunch Republican, said to his switchboard operator, "If Garfield is elected, light up that circuit. If not, do not light it."*[74]

Returns began to chatter through the laboratory's telegraph sounder soon after dark. Edison maintained full steam in the engine room, ready to trigger the line dynamo on command. When a swing toward Garfield became evident around nine o'clock, he gave the order for power. A mile of bamboo-filamented streetlamps lit up all the way from the depot to the barn behind his house.

They stayed on until nearly midnight, in the first use of incandescent light to salute the victory of an American presidential candidate.[75]

---

* James A. Garfield (R., Ohio) was running against Winfield S. Hancock (D., Pa.).

"NOT A WORD WAS SAID"

Edison had little else to celebrate that fall. He was under intense pressure from the board of the Electric Light Company to demonstrate the main elements of his proposed First District illumination plan to a delegation of New York City aldermen. But he was unable to do so until a new hundred-horsepower Porter-Allen steam engine he had ordered to power Menlo Park's enlarged plant was ready. It was still under laborious construction in Philadelphia. "Every little delay is embarrassing to us at this time," he wrote to its builder, "and we cannot wait longer."[76]

But he had to. Rumors multiplied in industry circles that after two years of announcing that he had solved the problem of subdivided electric light, the Wizard was defeated by its complexities. Meanwhile his senior glassblower, Ludwig Böhm, departed Menlo Park *mit Sturm und Drang,* saying in a resignation letter he was tired of being bullied by "the boys" in the laboratory. "Yesterday Mr. Batch and I had a disagreement from a cause not worth to be mentioned, which went so far that I had to hear that you were a dem side [*sic*] better if I were not here."[77]

Edison had been depending on Böhm to help him start up the world's first electric lamp factory—an elaborate, self-financed, $10,000 conversion of his old electric pen works down by the railroad. It had long been evident to Upton and Clarke, as they projected the labor costs involved in competing with gas illumination in New York, that some sort of molding machinery would have to be devised to speed up the production of bulbs. Deprived of Böhm's expertise and faced with multiple other start-up problems, Edison subcontracted with the Corning Glass Company to supply him with blank globes at five dollars per gross. They arrived every day in two freight cars, thirty thousand at a time, punctual as the morning milk.[78]

His principal challenge at the factory was the installation of 476 towering mercury pumps. For coordinated day-to-day operation, they could not be modeled on the delicate Sprengel-and-Geissler hybrid that Francis Jehl sweated over in the laboratory. It pumped well, if slowly, as gravity pulled the liquid metal down, each drip sucking the "atmospheres" out of an attached bulb blank. But it was so encumbered with extra gauges and tubes that it required constant maintenance and repair. Edison offered a prize to any employee who could design a simpler version. Not surprisingly, Jehl won. He was rewarded with a certificate of 1.6 Electric Light Company shares and a fatherly admonition from his boss: "Keep it under your cap, Francis."[79]

The efficiency of Jehl's prototype did not alter the fact that it too worked by gravity and therefore depended on a constant circulation of mercury. Edison saw that twenty-five tons of the liquid metal would have to be held in suspension at all times in his factory line. To attain this he invented what was, in effect, a superpump for the pumps on the Archimedean screw principle. Instead of using steam energy, he drove the entire evacuation system by means of an electromotor connected to his central station, in a pioneering step toward the industrial application of electric power.[80]

He also had to fabricate huge carbonizing and annealing ovens for the mass production of filaments. Batchelor designed beautiful molds for them. But the first test batch of ninety bulbs to come out of the factory had an average life of only 25.8 hours, compared to the 132 days racked up by Hammer's laboratory-made record breaker. This was less the fault of the ovens than the propensity of certain coarse-fibered bamboos to warp while incandescing. Some bent so far as to touch and melt the inside of the globe.[81]

Japanese bamboo remained Edison's wood of choice for filaments. But when he learned that there were over a thousand species of *Bambusca phyllostachys* growing worldwide, he gave letters of credit to six freelance explorers and told them to search the Caribbean, South America, and Asia for a cane close-textured enough to stand unlimited incandescence. It was a typical large gesture that over the next few years would cost him $100,000.[82]

"Edison said to be progressing towards the perfection of his Electric Light & may soon be a very wealthy man," the R. G. Dun credit agency reported. "He must have an income of a good many thousands a year but his constant experimenting eats up money exceedingly fast & it is thought he so far has not laid up much."[83]

In mid-November Edison heard that sixty bulbs designed by the inventor Hiram Maxim were working well in the Equitable Life Building in New York. Their light was said to be stronger than his own, albeit less steady, since it pulsed to the rhythm of a primitive generator. He remembered Maxim visiting Menlo Park earlier in the year and spending an entire day "looking over the whole place."[84] From what his spies told him, the new bulb was nothing but a copy of his own paper-fiber original, except that its filament was tweaked in the shape of an M.* But he could do nothing about a resultant whirlwind of competitive publicity, reversing the excitement he had whipped up in his own favor a year before.[85]

Much of it came from the professoriat. Henry Morton, in a paper read before the National Academy of Sciences, stated that Maxim's light was "more economical and efficient" than Edison's. The astrophysicist Henry Draper held a reception in his laboratory to coincide with the meeting and illuminated it with Maxim lamps. George Barker, the University of Pennsylvania physicist who had been so complimentary earlier in the year, told *The Evening Post,* "There is no doubt in my mind or in that of Professors Morton and Draper as to the value of Mr. Maxim's remarkable discovery. . . . I do not say that Maxim is a better electrician than Edison, but he has invented a lamp which surpasses, I believe, even Edison's dreams."[86]

Edison could afford to ignore Morton's criticism as that of a man with little real experience of electricity.[87] But Barker's hurt. The two of them had been friends long before they went west with Draper in '78 to observe the total eclipse of the sun. It was then that Edison had first conceived his idea of centralized electrical power, and Barker had become his most vocal academic supporter, praising him for having an "original and ingenious" scientific mind.[88] Pointy-nosed and sycophantic, especially when he wanted to borrow Menlo Park equipment for his public lectures, Barker now seemed ready to rat aboard a less heavily loaded freighter.

"I notice in last evening's NYork Post," Edison wrote him on 23 November, "what purports to be an interview with you & wherein you are made to say some things concerning my Electric Light work which I cannot bring myself to believe ever emanated from you. Will you be good enough to say if you even so much as supplied the re-

---

* The Maxim bulb also substituted a flame-suppressant hydrocarbon vapor, loosely sealed in, for Edison's high vacuum.

porter with a foundation upon which he could build such an interview."[89]

Before Barker could reply, Joseph Swan exhibited thirty-six linked,
filamentary incandescent bulbs to Britain's Society of Telegraphic Engineers. The meeting was attended by the flower of the English electrical establishment, including John Tyndall, Alexander Siemens, and
William Henry Preece. Swan was coy about the substance of his carbon, saying he had a patent pending, but convincingly—and ominously—he showed it to be wire-thin and hard, yet pliable. He again
claimed to have experimented with a carbonized-card conductor
twenty years before and made no reference to his transatlantic rival
except to recall the "non-success" of Edison's earliest platinum lamp.*
According to the official record of the proceedings, he was congratulated for the "beautiful steady light" of his demonstration display.[90]

"Not a word was said on your behalf," a sympathetic attendee
wrote Edison, without explaining his own failure to stand up.[91]

Barker, in turn, confirmed the substance of his remarks in the *Post*.
In a letter that was at once polite and patronizing, he said he had
come to the conclusion that Swan and Maxim had priority as inventors of practical incandescent lamps. He was especially complimentary about the latter, whose laboratory he had just visited. "I tell you
in all frankness, . . . the method he has for making his carbon loops,
consolidates them and gives them a wonderful resisting power and
durability. He has run them up to 60 candles for an entire month and
they are still good."[92]

Edison reserved judgment on the M-lamp. But he was aghast at
Barker's "ignorant" assertion that the coarse carbon stick that Maxim
had sought to patent on 4 October 1878, was anticipatory of his own
carbonized cotton filament, successfully held in incandescence a year
later. More disgraceful still, from a professional point of view, was
that Barker, who served as president of the American Association for
the Advancement of Science, seemed willing to embrace Swan's completely undocumented claims, as reported in *Chemical News*.[93]

---

* Swan's contempt for Edison at this time was plain in a letter he drafted on 24 September but apparently never sent: "I had the mortification one fine morning of finding you
on my track and in several particulars ahead of me—but now I think I have shot ahead
of you."

"This is a mean thing to throw at me at this late day," Edison complained, sending a copy of Barker's remarks to Henry Rowland. "Have you noticed lately the utter indifference of the technical press in giving credit [for] scientific work to 'previous or first publication and public exhibition.'" It was as if his past willingness to let any competitor visit Menlo Park and see and test his inventions counted for nothing in journals that were supposed to be objective about the empirical process. Apparently a summary of a lecture was as good as a patent. Only one magazine had come to his defense and said that "it would be interesting to know where Mr. Swan's labors may be found in printed form" previous to Edison's own publications and exhibitions.[94]

He was particularly bitter because he had tried hard in recent months to present an image of himself as a scientist as well as an inventor, going so far as to bankroll a new weekly, *Science,* out of his own pocket.* But the effort had been in vain. Ever since his invention of the phonograph, there had been a swelling chorus of establishment attacks on him as "the great successor of Barnum," an unschooled self-promoter greedy for money rather than the austere accolades of learned societies. The latest crescendo, joined by Barker, was so shrill he felt he was being penalized for "my criminal efforts to devise a subdivided electric light."[95]

Rowland was a scientist of impeccable probity, and for that reason he thought Barker had no business talking to reporters. "I was as much surprised as you were to see the statements about Maxim's lamp," he wrote Edison. "Of course it is only yours with a slight modification in the method of making it. . . . You alone will show the world what you have done and dispose of all these petty hangers on."[96]

The fact remained, however, that Maxim had filled a public building in Manhattan with incandescent light, while Edison was still rusticated in Menlo Park, with another winter coming on. When he heard that Ludwig Böhm was now blowing glass for Maxim's United

* Edison continued to finance *Science* through the fall of 1881, at a personal cost of $10,000, before declining to do so any longer. The magazine briefly failed and was restored to life by Alexander Graham Bell. It is still in publication.

States Electric Lighting Company, he put *zwei* and *zwei* together and decided to sue for patent infringement.[97] There was nothing he could do about Joseph Swan's claim to have lit up a paper carbon *in vacuo* somewhere around 1860 until that inscrutable inventor's specifications were made known.*

A GASLESS NEW YORK

Edison could have recalled all his bamboo explorers in December, because by then one of them, William H. Moore, had sent him an ideal variety for his filaments: *Yawata madake,* a giant timber from the Kansai forest of Japan.† Its long, steel-strong fibers were built up of eight-sided cells that carbonized with uniform density and stiffness, held the shape of the mold when electrified, and rated an average 2,450 hours of life.[98]

Francis Upton tried hard to prove on paper that a central station electric illumination franchise in Manhattan would be profitable, despite delay and the physical, political, and financial obstacles still to be surmounted before the first dynamos began to spin. Working with insurance maps and his beloved slide rule, he estimated that it would cost $150,680 to wire up an initial downtown district, plus $45,989 for patent rights and other expenses. If customers plugged in ten thousand lamps and ran them five hours a day (based on current gas consumption averages), receipts should total $136,875 a year—and then would surely increase at a compound rate, as more and more New Yorkers converted to the safety and economy of incandescent light. Upton therefore felt confident in recommending that the company capitalize the plant at around $300,000, with an expectation of

* The lighting historian Adam Allerhand points out that Edison could hardly have known about Swan's early lamp before Swan himself began to remember it publicly on 27 December 1879—the same day the "Village of Light" was illuminated for a press preview. From that date on, the British electrical establishment vigorously supported Swan's claim of priority, having not noticed it before. Swan admitted that his unpatented "invention" (one of twenty pre-Edison attempts to develop a workable electric light) had not shone for much more than a minute. As will be seen, in 1888 the London Court of Appeals ruled the first Swan lamp "a failure."

† An elegant monument "To the Memory of Thomas Alva Edison" stands at the Iwashimizu Hachimangu shrine between Osaka and Kyoto in Japan. The site was chosen for its proximity to the *madake* groves that supplied the Edison Electric Light Company with filament fibers for almost fifteen years.

a payable dividend of 30 percent and a 60 percent annual return on its investment.[99]

Edison's backers, however, had learned to look first and count twice before approving any scheme emanating from Menlo Park. On 17 December nine directors of the Edison Electric Light Company, led by Grosvenor Lowrey, formed a majority of the board of a new corporation, the Edison Electric Illuminating Company of New York. Its urgent mandate was to transfer Edison's operations to Manhattan as soon as City Hall could be persuaded to give him permission to start digging up the streets. Edison sensed from the board's composition, which notably represented the interests of some of the most powerful financiers on Wall Street, that he would lose much of his independence when he moved. In a vain gesture of protest, he declined to serve as a director. Lowrey slapped him down. "I shall not present your letter of resignation as Mr. Fabbri [of Drexel, Morgan] very strongly objects to your leaving the Board. His impression was that 'Edison's name is a tower of strength to us, and if he never attended a meeting, it would be a great loss if his name should not appear at all times among the names of the Directors.' "[100]

The Illuminating Company was duly organized and capitalized at $5 million. Five nights before Christmas Edison showed his ability to épater le bourgeois when a large party of municipal dignitaries, including eight aldermen, arrived at Menlo Park by special train.[101] The sun had just gone down, and some two hundred freshly polished streetlamps were already glowing up the hillside. A little tavern by the tracks stood ready to slake the thirst the visitors had worked up on their journey. But they were corralled without refreshment into the laboratory, where for two hours Edison, wearing a sealskin skullcap, explained the intricacies of multiple-arc circuitry, feeder-and-main distribution, metering by copper deposit, and the cold ohm resistance of various bamboos. The aldermen were less interested in these subjects than whether he would entertain them to dinner at an hour earlier than one of his famous "midnight lunches." Inexorable, he led them to a tour of the machine shop and the generation room, where the bed, if not yet the body of his new Porter-Allen engine sat on its massive foundation.

It was eight o'clock when he escorted his wilting guests back upstairs to the laboratory. The long room had been darkened during their absence, but as they crested the stairs, its thirty-seven ceiling lamps (one enclosed in a globe of shimmering water) burst into light, and a U-shaped dining table was revealed against the backdrop of Hilborne Roosevelt's pipe organ. White-gloved waiters stood ready to serve champagne. A banquet catered by Delmonico's ensued. Grosvenor Lowrey sat at the head of the table with Edison on his left and Chief Alderman John C. Morris, who was known to oppose the central station plan, on his right. The wine flowed copiously (Edison diluted his with liberal splashes of water), giving way to Kentucky bourbon as course followed course. By the time cigars were handed out, Morris had become an ardent advocate of municipal incandescent lighting. He told the table that Edison was "entitled to the thanks of the world for bringing this light to such perfection that it can now be made to take the place of gas."

The superintendent of gas, Stephen McCormick, allowed that electricity was a safer illuminant. It was too easy for a hotel guest in New York to blow out his lamp on retiring "and wake up dead." Parks Commissioner Andrew Green said that at last Central Park would have lights that did not burn foliage. Alderman John McClave waxed prophetic, seeing a gasless New York in 1900. "If at any time my voice or vote can be used to advocate the beautiful electric light which I have seen here tonight, you may count on me to use them."

Lowrey stood up and proposed a toast to the inventor. As the rest of the company reached for their glasses, Edison remembered he was still wearing his skullcap and awkwardly snatched it off. The toast, accompanied by loud cheers, was drunk standing.

NO GOING BACK NOW

On the following morning, New York newspapers announced that the Illuminating Company had a permit to bring incandescent electric light to fifty-one blocks downtown. Its First District, one of a projected twenty-six, would run from the East River to Nassau Street in the west, Wall Street in the south, and Spruce in the north. That square mile encompassed some of the densest real estate in the city, including the headquarters of several major financial institutions (notably Drexel, Morgan & Co.) and many townhouses and tenements.[102] Somewhere in the parcel, wherever Edison might find a lot that suited him, he could build his central station, and the streets were his to dig—subject to approval by the next board of aldermen, taking office in the new year.

Lowrey was uneasy about the permit, which could be revoked at any time. What he needed from the new administration was a formal ordinance, and knowing the ways of City Hall, it was bound to be expensive. But that was his problem, not Edison's. There could be no going back now on the revolution engineered at Menlo Park. After four and a half years of monastic seclusion and communal experiment, the inventor and his "boys" (many of whom had had their own champagne party at the tavern, with bottles purloined from the laboratory stash) were going to have to face the pain of diaspora—and with it, what amounted to the end of youth.[103]

Heavy snow fell on the twenty-seventh, whitening the little clutch of buildings that Henry Ford would one day resurrect in another state, in another century.

Menlo Park in the winter of 1880–81. Painting by Richard F. Outcault.

THE PATIENCE OF JOB

The advent of 1881, accompanied by a partial solar eclipse, portended great changes but found Edison in a melancholy mood. It was evident that the recent assaults on his reputation still stung. "I think I ought to have credit for what I have accomplished," he complained to a correspondent of the *Chicago Tribune*. "Only a year ago the subdivision of the current for lights was declared a practical impossibility. . . . Everybody was down on me, and now a fellow named Swan is making an exhibition in London of my incandescent lamps."[104] The possessive pronoun betrayed his fury that William Spottiswoode, president of the Royal Society and titular descendant of Sir Isaac Newton, should announce that Swan had at last solved the problem of the electric light.

Edison threw up his hands. "What's the use of a man trying to do anything anyway? If he keeps things secret and will not tell everything, he is denounced as a mountebank, and if he does things openly, they steal all his ideas."[105]

It did not occur to him that Old World sensibilities, attuned to the virtues of self-deprecation and proper procedure, recoiled from his American tendency to overshare—exactly what he was doing now— and his naïveté in assuming that every invention he boasted about, or let competitors borrow for testing, would not soon be imitated. To establishmentarians like Spottiswoode, a product of Eton, Harrow, and Oxford, Edison was an embarrassing example of the genus *Americanus egotisticus,* lacking Latin and even guile, which made him anybody's fool.

Guile was a quality Joseph Swan possessed in abundance. It had enabled him to climb in British society far above what Spottiswoode would call his "station," a working-class niche considerably lower than Edison's. He too had little formal education, having been apprenticed to a chemist in his teens and employed in a provincial pharmacy before he began experimenting with lightbulbs. This was, according to his first recollection, in 1855, a date that he and his family would progressively push back to 1848, the year of Edison's first birthday. Since then Swan had made all the right career moves, setting himself up in London as a gentleman-inventor and waiting twenty years to patent his filamentary lamp.[106]

"Talk about the patience of Job," Edison scoffed.[107]

LOVE OR MONEY

On the Feast of the Epiphany three kings of the New York financial world—John Pierpont Morgan, Egisto Fabbri, and Jacob C. Rogers—

paid a late-afternoon visit to Menlo Park to convince themselves that they were wisely investing in his system. The sight of five hundred lamps casting pools of orange-tinted light on the snow at a flick of Edison's wrist was all the evidence necessary. "I don't believe you could buy a share of this stock for love or money," one of them remarked.[108]

Edison had recovered his good humor by now, and drew an aide's attention to the sight of the great "J.P." leaning thoughtfully against one of the laboratory worktables and flicking his shoes with an ivory-topped umbrella. "Hammer, look at Morgan, you would not think he had $100,000 in this, would you?"[109]

From then on Drexel, Morgan & Co. acted as the Illuminating Company's bankers, promoted its interests abroad, and managed Edison's personal portfolio.[110]

As expected, the new mayoral administration of William Grace proved avid for tax money in return for its blessing on the First District lighting scheme. Its initial demand was for more than $1,056 per mile of street conduits, plus a 3 percent share of all gross receipts once the system commenced business. But Morgan's lobbying power was formidable, and the city eventually settled for a mere trench fee of five cents per linear foot. Its only other demand was that Edison reimburse the cost of having inspectors on site at all times during the installation period. (He was soon to find out that "at all times" meant a brief appearance on payday.) Otherwise he was free to start laying tubes as soon as the ground of lower Manhattan thawed.[111]

Before he transferred the bulk of his operation there, he wanted his systems analyst, Charles Clarke, to conduct a rigorous test of the whole Menlo Park system, to be sure that it could be economically duplicated to scale in New York.[112] It was not necessary to include the new Porter-Allen engine, which had at last been delivered but not yet set up in the machine shop. He had other plans for it. In the meantime his old eighty-horsepower Brown unit, linked to eleven dynamos, could be trusted to drive the test.

As the most mathematical intellect on Edison's staff, Clarke rejoiced in the algebra that thickened in his logbooks after the system powered up at 9:22 P.M. on 28 January.[113] The ciphers $772t$ $(W + wS + W's)$ signified to him that a calorimeter had obtained the full value of the economy of the lamps under observation, while a certain amount of energy was being lost in the conductors. Edison was happy to take his word for it, and even happier to accept Clarke's conclusion, at the end of the twelve-hour test, that all aspects of the system, from the "clear, free-burning egg coal" in the boiler to the last light in thirty-nine thousand feet of circuitry, were well coordinated. The most important figure in Clarke's final report was a ratio of 7.25 lamps per horsepower, substantially better than that of gas, and he did not doubt that with improvements in dynamo design it could be increased. Delighted, Edison told him, "After this we will make electric light so cheap that only the rich will be able to burn candles."[114]

YOUNG LADIES WHIRLED

January gave way to February, and the young men of Menlo Park, several of whom were now married, braced to hear which of them would shortly be ordered to find quarters in the city. For a year or two at least and perhaps forever, the Old Man was going to have to change his country hats for the bowlers and stovepipes of fashionable Manhattan. (Mr. Morgan wanted an Edison lighting plant in his house, and so did Mr. Vanderbilt.)[115]

Mary Edison—twenty-five years old, mother of three, universally liked for her sweet nature, if not her love of loud clothes—had as much reason as any to have conflicting feelings about the looming change.[116] Her big house was the village's social center, while she had only one good friend in New York. She was by no means a country girl, having grown up in Newark, and since she enjoyed spending money, her husband's intent to reside on Fifth Avenue in midtown Manhattan, not far from the couturiers and confectioners of "Ladies' Mile," sounded agreeable. But her working-class background might "show" more, in such a milieu, than it did in Menlo Park. Marion and Tom would no longer be able to roam freely about the countryside and make themselves pesky in their father's laboratory. They would require a governess in town, not to mention a nanny for William, aged two. Mary would miss the live-in companionship of her sister Alice, who was sweet on William Holzer the glassblower, and she would not be able to drive twelve miles down the turnpike to see her parents whenever she felt like it.[117]

Edison had no intention of cutting all local ties, as long as the lamp factory, railway, and machine shop kept running. The laboratory too could be maintained by a skeleton staff, until he found a substitute location in town. And the house with its adjoining green fields would make a pleasant summer retreat, remote as that prospect might seem in the middle of a particularly white winter.

Mary made the most of her last days at home by doing what she loved to do—dress up and entertain. A reporter from *The New York Herald*, sent out to view the laboratory's closing display of lamps, wrote a description of the little world she was about to leave behind:

> *Mrs. Edison's parlors were brilliant indeed. . . . You do not know what the Edison electric light in a house is until you have seen the pendant globes, spreading uninterrupted radiance on all beneath and around. There was a merry company, full of life and triumph. An Italian gentleman sang a Neapolitan impromptu to his own accompaniment. Young ladies whirled in the waltz. . . . We went down to the depot, and as the train came thundering by to bring us to the city, the jingle of sleigh bells rang over the snow from near the Professor's house, for there is no pleasure at Menlo Park like sleighing by electric lights when the public has gone away.*[118]

YOU SHOULD HAVE SEEN HER RUN

The Menlo Park diaspora began on 5 February with the departure of Charles Batchelor for Paris. He was charged with preparing an exhibit for the great Exposition Internationale d'Électricité, to be held in that city later in the year.[119]

Edison had hesitated before agreeing to take space in the show,[120] just when he should be establishing himself in New York and beginning the biggest practical task of his career. But it would be the first such event devoted entirely to the science and technology of electricity. Maxim and Swan were bound to be there, attempting to dazzle the public and the press with their imitation lamps. There would be demonstrations, medals, and worldwide publicity. Edison did not see how he could avoid participating—except to send the all-capable "Batch" to deputize for him.

With Batchelor gone, and Upton put in charge of the lamp factory—now incorporated as the Edison Electric Lamp Company—the question arose as to who would become the boss's new right-hand man. It did not remain unsettled for long. Two mornings after Batchelor sailed, Edison, bursting with energy and excitement, shouted across the laboratory at Charles L. Clarke, "Come on, Clarke; pack up at once and come with me to New York. We're going to begin business right off!"[121]

By noon they were walking into a four-story, double-width brownstone at 65 Fifth Avenue. Edison announced it as the new headquarters of the Edison Electric Light Company.

"The company has made you chief engineer," he said, rushing Clarke upstairs. "This is your office, the furniture will be here this afternoon. Furniture for your living room upstairs will be here too—I want you on hand all the time!"[122]

The brownstone quickly became known as "65." It stood on the east side of the avenue just south of Fourteenth Street, its sixteen tall windows unfurling striped awnings against the afternoon sun—which at this time of the year set, symbolically enough, over Menlo Park.[123]

Edison could have moved his family into a suite on the top floor. But he chose to use it as a laboratory and looked for an apartment elsewhere, pending a house rental somewhere in the neighborhood.* For the rest of that month Mary and the children remained in the country, while he supervised the transfer of staff and equipment across the Hudson.

As Edison's behavior with Clarke indicated, he was in one of his periods of cyclonic overcharge, excited by the project ahead of him much as a dynamo is "excited" by connection to a start-up machine. The comparison became actual on the twenty-eighth, when he and Clarke reunited at Menlo Park for an experiment that nearly became the last for both of them.

One integer of Edison's central station plan—the most important of all—was still unsatisfactory: its generation plant. Eleven bipolar dynamos had been able to handle the demands of the model system

---

* Some of Edison's bachelor employees were allowed to occupy bedrooms upstairs at "65."

and electric railway, but a plant many times more powerful would be needed to light up the First District. He had realized this since last spring, when he assigned Upton and Clarke, his two experts in electromagnetic theory, to build him a dynamo with sixteen times the capacity of any previously made. The spinning armature alone would weigh one and a half tons.[124]

It was for this leviathan that he needed his hundred-horsepower Porter-Allen steam engine. Clarke believed that the optimum rotor speed for the new dynamo should be 350 revolutions per minute. Edison, just to make sure, had asked Charles T. Porter to build a machine fast enough to drive a locomotive. Part of the delay in delivery lay in his extra demand that the engine be configured to couple directly with the dynamo by means of a mutual shaft, in a union of steam and electricity sure to incite coarse jokes among the "boys."

Hitherto all generators had been linked to their driving engines indirectly, through geared wheels and belting. Edison saw that much energy was lost that way. He was hoping that direct transfer, high speed, and low internal resistance might give him as much as 90 percent efficiency, rather than the 60 percent generally considered the limit any electrical engineer could expect from a dynamo.[125] But the imponderable was vibration—hence the two-foot depth of the Porter-Allen's cast iron bed, and the massiveness of its foundation in the machine shop.

The combined unit was now assembled and ready to test at Menlo Park.[126] Charles Porter, summoned from Philadelphia, was given the honor of operating his own engine. Feeling too nervous to do so at close quarters, he attached a chain to the throttle and backed away as far as possible before pulling it. Steam pressure built up slowly while Edison, stopwatch in hand, kept calling for more power. Then the governor took hold, and the dynamo accelerated at a compound rate until, in Clarke's words, "all the moving parts became a blur like that produced by flies' wings." Not only the foundation but the entire shale hillside began to shake underfoot. If any Wagnerian had been present, he might have called it *der Erdenton,* the bass note of all creation, but there was only Edison, stopwatch in hand, yelling "Hup . . . hup."[127]

Clarke could feel the hair rising on his neck. At Edison's signal he activated the speed indicator and found that the dynamo was spinning at 750 RPM. That being dangerously close to its disintegration point, Porter was allowed to throttle the engine down. Clarke was not happy with the performance of the armature, but Edison felt confident, now, that he had a prototype for six even larger dynamos to install in his central station. Years later he boasted to the editor of *Electrical Review* about the time he nearly maxed out a big Porter-Allen: "You should have seen her run! Why, every time the connecting rod went up she tried to lift the whole hill with her!"[128]

A PRIVATE SECRETARY BEYOND PRICE

Edison installed his family in the Chipman Boarding House at 72 Fifth Avenue on 1 March 1881. It was a date that coincided with the entry into his life of Samuel Insull, fresh off a steamer from England. Twenty-one years old, short and skinny and side-whiskered, with popping eyes and a humorless manner acquired from reading the motivational tomes of Samuel Smiles (*Self-Help; Character; Thrift; Duty*), Insull did not look like a youth destined to become one of the richest men on earth. However, he came highly recommended by Edward Johnson, who had known him in London and thought Edison could use someone with a bookkeeper's brain to take charge of his personal and financial affairs.[129]

Insull was well qualified, having worked as the factotum of Edison's chief European representative, Col. George Gouraud. During that time he had become a passionate subscriber to the Edison legend, and could have dreamed of no greater good luck than to be hired sight unseen by "one of the great master minds of the world." Johnson escorted him up the steps of "65" and introduced him to Edison in the bare back office. Insull's first reaction was surprise that anyone so famous would wear a seedy black three-piece suit and rough brown overcoat. But the face over the carelessly knotted white silk neckerchief was unforgettable. "What struck me above everything else was the wonderful intelligence and magnetism of his expression, and the extreme brightness of his eyes."[130]

That same night, Insull discovered that his new boss was something of a child about money. Edison pulled out his checkbook and revealed without embarrassment that he had $78,000 cash in the bank. Which of his European telephone securities, he asked, should he sell in order to capitalize three private ventures right away—a bigger lamp factory, a works for the production of dynamos, and a company to lay tubes under the streets of New York?[131]

Insull was able to answer on the spot, because he had made it his business to read all Edison's contracts passing through Gouraud's office. He had a photographic memory for stocks and shares, and told Johnson, who was returning to Europe to handle the transactions, exactly which ones to divest and where. Edison had, for example, a reversionary interest in the United Telephone Company of London worth around $100,000, and he might get as much again from a deal Gouraud was trying to swing with the Bell Company in the Far East. By four o'clock in the morning Insull had been through Edison's books and compiled a schedule of foreign patent rights as collateral against which further funds could be borrowed. If Edison was not yet assured by this performance that he had acquired a private secretary beyond price, then he was by Insull's ability to work through the night without apparent fatigue. A mutual contempt for the clock was to prove their strongest bond in the years to come.[132]

FACTORIES OR DEATH

Charles Clarke thought that Edison's impatience to start up three manufacturing adjuncts to the central station project, all independent of the Electric Light Company, was due to his "bull-like" overconfidence. He blamed himself for reporting so favorably on the Menlo Park system that the Old Man assumed it would work just as well when duplicated on a huge scale downtown.[133] By the same token, any small problems he had glossed over were likely to loom large.

Actually Edison already felt around his neck the "leaden collar" of corporate caution, personified by Sherburne B. Eaton, general manager and vice-president of the company. Eaton was a Civil War veteran who liked to be called "Major," and though small, occupied the largest office at "65." Even before they each moved in, Eaton had made it plain that his fellow directors believed their prime asset to be the patents they had acquired from Edison in 1878, in return for financing his development of the electric light. The time for experiments, Eaton's neat little goatee seemed to say, was now over, and the company's last great investment must be in construction of the First District.[134] If it was as successful as Edison promised it would be, cities around the world would clamor to replicate it, and his patents would become so priceless that he would never need lay another cable.

Consequently, most board members were opposed to getting into the manufacturing business, which they viewed as an unnecessary indulgence. They had put more than $130,000 into Edison's scheme without seeing so much as a cobble lifted downtown. Budget watchers at Drexel, Morgan failed to see why tubes and dynamos could not be bought instead of being expensively custom-made. Nor could they understand why he would want another lamp factory. The one at Menlo was blowing one thousand bulbs a day, and *madake* filaments were coming out of the ovens uniformly carbonized after being packed in with peat moss.[135]

Edison believed that it would be profitable in the long run to manufacture every part of the central station system. He was so convinced of this that he did not quail even when told that he would have to pay the Electric Light Company for the right to use his own patents. It was worth it in order to keep control of the whole operation. In any case, who but he could make things that nobody else had ever made—switchboards, regulators, current indicators, conduits, feeder-and-main junction boxes, connectors, meters, and house wiring, down to the very sockets that held his lamps?[136]

"Since capital is timid," he told Major Eaton, "I will raise and supply it. The issue is factories or death."[137]

Insull had no sooner snatched his first few hours of sleep in America than he found himself being hustled down to 104 Goerck Street, near the East River, to see the first of these facilities—an immense old iron-making shop now emblazoned EDISON MACHINE WORKS in letters three feet high. Edison had leased, refurbished, and equipped it for $65,000, contributing 90 percent of that sum himself, and Charles Batchelor was putting up the rest.*[138]

In view of what Insull already knew about Edison's finances, he uttered no protest when he heard that he was to be paid only $100 a month, half what he had earned in London. His discretion was rewarded. As Edison proceeded to organize subsidiary after subsidiary

---

* Neither man could guess at the time that they just co-founded General Electric.

in years following, Insull was appointed corporate secretary of all of them, and each paid him a salary to match. He even received, unasked for, a stock bonus of $15,000 after twelve months' service. "If you pushed Edison in money matters, he was as stingy as hell, but if you left the matter to him he was as generous as a prince."[139]

Another immediate start-up was the Electric Tube Company, to be run out of a shop on Washington Street by John Kruesi. It began to wire up subscriber buildings as soon as the city issued its ordinance of approval on 19 April. The franchise, negotiated by Grosvenor Lowrey, could hardly have been broader. It gave Edison the right to "lay tubes, wires, conductors and insulators, and to erect lamp-posts within the lines of the streets and avenues, parks and public places of the City of New York, for conveying and using electricity or electrical currents for purposes of illumination." He could do so not only in his chosen First District but in another one uptown if he wanted.[140]

In the same month two veterans of Edison's earliest days in Manhattan, Edward Johnson and the fabricator Sigmund Bergmann, partnered to form Bergmann & Co., with a contract "to produce electric-light fixtures"—switches, bulb-holders, panels, meters—that were too small for the mighty machines on Goerck Street. Edison contributed almost half the firm's capital, much more than Johnson's 12 percent, but as with the Tube Company, he took no titular credit—possibly to prevent his corporate backers from thinking he was spreading himself too thin.

However, when he put $5,000 cash down in early May on another vast complex in East Newark, New Jersey, reserving it for lamp production once the Menlo Park factory became too small, his pride in the basic bulb that had triggered all this expansion could not be suppressed. The factory—three massive wings linked with bridges, covering a whole city block—was acquired for $52,250 under the aegis of the Edison Electric Lamp Company.[141] As such it constituted the fourth and final arm of his new industrial empire.*

There remained only the purchase of a suitable structure to house his central station in lower Manhattan. Edison may not have been

---

* Edison, who had a gift for real estate, got an extraordinary bargain. The property's asking price was $136,000, but he bought it in receivership for $52,250. A number of years later he sold it for $1.08 million.

much of a money man, but he had a nose for real estate. It led him in more respects than one to the leather-trade corridor of Pearl Street. That part of the First District was especially aromatic in warming spring weather, because most of the ninety elevators that serviced its warehouses were horse-drawn. Drays ridden by boys could be seen on the flat rooftops, patrolling back and forth as bales of hides rose and fell. How the horses got up there was a question that Edison could investigate once he had made a deal to buy numbers 255 and 257, two conjoined four-story buildings on a lot five thousand feet square. He paid $65,000 on behalf of the Electric Illuminating Company, a pittance compared to what he would have had to shell out had Pearl Street been closer to the banking houses of Wall Street, five blocks away. But besides price, the site had the virtue of being—as required—central to the district, and the money he had saved buying it would enable him to reconfigure the buildings as extensively as he liked.[142]

He saw at once that the second floor of number 257 might collapse under the weight of the six dynamo-engine combinations he meant to install. Nor could he trust the north and south walls to support the heavy girders that would support them: what was necessary was for Clarke to build a complete internal wrought-iron bridge, almost as massive as an elevated railway. Beneath it, a battery of boilers would supply steam heat to the engines, and they would be fed by continuous conveyance of coal up from the basement. On the third floor he would place his voltage regulators, and on the fourth a bank of one thousand load-monitoring lamps would glow at all times. (The big advantage he had over the gaslight companies was that his power could be drawn on during the day, by sewing machines and the like.) The twin building next door at number 255 would serve for service, sleeping quarters, and storage.[143]

By 27 May Edison, working with manic energy, had started every project necessary to complete the First District within (he hoped) six or seven months. As yet all this peripheral activity—administrative planning at "65," lamp production at Menlo Park, dynamo assembly at the Works, Kruesi casting miles of conductor pipes in a shop on Washington Street, Bergmann rattling out auxiliary appliances on Wooster—was ragged and unconnected, like the slow start of a storm system. But momentum was building, and concentration would come, until everything converged on the switch he would throw—with luck, sometime in November—to begin the incandescent illumination of the world.

THE BALLS OF THE FIRMAMENT

"Boulevard St Antoine that damnable merchants of inhumanity Citenian wharfrats. Why Centenus dost run a line already greased from Sirus to Capella with angularity," Edison scrawled in the midst of a laboratory notebook otherwise full of engineering data, "whereon ten million devils slide down to the fathermost sag and piss into pendemonum."[144]

He was either recounting a dream, or teasing Charles Hughes, the main keeper of the notebook, into thinking that he had lost his wits, or—more likely—amusing himself by seeing how much nonsense his pencil would write before the point wore down. *"Tell me winged soldier of Hell if in the farthermost ends of infinity warted demons with carvernous mouths spit saliva on the balls of the firmament. . . ."*[145]

Some of the imagery in this stream of consciousness—the Rue Saint-Antoine, the wharf rats, the warted demons—pointed to the Paris of his favorite novelist, Victor Hugo.[146] The jets of saliva might as well have related to something of contemporary concern to him: a report that workers at the lamp factory were salivating excessively as they worked the pumps. That was a sure sign of mercury poisoning. Much else, including references to water closets in the underworld and men sleeping on telegraph poles, sounded like deliberate nonsense, unless it related somehow to the work he planned downtown.

Edison completed four pages, ending with an assertion that Thomas De Quincey, "had he a brain 300 miles in diameter full of opium," could never have comprehended the passion of the lovers in Longfellow's poem *Evangeline*—another of his favorites. Then he flipped the last page, scratched a tight, symmetrical zigzag, and left his screed for posterity to puzzle over.[147]

LA GRANDE GÉNÉRATRICE

As it happened, opium was a problem in his own family. Mary took so much of it, in the form of medicinal morphine, that her friends feared she might one day take too much. She was often plagued with neuralgia, that common complaint of housebound nineteenth-century women,[148] and when it struck, she could not rely on her peripatetic husband to nurse her.*

Edison was closemouthed about Mary in public, and only on the rarest occasions hinted, obliquely to intimates, that she was the girl he should have left behind. After meeting Kate Armour, the gifted young Canadian his attorney had married, he burst out with, "Why is it, Lowrey, that so few women have brains?"[149]

The question excluded Kate, whom he took to at once and presented with one of his souvenir calligraphed notes: "How do you do my dear Miss Armor? The Electric Light is a success, take my word for it." Lowrey was surprised and told Kate, "I never heard him refer to any woman the second time." He liked Mary but thought her ill equipped to be the helpmeet of a genius. "Edison's experience," he confided, "is of the slightest and poorest."[150]

For the moment Mary seemed well enough. She took full advantage of the stores along Ladies' Mile, wearing ever more brilliant outfits and parading Marion in party dresses of nile green or yellow satin, with hand-painted flowers.[151] She loved going to the theaters and music halls, and even bought tickets to the occasional society ball, but her husband invariably bowed out on account of his deafness, so she had to rely on the company of friends.

It was fortunate for Edison, busier than he had ever been,† that Mary had the house at Menlo Park to retreat to when the weather warmed in Manhattan. Her absence across the river this summer enabled him to supervise preparations at the Machine Works for the imminent international exposition in Paris. Ever since his decision to participate, Charles Batchelor had been working to fill two halls of the Palais de Champs-Élysées with a display of all his electrical inventions to date—the vote recorder, the duplex and quadruplex and oc-

---

* There is little doubt that Mary Edison's sufferings from neuralgia were genuine (see Part Six), no matter how they may have been related to emotional problems. They appear to have passed on genetically to her eldest son, who was afflicted with paroxysmal headaches all his life.

† Between 17 May and 25 June 1881 Edison applied for twenty-six U.S. lighting or power patents, only four of which were not granted. *Papers*, 6.4.

toplex telegraphs, the electric pen, the phonograph, the tasimeter, and dozens of others—all to be bathed in the incandescence of his latest and greatest. And the pièce de résistance was to be a dynamo even bigger than the one that had shaken the hillside at Menlo.[152]

Resistance, indeed, was key to its performance. Edison's theory of generator design was that the larger the armature, the fewer ohms would inhibit its flow of current. Accordingly he gave it a spinning core of laminated iron and heavy copper bars connected in pairs, fore and aft, to aureole-shaped copper "tits."[153] The field magnet, nearly six feet long, consisted of eight solid iron cylinders, each wound with more than two thousand turns of insulated copper wire.[154] Rotational power was applied directly, as to the dynamo's predecessor. But because the fast Porter-Allen engine had never worked well in a close embrace, causing dynamos to spark and build up heat, Edison ordered a 125-horsepower Armington & Sims unit that would run slower and, with luck, cooler.[155]

When all the components of this colossus were bolted together at the Works, its thirty-ton bulk inspired awe. It measured fourteen feet in length and towered taller than Francis Upton. Edison had spared no expense to perfect it, even gold-plating lugs and screws to lower resistance. Energy reduced to essence, it was a thing of brutal beauty, with all the basic forms of geometry massed around the invisible confluence of electrical and magnetic waves. But when tested at the end of June, it too got hot and sparked, with arcs cracking between adjacent induction bars. Edison began to lose hope that the machine would be ready in time for the opening of the exposition on 11 August. He ordered an emergency reconstruction and rewinding of the armature, deploying two shifts of fifty-five and sixty workers around the clock for eight days. While they dismantled the core, he filled twenty-three pages of his notebook with wiring diagrams of almost astronomical beauty.[156]

Eventually he settled on a combination of slimmer bars painted with zinc white, wrapped in japanned paper, and cooled by a fan blowing air through the interstices. Voltage dropped as a result, so he added two extra electromagnets to the upper field cores. This threw the circuit somewhat out of alignment, but restored tension to the point that the dynamo efficiently lit seven hundred lamps at 350 revo-

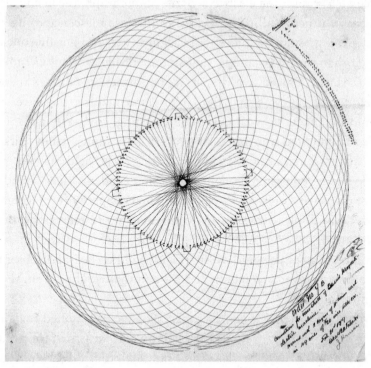

Shell winding for Edison's large magneto dynamo, February 1879.

lutions per minute.[157] But he made a mistake in amalgamating the commutator and brushes with an excess of mercury, to keep resistance to a minimum. Those surfaces oxidized in time and threw off such clouds of toxic vapor that attendants at the Works salivated as badly as their mates in the lamp factory.[158]

He solved the problem by reamalgamating often and polishing with the care of a silversmith.* That lowered resistance to less than one-hundredth of an ohm. But so many other "bugs" required fixing that Charles Batchelor, representing him at the exposition, had to fall back on two smaller dynamos to illuminate the Edison rooms. Opening day came and went. Visitors hoping to see *la grande génératrice d'Edison* were told they might have to wait another month before it could be exhibited.[159]

---

* Edison's fanatically detailed, 8,700-word letter of 8 September 1881, consigning this dynamo to Batchelor in Paris ("I will write you further if I have omitted anything"), is a good example of how closely he dominated his subordinates.

By the end of August it seemed ready to ship. Edison had to go west to pick up Mary, who had been taken ill on a family visit to Port Huron.[160] Some instinct prompted him to test the dynamo one more time before he left. No sooner had it powered up than the crankshaft shattered and flew across the room. Fortunately no one was killed. He examined the broken iron, cursing, and was amazed to see that Armington & Sims had failed to anneal it.[161]

When he got back from Michigan early in September, the dynamo was working again with a new steel shaft and was at last ready to ship. It was booked on the French liner *Canada,* departing 7 September for Le Havre. With only four hours to go before the hold closed, sixty Machine Works employees disassembled the dynamo and packed it into 137 crates, while Edison leaned on Tammany Hall to give his delivery trucks quick passage to the loading dock. Police held up traffic, and a fire bell cleared the way as the horses sped across town. Stevedores were waiting at the Compagnie générale transatlantique pier. The last box was taken aboard with an hour to spare.[162]

"FIVE GOLD MEDALS"

As things turned out, he need not have hurried. The Paris exposition, plagued by power problems, had not so much opened as half-opened, offering only dull or dark displays at first, except for a bluish mix of arc and incandescent light near the ground-floor entrance. Edison's lamps at least shone steadily, if not strongly, while Batchelor and William Hammer coaxed as much current as they could out of the main line at their disposal. The arrival of the great dynamo from America on 23 September caused widespread excitement, since it was four times the size of any generator yet seen in Europe. Edison had given it the model mark C, to distinguish it from his smaller bipolars, but because of its elephantine proportions, it soon acquired the nickname "Jumbo."[163]

When it brought Edison's exquisite lamp displays to full brilliancy, a tone of rueful admiration crept into the commentaries of French scientific writers, who for the last couple of years had vied with their British colleagues in mocking the promises of *le solitaire de Menlo-Park*. They could hardly avoid seeing that he had put together all the elements of a complete lighting system—as opposed to Swan and

Edison's "Jumbo" dynamo at the Paris Electrical Exposition, 1881.

Maxim, who exhibited lamps and chandeliers only. "Edison is not a myth," *Le Figaro* had to admit. Henri de Parville wrote in *Le Journal des débats*, "Times have certainly changed. All doubts are gone. Those who want physical evidence, like Saint Thomas, can see his lamps now with their own eyes."[164]

Perhaps the most influential of these skeptics was Théodose du Moncel. He published a long article in *La Lumière électrique* retracting his former dismissal of Edison as a "pompous" poseur—as well he might, because the Electric Light Company was now paying him a thousand francs a month to represent its interests in Europe. Nevertheless, an international panel of examiners found in mid-October that Edison's bulbs, boosted by his three-thousand-watt power plant, had an efficiency rating of 12.73 lamps per horsepower. Swan's rated 10.71, little better than those of his compatriot George Lane Fox at 10.61 and Maxim's, at 9.48.[165]

On 22 October Grosvenor Lowrey, who was in Paris representing the interests of the Electric Light Company, cabled Edison in New York:

OFFICIAL LIST PUBLISHED TODAY SHOWS YOU IN THE HIGHEST CLASS OF INVENTORS. NO OTHER EXHIBITORS OF ELECTRIC LIGHT IN THAT CLASS, SWAN LANE FOX AND MAXIM RECEIVE MEDALS IN CLASS BELOW. THE SUB-JURIES HAD VOTED YOU FIVE GOLD MED-

ALS BUT GENERAL CONGRESS PROMOTED YOU TO THE DIPLOMA
OF HONOR CLASS ABOVE. THIS IS COMPLETE SUCCESS THE CON-
GRESS HAVING NOTHING HIGHER TO GIVE.[166]

Almost simultaneously another cable arrived at 65:

EDISON N.Y.—YOU HAVE RECEIVED THE HIGHEST AWARD THE
JURY HAD TO GIVE. I CONGRATULATE YOU.
JOSEPH W. SWAN [*][167]

BOTH OF THEM DIED

Edison received the news of his five medals without comment. He was
at work on another generator—"Jumbo No. 2"—even bigger than
the one in Paris. Edward Johnson had ordered it for an exhibition to
be held at the Crystal Palace, London, in January 1882. A *New York
Times* reporter was given the honor of a private demonstration in the
Machine Works. It took place at four in the morning, an hour more
convenient to his host than to himself.[168]

"You are seeing what nobody else ever witnessed before tonight,"
Edison said, rubbing his hands with glee as a rheostat turned and row
after row of lamps ignited on the test room's high ceiling. "A thousand electric lights, all from one dynamo." The armature accelerated
to 360 RPM, flickering with an electric nimbus so strangely colored
that the *Times* man could only describe it as "indescribable."[169]

It would be equally hard for any chronicler to find words for the
blur of energy that Edison himself had now become. Thirty-five and
at his mental and physical peak, he was everywhere and nowhere to
those who tried to keep up with him or merely corner him long
enough to get a recall of the old Menlo Park experimenter, always
willing to stop and chat, doodle out tunes on the organ, and swap
lunch boxes—even, on occasion, shutting up shop, renting a boat,
and taking the "boys" out fishing off Sandy Hook. Over and above
medals, he had gotten what he most wanted from the exposition—
international respect.

---

* Swan also conceded privately to Lowrey that "Edison is entitled to more than I. . . .
He has seen further into this subject, [more] vastly than I, and foreseen and provided for
details that I did not comprehend until I saw his system."

This, however, did not help him much in downtown New York, where Kruesi's Electric Tube Company was finding the work of completing the First District's distribution system almost prohibitively slow. The businessmen and householders who had signed up for electric lighting were wired up and waiting to see and smell the last of their gas mantles. But the city would not allow the laying of mains and feeders under the streets during the day, and the delayed delivery of copper cable and parts had prevented a start to nighttime excavations until the fall. A troop of Irish navvies was racing to dig as many trenches as possible before the subsoil froze. Edison saw that he would have to abandon his dream of illuminating the District in November, and that he would be lucky to do so within the next year.[170]

The navvies had fifteen miles of iron pipes to connect, twice that length of half-moon conductors to thread, hot insulation compound to pour, and heavy junction boxes to bolt down, toiling beneath harsh arc lamps and enduring the ire of pedestrians who wanted to know why current could not be distributed via overhead wires. It was difficult for them to understand that something as bodiless as electricity needed protection. The work was filthy and dangerous, with accidental gas leaks and at least one short-circuit that lifted a passing horse off the wet cobbles.[171]

Edison often helped out in the trenches, as if his own muscle would accelerate the Tube Company's forward progress of no more than a mile a week. He rejoiced in hard labor and often did not go home to sleep. Instead he napped on the spare tubes that Kruesi stored in the cellar at number 255, half-fulfilling his fantasy of men who slept on telegraph poles. He did not seem to care that the iron rounds were tarry and striped his overcoat. Nor was he bothered by the damp as winter approached. "I had two Germans who were testing there," he reported, "and both of them died of diphtheria."[172]

In contrast to these nights downtown, he enjoyed cerebral evenings at Delmonico's with a new friend, the great Hungarian violinist Edouard Reményi. They could hardly have come together from cultures farther apart, but to Reményi, Edison's technological talk was a new kind of music. "Since I was with Victor Hugo and Liszt," he wrote after one of their dialogues, "I was never so much in intellectual heaven."[173] He jokingly appointed himself "court musician" at the

Machine Works, as he had once been at Windsor Palace, and treated Edison to several private recitals there and at 65, weeping as he played. When Edison asked why, he said, "I always weep when I hear really good music."*[174]

By November the Works had geared up to the extent that it had a backlog of well over 130 smaller generators for sale, along with successors to the giant Paris and London machines. This contrasted with a double decline in productivity and quality at the Lamp Company in Menlo Park. Edison resolved on "a grand bounce of the bugs" before he moved that facility to its new quarters in East Newark. He recrossed the river and within eight days had lengthened lamp life from four hundred hours to six hundred. "I had just 18 hours sleep that week without my boots being off." Not trusting Francis Upton to improve on his improvement, he elected to stay at the Park all winter if necessary, until his bulbs were twice as energy-efficient as they had been.[175]

Christmas marked the tenth anniversary of his marriage to Mary. She celebrated by reopening their house and hosting a dance party. It was elaborate enough that Insull asked the Pennsylvania Railroad to make special arrangements for guests traveling back to New York in the small hours.[176] Edison's deafness prevented him from enjoying such occasions as much as she did. But the Patent Office had a gift for him on 27 December: the award of a patent, U.S. 251,545, on his electrolytic meter, a coilless device so simple that it measured current without needing any to operate itself.[177]

Much as his future lighting customers were going to dislike it, the meter was, with the possible exception of his big dynamo, his most important invention of the year. Without a reliable tally of power consumed or power saved month by month, the Edison Electric Illuminating Company of New York could never function profitably. There was no shortage of other devices almost as ingenious, such as an "electrical knockdown chandelier" that could be pulled apart and reassembled without tortuous rewiring. Edison was in the midst of a phenomenally fertile period, executing, on average, one new patent every four days.[178] His total of successful applications in 1880 had

---

* In 1898 Edison served as a pallbearer at Reményi's funeral in New York.

been fifty-nine; this year would see another ninety, and next year well over a hundred. Over the entire decade he would average one patent a week (starting with the electric light and ending with a hydraulically regulated phonograph), while combining the duties of manufacturer, engineer, entrepreneur, publicist, plotter, executive, and family man in a torrent of hyperactivity to be stopped only twice, and temporarily, by *das Ewig-Weibliche*.

DESIGNED TO FLASH DAGGERS

When Chief Sitting Bull, one of the celebrities who chose to drop in on Edison unannounced, saw the jumbo dynamo destined for London, he allowed that it was "damn big." That was also the opinion of William H. Preece, consulting electrical engineer to the British Post Office, who had boggled at its predecessor in Paris. In an address at the Royal Society for Arts, he informed his colleagues that "those who are interested in this machine, and everyone should be, because it is a decided step in advance—will soon have an opportunity of seeing it at work at 57, High Holborn."[179]

Preece was referring to the Edison central station system that Edward Johnson was installing in London as a curtain-raiser to the opening of the Crystal Palace Exhibition on 25 February 1882. Although the station was not intended to be permanent (it was part of a test incandescent-lighting project organized by the London County Council) and would illumine only half a mile of the Holborn Viaduct, there was no question now that it, and not 257 Pearl Street in Manhattan, would be the true cradle of incandescent street lighting.[180]

It could be finished quickly because Johnson did not have to go underground to wire up the buildings lining the viaduct (despite its name, nothing more than a broad thoroughfare elevated above Farrington Bridge Road). All he had to do was string his mains and feeders under the supporting stonework, along conduits already hollowed out by the city's gas utility.* John Kruesi had no choice, meanwhile, but to wait for the subsoil of downtown Manhattan to thaw and permit the completion of his distribution system. If he could get that

---

* Today the conduit access door at Farrington Bridge is still labeled "North Thames Gas."

done by midsummer, there was a good chance that the First District could be lit up before the fall.

Johnson was an eager, honest, torrentially garrulous promoter of whatever business he happened to be in at any given time. He had begun his career selling telegraph equipment out west and would end it selling milk cartons in upstate New York; currently he was working with absolute devotion on behalf of Edison's telephone and lighting interests in Britain. "There is but one Edison," the London *Daily News* remarked, "and Johnson is his prophet."[181]

With the help of Hammer and Jehl, sent over by the Electric Light Company as consultant engineers, Johnson literally dazzled the British press on 19 January with a coruscation of four hundred Edison lights along the viaduct, and 250 more at a black-tie dinner in the Crystal Palace. It went without saying that any lamp display in that building was bound to reflect in many directions. But the chandelier Johnson hung in the concert room, with its own crystals multiplying the bulbs ten times over, was designed to flash daggers into the heart of any gas industry executive present. Both installations were expanded in the weeks that followed, as extra dynamos ramped up their voltage. The viaduct system eventually reached a capacity of three thousand lamps, and the Crystal Palace one thousand—some of which were rigged by Hammer to spell out the letters E-D-I-S-O-N, making his boss the first man ever to have "his name in lights."[182]

Edison's hopes of illuminating all of London, however, were dashed when Parliament adopted an anticommercial Electric Lighting Act, effectively discouraging central franchises. This did not prevent him from forming a British subsidiary, the Edison Electric Light Company, Ltd., in March, to join a number of European start-ups, proliferating like branch feeders in the afterglow of his Paris triumph. In France alone he organized the Société industrielle et commerciale to manufacture lamps under the management of Charles Batchelor, the Société électrique Edison to build central stations locally, and the Compagnie continentale Edison to do the same across Europe. One of the most successful of these licensed plants was the Deutsche Edison-Gesellschaft in Berlin. Edison lamps shone in the railroad station at Strasbourg and the grand foyers of the Paris Opéra and La Scala in Milan. Teams trained by Batchelor installed isolated systems as far away as Finland.[183]

Joseph Swan, competing strongly, put a system of his own into the Savoy Theater in London, well before Francis Jehl wired up a municipal theater in Brünn, Bohemia. However derivative Swan's new lamp might or might not have been of Edison's—a question that could be settled only in a court of law—it was equally efficient if not superior with its filament of parchmentized cotton thread, even smoother and harder than *madake* bamboo. Johnson thought the best thing for both inventors would be to merge their British interests. Edison would not hear of it. Swan's gentlemanly concession of victory to him at the Paris exposition left him unmoved: "My own private opinion is that he tries to claim other peoples work & carries to extreme the idea of enormous respectability while being at heart what his compatriots call a 'bloody liar.' "[184]

RED, WHITE, AND BLACK

Mary Edison suffered during her husband's hyperactivity that winter with uterine troubles and an attack of depression. "She seems very nervous and despondent and thinks she will never recover," the family doctor wrote Edison. "She seems so changed physically and mentally of late that I think something ought to be done."[185]

He suggested she be taken to Europe for a few months. But the best Edison could do, with resumption of work at Pearl Street looming, was escort her and the children to Florida for four weeks in March. He was exhausted himself, after a spell of seventy-two-hour working jags, and also under doctor's orders to get away. It was his first visit to the Sunshine State. The strawberry season had begun, and the sulfurous waters at Green Cove Springs, a resort on the St. John's River in Clay County, were therapeutic. Mary was in no hurry to go back north. Insull, who enjoyed writing letters in Edison's name, was pleased to hear nothing from "the Great Mogul" until the twenty-eighth, when he suddenly announced his return home.[186]

For the moment that meant Menlo Park rather than Manhattan, Edison not having relinquished his takeback of the Lamp Company. He felt that it was now ready to relocate to the enormous plant he had bought for it in East Newark. The transfer, supervised by Francis Upton, began on the first of April, and more than a hundred local jobs melted away.[187] For as long as Edison needed his laboratory and electric railway for experiments, and Mary the house as their country retreat, Menlo Park would retain some signs of life. But with the Uptons and the Batchelors and Kruesis gone, and Mrs. Jordan's boardinghouse in need of guests, and the lamp works standing empty, its ghost days were near.

Edison returned to Manhattan and rented a suite in the Everett House, a luxury hotel on Union Square.[188] It was to be a base for him during what promised to be the most urgent summer of his life. If he took much longer to light up the First District, after Batchelor and Johnson had won such cheers for him last autumn in Paris and London, he could expect only ruder noises from the citizens of New York.

Already, *The New York Times* reported, there was "grumbling" by Electric Light Company subscribers in the First District, tired of seeing dead wiring hanging out of their walls. Trench work had resumed with the spring thaw, but there were still seven miles to go, at one thousand feet a day. The paper sent a representative to ask Sherburne Eaton if he had set a completion date, and his replies made plain the pressure on Edison to deliver.

A We can fix no limit whatever. We should have completed the work before the frost came last fall, if the parties who furnish our material had not failed to keep their contracts.
Q Can a very distant limit be fixed, say four months?
A Not definitely.
Q Will it be concluded in a year?
A I can't positively fix any limit whatever. Our contractors may disappoint us again about material.
Q Do you expect to be delayed again by frosty weather?
A. You can judge that as well as I can. . . . We are doing our best to get the wire laid, immediately after which we should be able to light the lamps.[189]

Edison was further driven by the desire of his backers to develop a lucrative and less costly alternative to the construction of central stations—isolated systems for private customers like J. P. Morgan, or suburban factories and small towns.[190] That switch of interest signified that financing for a Second District in New York might be a long time coming or never at all if he did not finish the First before another winter came round.

Four 240-horsepower boilers were installed at 257 Pearl in the spring, along with three jumbo dynamos on the second floor and a mini-avalanche of auxiliary fixtures next door. The first dynamo, directly shafted to its Porter-Allen engine, whirred into life on 5 July. Three days later it was connected to the monitoring panel on the top floor of number 257.[191] A vision of a future unimaginable even by Edison materialized when the wall-mounted oblong lit up: a thousand bulbs packed close in rows, their brightness flickering at different strengths according to the current that fed them.*

The pace of pipe laying increased during July, Edison helping out as before. It did not stem his flow of patent applications, more tumultuous now than it would ever be again. Among them were a coal conversion method of power generation that essentially presaged fuel cell energy, and a 330-volt overhead "village" distribution network that reduced the already economical copper quotient of the Pearl Street system.[192] He simultaneously and brilliantly invented, but did not have time to caveat, a three-wire branch circuit that interposed a neutral conductor between two "hot" ones of 220 volts each, permitting independent operation of multiple lamps at 110 volts. It too looked forward to a time when red, white, and black wires would be standard equipment in American homes.†[193]

His total of fifty-three successful patents that spring and summer did not include foreign ones, or the seventy-eight applications lost or stolen from him by his alcoholic patent attorney, Zenas Wilber. "I am free to confess," Edison said in later life, "that the loss of these 78 inventions has left a sore spot in me that has never healed. They were important, useful, and valuable."[194]

While oddly forgiving of Wilber, he took the precaution of hiring a

---

* When the black-and-white image of this panel printed in *Scientific American,* 27 August, is viewed on a modern computer screen, the rows of lights suffuse with patterns of refractive color. See https://babel.hathitrust.org/cgi/pt?id=pst.,000062999472;view=1up;seq=137.

† Edison delayed executing his patent on this invention until 27 November 1882, allowing John Hopkinson to file a similar application in Britain well before that date. The Edison system was awarded priority in the United States on 20 March 1883 (U.S. Patent 274,290), but by then Hopkinson already had his British patent.

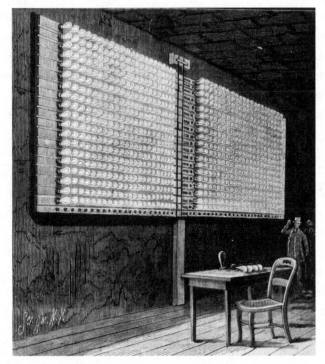

Power monitor panel, Edison Pearl Street station, 1882.

new young intellectual property lawyer, Richard Dyer, and gave him the task of organizing his cumulative total of letters patent, now numbering more than four hundred in the United States alone.[195]

August was a month of rapid progress for the Edison Light, Lamp, Isolated Lighting, and Electric Illuminating companies and especially for the Electric Tube Company and the Machine Works. All of them were aware that the consummation of their effort was in sight, with a momentum that seemed determinate now rather than willed by any manager. Even Edison was borne along. Lamp production in Newark rose to fourteen hundred a day, with an estimated capacity of thirty times that output. The factory had so many dynamos to build, thanks to orders coming in from Edison's foreign enterprises, that it had to put them on hold while it finished the six jumbos he needed for Pearl Street.[196]

Finally the full column of paired engines and dynamos stood ready at 257 Pearl. Kreusi paved over the last of the First District's feeders

and mains and completed connections to the premises of major system subscribers—most notably, the *New York Times* building on
Park Row. So far the Electric Light Company had 946 customers with
well over fourteen thousand lamps installed. Property owners and
tenants who had not yet signed up were wooed with circulars promising no charge for installations unless "you ultimately decide to adopt
our light permanently."[197]

Scientific American published the first detailed description of "the
Edison Electric Lighting Station" system on 26 August. It was illustrated with exquisite technical engravings that conveyed, better than
any screened photographs, the radiance of the high monitor (every one
of its thousand bulbs limned), the frightening mass of the dynamos,
the ocean liner dimensions of the stoke hole, with its sixteen furnaces
and attendant *Nibelungen* (all wearing neat black bowlers), and the
precision engineering of the two dozen street conductors attached outside, beneath the Pearl Street sidewalk, in a service hall bright with
reflector sconces, unknown to the pedestrians clomping overhead. The
text explained how all the moving devices worked, from the coal conveyors to a giant switch, something like a triple-bladed guillotine, that
could be used to slice any dynamo out of circuit.[198]

"We have no doubt," the editors wrote, "that before this paper
meets the eye of the reader, the district will have been illuminated."[199]

In the days immediately following, gas company workers were seen
removing their globes from streetlamps around Pearl and trundling
them away in wagons.[200]

AN END TO THE DOMINANCE

Another issue of *Scientific American* had gone to press before Edison
felt ready to activate his system on Monday 4 September. He did so
with some dread, and none of his usual theatrical flair, powering up
just one dynamo at three in the afternoon and directing its current to
a scattering of customers across one-third of the district.[201] It was as
if he hoped daylight would blank out the failure of any of his "luminous horseshoes" to incandesce. Workers at *The New York Times* did
not notice until dusk that there was any change in the light they were
used to. They turned the thumbscrews on their office walls and, instead of the flickery glare of gas mantles, found themselves bathed in

a soft glow that remained pleasantly steady. Commuters walking down Fulton Street to the Brooklyn Ferry noticed circles of the same light on the sidewalk. Those who glanced up saw pear-shaped globes with porcelain shades hanging from iron crooks, each filament leaving a tiny imprint on the eye if stared at too long.[202]

By seven o'clock it was dark enough[203] for reporters seeking Edison out to be surprised that the lamps of Pearl Street had not been cut in, although the station itself was radiant. He was found on the second floor at number 257, looking gleeful and as slovenly as ever in a high-crowned white hat and collarless shirt:[204] "I have accomplished all that I promised." Asked why he had not illuminated the whole district, he said that he would have, but for the insistence of the New York Board of Fire Underwriters that the city should sign off on every wired building. He rejoiced in those that had already been connected: "The lights in the office of Drexel, Morgan & Co., half a mile away, are burning as brightly as the lights here."[205]

Brightness, however, was not a novelty anymore. The sixteen-candlepower average of the eight hundred lamps aglow that evening was pallid compared to the intensity of arc lights in public places elsewhere in the city. The revolution Edison had wrought was so unobtrusive and at the same time so world changing that few, if any, of the people who experienced it realized what had happened: an end to the counterbalance of night and day that had obtained for all of human history, mocking the attempts of torchbearers and lamplighters and gas companies to alter it with their puny waves of flame.

A TUSSLE BETWEEN ELEPHANTS

The only morning papers that paid attention to the inauguration of central station service in New York were, not surprisingly, the two with the most Edison lamps in their newsrooms: the *Times* and the *Herald*.[206] There were a few other brief reports across the country, none conveying much excitement. The best that could be said for British comment was that it was respectful, perhaps because Edison had for once achieved something without boasting about it. An editorial wisecrack in *The Boston Globe* came nearest to the truth of the matter: "Chevalier Tom Edison has had an 'opening night.' His aim is to open night until it shall be as day." *[207]

For a week Edison slept on a cot at the station, determined to keep its output up as subscriptions to the First District network multiplied. Foreseeing a time when he would need his full battery of six generators to supply the demand of sixteen thousand lamps, he experimentally yoked two jumbos together, in a union that he thought would make for a smooth aggregation of power. As things turned out, he could not have been more mistaken.[208]

> The moment we threw in the second engine the first engine slowed way down and the second engine jumped up to speed almost in an instant, and then went two or three times its speed, until we thought the building would collapse. Then the other engine would speed up, and they would see-saw, from 50 revolutions a minute to 800 revolutions a minute. Nothing of steel or iron could stand it. The commutator brushes burned and red-hot globules of copper flowed down on the floor and began to burn the wood. Smoke poured all over. The building was apparently going to come down, and everybody made for the stairs. Finally I yelled to shut down, and two of the men jumped in and closed the throttle.[209]

Edison had encountered the phenomenon of "hunting"—the contrary torques of a rotary machine searching for a balance between the mechanical power applied to it and the electromagnetic forces inside it. But stability is key. The boards beneath the Pearl Street generators rested on a bridge of cast iron. Strong as it was, it stood free of the walls, transmitting the vibrations of each dynamo downward, in the same direction as gravity operated the Porter-Allen engine governors. They were extra sensitive and, confused by the electrical connections between dynamos that let one run as a generator one moment and as a motor the next, began to hunt wildly for equilibrium. The result was changes of speed in one dynamo that conflicted with changes in the other. It was a tussle between elephants no trainer could handle, com-

---

* In 1878 Edison had been appointed a *chevalier* of the French Legion of Honor for his work on the phonograph.

plete with deep groans and shrieks. Edison was lucky not to lose his station.[210]

For once in his life he needed a slug of liquor and went across the street with Edward Johnson to get one. "Am I to drink the whole of that?" he asked as he watched the glass being topped up.

"Yes," Johnson said.[211]

The next issue of the Electric Light Company's promotional bulletin made no mention of the near-catastrophe but admitted that there had been problems at Pearl "of a peculiarly mechanical nature relating to the imperfect regulation of the engines." Edison solved the problem by devising a tubular connecting shaft, full of trapped torsion, that brought the Porter-Allens into sync, but he thought it wise to order new engines with more centrifugally weighted governors from Armington & Sims. Meanwhile the station continued to operate, as it would do, with only two short service breaks, for the rest of the decade.[212]

NOW THAT IT LIVES

The transfer of a personal library from one home to another is always, for an intellectual, a sign of irreversible change, and for an inventor, a transfer of test tubes and precision instruments amounts to the same thing. Edison accomplished both at the end of September 1882, taking a two-year lease on a gray stone townhouse at 25 Gramercy Park and opening a new laboratory on the top floor of the Bergmann factory at Avenue B and Seventeenth Street in Manhattan. For the time being he held on to his country house but said that "because of the women constantly bothering him," he would henceforth operate out of New York.[213]

Mary Edison was no doubt a member of this female lobby, along with his daughter and Mary's younger sister Eugenia, a recent addition to the Edison ménage. Marion, at nine and a half, had endured a year in boarding school and looked forward to the more pleasant prospect of living with her parents in the most fashionable quarter of the city, while she and Tom attended Mlle. de Janon's "English and French School for Young Ladies and Children" nearby.*[214]

* Thomas Edison, Jr., attended the same school.

Edison grumbled to Insull about preferring life across the river, but the young man was not fooled. "Johnson and myself are of the opinion that it is six of one and half a dozen of the other," he wrote Charles Batchelor, "and that he wants to come in just as much as the women do."[215]

The lease on Edison's new home included furniture and fittings but not, apparently, many books. To supplement those he brought from Menlo Park, he ordered sets of the novels of Dickens, George Eliot, Hugo, Cooper, and Hawthorne, as well as *Don Quixote, Gil Blas,* Longfellow's complete poems, Macaulay's essays, and a number of other volumes—"good solid binding only nothing fancy." The house had been Mary's choice out of nineteen other properties available, but he felt a sense of grand design when he explored the attic and found the private diary of Samuel Morse, inventor of the telegraphic code that had once been—still was—his second language.[216]

The last woman to "bother" him in 1882 he scarcely knew and could not have cared less about, except that a New Jersey state court found on 18 December that he owed her $5,065. She was Mrs. Lucy Seyfert, and she based her claim on a promissory note he had written six years before while raising capital for the Automatic Telegraph Company. Edison remembered the note but also that it had not then been held by her. The blood of old Sam Edison, a compulsive litigant, arose in his veins, and he declined to pay. As a result, the case was referred for trial at the New Jersey Supreme Court.[217]

Except for that unpleasantness, promising more discord later, the year rang out happily for Edison. It would be a while yet before all the lighting enterprises he had started became profitable or even proved themselves individually viable. The Pearl Street project in particular had years to go before it would return a dime of its $600,000 capitalization. Its growth rate from four hundred subscriber lights in September to nearly five thousand in December looked impressive, but offering people power for free at first had much to do with their willingness to wire up. Still, there was no mistaking the admiration with which Britain's newspaper of record recognized the "constant and equal" amperage put out by the central station and the superior effulgence of Edison's bulbs over any manufactured in England. The success of his system, declared *The Times,* "is now beyond question."[218]

Another money loser for the moment was the Edison Electric Lamp Company. Its hugely expensive factory in Newark was not yet at the break-even point of producing fifteen hundred bulbs a day. Each one had to go through two hundred delicate processes before shipping and sale, at forty cents apiece—a price that was sure to come down as output cranked up.[219] Lamp life and lamp quality were steadily improving. If the plant ever reached its intended capacity of forty-two thousand bulbs a day, the rate it was designed for, it might well relabel itself the First Edison Bank of New Jersey.

The Machine Works (which Batchelor had successfully duplicated at Ivry-sur-Seine in France) had the contrary problem of being so productive that it was warehousing seven unsold jumbo dynamos—"very heavy stuff for us to carry," Insull complained, with his English tendency to make lame puns. Despite this $140,000 liability, it was at least in the black and had already paid Edison $38,000, the first decent stash he had pocketed all year.[220]

By far the most lucrative of his start-ups was the Edison Company for Isolated Lighting. One hundred and thirty-seven domestic or small-business plants were now on its books, all running Edison dynamos and burning Edison lights. The most prestigious of these was in J. P. Morgan's midtown mansion on Madison Avenue, where it short-circuited frequently, set fire to the mogul's desk, frightened his horses, and drew noise complaints from neighbors, but otherwise gave complete satisfaction.[221] Better reports came from distant locations, none more remote than a sawmill at Yväskylä, Russia, north of the sixty-second parallel. The town councilors were so pleased with its light that they had voted to upgrade to a central station. In the shipyards of Glasgow, Scotland, crowds queued in drenching December rain to tour a new Edison-illuminated steamer. Even its shaft tunnelway was aglow.[222]

Edison foresaw many corporate problems in the immediate future, as he sought to lean on the Electric Light Company directorate for more central station capital and they looked for easier money by offering "help" with his independent companies. He wrote in a draft memorandum to the English Light Company, "I have nursed the baby so far & I believe I can continue to do so without any extraneous aid, especially from those who said the baby would never be born & when born would never live, & now that it lives wants to change the manner of nursing. If I should fail in any particular it will then be time to call in other inventors."[223]

IS NOT A PATRON

Except for another Florida vacation with his family, which he cut short before February was out, Edison was not seen much in public in the early months of 1883. He luxuriated in the spaciousness of his sixth-floor laboratory in the Bergmann building, discovering, typically, that the elevator ride up there lasted as long as it took him to wind his watch. When he held the grilled stem against the shaft column as he rose, he saved his thumb and forefinger much labor and arrived at the top fully wound.*[224]

He had become so identified with electric light innovation since 1880 that occasional news reports noted, almost with surprise, that he was still capable of inventing other things, such as a horse-drawn truck that scooped up snow, compressed it by 90 percent, and deposited it in the form of neat ice blocks that left the rest of the street clear. He also toyed, for reasons best known to himself, with the idea of vacuum-packing bran. Much to his regret, he had to give up on another device that he had patented out of farthest left field two years before. It was a magnetic iron ore separator, designed to refine the sheets of black sand that covered the beaches of Quogue, Long Island, and Quonochontaug, Rhode Island.[225] Major Eaton had become as excited as he at the ferric richness of those deposits, in some places twenty feet deep. Yet the Edison Ore-Milling Company they formed together had never flourished, not least because the sea that had washed up the deposits in the first place kept reclaiming its own. As Edison later groused, "It was too much like taking out a mortgage on a school of herring."[226]

The winter somewhat strained his relations with Grosvenor Lowrey, hitherto his best friend on the board of the Electric Light Company. A Wall Street man at heart, Lowrey waved aside the complaints of Edison, Upton, and Johnson that the company was greedy in taking ten cents for every lamp it sold, at cost, for forty cents. Major

---

* Sam Edison, just as typically, delighted in climbing all the stairs whenever he visited the factory. At eighty-two he had the legs of an elk and a chest expansion that his son was proud of: "I think it was five and one-half inches."

Eaton, who was now president, darkly hinted that the autonomous manufacturing shops they represented—the Machine Works, plus the Lamp, Tube, Isolated, and Bergmann companies—would be better off under corporate patronage.[227]

Every hackle Edison possessed rose at this takeover threat. He and his partners signed a joint nonnegotiable declaration of independence, reminding Eaton that the Electric Light Company had been ungenerous to them back in the days when they were trying to capitalize their shops. The letter was drafted by Johnson, who may have recalled the cry of another Johnson, to another plutocrat, 127 years before: "*Is not a Patron, my Lord, one who looks with unconcern on a man struggling for life in the water, and, when he has reached ground, encumbers him with help?*"[228]

AMERICAN CERTAINTY

"How your genial—delightful light spreads—and it spreads—and spreads and spreads," Edouard Reményi wrote Edison from Lincoln, Nebraska, on 25 April. He had just attended a banquet where he had basked in the double glow of Edison lamps and his own intimacy with the man who had made them: "I—old fiddler I brag that I am your faithful and affectionate friend and your court musician."[229]

Edison was aware of the spread himself. He had already taken steps to go after the "Village Plant Biz" by forming the Thomas A. Edison Construction Department, an independent company that would install cheap, overhead-strung central lighting systems in the provinces. Although the name of his new venture was strange—how could a department be a company?—and seemed contrived to disguise its purpose, he headquartered it at 65, as if to reprove Major Eaton for not being more interested in central stations outside New York.[230]

He wrote to tell Edward Johnson, who had gone back to London to straighten out the tangled affairs of the local Edison Electric Light Company, that it would be as well to let the management there muddle along and come home for good. "Just now we are doing all we can to rush the Village business. There is immediate money and plenty of it in that. . . . Here we can get things done just as we say, and I therefore think that it is better to concentrate our efforts on American certainty rather than an English possibility."[231]

Johnson was respected in London and did not want to be seen there as a quitter. He needed a couple of months to finish his task before joining the Construction Department. Besides, he wanted to help a London barrister argue a patent infringement injunction suit Edison had insisted on, against Joseph Swan in the High Court.[232] The prospects for a favorable decision were not good, considering that the court building was illuminated with Swan bulbs.

Edison appointed him a partner in absentia, along with Batchelor, Insull, and Eaton, who could hardly be left out. The little major was a good-natured, if cautious person, the sort of gray bureaucrat who burrows molelike into positions of great power. Somehow he had become president not only of the Edison Electric Light Company, but of the Isolated Company too. The cooperation of each was necessary if Edison was to use his own patents for Construction Department projects. In return, the parent company would have a share of the income from every new station that joined the spread of his "delightful light."[233]

For Insull—another burrower, but much more aggressive and devious than Eaton—his partnership amounted to a reward for two years of self-sacrificing service to Edison. It came with the proviso that he must handle the new Department's finances, which were bound to be complex, as well as run its head office, while Johnson functioned as sales director and the Old Man supervised designs and installations. Still, the more power he got, the happier Insull was—especially when Edison agreed to pay him an annual minimum of $2,400 on all regional plant profits, plus a generous 20 percent of the rest. On 3 May Insull also took financial control of the Machine Works, potentially the most lucrative of all Edison's businesses.* The manager of that enterprise "kicked," he wrote Johnson, but "Edison supported me in a bully fashion & I came out top of the heap."[234]

BUSINESS PURE AND SIMPLE

Except for a highly successful, nontaxpaying enterprise undertaken at age twelve, to do with the sale of candy, fruits, and newspapers, Edison had never before acted alone as a businessman. The Construction Department was his own commercial conception and responsibil-

---

* As indeed it was, when it became General Electric.

ity, launched with $11,000 of his own money and likely to enrich him, or impoverish him, to a far greater extent than it would his partners. For every twenty dollars they won or lost, he would gamble sixty.[235] Consequently he had to forsake what he loved most—experimenting and doodling in notebooks—and take on a new identity that shocked many who knew him.

"What has happened to 'the wizard of Menlo Park?'" a *Brooklyn Daily Eagle* correspondent wrote on 29 July. "The last time I saw Edison he had grown very stout, and no longer wandered around with a misty far off look in his eye and a battered felt hat on his

Mary Edison and feathered friends, 1883.

head. Instead he wore a shiny beaver, gold eye glasses, and looked fashionable. . . . Perhaps too much prosperity—for Edison has made a vast fortune—has driven all ideas of inventing out of his head."[236]

There were elements of caricature in this description, although Mary Edison had certainly discovered that her abstemious husband had a weakness for pie. She was partial to it herself, as well as to expensive Huyler chocolates, which she ate by the pound.[237] Her gowns grew larger and more elaborate by the season. She posed in one brocaded creation for a studio photographer, who needed all the focus he could get to delineate the dead, stuffed, red-and-black birds pinned to her breast and thigh.*

Edison confirmed in an August interview with *The Evening Post* that

---

* The number of cardinals on this dress, designed by Madame Anna Duval, was seven. It cost $391.90, or $9,500 in 2019 dollars.

he was taking "a long vacation" from his workbench. "I am going to be simply a businessman for a year. . . . I won't go near a laboratory." Sounding more like Insull than himself, he ran off a long list of the contracts the Construction Department had already signed with regional municipalities: "Sunbury, Pa., where we are putting in 500 lights; Shamokin, Pa., 1,600 lights; Brockton, Mass., 1,600 lights; Lowell, Mass., 1,200 lights; Lawrence, Mass, 4,000 lights." Before he ran out of breath, he had strung imaginary wires across Ohio, Wisconsin, and Minnesota as far west as Davenport, Iowa. "I am so convinced of the system's success that, as I said before, I have given up inventing and taken to business pure and simple."[238]

The trouble with such interviews was that they were read by Edisons less fortunate than he, such as his fifty-two-year-old brother Pitt, a farmer in Port Huron:

Dear Bro
I see by the papers that you are agoing to be a buisness Man for a
year . . . I keep a good man on the farm so it is not nessesary for
me to be thare much of the time for a year at least now al can't
you place me somewhere for that time thare is lots of work for me
yet I would not care whare I was placed in New York or any
whare Else[239]

Edison replied, "I think the best thing that you can do is to look out for something where you are."[240]

Another reader of the *Evening Post* article may have been Henry Rowland, the Johns Hopkins professor who had praised the efficiency of his lightbulbs three years before. Edison's worldly success since then, together with speculation (incorrect, as it turned out) that he was now a millionaire, was evidently on Rowland's mind when he delivered an impassioned "Plea for Pure Science," at the August meeting of the American Association for the Advancement of Science in Minneapolis. Refusing to dignify such "conveniences" as telegraphs and electric lights by the name of science, he said that money-seeking, manufacture, and the pursuit of fame were hindrances to intellectual progress. "It is not an uncommon thing, especially in American newspapers, to have the applications of science confounded with pure science; and some obscure American who steals the ideas of some great mind of the past, and enriches himself by the application of the same to domestic uses, is often lauded above the great originator, who might have worked out hundreds of such applications, had his mind possessed the necessary element of vulgarity."[241]

As Rowland proceeded with his address, it became clear that he was pleading, not for pure science per se, but for more funding for university laboratories—a complaint that would be renewed a century later.[242]

BOUND BY PRIDE

By now Edward Johnson was back from London. As he expected, the High Court had rejected *Edison v. Swan* as a suit of no value. Mr. Justice Chitty held that the plaintiff had failed to show any fundamental dissimilarity between his filament and the defendant's spaghetti-thin carbon rod. Edison's case would have been stronger if he had not carelessly neglected to describe, in his own application for a British patent, the unique "running on the pumps" method he had devised to suck occluded gases out of a bulb when the carbonized element was first heated *in vacuo*.[243]

Edison felt that any judge able to see past his wig flaps should understand that there was a difference between a flexible black fibrous hoop that incandesced cleanly, and a brittle stick that smoked up its bulb in less than an hour. (Swan's more recent parchmentized cotton filament was not part of the priority issue.) Johnson repeated his urgent suggestion that there should be a merger of the Edison and Swan United Electric companies.[244]

Reluctantly, Edison agreed. But he made it as difficult as possible for Swan to agree too, by insisting that the joint concern "shall be distinguished by my name in its title solely." He disclaimed, with an ingenuousness sure to make Swan's representatives groan, "any such feeling as might naturally be imputed to me of wishing to gain in this way a concession as to the disputed claims of Mr. Swan and myself concerning lamp patents." On the contrary, he would be happy to drop out of the British market altogether, were it not for the importance of having his rights to all other aspects of his system recognized around the world.

I remain in this country, and wherever else I can, as large an owner as possible in my inventions. I have never parted with any of my holdings except when compelled to in order to carry on my various works. . . . I am bound by pride of reputation and by pride and interest in my work, to remain interested in the business. I expect to be a large owner in all companies employing my inventions after most persons now interested shall have sold out and retired with their profits.[245]

Before signing off on a letter that was supposed to be a sober statement of terms, he could not resist adding a sarcastic suggestion that if Swan had contributed as much to the science of lighting as he had, "then his friends may with equal force say what I have said."[246]

Johnson could only allow him to send it off as Exhibit A in what was sure to be a legal battle. Then he, Insull, and Edison turned their attention to the more immediate exigence of making the Construction Department viable.

OUT OF THE DAM-D HOLE

They found that there was a budgetary penalty to Edison's public offer to plan, wire, and light any town or village "within sixty days." Before a contract could be drawn up, the distribution area had to be surveyed and canvassed street by street, so that Insull could figure how much revenue to expect, and how many physical obstacles to overcome, in arriving at a quotable price. This cost serious money. More often than not, the outlay had to be swallowed when towns (including several Edison had boasted about) chose not to proceed. If they did, he was charged with the initial costs of manufacturing the necessary hardware, plus providing transportation and on-site labor— not to mention teaching the local illumination companies how to maintain their systems. Since everything about central station technology was so new, capable engineers were almost impossible to find. Edison therefore needed to establish a school at the Machine Works to train men for the job and persuade them that life in places like Canyon, Arkansas, was just as interesting as life in New York.[247]

There followed the difficulty—often the impossibility—of getting paid in cash for each central station installed. As early as mid-

September, five regional utilities owed Edison $43,000. Others were too poor to pay at all, and he had to accept their promises of stock dividends. It was better than nothing, which he often had to settle for. He spent $10,400 to canvass eighty cities and towns, only twelve of which ordered systems.[248]

The only encouraging aspect to his enterprise was the steady increase in customers once a community lit up. But that meant future, not present income, and meanwhile most of the Construction Department's assets and liabilities accrued to his personal account. He was rich—incalculably rich—in expectations, and poorer by the day in reality.

An expense he could cut was that of his posh townhouse in Gramercy Park. He had been falling behind on rental payments for several months. "Would get out of the dam-d hole if could," he scrawled on one of James Pryor's bills, referring it to Insull for inaction. (One of the reasons Edison liked his secretary was that Insull was a virtuoso prevaricator.) Mary's invalidism gave him an excuse to beg for cancellation of his two-year lease. "I very much regret to say that owing to the illness of my wife it has become imperative that she should give up housekeeping in accordance with the Doctor's instructions."[249]

Pryor declined, with contempt for this hiding behind skirts, but agreed to let him sublet the house and "lighten the burden under which you find yourself placed." Edison removed his family to a luxury annex of the Clarendon Hotel, which was twice as expensive but at least had a more corporate tolerance of credit.[250]

He was not as short of money as Insull made out, since he repaid two loans totaling $42,806 to Drexel, Morgan in December. At the same time he instructed his counsel in the Lucy Seyfert case to postpone "just as long as possible" any settlement of the judgment against him. He preferred to risk the wrath of the Supreme Court of New Jersey than to lose the goodwill of Wall Street.

When Christmas came, Mary let no financial considerations cramp her style. She loved giving presents, some of which—for lack of any later ones—Marion remembered with especial vividness: "a first edition of La Fontaine Fables, with beautiful etchings, a ring with diamonds and turquoise, a Le Maire mother-of-pearl Opera Glass which was in a blue velvet case."[251]

Edison's personal gift to himself was the knowledge that he now had 12,843 lamps shining around Pearl Street, plus a further 64,856 nationwide.[252]

YOURS DEVOTEDLY

Around this time he became aware of a twenty-three-year-old Scotsman with a ridiculously long name hanging around the testing room at the Machine Works. "W. Kennedy Laurie Dickson," as this engineer signed himself over a double curlicue, had been hired by Insull several months before, on the strength of a recommendation that spoke to his electrical training and mastery of French and German. Dickson was also a master of flattery, to which Edison was not immune. "If you only knew how I am heart [and] soul in all your inventions & all you do," he wrote, in a note attached to some lamp designs, "you would now & then stoop to assist & better my prospects in life."[253]

Edison ignored the lamp designs but gave Dickson two of his own to test. They looked like regular T models except for an unusual tongue of platinum inside the hoop of the carbon. It was separately wired and, when the filament incandesced, deflected a galvanometer needle. This indicated a ghost flow of electromagnetic force within the lamp's vacuum and was further proof of thermionic emission, the famous Edison Effect. Its discoverer, apparently forgetting that he was supposed to be a businessman only, wanted to patent it as an "electrical indicator" that would gauge and regulate the voltage of lamps connected in multiple arc.[254]

Dickson reported favorable results to Edison one night in the cavernous testing room. The scene registered in his photosensitive memory in such detail that he could draw it forty years later: bare brick walls ascending to a galaxy of pendant bulbs, a German silver (as in French door: an alloy, not silver) shunt in one corner, two assistants pottering, a central stove discharging heat. Edison, his hair disheveled, sat tilted back in a Windsor chair, one foot up on the work table, idly playing with one of the test lamps while Dickson talked to him.[255]

Shortly afterward he executed his patent for the indicator, U.S. 307,031. The device did not work well, and he was too busy with other projects to develop it. But in its use of thermionic emission for a practical purpose, it was technology's first attempt at what would one day be called electronics.*[256]

As for Dickson, he had at last been noticed. He was soon put in charge of the testing room and began his long rise to obscurity.[257]

"THE FOLLOY OF HIS WAYS"

By the new year of 1884 Edison was back in his laboratory on the top floor of the Bergmann building, trying to develop synthetic filaments from various gelatines and researching electrodeposition techniques that he hoped might precipitate gold foil.[258] Such absorption in experiment, with his hands moving and his deafness muffling outside noise, was usually a sign that he was tired of pretending to be what he was not: a man of the world comfortable in society, savvy about money, adept at boardroom maneuverings, and interested in politics, women, and children.

The Construction Department had been his attempt to show Grosvenor Lowrey, Sherburne Eaton, and other directors of the Edison Electric Light Company that the traits that distinguished him as an inventor—contrary thinking, obstinate repetition, daydreaming, delight in difficulty—would bring about the demise of the gas industry much faster than their insistence on cautious progress. He had expected them to admire him for his courage in daring to launch a new enterprise on his own nickel. But although the Department had signed up many towns and cities, its expenses were outrunning its income at a compound rate. Moreover, the quality of its work—rushed through as fast as possible, to collect fees—was often shoddy. There was a sour joke going around about "Edison's Destruction Department," and he now faced the prospect of having to ask Sherburne Eaton if the Electric Light Company would defray his personal deficit of $11,000. Since Eaton was likely to say no, his high laboratory offered both refuge and solace.[259]

Eaton did turn him down. Edison's hurt anger (the Light Company, after all, stood to gain hugely from his patents) was gratifying to Insull. Daily more powerful as he acquired financial and administrative authority, the secretary saw a corporate crisis coming that he could turn

* See Part Four.

to his own advantage. "There is no one more anxious after wealth than Samuel Insull," he admitted. Edison made no effort to restrain him. He was grateful for the icy efficiency with which Insull kept creditors at bay, while always coming up with whatever cash he and Mary needed.[260]

Insull's secret plan was to relieve Edison, as tactfully as possible, of responsibility for the Construction Department. He intended to amalgamate it with the most successful of his boss's ventures so far, the Edison Company for Isolated Lighting. At the same time he wanted to bring down the man they both now saw as their corporate opponent—Sherburne Eaton—and Grosvenor Lowrey too. His target date for this coup was 29 October, when Eaton would preside over the parent company's annual board meeting. That gave Insull most of the year to gather enough shareholder support to force the election of a new president, who would be beholden to him rather than to Drexel, Morgan & Co.[261]

Eaton was a courtly man who had never gotten used to Insull's brash lack of manners. He made a mistake in trying to be acerbic when he sent Insull a memo on 18 February, drawing attention to one of Edison's Construction Department shortfalls. "I have no doubt that he will see the folloy [sic] of his ways after having learned experience at heavy and unnecessary cost." Insull may have been self-serving, but nobody ever criticized his adored boss without penalty.[262]

THE DEMON SHARK

Edison was then in Clay County, Florida, letting Insull act for him while he and Mary treated themselves to their most extended vacation yet. This time they left the children behind, traveling instead with Mary's good friend Josie Reimer and her husband. Edison had $1,500 in Construction Department funds in his pocket, charged up, truthfully enough, to "Expenses South."[263]

Mary, her daughter noted, was never happier than when she had Edison to herself in Florida—enjoying white glove service at the Magnolia Springs Resort Hotel, lolling in the warm baths, and cruising with him down the St. John's River between palms and sour-orange plantations, while he scribbled laboratory ideas in his pocket notebook.[264]

"Stay away as long as you feel like it," Insull wrote Edison on "Birthington's wash Day," one of his occasional attempts at humor.[265] "At least give me till 1st April before you show your face in New

York. I am conceited enough to want to try & get some work single handed for Const Dept."[266]

Edison was not sorry to be relieved of that responsibility for a while, and did what he was told. He was at last able to accept that his affairs were too complex for him to manage alone, with new lighting systems, wholly owned or affiliated, opening up almost weekly around the world, and competition harder and harder to restrain. For that reason he agreed, after all, to let Joseph Swan join names with him in the union of their British interests, henceforth to be known as the Edison & Swan United Electric Company, Ltd. He also consented to the incorporation of his Lamp Company and Machine Works, while continuing to resist Eaton's desire to add them and the Tube Company to the portfolio of Drexel, Morgan. Of Insull's private intent to the contrary, he had, as yet, no idea.[267]

He continued to fill his notebook with mostly electrical notions but omitted to include one that instantly became one of Florida's taller fishing yarns. In the last week of March he was seen by a reporter escorting Mary ("a superb blonde") aboard a yacht in the harbor of St. Augustine, accompanied by the Reimers and a small black boy toting a basket of what onlookers assumed was champagne.[268] The yacht pulled out to the fishing ground off the lighthouse, where for years a "demon shark" had consumed multiple blackfish and bass and, reportedly, one or two swimmers. Lines went overboard, one of which unspooled from the basket and proved to be a regular telegraph wire insulated with gutta-percha. It was attached to a powerful battery, and baited at the other end with an electrode. Within fifteen minutes Edison and the captain were hauling in a mortally shocked seven-hundred-pound shark. It ended up on permanent display in the local Vedder Museum, labeled:

THE DEMON SHARK.
CAUGHT BY T. A. EDISON,
WITH ELECTRIC BAIT.[269]

HUBRIS

Sharks of another kind (or so Edison chose to see them) gathered when he resumed work in New York at the beginning of April. Dur-

ing his absence Major Eaton had menacingly attempted to glean full
details of the finances of his profitable manufacturing shops. The
Electric Light Company derived no proceeds from them, whereas
Eaton kept receiving bills for "sundry" Construction Department expenses that Edison seemed to think he should pay. Eaton was not sure
that he would, and pressed his demand for information about the
shops, pointing out that they were, after all, "connected with our
business."[270]

Edison replied that he would prefer not to comply "until I have had
an opportunity of discussing the matters in question with Mr. Villard," referring to the one Light Company director he had always
been able to count on for moral and monetary support. But Villard
was a broken man now, having pushed his Oregon & Transcontinental Railroad too far and too fast toward the Pacific, and caused both
it and himself to collapse. He could not suggest anything to save Edison from similar hubris, trivial as the latter's entrepreneurial difficulties were compared to his own.[271]

On 24 April Edison wrote Eaton to say that he had been unable
to win any new contracts for the Construction Department, and
could not coax any more cash out of the ones he had acquired. "I
find myself in the position of being obliged to immediately disband
my organization, as the expenses in connection with it are too large
to allow of my continuing it." He would therefore allow the Electric
Light Company, "as it has been suggested," to take over the Edison
Company for Isolated Lighting, along with all his current construction projects. The sooner this was done the better, because he had
recruited some of the best men in the electrical industry over the
past year, and it would be shortsighted to let them go for lack of
pay.[272]

Actually Charles Clarke and Frank Sprague, a brilliant (if obstrep-
erous) young engineer brought in by Edward Johnson, had already
walked, early refugees from a business empire widely perceived to be
in trouble. The perception was inaccurate. Pearl Street was pouring
out more power by the month and looked sure to become profitable
sometime soon. Planning had begun for Manhattan's second central
station, and John Kruesi had taken extra space for the Tube Company

in Brooklyn. But negativism was the prevailing mood on Wall Street these days, due largely to Villard's fall. It was a gloomy time for Edison to have to admit his own failure to push a grand project, even in another man's handwriting.[273]

The Electric Light Company board accepted his proposal and reaffirmed its interest in buying his shops, while Insull, emulating Brer Fox, lay low. Then in mid-May a liquidity crisis hit the nation's banks. Commerce froze, and Edison in a panic fired his engineering staff (retaining only Dickson and an assistant). He also closed down the Machine Works for "maintenance." Several jumbo generators sat unsold on the factory floor, and he unsuccessfully tried to get his British company to buy them.[274]

For Mary, too, the spring was bleak. Her adored father died, and she felt less well than she should after a long vacation. Besides being caught up in funeral preparations, she found that for budgetary reasons she must transfer her husband and children out of the Clarendon Hotel and back into the house in Gramercy Park. Edison's sublessees had come up short, and the lease there would not be up until the first of October.[275]

Lucy Seyfert's legal team chose this moment to inform Edison that the New Jersey Supreme Court had again validated his debt to her, now amounting to $5,349 exclusive of fees. In view of his obstinate refusal to pay, the Middlesex County sheriff had authority to seize his holdings in Menlo Park.[276]

Edison was convinced against all reason that Mrs. Seyfert would settle for $300, the original value of the note she held, if he continued to stall. He ordered his lawyers to assert that the Electric Light Company was the owner of his former laboratory and auxiliary buildings. Everything in the house on Thornall Avenue should be put in Mary's name and out of the sheriff's reach. The house itself belonged to him but was heavily mortgaged in New York, which meant that the plaintiff would have to cross state lines, and celebrate several more birthdays, before she got satisfaction on that score. He presented Mary ("Duck—please sign your name below") with an affidavit claiming title to all the goods and chattels they still possessed at Menlo Park, from a six-piece marble top suite in the master bedroom to a gray horse, three cattle, two pigs, and a "lot of manure" upwind in the yard.*[277]

HIS PASSING ALONG

Olive Harper, a roving reporter on women's issues, was impressed with the splendor of Mary's home in Gramercy Park, when she interviewed her for a profile published in *The World* at the beginning of June. The pale blue satin furniture and Chickering piano had come with the lease, but Mary had rearranged the first-floor parlors, laid down some extra Persian rugs, and crammed in many bits of painting and porcelain that were evidently her own artistry.[278]

Miss Harper seemed to want to write a mainly descriptive article, noting that "Mrs. Edison has been called the most extravagant woman in New York as to personal adornment," and estimating her weight at 160 pounds. But Mary was wearing nothing but black in mourning for her father. She wanted, in her first and last chance to speak to the press, to correct a story about her marriage that had irritated her for more than five years.[279]

"In the first place," she said, "I never worked in any factory, not for Mr. Edison, nor anybody else in any capacity, and therefore all the stories about his passing along where I was at work Monday evening and proposing to me and setting the wedding for Tuesday morning hasn't a word of truth in it."

She confirmed that she had been fifteen and a half when he first set eyes on her, ducking out of the rain into his works on Ward Street in Newark. And "very handsome" eyes they were, although the rest of him had been so grimy and oily. "I'm a little in love with my husband's eyes—yes, in fact, a good deal." He had been the most gentlemanly of courtiers, gradually winning her father's trust.

Mary became so sentimental as she rambled on about Edison and her children that she forgot the main rumor she had wanted to deny—that he had gone to the laboratory on their honeymoon night and forgotten to come home. "I have been very happy with him, and I expect to be as long as I live."[280]

---

* Mary was by no means passive in signing this affidavit. She attached a covering letter warning the sheriff, "You will interfere with same at your peril."

QUIETNESS AND COUNTRY AIR

Menlo Park was a stripped and saddened place for Mary to return to that June, with none of the "boys" and their wives remaining, the laboratory emptied out, the electric railway grassed over, and the famous streetlamps dead. She would be seeing little of her husband, now that his work was concentrated in New York, and Mr. Batchelor back at last from France. She had her mother, her sister Jennie, and her children to talk with during the day. But with no man in the house at night, and hoboes squatting in the old lamp factory nearby, she slept with a gun under her pillow.[281]

The quietness and country air at least were good for her in her uncertain state of health—and for eight-year-old "Tommie" as well. He was like his father with his bright eyes and large head, but he was like her too, being prone to fainting spells and mysterious headaches. Little "Willie" was sturdier and stronger, and Marion, now in her twelfth year, even more so, with long blond hair that reminded Mary of herself as a girl.[282]

A DAMN GOOD MAN

It was a comfort for Edison to have Batchelor at close quarters again, as they worked together on dynamo improvements in the shut-down Machine Works. Edward Johnson (whom Insull was plotting to put on the board of the reorganized Light Company) was the only aide who had served him as long and with as much devotion. But whereas Johnson was an excitable, affectionate dog, constantly pulling to the next pole ahead, Batchelor was a cat who kept his own counsel. He had wisely invested the bonuses, stock certificates, and other pourboires Edison had given him over the years, in moments of shared elation over some triumph at the workbench. As a result, he was by now a wealthy man, and would have been shocked if he knew that his employer currently had little more than twenty dollars in the bank.*[283]

Edison appointed him general manager of the Works and accepted

* The editors of *The Papers of Thomas A. Edison* found that on 1 June 1884 Edison had only $18.64 on deposit at Drexel, Morgan, and $3.80 at the Bank of the Metropolis. They point out, however, that he always regarded banks as clearinghouses: "Large sums flowed into his two checking accounts, [and] flowed out just as quickly." *Papers*, 7.575.

his recommendation to hire Nikola Tesla, a phenomenally gifted young Serbian engineer, just off the boat from France. Batchelor had discovered Tesla in Paris the year before, and been awed by his understanding of electricity, as well as the voracity of his appetite for steak.* On both counts, America was clearly where Tesla should be. It had not been difficult for Batchelor to persuade him to cross the Atlantic and become the newest of Edison's "boys."[284]

He at once solved a dynamo problem that was preventing an Edison-lighted steamer, the *Oregon,* from leaving New York Harbor. Having stayed up all night, he then reported to the Works for another assignment. Edison murmured to Batchelor, "This is a damn good man."[285]

Tesla was reciprocally impressed: "The effect that Edison produced on me was rather extraordinary. When I saw this wonderful man, who had had no training at all, no advantages, and did it all himself, and [saw] the great results by virtue of his industry and application, I felt mortified that I had squandered my life . . . ruminating through libraries and reading all sorts of stuff."[286]

ALL TWENTY-ONE LOTS

Mary's affidavit claiming possession of everything in the house at Menlo Park had not persuaded her husband's lawyers that the sheriff of Middlesex County would be put off by it. Nor was he likely to be intimidated by the note she had added for good measure: "You will interfere with same at your peril." Frail as she might be at the moment, with gastritis complicating her chronic neuralgia, Mary was a fighter.[287]

So, but in a less emotional, more vengeful way, was Edison. Neither of his ploys to frustrate Mrs. Seyfert's suit passed legal muster. The mortgage argument admitted of no postponement under New Jersey law, while Mary was unable to show a transfer deed that proved she was the rightful owner of the house's chattels. Edison, who had excellent credit, could have borrowed money to honor the obligation imposed on him by the state's highest court. But he de-

---

* According to T. C. Martin, writing in the February 1894 issue of *The Century Magazine,* Edison wondered aloud if Tesla was a cannibal.

clined to do so, and the sheriff thereupon announced that his entire property at Menlo Park would be auctioned to satisfy the judgment, at two P.M. on 22 July.

Nobody involved in the action chose to inform posterity where Mary was that day or what she felt about strangers bidding low for things she held dear. However, a bidder comfortingly familiar to her won out. All twenty-one lots were gaveled down to a Mr. Charles Batchelor, of New York.

The total price paid was only $2,750, reflecting Menlo Park's desuetude as much as the constriction of the economy. Batchelor acted only as a front for Edison, who had arranged to reimburse him later. But $2,852 was still owing on the judgment, and as far as Edison was concerned, the sheriff could sing for it.[288]

Mary moved back into the house pending further court action. Simultaneously, Edison vacated his office at "65," on the ground that he was now an inventor again and could safely leave the reorganization of his corporate affairs to Insull. He returned full time to his laboratory in New York—and at once blew out all its windows in an attempt at the direct conversion of coal to electricity.[289]

He had only just settled in when, without explanation on Thursday 7 August, he left for Menlo Park. His train arrived there before sunset. Two nights later, in the small hours of the morning, Mary died.[290]

SORROW AT MENLO PARK

Edison had suffered no major bereavement before, except the loss of his mother in 1871. She had ailed for several years with dementia, so he had had time to brace for her death. Mary—still only twenty-eight, and usually able to bounce back to fun-loving health from her spells of illness and depression—departed with such suddenness that for the only recorded time in his life, he cried uncontrollably. When he broke the news to Marion, he was shaking and sobbing and hardly able to speak.[291]

The immediate question to be asked was what had killed Mary Stilwell Edison. Half a century later her sister Alice told an Edison biographer, "The cause of death was typhoid fever."[292] If so, Mary's prostration was remarkably rapid, bringing the gigs of country doctors to the house at a gallop even as Edison took his train from New York. Her death certificate, and a terse report issued by the Electric Light Company, cited "congestion of the brain," which in contemporary parlance could mean anything from meningitis to apoplexy. Or it could mean the alternate dilation and contraction of cranial arteries stimulated by morphine.[293] An unsigned article, "Sorrow at Menlo Park," in *The World* on 7 August (reading as if written by Olive Harper) went beyond circumspection in suggesting that Mary died of opioid abuse.[294]

> She suffered from obstinate neuralgia that refused all manner of treatment. The best physicians were called in, but their remedies were useless. At last for temporary relief she tried morphine, and soon learned the great palliative powers of the seductive drug—a ready dose of which was always at her side—and when the premonitory symptoms of an attack came on she knew the value of her white powder.
>
> At the request of Mr. Edison she took a trip to Florida last winter. Instead of obtaining relief she fell victim to gastritis, due to the peculiar atmosphere or perhaps the long acquaintance with morphine. She returned to Menlo Park in a more troubled condition. Her pain intensified, and at times she was almost frantic. Morphia was the only remedy, and naturally she tried to increase the quantity prescribed by the doctors. From the careless word dropped by [a] friend of the family it was more than intimated that an overdose of morphine swallowed in a moment of frenzy caused by pain greater than she could bear brought on her untimely death. The doctor in attendance said she died of congestion of the brain. When a reporter put the question to him he positively asserted that it was the immediate cause, but about the more remote causes he preferred to remain silent.*[295]

So did Edison. Like Henry Adams and Theodore Roosevelt and other dumbstruck widowers of the time, he honored the dead by keeping his grief to himself. Except for a brief reference to "my poor wife" in an interview at the end of the month, he rarely mentioned Mary again. Because she was soon to be replaced, she was edited out

---

* Mary Edison's death certificate was unsigned by any doctor, and the space for registering the cause of death was left blank. The original, in the New Jersey State Archives, shows signs of being mutilated. As a family friend remarked at her funeral, attended by four hundred mourners, "She is dead now, poor thing, but no one will ever know what she died of."

of the Edison family's later history—except among her children, and only the eldest of them had much of her to remember. In the mythology of the Stilwells, Mary became a limp, naked figure being lifted from her bath by Grace-like vestals, or a ghost walking in the front yard of Edison's house, rising as he ran forward to clutch at her white summer dress, which dissolved in his fingers like a cloud.[296]

DAMON AND PYTHIAS

The barrage of emotional blows that made 1884 Edison's *annus horribilis*—his business folly, his near bankruptcy, the humiliating sale of his house, the abruption of the mother of his children—drew him closer to Marion than to either of his bewildered boys. At eleven and a half, she was old enough to feel another's grief as well as her own. He took what consolation he could from her girlish company, addressing her as "Miss Marion Edison, sweetest of all." In September "Grammach" Stilwell, Mary's mother, looked after Tom and William at Menlo Park while Edison took Marion to the International Electrical Exhibition in Philadelphia. It was a grown-up treat for her before she went back to school in New York. There was a nice new apartment waiting for them on East Eighteenth Street. They would not live again in that stone house on Gramercy Park, with its pale blue satin furniture and mirrors full of memory.[297]

Father and daughter made a touching duo as they toured the Philadelphia show hand in hand, gazing up at a Doric column of more than two thousand lamps that dazzlingly spelled out his name letter by letter over spirals of colored light. As if that were not apotheosis enough, an electrified bust of the inventor represented him at the moment of perfecting his first carbon bulb, with a halo of incandescence encircling his brow.[298]

"As soon as I go the laboratory again I'm going to work on several new things," he said to a reporter. "I haven't been doing work on anything much but light."[299]

In an encounter that would pluck him from the slough of despond, he met an old colleague from his days as a wandering telegraph operator. Ezra Gilliland was a humorous, loose-mouthed electrician from upstate New York who had helped him promote his phonograph six years before. In the interim since then, Gilliland had married well, and was now working for the research arm of the American Bell Telephone Company in Boston. He had acquired a healthy paunch and a beach house on the North Shore.[300]

Gilliland dabbled in invention and owned a share in several communications patents. When Edison asked him "what would be a good thing to take up next," he suggested they collaborate on a long-distance telephone transmitter for American Bell. Edison was immediately interested, having himself, seven years before, invented the carbon button that made Bell telephones audible.* After returning to New York, he at once reverted to acoustic technology, and as early as 24 September executed a patent on a xylophone-like signal receiver that chimed at different pitches, depending on who was being called.[301]

He felt at liberty to work for an outside client now, especially after Insull (who had gone into his own private depression over Mary's death) succeeded in reorganizing the Electric Light Company in October as promised. Eaton was out as president, replaced by a more compliant Eugene Crowell; Edward Johnson was vice-president, and Lowrey dropped from the board.† The independence of the manufacturing shops was preserved, and the power of Drexel, Morgan to block innovation nullified. "I have got mine at last," Insull exulted. Edison, less vindictively, expressed relief at being free of corporate restrictions. "I have worked eighteen and twenty hours a day for five years, and don't want to see my work killed for want of proper pushing." With Johnson in control of the Light Company and other Menlo Park alumni running the shops, he returned to the study of telecommunications, his once and future passion—and the best therapy imaginable for a man in mourning.[302]

Edison and Gilliland cemented their professional relationship at the beginning of December with a joint application for a patent on the prevention of electromagnetic interference in speech transmission.[303] For the rest of the winter they worked together in New York and Boston, staying in each other's apartments and recapturing the intimacy that had linked them as youthful wire gypsies. It burgeoned with un-

---

* See Part Six.

† Lowrey was deeply hurt by Edison's apparent acquiescence in the purge, but persuaded himself that Insull alone was responsible.

usual speed, since Edison had a widower's need for company at night. His new laboratory did not offer the rough camaraderie of the old, with only a couple of mechanics and a boy to join him for midnight lunches, if they could be persuaded to stay so late. Gilliland and his birdlike wife, Lillian, were childless. They compensated with many entertainments, to which the teenage daughters of their friends, many of them students in Boston's private academies, were always welcome.[304]

On 20 February 1885 Edison and Marion—delighted to play hooky from her own school in New York—set off with the Gillilands on a marathon railroad tour. They headed first for Adrian, Michigan, where Gilliland's father lived and where Edison had worked as a sixteen-year-old night operator on the Lake Shore & Michigan Southern Railroad. A blizzard slowed their progress. The two men passed the time discussing a patent that Gilliland co-owned, for sending wireless telephone waves from a moving train by means of electromagnetic induction. They believed that the technique could be refined for telegraphy by using a vibrating reed to compress the dots and dashes of Morse code into rapid pulses—as many as 250,000 a second, Edison calculated—which would then be "jumped" to wires running alongside the track for transmission to stations along the line.[305]

The idea grew in their minds as they proceeded south via Chicago and Cincinnati, where they had once worked together for Western Union, and where Edison had conducted his first experiments in multiple telegraphy. At the end of the month they attended an industrial fair in New Orleans before moving east into Florida. They installed Mrs. Gilliland and Marion in St. Augustine's luxury hotel, the San Marco, then crossed over to the considerably wilder Gulf Coast. The tarpon fishing was said to be especially good off Punta Rassa, so they rented a sloop at Cedar Key and sailed south to that cow town on the mouth of the Caloosahatchee River. They checked into the Schultz Hotel, which was in all respects the opposite of the San Marco and therefore entirely to Edison's taste.[306]

One day he became curious about Fort Myers, a village twelve miles upstream, after hearing that bamboo grew seventy feet tall there. He still had a bamboo explorer on his payroll, scouting the world for splints, but had not thought of Florida as a possible source of supply. On 20 March he and Gilliland took the sloop and, leaving behind Punta Rassa's miasma of fish parts and cattle dip, sailed inland into the fragrance of orange trees and fan palmettos blooming.[307]

A white road of crushed oyster shells paralleled the left bank of the river, half-screened by live oaks, tamarinds, date palms, and native cinnamon. It was besplattered with dung, testifying to its function as a cattle conduit to the southern part of the state. At the head of the road Fort Myers came into view: a straggly settlement consisting of a few dozen houses, a tiny telegraph office, a drugstore, a hotel, a schoolhouse, a church, and that feature of all American outposts, a real estate office.

Although the bamboo growing nearby at "Billy's Creek" did not compare with Japanese *madake* for hardness, Edison was charmed enough with the little town to request a tour of a thirteen-acre riverfront property advertised for sale a mile down the white road. It was unfenced and overgrown, but the panorama of the river, one and a half miles wide, was magnificent. Before he sailed back to Punta Rassa the following day, he had contracted to buy it for $3,000.[308]

Separately, Gilliland agreed to pay a quarter of the price of the estate. Although that gave Edison most of it, they planned to build twin winter homes there, in a grove facing the water. When they journeyed back north a few days later with their female companions, they had something other than "jumping" train telegraphs to discuss. Punta Rassa was becoming fashionable and an investment upriver seemed worthwhile, although as Edison admitted, "it will make a savage onslaught on our bank account." United now in business and at least partially in domicile, they began to call themselves Damon and Pythias—Gilliland assuming the former identity, with its Greek connotation of readiness to die for his best friend.[309]

Pythias had no sooner returned to his laboratory than he applied in Damon's name for two patents on a wireless train communications system using induction telegraphy. Possessive as Edison always was of his own patents, he was scrupulous in recognizing the antecedence of other inventors—even though the claimant in this case had merely bought his way in. But on legal advice, and recognizing that he was to be the major developer of the system from now on, he added his own name at filing time.[310] Week by week, signature by signature, the Pythagorean duo was being yoked closer together.

A SORT OF RAPHAELIZED BEAUTY

Among the young women who prettified the Gilliland apartment in Boston was a Miss Mina Miller, the nineteen-year-old daughter of a wealthy businessman in Akron, Ohio. She was a student at a finishing school on Newbury Street, where she had become fluent in French and well trained in fine and domestic arts. Her piano teacher, however, had been unable to match Mina's love of music with anything resembling musicality. Edison's first reaction when she sat down at Mrs. Gilliland's piano was a mixture of surprise and curiosity. "I could not help being interested immediately in anyone who would play and sing without hesitation, when they did it as badly as that."[311]

Mina performed not because she was vain, but because she had been asked to. It was her nature to oblige. She doubted that she would see Edison again. He was, of course, famous and "a genial, lovely man" to boot.[312] But he was twice her age, with gray streaking his hair and a habit of cupping his right ear in conversation. For all her study of English literature, it did not occur to her that such a man, single and in possession of a good fortune—not to mention three motherless children—must be in want of a wife.*

All he was aware of, at first, was a pair of "great dazzling eyes." On subsequent visits to Boston to do business with Gilliland and American Bell, he could not fail to notice that Miss Miller had other double attributes, agreeably arranged elsewhere on her sturdy person. Had she been fair instead of darkly brunette, she might have reminded him of an even younger schoolgirl, ducking out of the rain and into his life fifteen years before.† Except that this one had an easy sophistication poor Mary never attained. Her four elder brothers were all college men, and her two younger ones were destined for Yale. Her older sister was as polished and well traveled as she, and the other two were at or put down for Wellesley. Her millionaire father, Lewis Miller, was a pillar of Akron society, an elder of the Methodist Episcopal Church, and co-founding president of the Chautauqua Institution. From her equally pious mother Mina had inherited a certain

---

* Edison had confessed this need to Lillian Gilliland during their railroad trip to Florida, and asked her to introduce him "to some suitable girls."

† See Part Six.

dourness that was less attractive to Edison than those big eyes and—was fate again speaking to him?—the delightful fact that she was at home in a workshop.[313]

It turned out that Mr. Miller was also an inventor, with a hundred farm implement patents to his credit, so she had an understanding of technology and was not likely to be bored when Edison talked to her about electromotographic mirrors. Luxuriating in company with her and other "fresh invoice[s] of innocence and beauty" at Woodside Villa, Gilliland's beach house in Winthrop, Massachusetts, Edison became so infatuated that he sounded like a teenager himself, writing

Menlo Park N.J.

Sunday July 12 1885

Awakened at 5.15 AM. my eyes were embarrassed by the sunbeams – turned my back to them and tried to take another dip into oblivion – succeeded – awakened at 7 AM. thought of Mina, Daisy and Mamma G— put all 3 in my mental kaleidoscope to obtain a new combination a la Galton. took Mina as a basis, tried to improve her beauty by discarding and adding certain features borrowed from Daisy and Mamma G. a sort of Raphaelized beauty, got into it too deep, mind flew away and I went to sleep again. Awakened at 8.15 AM. Poweful itching of my head, lots of white dry dandruff – what is this d—mnable material, Perhaps its the dust from the dry literary matter I've crowded into my noddle lately Its nomadic. gets all over my coat, must read about it in the Encyclopedia, Smoking too much makes me nervous – must lasso my natural tendency to acquire such habits – holding heavy cigar constantly in my mouth has deformed my upper lip, it has a sort of Havana curl. Arose at 9 oclock came down stairs expecting twas too late for breakfast – twas'nt. couldnt eat much, nerves of stomach too nicotinny. The roots of tobacco plants must go clear through to hell. Satans principal agent Dyspepsia

A page of Edison's diary, summer 1885. *

* Edison's references are to Grace "Daisy" Gaston, one of the other girls visiting with the Gillilands that summer; Lillian "Mamma" Gilliland; and the English geneticist Francis Galton (1822–1911).

Insull at the end of June, "Could you come over here to spend 4th at Gills—there is lots [*sic*] pretty girls."[314]

Around this time someone in the Gilliland circle suggested that they all start keeping diaries, full of as many personal details as possible, to be shared for common amusement. Edison began his on 12 July in Menlo Park, where his children were being looked after by Mrs. Stilwell.[315]

When he returned to Woodside Villa with Marion in tow, Mina had left to join her family at Chautauqua. Lillian Gilliland, who was openly seeking a mate for him, offered Louise Igoe from Indiana for his consideration. "Miss Igoe," he wrote, "is a pronounced blonde, blue eyes, with a complexion as clear as the conscience of a baby angel." But he could not shake Mina's darker charms from his mind. During a Boston book-buying excursion, he "got thinking about Mina and came near being run over by a street car—If Mina interferes much more I will have to take out an insurance policy."[316]

There followed the laziest, most ruminative period of his life, a sun-soaked, ozone-flavored, female-graced, and oddly frenchified interlude so unlike what he was used to that the days seemed to blend into a prolonged raptus that was more dream than reality. Missing from it only, yet permeating his diary through and through, was "the Maid of Chautauqua," whose remoteness in western New York State he was determined to make temporary. In the meantime Mina could be pleasurably associated with Madame Récamier, Lucien Bonaparte's scantily dressed muse and the embodiment of unattainable, sophisticated sexuality. Although Edison spoke little French, he had always been drawn to French literature, and one of his book purchases in Boston was the autobiography of *la divine* Juliette—"I should like to see such a woman."[317] Absorbed in it, he drowsily envisioned the jealous tyrant who had sent her into exile.

After breakfast laid down on a sofa, fell into light draught sleep dreamed that in the depth of space, on a bleak and gigantic planet the solitary soul of the great Napoleon was the sole inhabitant. I saw him as in the pictures, in contemplative aspect with his blue eagle eye, amid the howl of the tempest and the lashing of gigantic waves high

Mina Miller dressed as a gypsy, at about the time Edison first met her.

up on a jutting promontory gazing out among the worlds & stars that stud the depths of infinity Miles above him circled and swept the sky with ponderous wing the imperial condor bearing in his talons a message. . . .

Then my dream changed—Thought I was looking out upon the sea, suddenly the air was filled with millions of little cherubs as one sees in Raphaels pictures each I thought was about the size of a fly. They were perfectly formed & seemed semi-transparent, each swept down to the surface of the sea, reached out both their tiny hands and grabbed a very small drop of water, and flew upwards where they assembled and appeared to form a cloud.[318]

Clearly Edison was in the grip of an emotional turmoil that had him struggling to keep his balance amid the vast stability of a universe governed by immutable natural laws or (if Mina insisted) by God:

Went out on Veranda to exercise my appreciation of Nature. Saw bugs, butterflies as varied as Prang's Chromos,* Birds innumerable, flowers with as great a variety of color as Calico for the African market. . . . What a wonderfully small idea mankind has of the Almighty. My impression is that he has made unchangeable laws to govern this and billions of other worlds and that he has forgotten even the existence of this little mote of ours ages ago. Why cant man follow up and practice the teachings of his own conscience, mind his business, and not obtrude his purposely created finite mind in affairs that will be attended to without any voluntee[re]d advice.[319]

Marion, not to be outdone by her father in imaginative expression, wrote an outline for a novel about "a marriage under duress." When she read it to him, he said, in words that might have told her something about his own past experience, "Put in bucketfulls of misery."[320]

It was the only sour note sounded in what was otherwise, for him, a period of ecstatic expectation that Mina (who had turned twenty on the sixth) would be receptive to his advances. If by heading west she was playing hard to get, then he would pursue her, even into the bosom of her family. That meant he would have to charm as many as ten other Millers, not to mention Mina's countless cousins, and especially win the favor of her father, the largest frog in the small pool of Chautauqua Lake.

Edison's best chance there was to ingratiate himself with Lewis Miller as a fellow inventor, while staying off the subject of churchgoing. "My conscience seems to be oblivious of Sunday," he wrote in his diary. He learned what he could about the great man by reading a collection of business profiles encouragingly entitled *How Success Is Won*. There was little he could do about his other possible liability—middle age—except to strive for an elegant appearance by buying a pair of uncomfortably tight, French-made *chaussures de monsieur*.[†] "These shoes are small and look nice," he noted in his diary. "My

---

* The colored "chromolithographs" of Louis Prang, popular in late-nineteenth-century America.

† These shoes cost Edison fourteen dollars, or $355.60 in modern money.

No. 2 mind (acquired mind) has succeeded in convincing my No. 1 mind, (primal mind or heart) that it is pure vanity, conceit and folly to suffer bodily pains that ones person may have graces [which are] the outcome of secret agony."[321]

He also brushed up intellectually by delving into his pile of beach books—Rousseau's *La nouvelle Héloïse*, Disraeli's *Curiosities of Literature*, Hjalmar Boyesen's *Goethe and Schiller* ("a little wit & anecdote in this style of literature would have the same effect as baking soda on bread"), Johann Kaspar Lavater's *Essays on Physiognomy*, Hawthorne's *Passages from the English Note-Books*, Rose Cleveland's *George Eliot's Poetry and Other Studies*, Goethe's *Wilhelm Meister* and *The Sorrows of Young Werther*, Thomas B. Aldrich's *Story of a Bad Boy* ("very witty and charming"), and Longfellow's *Hyperion*. As if this were not enough erudition, he reveled in the epigrams of Sydney Smith and reminded himself, "I must read *Jane Eyre*."[322]

On 10 August Edison presented himself at Chautauqua. Marion accompanied her father, not altogether congenially. He had sensed, while extolling Mina's "perfection" to the Gillilands, that his daughter was becoming jealous. "She threatens to become an incipient Lucretia Borgia."[323]

Lewis Miller turned out to be a pleasant surprise. Reciprocally, he took to Edison at once as a man who, like himself, made large amounts of money for the benefit of humanity. Although the light he sought to bestow was more spiritual than electrical, he was at the same time an earthy, good-natured, receptive personality, more willing than his austere wife to laugh at cornball jokes—which Edison was careful to scrub clean of words likely to shock them. Even Mary Valinda Miller admitted that her daughter's new suitor had a winning way with him. She had previously hoped for a match between Mina and young George Vincent, the son of Chautauqua's other co-founder, but neither she nor Lewis could help being flattered that one of the most famous men in the world was calling at their cottage, hat and heart in hand.*[324]

They jibbed, however, when Edison asked permission to take Mina

---

* The Miller Cottage still stands at Chautauqua.

off on a tour of northern New York State and New Hampshire. Decent girls did not go on such jaunts, even with respectable widowers. Ezra and Lillian Gilliland offered themselves as chaperones, and Louise Igoe (who was sweet on Mina's brother Robert) volunteered further company. They all, in Mina's words, "made it so attractive [that] father at last consented."[325]

The six travelers set off on 18 August. Their rail and pleasure boat itinerary took them via Niagara Falls and the Thousand Islands of upstate New York to Montreal, whence they swung south into the White Mountains of New Hampshire and stopped at the grand Maplewood Hotel. Here, Mina primly recalled, "things got a little warmer" between herself and the inventor of the quadruplex telegraph.[326]

> One evening after spending the day on top of Mount Washington, we were sitting around the hotel in the foothills. Mr. Edison wrote down for me the Morse code characters and by next morning I had memorized them. A short time later he slowly tapped a message to me which I was able to understand. Just what the message said I consider too sacred to repeat.[327]

Marion claimed to have witnessed the tapping, which took place on Mina's hand, as well as the *dash-dot-dash-dash, dot, dot-dot-dot* response. Having often hung about her father's laboratory in Menlo Park, she may have learned Morse code too. More likely her memory of Mina's story transmuted over time into an imagined recollection. At any rate it was an end to the happiest year she ever spent,[328] and for Edison and Mina, the beginning of a union till death.

CUPID'S DEMANDS

On 30 September Edison wrote Lewis Miller, formally asking if he could marry Mina.

> *My Dear Sir*
> *Some months since, as you are aware, I was introduced to your daughter, Miss Mina. The friendship which ensued became admiration as I began to appreciate her gentleness and grace of manner, and her beauty and strength of mind.*

That admiration has on my part ripend into love, and I have asked her to become my wife. She has referred me to you, and our engagement needs but for its confirmation your consent.

I trust you will not accuse me of egotism when I say that my life and history and standing are so well known as to call for no statement concerning myself. My reputation is so far made that I recognize that I must be judged by it for good or ill.

I need only add in conclusion that the step I have taken in asking your daughter to intrust her happiness into my keeping has been the result of mature deliberation, and with the full appreciation of the responsibility I have assumed, and the duty I have undertaken to fulfil.

I do not deny that your answer will seriously affect my happiness, and I trust my suit may meet with your approval. Very sincerely yours

Thomas A. Edison[329]

He gave his New York laboratory as his return address, and was good for nothing connected with business until he heard his fate. Edward Johnson tried to get his attention on a matter of some urgency—what to do about a new rival to the phonograph, the "graphophone"—but had to postpone further discussion, "for the simple reason that he is in love and don't want to make any appointments in advance that might possibly conflict with Cupid's demands."[330]

Miller replied promptly and with equal formality, inviting Edison to visit him *en famille* at Oak Place, Akron, early in October. Mina's towering home was defended by a profusion of deer, horse, and dog statuary, and Edison was not made entirely welcome by her mother. Mrs. Miller doubted his assurance that he would be a churchgoer, were it not for the unfortunate problem of his deafness. Mina had some concerns on that score herself, but Lewis had none at all. He liked Edison enormously and gave the lovers permission to marry under his roof on 24 February 1886.[331]

In the interim, Edison had some major real estate decisions to make. On a bright night by the sea the previous summer, indulging fantasies of Mina far away, he had taken an imaginary triangulation of the moon, "the two sides of said triangle meeting the base line of the earth at Woodside and Akron, Ohio." His calculations had got him to the latter point precisely as planned. Now he had to plot a series of other extensions, which would become the geometry of his future life. First, from Oak Place to Fort Myers, where he wanted to take his bride on honeymoon, and where he and Gilliland were building twin houses and a winter laboratory to share; then back to wherever in the New York area Mina wanted to settle (he would give her the choice of city or county);[332] then the shortest possible connection from that base to the location of a new superlaboratory that would erase all memories of Menlo Park.

Mina chose Llewellyn Park, an exclusive, gated, hillside enclave in West Orange, New Jersey. It was far enough from the railroad station in downtown Orange to be considered rural, yet close enough to merit municipal horsecar service. Glenmont, the estate's premier residence, was listed for sale fully furnished, thanks to the downfall of the owner, Henry Pedder, in a million-dollar embezzlement case. It was bigger than the house she had grown up in, a many-gabled, twenty-three-room Queen Anne mansion, red of brick and exterior framing,

Glenmont in Llewellyn Park, soon after Edison's purchase of it for Mina.

almost new and built as solidly as a bank, with a mahogany central staircase, a billiard den, a music room, and an immense curving conservatory that caught the morning sun. It had hot running water in all bathrooms and fireplaces in all bedrooms, central heat, hand-stenciled ceilings, oil paintings, statuary, a huge service of Tiffany silver tableware, and a library of leather-tooled books Mr. Pedder had been unable to take with him when he skipped the country for St. Kitts.[333]

Glenmont was so named because it had a valley view, facing east across the Oranges toward New York, only twenty miles away. It was surrounded by eleven acres of shaven lawns and plantings designed by Nathan Franklin Barrett, the nation's foremost landscape architect. Behind and to the north, fragrant woodlands soared toward the Eagle Rock reservation, which offered skating in winter and a refuge from the state's mosquitoes in summer. All Mina needed to acquire this paradise from its receiver was a fiancé able to unbelt $125,000— well under a third of the estate's estimated value.[334]

Edison did not disappoint. He was beginning to be flush again, with profits surging in from his manufacturing shops and other enterprises. Pearl Street was in the black, having paid off its start-up costs in a year, and already issued its first dividend. The Edison Illuminating Company of New York was poised to begin construction of a much larger Second District, extending as far north as Central Park. Nationwide, fifty-eight Edison central stations and 520 isolated plants had more than three hundred thousand lamps in circuit.[335] The Machine Works had so outgrown its cramped Lower East Side neighborhood that it would soon have to move out of Manhattan or even upstate. Not for some years yet might Edison count himself as rich as some of the other mansion dwellers in Llewellyn Park, but for the first time in his life he felt wealthy enough to match his spending to the scope of his ambitions.

He signed for the purchase of Glenmont on 10 January. Not wishing the pesky Mrs. Seyfert to place a lien on the property, he settled her suit for $6,134—more than twenty times his original debt to her husband. In another severance, he ordered a Newark florist to discontinue the placement of flowers on Mary's grave. He treated his ninety-one-year-old father to a three-month tour of Europe. He wrote his real estate agents in Fort Myers to notify them he was coming there soon (without mentioning it would be on honeymoon) and expected that he and Mr. Gilliland would be able to move into their completed houses.* He dispatched two schooners loaded with heavy equipment for the laboratory, and when one of them was destroyed by lightning en route, he sent another with duplicate cargo. Looking further ahead, to the work he would be doing after settling into Glenmont, he doodled several elevations of his northern laboratory and conjoined works, massively built in beaux arts style around a quadrangle and anchored, to the right of its gates, by a library.[336]

Edison's Magritte-like sketch of Mina as an airborne clock.

Ezra and Lillian Gilliland opted to go to Fort Myers in advance of the wedding, to prepare the compound for his arrival. They took Marion with them. Tom and William stayed in school.† On 20 February the Menlo Park "boys" gave the Old Man a stag party in New York at Delmonico's. Three days later Batchelor, Johnson, Insull, and a few others entrained in a private car to Akron, where at three P.M. on the twenty-fourth Edison stood under a wishbone of roses in the parlor at Oak Place and waited for Mina to marry him.[337]

---

* The houses were being constructed from kit parts shipped down from Maine.

† On 1 March Tom wrote his father, "I am in long division and willie is in subtraction."

TO LOVE AND TO CHERISH

She made a glittering prize in white satin, with a diamond and pearl necklace he had bought for her. Heaps of other gifts of diamonds, rubies, sapphires, and silver were on display in the great room, arrayed around a column of solid onyx, capped with a gold capital, confirming (along with a bronze dog outside, garlanded for the occasion) that Edison had been accepted into the ranks of the nouveau riche. He made a stab at elegance in a black Prince Albert coat, but declined to wear gloves.[338]

This omission was widely noted in the nation's press. It served as a signal that for all his new wife's social aspirations, he was still a man who worked with his hands. And his haste to leave for Florida that same evening indicated a desire to return to inventive engineering that was at least as urgent as sex. No sooner had they begun to pass through Georgia plantations than he figured out the mechanics of an automatic cotton picker, with air-blown depilatory spindles.[339]

Peach trees bloomed on the approach to Florida. When "Mr. and Mrs. Edison" checked into the St. James Hotel in Jacksonville, Mina found herself the object of such avid public scrutiny that she was overcome with shyness. She took refuge in their room while her husband, used to fame, went sightseeing. Like many a young bride suffering the anticlimax of honeymoon, she found she had given herself to someone not wholly congenial. Her husband's irreverence bothered her. Having knelt beside him before a white altar in her father's house and heard him vow "to love and to cherish, till death do us part, according to God's holy ordinance," she hoped that he might be receptive to the Methodist Episcopal Church's 223-page *Doctrines and Discipline,* a copy of which Lewis Miller had stuck in his coat pocket for light reading on the train. "He intends to study it well," Mina wrote her mother, sounding doubtful. "He wanted to know the other day if I married him to convert him."[340]

Edison tried to make her understand that he needed hard evidence, or at least logical argument, to believe anything, and that religion was deficient in both respects. He had no wish to convert her to agnosticism and was willing to admit that he might be wrong to shrug at faith. But he could not help the way he felt: "Everyday life must be the convincing power."[341] On that, at least, they agreed.

"XYZ"

If Mina was aware how much Edison's creativity had diminished in the eighteen months since Mary's death, she might have given herself inspirational credit for the volcano of ideas that erupted out of him as soon as they got to Fort Myers.[342] His laboratory there—a plain prefabricated shed—was still not ready, and $16,000 worth of equipment had yet to be installed. But that did not stop him from filling six notebooks with enough drawings and specifications to keep a research team busy for the rest of the century. One or two sketches prefigured surrealism, such as a piano that produced speech instead of music by means of keys "playing" a rubber larynx, or a bust of Mina, bareshouldered, hanging upside down from an airborne clock. But most of the entries were so precisely conceived, dated, and signed that it was clear he had set up a laboratory in his mind.[343]

Notebook number one began with three images that on any other honeymoon might be considered phallic—

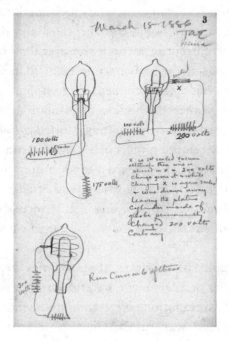

—but for Edison were just variations on the endlessly fascinating theme of incandescent light. Before that day's jotting was over, he had run eleven mental tests on carbonizing solutions, purified and desiccated "city gas" by passing it through tubes of finely divided copper, experimented with a foil balloon for long-distance electric signaling, and converted natural gas into lampblack. There followed, over the next six weeks, more than four hundred inventions, including fluid prisms, a phonographic siren, a motorized "cash carrier" for department stores, a metal fatigue detector, sonar depth sounders, a squirter of artificial silk, and a pneumatic device to suck turpentine out of trees.[344] Amid this array of minor notions there emerged two concepts that he considered to be of major importance: an electromagnetic theory of gravitation (influenced by his readings of Faraday) and conversion of light or heat into electricity. The former made him see the solar system as one giant centrifugal dynamo, or in universal terms as a molecule among billions of others whirling in the cosmos. The latter derived from an idea that had tantalized him for years, that there existed in frequencies above and below the limits of human reception a type of energy so new he could only call it "xyz." He sensed it again now, when he mentally projected beams of light or heat through liquids, or theorized an opposing relationship between the current in a condenser and the lines of force in a magnet. What was the signal-emitting train that he and Gilliland were working on but a giant condenser, "jumping" electrical energy through air, in defiance of the law of insulation? He drew, three-dimensionally, various ways a rotating, slotted cylinder could throw slices of light between the prongs of a magnet "straight and at right angles." In his brain's ear he heard a tone emitted by a rapidly spinning magnetized wheel, audible through an attached telephone as a kind of spectral music. "Now if this disturbance is created without the production of electricity or magnetism, then we have a new form of energy."[345]

The orthography of his notes at such times showed him being swept into a fever of excitement, careful script degenerating into a sprawl, as if his speculations were running ahead of the pencil in his hand.[346] Edison was rushing into realms of thought where even pure scientists feared to tread, and he knew that he was unqualified, but as when arguing theology with Mina, he could not help himself.*

She in turn could not help feeling lost in the primitive environment of a riverside estate stripped of jungle growth and only partially replanted. The twin houses were attractive enough, in a raw-looking, just-carpentered way, and it was comforting to have Lillian Gilliland at hand to help her deal with Marion, but she looked askance at the cowboys and colored people—"nearly every one of the darkest shade," she wrote home—that constituted the lower ranks of Floridian society. They were not what her schoolmates in Boston might call *de notre monde*.[347]

Mina's doubts about having married Edison were compounded by his incessant jocularity.[348] She was devoid of humor herself and flinched at the way he teased both men and women, sometimes in a rough way that made her wonder what kind of language he used when closeted with men only. Just as hard to get used to was his need to control everyone and everything around him. Even so personal a task as planning the gardens around their winter home had to be executed to the last detail by himself:

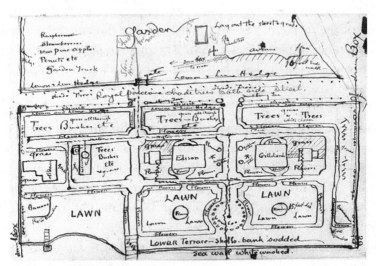

Edison's plan for his Fort Myers estate, spring 1886. Laboratory to the left of the twin houses.

* The editors of *The Papers of Thomas A. Edison* have concluded that Edison was familiar with the nonmathematical aspects of James Clerk Maxwell's classic *Treatise on Electricity and Magnetism* and with all of Oliver Heaviside's papers on "Electromagnetic Induction and Its Propagation," recently published in *The Electrician* (UK).

Edison presented this design to their gardener/caretaker with fifteen hundred words of precise instructions, informing him that 280 boatloads of topsoil would be needed to cover the eight-acre riverside parcel four inches deep. He ordered ninety different fruit trees, including figs, mangoes, mulberries, alligator pears, plums, peaches, apricots, persimmons, and "as many orange trees of best variety as will go on end of the House plot." There should also be a banana bed twenty feet square, a thousand pineapple plants, and a lemon hedge. ("If you cannot procure the regular Italian Lemon seed elsewhere . . . raise the slips yourself from seeds found in the lemon.") He also authorized the purchase of eight tons of fertilizer: "We propose to have our ground the best manured in Florida. . . . I think you should go back from the river & look for black muck fresh water muck. . . . It evidently wants some fine decayed fibrous spongy matter like they are putting in the Coconut holes to hold the manure & prevent it going clear through to China."[349]

Mina learned, like Mary before her, that she would never have the whole of her husband, nor even the best of the balance. Perhaps sensing her feelings of inadequacy, Edison asked her to copy and countersign many of the inventions in his notebooks. But the only experiment he permitted her to perform with him was an attempt to shock an oyster into opening its shell. They had to write it off as a "dead failure."[350]

BILLY, GEORGE, TOM, AND WILLIE

Edison was in no hurry to resume normal life after his honeymoon ended in late April. He and Mina spent the early days of May in Akron before moving into their mansion—with apprehension on her part and false modesty on his. "It is a great deal too nice for me," he told a reporter, "but"—touching Mina's arm—"it isn't half nice enough for my little wife here."[351]

Her problem was not its grandeur but the intimidating prospect of having to run such a large establishment, with many servants and three stepchildren looking to her for authority. Edison could be relied on not to supply that, once he was back at work full-time. Domesticity, including the balancing of household accounts, was a woman's work. Her chances of civilizing him were about as slim as the Widow

Douglas's with Huck Finn, although for a while friends were amused to see that his pants were now pressed, his shoes shined, and his jacket buttons inserted into the correct holes.[352]

Edison began to mind his own business again, moving decisively to punish a strike for union recognition and higher pay at the Edison Machine Works on 19 May. If the "communists" of New York City thought they could organize one of his shops, he was happy to move the entire plant to Schenectady, where Insull had providentially found an old locomotive factory. "Do it big, Sammy. Make it a big success," he said. "Or a big failure."[353]

Insull took the word *big* to heart and established beside the Erie Canal the future world headquarters of General Electric.*

With him and John Kruesi gone, Edison was freer of corporate encumbrances than ever, and able to indulge his most consuming current interest, wireless telegraphy. The "jumping" train communications device he had developed with Gilliland the year before, now nicknamed the "grasshopper" system, was being tested with only partial success on the Milwaukee & St. Paul Railroad. He had much more faith in an alternative device he had patented himself, the phonoplex. It gave depots along the track the ability to telegraph one another via multiple lines, without interfering with terminal communications. When in a moment of inspiration he devised a weighted diaphragm to enhance its acoustics, it performed so well along the Grand Trunk Railroad—the very route he had run as a newsboy—that his tester, Alfred Tate,† reported, "There is no 'frying pan' induction or 'morse hash' to drown the writing of the phone key."[354] Translated from telegraphese, that meant that the phonoplex sounded clear, with no crackly interference or blurring together of dots and dashes.

The "grasshopper" system was eventually taken over by the Consolidated Railway Telegraph Company and never flourished, but the Edison phonoplex was adopted by the Baltimore & Ohio in July and became a staple on American railroads well into the next century.[355]

---

* In two years Insull quadrupled Machine Works sales, and in six he increased its payroll from two hundred to six thousand employees.

† Tate had succeeded Insull as Edison's secretary-factotum in 1883.

That summer Edison transferred his laboratory from New York to the Lamp Company in East Newark, which was closer to home. He continued to involve Mina in his experimental work, taking her there as an assistant and—since she was being allowed into a man's world—calling her by the pet nickname "Billy." When on occasion Marion joined them, she was likewise "George." In mid-August he joined Billy, George, Tom, Willie, and numerous Millers at Chautauqua, and basked for as long as he could stand it in what William James called its "unmitigated goodness."[356]

For the rest of Mina's life this annual pilgrimage, plus churchgoing, was to offer her spiritual relief from her husband's material preoccupations. Chautauqua, however, never quite alleviated her tendency to melancholy; nor did the sermons she heard at the 1886 assembly help her deal with the problem of a jealous stepdaughter. Marion, with adolescence coming on, could not adjust to being displaced from Edison's affections by a woman only six and a half years older than herself. The boys were difficult too. Mina unloaded her angst in letters to her mother, who replied in comforting Chautauquese. "Try and love them and they will love you and Mr Edison will be perfectly happy."[357]

But Edison was so already. In October his good friend Gilliland joined him in East Newark for a new round of experiments, and Mina found her services there were no longer needed. About the same time, she became pregnant.[358]

FOR SHOCKING PURPOSES

On 2 November the U.S. Patent Office issued one of its most unpronounceable patents to Károlyi Zipernowski, Ottó Bláthy, and Miksa Déri, of Budapest, for an induction coil transformer to provide the high voltages required to distribute alternating current economically over distances far beyond the limit of Edison's direct current system.[359] The device became popularly known as the "ZBD" transformer, and its power as "AC," while Edison was positioned as the defender of "DC." Thus began the competition that would develop, over time and in conspiratorial myth, into the "war of the currents" between Edison and an opponent always identified as Nicola Tesla but whose real name was George Westinghouse.[360]

At first it was not so much war as a research effort by Edison to see if AC technology—which Frank Sprague predicted was "going to be a formidable rival to the system of direct supply"—could be integrated with his own. He had long been aware that DC power, suited as it was to a compact urban area like the First District of New York, was not suitable for long-distance transmission, because the farther it was extended, the thicker and costlier its copper conductors would have to be. His "three-wire" distribution system was an ingenious answer to that problem, but again best served a close-spread circuit. DC power flowed in one direction, steadily and at moderate voltage from dynamo to lamp. AC power zigzagged back and forth as it flashed along the surface of any wire, alternately swelling to maximum and dropping to zero pressure, forced by tranformers to as many as three thousand volts, then using magnetic induction to reduce them, transformer by transformer, to levels that would not melt a filament. It used little copper and went as far as any supplier needed to send it. Until the perfection of the ZBD, however, high-voltage AC had been too unsteady for reliability.* But the Hungarian transformer smoothed it out so effectively that Sprague warned Edward Johnson, president of the Edison Electric Light Company, "You cannot too soon take steps to prevent someone getting into the field ahead of you."[361]

Johnson accordingly bought American rights to the ZBD.[362] It availed him little. Westinghouse had already formed his own eponymous Electric Company, financed an AC system in Great Barrington, Massachusetts, and acquired an even more sophisticated transformer, recently designed by William Stanley. Meanwhile Edison, acting out of curiosity rather than combativeness, applied for nearly a dozen AC-related patents of his own. Experimenting with it gingerly, he convinced himself that the system's invisible, flickering, trapped lightning was likely to kill careless utility workers, not to mention householders who tinkered with the wrong outlet or al-

---

* Edison had tried and dismissed alternating current as early as 1882, saying it was good for nothing except the euthanasia of animals. However, he looked the other way when it was successfully installed on his isolated systems in Europe, for example at La Scala in Milan, May 1886.

lowed the insulation around their feed wiring to fray. He summed up his findings and his feelings in a lengthy, highly technical memorandum to Johnson: "As the [back-and-forth] wave cannot start instantly and stop instantly . . . it will require 130 volts or thereabouts to produce the equivalent of 100, thus we have for shocking purposes a reversed intermitting current of unlimited amperes as far as the body is concerned (!) and [at] a difference of 260 volts it will certainly be unpleasant."[363]

ALL REFERENCES TO CRIBS

Telling jokes coatless outside his East Newark facility, one icy day around the turn of the year, Edison contracted pleurisy and for many weeks lay dangerously ill.* It was the end of January before he could sit up in bed[364] and attend to his first order of business in 1887: the purchase of fourteen acres of property at the corner of Valley Road and Lakeside Avenue in West Orange, about a mile from his home in Llewellyn Park.

As soon as he was mobile again, he wanted to resume work at the lamp factory and develop a squirted-cellulose filament for the Edison lightbulb. He believed that if he could double the reach of his system by such advances, it would beat back the challenge of AC power. But his doctors insisted he recuperate further in Florida, sending him there so early in February that he was obliged to celebrate his fortieth birthday on the train. A press rumor that he had tuberculosis and would never return north flashed across the country. Edward Johnson attributed it to the Westinghouse Company. Edison's recovery in Fort Myers was slow and complicated by an ear abscess that had to be lanced on 24 March. A reporter from *The World* found him a few days later at work in his wooden laboratory, healthy but harder of hearing. "He certainly was a sick man at one time," Ezra Gilliland admitted. "The trouble somehow reached the heart, and it was found necessary several times to administer hypodermic injections of morphine."[365]

Mina had her own problems, being by now visibly pregnant and worried after a year of marriage that she did not have Edison's "full

---

* Edison may also, or alternatively, have contracted pneumonia.

affection."[366] She preceded him north in mid-April, taking Marion with her, leaving the field clear for her father to visit with Edison alone and subject him to Methodist scrutiny. Lewis Miller came, stayed, and was beguiled. He listened in vain for improprieties in his host's storytelling and for any remarks that might indicate disappointment in her as a wife.

"The more I see him," he wrote Mina, "the more I am impressed with his greatness and genuine good heart. I am thoroughly convinced that he is true to you and true to what he appears to be. And socially he is superior to any man I know."[367]

However cheering this was to her, she had to deal sometime that spring with a trauma more real than any imagined marital strain. All references to cribs and specially measured dresses vanished from her correspondence.[368]

HILLS AND DALES

Edison had Mina work with him in his East Newark laboratory for a while in May, but soon got caught up in a project so technical that only Batchelor and Gilliland could help. It was the development of a dictating machine to counter the cylinder recorder that Charles Sumner Tainter and Chichester Bell had patented last year, when he was too lovesick to pay it much attention. Now he had to. As if calling their device the "graphophone" were not cynically evocative enough of his favorite invention, it looked so like the phonograph, with its voice funnel and helical grooves and hand crank, as to trick casual shoppers into thinking he must have licensed it.* The crucial difference between the two instruments was that the graphophone stylus incised a wax sleeve, instead of indenting a tinfoil wrap. This was a definite selling plus—on top of the fact that Tainter and Bell were ready to market their device, while Edison had little more than a rough sketch of an "Improved Phonograph" to offer in competition.

He was further lumbered by a mistake made in 1878, when he patented an extraordinary number of potential phonograph develop-

---

* The "graphophone" that Tainter and Bell caveated at the Patent Office in a sealed and dated box on 20 October 1881 actually *was* an Edison phonograph, with a coating of wax applied to the metal cylinder. The deception was not uncovered until 1937.

ments in Britain, rather than in America. Among those innovations (itemized in sixty-seven descriptive drawings) had been all the features that Tainter and Bell now claimed to have come up with independently. But the Patent Office refused to award him priority over them at home, on the grounds that protection overseas was all that he was entitled to. Whatever new instrument Edison designed for the American market must, by maddening irony, not infringe on the graphophone, whose technology he had anticipated nine years before.[369]

His sketch, dated 7 May, accordingly showed a phonograph with an electric motor drive and rubberized acoustic tubes for recording and playback. The round of the cylinder was left enigmatically blank, and as he and Batchelor built what became known as model M, it was clear that they acknowledged the advantage of a recording surface of wax coating cardboard over one of foil laid on iron.[370] When incising, the stylus cut the wax cleanly and lightly, with minimal drag. When indenting, it needed weight to press the foil into hills and dales that were less well defined. The contrary problem with wax was that it was softer than foil, so sound quality deteriorated with each replay.

Hence, Edison's challenge was to formulate a wax that was hard enough to wear well yet receptive to the highest, least incisive frequencies, contained in such sibilant words as *sphynx*. (Clarity of speech was essential to this machine, since both he and Tainter were aiming at the dictating market, in a boom time for American business.) The wax should not clog the stylus by curling up behind it as the cylinder rotated; ideally what was etched out should float into the air, dust from the sonar landscape. That meant, of course, that the stylus itself should be as durable as it was sharp—yet not so sharp as to erode the very slopes it was shaping, when it traveled over them again.[371]

The search for an ideal counterbalance between all these requirements, involving chemistry as well as physics and electrical engineering, was to preoccupy Edison for the next thirteen months. At the same time, and for almost as long, he had the equally challenging task of planning, building, equipping, moving into, and staffing his huge laboratory on Valley Road in West Orange—the ground for which was broken on 5 July.

Both projects afforded him deep pleasure, the first being the kind of experimental marathon that had produced his best inventions. He had at least one seminal idea, a playback needle that "never touches the surface of the record but is itself electrified," in essence a magnetic pickup. For some reason he abandoned it, postponing the advent of electrical recording for thirty-three years.[372]

There were three other distractions to deal with during this period: a corporate squabble about the future of the long-defunct Edison Speaking Phonograph Company, a second pregnancy for Mina commencing in the fall of 1887, and a visit from the photographer Eadweard Muybridge later on that would profoundly affect his future.

Edison had hardly started work on model M when Edward Johnson and Uriah Painter, co-founders of the old phonograph company, lobbied him to accept a move by Tainter and Bell to combine all his patents with theirs, and form a new company that would practically monopolize the dictating machine market. To Edison, any such merger, ludicrously scrambling together phonograph and graphophone and Painter and Tainter, was unthinkable. It would compromise his primacy in the recording field, and he could see that what the Bell interests really wanted was his all-embracing British patents. "Under no circumstances will I have anything to do with Graham Bell with his phonograph pronounced backward graphphone [sic]," he wrote his London agent, Col. George Gouraud. "I have a much better apparatus and am already building the factory to manufacture [them]."[373]

Alexander Graham Bell was involved with his brother Chichester and Sumner Tainter in the sense that all three of them comprised the newly formed American Graphophone Company of Washington, D.C. With increasing obstinacy, Edison rejected their repeated efforts to do business with him and, rather than revive the Speaking Phonograph Company, decided to form a new one, the Edison Phonograph Company. He incorporated it on 10 October 1887, establishing its capital at $1.2 million and waving aside the anguished protests of Johnson and Painter that he was trampling on their rights as shareholders of the old firm. They declined to accept his offer of a 30 percent stake in Edison Phonograph as dishonorable and inadequate, since they would have profited to the extent of a half interest if he had accepted the Bell offer.[374] That merger, in retrospect, looked all the more sensible a month later, when Emile Berliner, a German immigrant, patented a recorder that played disks instead of a cylinder.

In several ways, Berliner's was a revolutionary device, but he was years away from making it commercially feasible. Edison had an advantage over him and the Bell interests too, being able to boast the imminent completion of "the best equipped & largest Laboratory extant." It would loom three stories high, total 37,500 feet of floor space, and employ a large staff of scientists, specialist engineers, and craftsmen—in all, a multidepartmental facility "incomparably superior to any other for rapid & cheap development of an invention." He intended to occupy it by Thanksgiving and have it fully operational by the end of the year. Numerous outbuildings would surround it, each with its coordinated research or development function, including a gigantic Phonograph Works that would ship (by means of the Erie Railroad, curving right past his back door) so many perfected talking machines that Graham Bell would wish he had stuck to telephone design. In its integration of innovation and manufacture, the West Orange facility would amount to an apotheosis of Menlo Park. "In fact," Edison wrote, as if the plant were already in operation, "there is no similar institution in existence. . . . Can build anything from a ladys watch to a Locomotive."[375]

During much of the time that he was jockeying with corporate lawyers and working out the mechanics of the M phonograph, Ezra Gilliland found it convenient to be ill and away from work. He thus avoided having to take sides in what became a painful showdown for Johnson particularly, as Edison's oldest friend and passionate promoter of the first phonograph. Johnson was for the first time in his life rich, having risen to the presidency of both the Edison Electric Light Company and the Edison Company for Isolated Lighting. He also enjoyed royalties from a personal invention—twinkling, colored Christmas tree lights—was living in a baronial Connecticut mansion, and had looked to the resurrection of the old phonograph company as the clincher of his fortune. But now, instead of sympathetically choosing him as "general agent" of the new one, Edison appointed Gilliland. The terms of the latter's contract were more than generous, assuring him an income of around $160,000 during the first full year of production.[376]

There was little Johnson could do but wonder at Damon's preferment at the hand of Pythias. He was a sensitive soul, and cringed when Uriah Painter tried to bully him into a shareholder revolt that would punish Edison for trading away their former rights. "Can we not get together & straighten this out?" he wrote Edison. "It is not a matter of money, but of wounded pride—Upon receipt of your ans[wer,] I shall take such action as will forever remove me from my present unpleasant position. . . . The burial of all my long cherished ambitions in this Phono. matter will cost me no slight regret."[377]

Edison replied with the written equivalent of a shrug. "Its not your fault in any way that the present condition of affairs have [*sic*] come about. I'll take care of Mr U. H. P. after he gets through his outburst of temper & damfoolery."[378]

Gilliland quailed at the unattractiveness of the "improved" phonograph Edison expected him to sell. It was still only a dictating machine, for one person to speak into and another to listen to at close range. The best that could be said of it was that the sound quality was remarkable. It was small but grotesquely complicated, with two diaphragms, exposed electric coils, and an array of studs, slides, knobs, and screws likely to cause mass resignations in the stenographic industry.[379]

He worried that Edison, who had recently won court victories that gave him almost total patent protection for his lamp design worldwide, was beginning to think that anything he invented obviated the work of other inventors, including men as gifted as Tainter and Berliner. The graphophone, operated by a treadle, like a sewing machine, struck Gilliland as a simpler and better device than his friend's prototype M. Early that fall Edison bragged to a *New York Post* reporter that although his Phonograph Works was still under construction, he would have five hundred of the new machines on sale by the end of January 1888.[380] He also showed his mastery of mathematics by figuring that four of its detachable cylinders, each with a capacity of six hundred words, would be enough to record all of *Nicholas Nickleby*.*

---

* Edison wildly overestimated. Dickens's novel totals 263,520 words.

BUILDING 5

Fortunately for all concerned, he forgot about these promises in the excitement of opening his new plant on Valley Road in the new year. An imposing gatehouse admitted nobody to the campus without a pass. The redbrick complex beyond, designed by Joseph Taft, was a considerable expansion of Edison's original 37,500-square-foot concept, with four experimental longhouses servicing the main laboratory. They were respectively devoted to physics, chemistry, chemical storage, and metallurgy. The first building was completely nonferrous, so its galvanometers and other delicate instruments would relate only to the earth's magnetic field. The second, sure to be Edison's favorite retreat, had a concrete floor inclined and drained for toxic spillage. Among his personal stash of supplies, there were a few platinum cathodes remaining from some old Grove batteries he had broken open as a boy telegrapher.[381]

Building 5 opposite presented three great arched windows to passing traffic—as yet mainly buggies, in an otherwise rural landscape. Behind the glass was Edison's library, double-storied, galleried, and paneled in yellow pine that would take years to darken to a more studious shade. The first aide he intended to install there, beside Alfred Tate, was a linguist able to translate his subscription list of German,

Edison's new laboratory in West Orange. Phonograph Works in background.

French, and Italian technical periodicals, not to mention the jargon of scientists purportedly writing in English. His executive bathroom adjoined, gleaming with porcelain and Italian marble. (Elsewhere in the laboratory complex, galvanized iron was the noisy surface of choice.) Farther back was a house-size stockroom under orders to acquire, catalog, and index every nonperishable substance in the world, from hardwoods, graphite, waxes, drugs, and gems to sheet glass, silk, meerschaum, seeds, bone, aromatic oils, and the hair of the red deer, which Edison had found more delicate than camel skin to clean the grooves of his cylinders. The rest of the immense structure was given over to light and heavy machinery shops, plus a third-floor warren of research rooms whose walls were movable and whose functions would change as his interests changed. A tall-chimneyed powerhouse was annexed to the rear, its output of DC current set to handle the demands of all the laboratory buildings as well as the Phonograph Works—as yet little more than a frozen field alongside Alden Street, awaiting the spring thaw. Edison planned to run a branch feeder to his house in Llewellyn Park, a mile away.[382]

By the end of January he had recruited or transferred from New York seventy-five laboratory assistants, a payroll that steadily increased. So did the number of extra hours he seemed to expect everybody to put in. One employee defined the Old Man's idea of a basic schedule: "Saturdays the laboratory closed at five o'clock instead of six . . . and holidays were celebrated with work." When he complained of having no life of his own, Edison said to his mystification, "There's just as much time coming as going, young man."[383]

PICTURE AFTER PICTURE

On 27 February the photographer Eadweard Muybridge visited Edison after giving a demonstration of "zoopraxiscope" images in Orange. Having spent years encouraging racehorses and naked men and women to run, jump, and stroll past a long row of cameras—each one synchronized to capture a moment of apparent stillness that was actually a moment of motion—he had a suggestion that he thought would appeal to a man who had been working just as long to record sound and project light. Would Edison consider applying himself to the invention of a machine that would show moving, talking pictures?[384]

What Muybridge had in mind was, ironically, a system that reversed his own concept of photography. Whereas his multiple cameras—as many as twenty-four in a row—exposed only one frame each in a broken sequence, Edison's imagined single camera must expose hundreds, even thousands, in smooth succession, and reproduce them in much the same way as his phonograph played back linked sound waves. The impression of continuity would of course be an illusion—picture after picture, each slightly different, succeeding one another so rapidly that the eye could not register the blanks between them.

Edison was interested, saying he would look into the idea when he had time. At his request, Muybridge sent a selection of his *Animals in Motion* plates to West Orange for exhibit in the laboratory library,[385] and left their haunting sequences to float in Edison's subconscious.

NOT BAD FOR A FIRST EXPERIMENT

In March, just as Gilliland was denying yet another rumor of the amalgamation of the Edison Phonograph and American Graphophone companies, a Pittsburgh multimillionaire, Jesse Lippincott, made a move to take over the latter firm. Lippincott's fortune derived from glass. He knew nothing about the dictating machine business except that it looked like a good investment at a time when the American economy was booming. He also coveted the marketing rights to the "improved" phonograph, despite its nonappearance since being announced. With a capitalist's wolflike alertness to any whiff of financial vulnerability, he sensed that Edison had overextended himself in West Orange and might be amenable to a sale that would give him titular preeminence over Tainter and the Bell brothers.[386]

Lippincott guessed correctly. Edison was indeed short of cash, having sunk $140,000 into his new laboratory and budgeted another $250,000 for the Phonograph Works, on top of his expenditures on personal real estate. He had covertly tried and failed to get Henry Villard to finance the West Orange plant, not pausing to think that he might be encouraging another wolf to start prowling around him.[387]

Insull, who still managed Edison's financial affairs, wrote Tate in late May to say that he had heard the Phonograph Company was not paying its bills. He worried that this would reflect on the credit of the Machine Works in Schenectady. "If you people at Orange are going to abuse your credit you will cripple us."[388]

This plaint coincided with a disastrous demonstration Edison put on at the laboratory, hoping to impress another group of financiers with the latest refinements to his prototype M phonograph. Unknown to him, Fred Ott had put a stylus into the machine's reproducer that was broader than the recording point. The result in playback was a prolonged hiss, bewildering Edison, and the money men retreated to New York with checkbooks intact.[389]

A similar humiliation had befallen Edison once before, when he was experimenting with his first lightbulbs at Menlo Park. Now as then and at other crisis times in his life, he marshaled a team of his best men—Arthur Kennelly the mathematician, Walter Aylsworth the chemist, Franz Schulze-Berge, and Theo Wangemann, German-trained acousticians—and plunged into a prolonged blitz for mechanical perfection. The phonograph was a much more sophisticated instrument now than it had been in January, attuned to music as well as speech, but it was still not ready to manufacture or market. Edison's urgency was prompted by demands from Colonel Gouraud in London for a machine to recruit British investors, as well as by a need to make the stock of the Phonograph Company as attractive as possible to Lippincott—for by now that entrepreneur had made a direct purchase offer of $500,000, and Gilliland thought it was too good to pass up.[390]

Absorbed in his laboratory work, Edison did not notice, or did not sufficiently heed, a clause in the offer awarding Gilliland $50,000 in cash and $200,000 more in new-company shares, for his "general agency" contract. What was more, the shares could be redeemed for cash as soon as the purchase was consummated.[391]

On 16 June Edison declared that he was through, for the time being, with improving the "improvements" and perfecting the "perfection" of his new recorder. He shipped a handmade model on the next steamer to London, along with what he called a "phonogram"—history's first audio letter. It was recorded by himself on a detachable spool, and contained the news that he had just become a father again.[392]

> *Ahem! In my laboratory in Orange, New Jersey,*
> *June 16, 1888, 3 o'clock,* A.M.

> *Friend Gouraud—Ahem! This is my first mailing phonogram. . . .*
> *I send you by Mr. Hamilton a new phonograph, the first one of*
> *the new model that has just left my hands. It has been put together*
> *very hurriedly and is not finished, as you will see. I have sent you*
> *a quantity of experimental phonogram blanks, so that you can*
> *talk back to me. . . .*
>
> *Mrs. Edison and the baby are doing well. The baby's*
> *articulation is quite loud enough but a trifle indistinct. It can be*
> *improved but it is not bad for a first experiment.**

> *With kind regards,*
> *Yours,*
> *Edison.*[393]

The organization by Jesse Lippincott of the North American Phonograph Company on 14 July ended the strife between the Edison and Tainter-Bell interests and pooled all their patents. It enriched everybody concerned, especially Gilliland, who hastened to cash in the rest of his quarter-million-dollar bonus and depart posthaste for Europe. He claimed to be exhibiting the phonograph there, but when Edison found out about the agency sale clause and heard that it had been negotiated by his own personal attorney, John Tomlinson, it was as if Damon had slipped a knife between the ribs of Pythias. "I have this day abrogated your contract," he cabled Gilliland in London, "and notified Mr Lippincott of the fact and that he pay any further sum at his own risk. Since you have been so underhanded I shall demand all the money paid you."[394]

Gilliland cabled back, "Sale made to Lippincott exactly as presented and had your approval. . . . You certainly are acting without

---

* Madeleine Edison was born on 31 May 1888. H. de Coursey Hamilton was one of Edison's paid globe-trotters.

knowledge of facts and are doing me great injustice." A court of law agreed when Edison, in one of his litigious furies, sued him for breach of contract. But their friendship, which had yielded them both such dividends—financial, professional, and in Edison's case romantic— was over. For as long as the Gillilands continued to winter in Fort Myers, the Edisons stayed away, depriving themselves of vacations in the sun for the next fourteen years and making sure that no electric power from their generator or water from their windmill pump would ever cross the space between the twin houses.*[395]

THINGS IN MOTION

On 8 October Edison, free now of another corporate responsibility, sketched a device that at first sight looked like a phonograph.

But M was not a stylus or a speaking tube, and N was not a wax cylinder. Nor were the dots on its surface indentations or incisions: they were too widely spaced for that. Actually he was thinking of microphotographs—Muybridgean images reduced down to $\frac{1}{32}$nd inch wide, spiraling past a tiny telescope at a rate of twenty-five exposures per second. He calculated he could embed forty-two thousand such images on a cylinder of plaster of paris that, rotating at phonograph speed, would present a moving-picture show twenty-eight minutes long. If the drive shaft P was geared to that of an acoustic cylinder in playback mode, sight and sound would blend.

"*I am experimenting upon an instrument which does for the Eye what the phonograph does for the Ear,*" he wrote, in the beautiful script he reserved for important documents, "*which is the recording and reproduction of things in motion, and in such a form as to be both cheap practical and convenient. This apparatus I call a Kineto-scope 'Moving View.' In the first production of the actual motions that is to say of a continuous Opera the Instrument may be called a Kinetograph but its subsequent reproduction for which it will be of most use to the public it is properly called a Kinetoscope.*"[396]

He went on to describe how the recording version of his device

* Ezra Gilliland died childless, of heart disease, on 13 May 1903. In extreme old age his widow wrote Edison, "He loved you very dearly, Edison, and regretted all those misunderstandings."

Edison's first Kinetoscope caveat, 8 October 1888.

would be a camera big enough to contain a photosensitive cylinder, or even a reel of film, either of which would advance in a stop-start motion so rapid as to seem continuous. The essence of his invention was deception: each unbelievably short stop would be enough to photograph a slice of action, and each unbelievably fast jump forward enough to expose another frame to the light. That would necessitate a shutter just as kinetic, or mobile, as the advancing cylinder, able to snap at least eight pictures a second (but preferably twenty-five) in order to flow them past an enlarging lens later, and pull off the con known as *persistence of vision.*

Edison thus articulated for the first time in history the idea of motion pictures synchronized with words and music. He was never to achieve the combination practically, and he was wrong to imply in his caveat that a spiraling cylinder was preferable to unrolling film. Nor—even as he datelined the document "Orange N. J. Oct 8 1888," with the initial O inked in a perfect circle—did he know that Louis Le Prince, an obscure French camera inventor working in Leeds, England, was six days away from doing precisely for the Eye what the phonograph did for the Ear.[*397]

"Rush this I am getting good results," Edison wrote later that day, dispatching the Kinetoscope caveat to his patent attorneys for submission to the Patent Office. It was followed in due course by two others, all attesting in their force and specificity to the power of the idea that had seized him. But for the next nine months he was so busy with electric power and phonograph promotions, in anticipation of the epochal Paris Exposition Universelle in 1889, that he left the technical work of building his cylinder-camera to W. K. L. Dickson. The young man was not only a gifted engineer but an expert photographer. Fluent in French and German, he was able to browse foreign periodicals in the laboratory library and keep track of experiments in motion photography overseas.[†398]

Le Prince's supreme achievement with a single-lens camera was unpublished news and in fact would not be recognized for almost half a century. If Dickson was ignorant of it in 1888, he certainly knew about other pioneers in France and Germany, and lived to see the day when Edison's own pretension of primacy in the history of cinema (a word not yet coined) was under justified attack by their proponents. As an old man, Dickson would again and again notch his memory calendar back one year, to prove beyond doubt that the Kinetoscope was the mother of all motion picture cameras, and that he and Edison had started working on it in the fall of 1887. He wasted much ink in refusing to acknowledge that there was such a thing as a *Zeitgeist*, as his revered boss did in 1912, in words as modest as they were true:

* Le Prince's *Roundhay Gardens*, filmed on 14 October 1888, is now generally regarded as the first motion picture. *The First Film*, a 2013 documentary by David Nicholas Wilkinson, is available at https://vimeo.com/ondemand/thefirstfilm/181293064.

† In some respects the contemporaneous Electrotachyscope of Ottomar Anschütz and the Chronophotographes of Étienne-Jules Marey were more sophisticated than Edison's, the one employing projection and the other intermittent action.

"My so-called inventions already existed in the environment—I took them out."[399]

COMPARATIVE DANGER TO LIFE

In November Edison was drawn into the most disagreeable controversy of his career, when a body calling itself the Medico-Legal Society advised the New York State legislature that the quickest, most painless way to execute criminals sentenced to capital punishment was to subject them to three thousand volts of direct or alternating electric current.[400]

Earlier in the year Gov. David B. Hill, impressed by this argument, had signed a bill to abolish hanging as the state's standard execution method, in favor of death by electricity—or electrocution, as it became known, to the distress of word purists. Few understood the implication of the decision better than Charles Batchelor, who nearly killed himself fixing a light on the Edison laboratory's direct current system. Had the current been alternating, instead of flowing directly through him, he would unquestionably be dead at forty-two, mourned by all.* The savage sawing motion of AC, at hundreds of reversals a second, would have shredded every cell in his body. Or so Edison persuaded himself, on the basis of animal tests conducted on his own premises by Batchelor, Arthur Kennelly, and Harold P. Brown, an independent, passionate proponent of DC power.[401]

Brown was responsible for perverting what had been a reasoned debate between the Edison and Westinghouse interests as to which system was preferable for most lighting purposes. In all respects except safety, the so-called "war of the currents" was now resolved in favor of AC. During the last month, George Westinghouse had received more orders for central station lights than the Edison Company had in the previous year.[402] Local utilities simply found his thin-wire, high-tension system cheaper to install and operate, as well as extensible to suburban areas where affluent customers lived.

Brown's only recourse was to stigmatize AC as the "executioner's

---

* Experiments conducted in the 1970s confirmed what Edison had believed, but been unable to prove, in the 1880s: that AC current is two and a half to three times more lethal than DC. There is no evidence, however, that death by either is painless.

current," better used on death row than in the home. Although his professional credentials were slight, he was possessed, to a near-pathological degree, with desire to prove his point by electrocuting dogs, calves, and horses both at West Orange (where Arthur Kennelly meticulously noted their convulsions) and at venues as public as Columbia University. His evangelism on the subject was based on industrial deaths he had observed during five years as an electrician working for Brush Arc Light Company.[403]

Edison witnessed several of the electrocutions, and gave no sign of being disturbed by them. "I have taken life—not human life, in the belief and full consciousness that the end justifies the means." He had opposed capital punishment in the past, but found his moral attitude toward it wavering now that it had become a concern of electricians. If criminals were to die for their sins, then he would prefer them to be dispatched by a Westinghouse dynamo. "Electricity of a high tension must be used," he told a reporter who questioned him on the subject, "and an alternating one rather than a straight one."[404]

He was, nevertheless, sincere in his belief that DC power of around three hundred volts was safer for common distribution than the lightning bolts Westinghouse was sending around cities like Pittsburgh—via overhead wires that were liable to tangle disastrously with those of telegraph and telephone companies. Had he given the word, the Edison Electric Light Company could have used its own AC patents to compete on both the high- and the low-tension levels. But he would not consent, to the company's great cost.[*][405]

In December the Medico-Legal Society formally recommended the installation of an AC system at Sing Sing Prison, to excite the electric chair being built there by Edwin F. Davis, the New York state executioner. *The New York Times* published an approving article headlined "SURER THAN ROPE." At this, George Westinghouse, who had maintained a dignified silence in the controversy so far, publicly accused Harold Brown of being a paid Edison Electric Light Company stooge and a cynical scaremonger. Brown responded by challenging Westing-

---

* With the advent of sophisticated new transformer stations in the early twenty-first century, high-voltage DC power has been rediscovered as a superior force for, e.g., underwater cable systems.

house to an electrical boxing match, wherein they would each submit themselves to punches of their preferred current. The punches would increase by fifty volts, round by round, until one of them was forced to admit defeat. He was pretty sure it would not be him.[406]

To general disappointment, Westinghouse did not take up the challenge. Brown expanded his campaign in the new year, calling for a nationwide ban on AC distributions of more than three hundred volts. He published a booklet, *The Comparative Danger to Life of the Alternating and Continuous Currents,* which thanked Edison for giving him the space and the power to conduct his experiments. The authorities at Sing Sing rewarded him with an order for three Westinghouse dynamos, which were secretly provided by Charles A. Coffin of the Thomson-Houston Electric Company. A murderer on death row, William Kemmler, was given the honor of being the first man to test the effectiveness of the prison's electric chair.*[407]

YOUR INSTRUMENTS TAUNT ME

Edison paid only sporadic attention to the protracted and worsening "war." It was primarily the concern of Edward Johnson,[408] president of the Electric Light Company and Brown's main backer. Nowadays he was more interested in the applied sciences of sight and sound. He wrote two further moving picture caveats as stimuli to W. K. L. Dickson, whom he had charged with the construction of a cylinder-spinning, microphotographic camera.[409]

Production, meanwhile, of the improved phonograph began at West Orange. It reproduced so clearly that Edison changed his mind about marketing it only as a business instrument. He opened a soundproof studio in the laboratory and began to engroove a long series of musical and spoken-word performances on spools coated with a hard-wax formula that he kept as secret as Babbage's solution to the Vigenère cipher. It consisted of 80 percent burgundy wine, 25 percent frankincense, 9 percent colophony (a rosin derived from spruce), 8 percent beeswax,

---

* Kemmler's execution on 6 August 1890, was a notoriously agonizing disaster, with two long applications of current being necessary to kill the prisoner. Edison agreed that an account of it was "not pleasant reading," and suggested that Kemmler would have died much more quickly if the executioner had dunked his hands in cans of water added to the circuit.

Edison with his microphotographic camera. (Photograph by W. K. Dickson, 1888.)

4 percent olive oil, and 4 percent water, heated at 110 degrees until it steamed solid and was left to cool in molds.[410] Record production was a simple matter of remelting the wax, so that blank spools of plaster of paris could be dipped in it and rotated for an even coating.

Before the decade was out, Edison would imbue this dark red medium with the sonic presence of some of the great names in classical music, including Johannes Brahms, Hans von Bülow, Josef Hofmann, and Johanna Dietz, as well as such celebrities as Mark Twain, William E. Gladstone, Lord Alfred Tennyson, Florence Nightingale, Sir Arthur Sullivan, Prince Napoleon, Otto von Bismarck, and the aged Count Helmuth von Moltke—whose voice had first been heard in the eighteenth century. Von Moltke recited some lines of Goethe that spoke forward through time: *Ihr Instrumente freilich spottet mein / Mit Rad und Kämmen, Walz' und Bügel* ("Your instruments taunt me / With cylinders and levers, wheels and cogs").[411]

Most of these records were cut by Theo Wangemann, whom Edison appointed manager of his new Phonograph Experimental Department and sent abroad as his musical emissary. But he also dispatched another sound technician, Julius Block, to Russia, whence a high-pitched Peter Ilyich Tchaikovsky could be heard giggling with excitement over what he called "the most surpassing, most beautiful, most interesting invention of the nineteenth century."[412] Block was unable to record Tsar Alexander III and Leo Tolstoy, but they joined Tchaikovsky in sending good wishes and *"gloire au grand inventeur Edison!"* *

These and other plaudits from Europe, where his isolated lighting systems were proliferating, set the tone for Edison's looming reception at the 1889 Paris Exposition Universelle. He had decided to attend in August, partly because his advance man on the site, William J. Hammer, was putting together a complete retrospective of his career so far, and partly because Mina deserved a treat after two years of adjustments to marriage and stepmotherhood. (Marion had become such a resentful problem that they had allowed her to leave school and sent her to France ahead of them, accompanied by an aunt.)†[413]

Besides, as the pianist Hans von Bülow noticed in April, Edison was an exhausted man. He had patented thirty-eight phonograph refinements in the last year, as well as doubling the luminosity of his lamps, inventing the Kinetoscope, embarking on a new career as a record producer, fighting a hopeless breach of trust case against Ezra Gilliland, and nearly blinding himself in a chemical explosion that had his face swathed in bandages for almost a month. Throughout this draining period he had to stand by, half regretful and half relieved, while Henry Villard reorganized all his lighting companies and shops (excluding only the Phonograph Works in New Jersey) into the Edison General Electric Company, capitalized at $12 million.‡[414]

For good measure, Villard included Frank Sprague's highly success-

* Tolstoy did record some cylinders after Edison sent him a gift phonograph in 1908. They can be heard on YouTube, e.g. at https://www.youtube.com/watch?v=6310hAtdl6k. The machine is on exhibit at Yasnaya Polyana.

† Once in France, Marion was placed in the care of a governess, and she received the rest of her education in Europe.

‡ Villard had previously and successfully combined all Edison's European lighting companies.

ful Electric Railway and Motor Company, which furnished the Machine Works in Schenectady with two-thirds of its business. Sprague had created the world's first electric streetcar service in Richmond, Virginia, in 1887, obliterating the memory of the little track and train Edison had built six years earlier. Edward Johnson and the other Menlo Park veterans running Edison shops were sullen about Villard negotiating away their autonomy. Edison pretended to sympathize, but he saw that the combination made business sense. Better even than its enormous profit to himself—at about $1.25 million in cash and stock—was the feeling, when he signed the incorporation papers on 24 April, that the "leaden collar" of company ownership had finally fallen from his shoulders.[415]

For twenty years he had had to find capital to keep his various enterprises going and over that span had been compelled to shelve many inventions, for lack of time or money to develop them. The hundreds that had come to him on honeymoon were only the latest examples of these suppressions. If he had nevertheless managed to invent a pyromagnetic motor and glassmaking machine since then, along with the Kinetoscope, how free would he now be to develop new ventures! For one thing, he had developed a compulsive interest in mining. When the exposition was over, he might switch to that entirely.[416]

GO AND BE PART OF A ROSE

Edison's five-week visit to Paris in the summer of 1889 was a succession of dazzling social and professional triumphs, culminating in near apotheosis when the French minster of foreign affairs inducted him into the highest rank of the Légion d'honneur.*[417] Instead of his usual loose black bow tie, he had to wear the red ribbon of a *commandeur* under his collar, plus a dangling, enameled grand cross. Save for two superior "dignities" usually awarded to statesmen, it was the greatest civil honor France could bestow, recognizing his "eminent merits" as a benefactor of civilization.

Wherever he went in the city (which he had mapped in his head before arrival), crowds pressed close to stare at him, inventors waylaid him with devices under their arms, and sycophants hailed him as

---

* The ceremony was performed at the Élysée Palace in September.

"*Maître,*" "*Sa Majesté Edison,*" "*le roi de Paris,*" and less formally, "*le papa du phonographe.*" Two hundred letters a day poured into his hotel suite on the Place Vendôme. The conservative *feuilletoniste* "Caliban" went to extremes of Gallic metaphor in predicting that Prometheus must soon wreak vengeance on Edison, out of "divine jealousy" for his success in "shackling lightning" and "externalizing sound." He was serenaded by Charles Gounod *au piano* and by Emma Eames, the Opéra's latest ingenue, who sang him Liszt's setting of "*Comment, disaient-ils,*" a poem by his favorite Victor Hugo. Louis Pasteur showed him around his institute. Alexandre Dumas *fils* begged him to come to Puys so that they could hold hands.

He was twice received at the Élysée Palace by President Sadi Carnot, and was twice treated to champagne luncheons in Gustave Eiffel's vertiginous new tower.* The hurtling elevators, the *poulet braisé aux truffes* and *langoustines et écrevisses au buisson,* the clamor of French conversation in his muffled ears, agreed with him no better than an "American breakfast," hosted by Buffalo Bill Cody, at an outdoor table loaded with clam chowder, cornbread, pork 'n' beans, "grub stake," hominy, and two kinds of pie. Long before the biggest celebration of all, a seventeen-course banquet staged in his honor at the Hôtel de Ville by the municipality of Paris, opening with sherry-infused soup and sluiced down with Château d'Yquem '75, Edison's shrunken stomach rebelled, and he was pale with dyspepsia.[418]

At all public functions, he remained obstinately mute in his refusal to acknowledge the countless toasts tilting his way. Yet he was accessible as usual to reporters, telling them with a straight face that he was designing a telephone that would allow parties to a call to see each other. He said nothing about the moving picture machine that he and Dickson would be working on when he returned home. But Étienne-Jules Marey, whom he met at a dinner commemorating the fiftieth anniversary of the daguerreotype, divined enough of that secret to give him a private demonstration of his own *zoetrope électro-photographique,* a rolling film device that could shoot twenty images a second of birds in flight. The difference between them, as Marey

---

* To this day, a wax Edison may be seen visiting with Eiffel in the latter's office atop the tower.

afterward acknowledged, was that he sought the illusion of moments of stillness in motion, to elucidate avian or animal mechanics, whereas Edison sought the illusion of movement, by streaming stills so rapidly that the eye could not "seize" on their separation.*[419]

His exhibit was by far the largest at the exposition, attracting thirty thousand visitors a day. It covered almost an acre of the *Palais des machines* and centered on a lamp of lamps—twenty thousand bulbs clustered into the shape of a single bulb forty feet high. At periodic intervals they were unlit, irradiated only by a giant concealed carbon. Then crowds thronging the gallery gasped as a wave of light ascended from the base, transforming opalescence into incandescence, until the whole thing shone like an effulgent balloon about to take off. Its glow fell on 493 of Edison's inventions, dating back as far as his 1869 vote recorder. Polished and carefully positioned, they gleamed and throbbed while phonographs cranked out "cheerful American songs." The music played only tinnily through speaker diaphragms (the problem of amplification had not yet been solved), but when Parisians listened through attached white rubber tubes, they were amazed that so much sound could emanate from wax. Not everyone liked what they heard. The bone earbuds scrubbed words and music clean of the "varnish" of ambient noise. It was the sound of a harsh new age of talking machines and artificial sunshine and *le Dieu moderne,* technology as God.[420]

Unknown to Edison as he went his rounds of the city, a French writer who had long obsessed about him lay dying in the care of nuns. Auguste Villiers de l'Isle-Adam was the author of *L'Ève future,* a visionary novel that imagined Edison's creation of an android woman. Its opening pages, written in 1877, depicted *le Sorcier du Menlo-Park* as a reclusive, Faustian figure, frustrated in his attempts to record the voice of the Almighty. Instead, this fictional Edison had applied his mechanical magic to the creation of a "new Eve," who could be the progenitor of a whole new race of synthetic beings, untroubled by morals, dedicated only to the advancement of science. At first he

---

* Marey sent Edison an advance copy of his magnificent *Le vol des oiseaux* in the fall of 1889. It included the specifications for all his cameras. Edison's subsequent Kinetograph and Kinetoscope were, as Marey dryly noted, "not without resemblance to my *appareil.*"

shaped her out of sound, as a singer heard but not seen, but then he made her a moving image, a singer-dancer whose seductive vivacity derived from the synchronization of audible and visible effects—just what the real Edison dreamed of achieving now with his Kinetoscope-Kinetograph. "I am experimenting upon an instrument which does for the Eye what the phonograph does for the Ear." *[421]

Villiers's death was reported on the front page of *Le Figaro* on 20 August. A few days later "Caliban" wrote in the same paper:

> It is clear that Edison has never read *L'Ève future,* and without doubt the name of [its author] is totally strange to him. Maybe he'll learn from this article that, during his stay in Paris, he passed a hundred paces from the hospice where his prophet lay in agony.
>
> I don't subscribe, like poor Villiers, to the philosophy that holds up the dressed crucifix against the invasion of the scientific barbarian. . . . But, being a man of the sort menaced by the American, who resembles Napoleon and is deaf like Beethoven, I find he plunges me into unspeakable melancholy, because I know well that he holds the future in his fob pocket.[422]

Before returning to America, Edison spent two scientifically oriented weeks in Germany, Belgium, and London, being further lionized by the likes of Hermann von Helmholtz, Werner von Siemens, Heinrich Hertz, and Sebastian de Ferranti. He revisited Paris just once to receive his final awards, pack a major souvenir—Aurelio Bordiga's white marble statue of the "Genius of Electricity"—and say goodbye to Marion, who was remaining behind in a *pensionnat* on the Champs-Élysées to complete her European studies.[423]

By the time he sailed from Le Havre on 28 September, he had picked at so many banquets and pretended to hear so many toasts that he did not care if he never donned a dress suit again. Across the Atlantic the prototype Kinetoscope awaited his inspection, and great tracts of Appalachian ironland lay ready for prospecting by a new company he had formed, the New Jersey & Pennsylvania Concentrating Works. Pearl Street—by any account his supreme achievement—

---

* Life imitated art on 9 September, when Edison went to see Léo Delibes's ballet *Coppélia,* about an old doctor's creation of a life-size singing and dancing doll.

was aglow with 16,377 lamps, just one of thousands of imitative constellations around the world.[424]

Edison had had enough of light, and enough of fame.[425] A new decade beckoned, in which he meant to recover his old identity as an empirical inventor, feeling his way by hand and intuition toward fresh fields of discovery—perhaps that of submolecular science, which suggested the communality, and maybe interchangeability, of all matter at the most basic level. What a great thing it would be to have every microscopic unit of his own body under control, detachable and adjustable at will!

"I would say to one particular atom in me—call it atom No. 4320—'Go and be part of a rose for a while.' All the atoms could be sent off to become parts of different minerals, plants, and other substances. Then, if by just pressing a little push button they could be called together again, they would bring back their experiences while they were parts of those different substances, and I would have the benefit of the knowledge."[426]

PART SIX

Sound

1870–1879

A T TWENTY-THREE, EDISON landed his first major contract as an inventor. The Gold & Stock Telegraph Company of New York City paid him $7,000 to lease a shop in Newark and develop a small, fast, one-wire printing telegraph. He was given an extra $400 to buy tools and equipment and hire a machinist for six months. When (the contract was respectful enough not to say *if*) his instrument was "clearly patentable," he would receive a salaried appointment as the firm's consulting electrician. And in further recognition of the ingenuity he had displayed last summer in Manhattan, Gold & Stock promised him a bonus of $3,000 if he could invent a facsimile telegraph that transmitted shapes, not just dots and dashes. Wealth—at least in terms of his absolute penury not so long ago—stared Edison in the face.[1]

It was an unusual face for a young telegrapher at the beginning of the Gilded Age, when beards were as obligatory as bowler hats: smooth-shaven, large, and pale, blank during moments of thought, yet animated and focused when speaking. Edison looked like—and was—an open, engaging personality, unusual only in his lack of interest in food and apparent habit of sleeping in his suits. He compensated for being unable to hear much of general conversation by dominating it himself. His near-cataleptic concentration on any technical problem could have been that of a recluse, yet upon emergence from it, he was as gregarious as an actor, eager for approval of his sometimes labored jokes. The fact that he thought himself funnier than he really was betrayed a certain detachment from society. His frequent acts of generosity and kindness contrasted with an apparent inability to care about, or even notice, the emotions of other people.

Victims of this indifference put it down to his haste to outrace every-body, whereas in fact it often hampered Edison's own progress.

He constantly sought to widen his web of business and social rela-tionships, while detaching himself with difficulty from those of no use or interest. Even now, as he put up the shingle of "Newark Telegraph Works" at 15 Railroad Avenue, he was still a partner in Pope, Edison & Co., electrical engineers of New York, and a co-founder of the Fi-nancial & Commercial Telegraph Company, directly competing with Gold & Stock. He had a telegraph patent pending for the former firm, and was about to execute two more for the latter, one in his own name and another co-signed by Franklin Pope—in whose mother's house he was boarding, under increasingly strained circumstances. Pope and he were further allied with James Ashley, editor of *The Telegrapher,* a New York journal always willing to promote new devices in which Ashley was personally interested. Back in Boston there were even older associates with varying claims on him, one owning the rights to Edi-son's first patented invention, the electric vote recorder of 1868.[2]

Had he been less of a confidently self-centered spider, adept at tweaking all these strands, he might have quailed at acquiring yet another partner. But in late February he took on William Unger as his machinist and gave him a large share of the Newark shop. It was an impulse he would soon regret—unlike the friendship he at once struck up with John Ott, a twenty-year-old mechanic whose ability to con-struct any device on order amazed him. "Here, I tell you what I want: you come and take charge of this place for me." Ott would remain on Edison's payroll for more than half a century, until they died within hours of each other.[*][3]

YOU CAN DRAW ON ME

The Newark Telegraph Works, soon to be known as Edison & Unger, was that city's first communications laboratory. Edison intended forthwith to manufacture his own inventions, but as Ott noted, he was not a man to waste time in manual labor.[4] He liked to use his hands creatively—spilling just the right number of drops from one

---

* John Ott suffered a crippling stroke in 1895. Edison continued to employ and support him for life.

test tube into another, sketching rapid diagrams of telegraph circuitry (clear to the last relay, for all their speed), or calligraphing important letters with graceful curlicues. For the next several years, while bending his brain to complex theories of information exchange, he would devote himself with increasing fascination to the mechanics of recording—printing, perforating, engraving, motographing, mimeographing, and other still-undreamed-of ways of capturing words on the fly.

It took him three months to perfect the device specified on his first contract, a stock printer that would be smaller and at least as fast as the telegraph industry's standard Calahan ticker.[5] Gen. Marshall Lefferts, president of Gold & Stock, was impressed with it. He began to keep an admiring eye on Edison, even though the "Autographic or Fac Simile Telegraph instrument" he also wanted would not materialize for another eleven years.[6]

Edison's delightful feeling of being affluent for a change was buttressed in the spring when Gold & Stock paid $15,000 for rights to a ticker he and Pope had co-patented. They called it their "gold printer," although it was neither the first nor the last instrument Edison designed to work on circuits reporting gold price fluctuations on the New York Stock Exchange. The high value Gold & Stock assigned to it derived from his innovative use of electricity to operate a "unison stop" that synchronized all the printers subscribing to the circuit. James Ashley had contributed nothing to the patent, but as a partner in Pope, Edison & Co., was happy to accept one-third of the purchase price.*[7]

Edison was less happy to have either man profit from a machine he regarded as largely his own. He looked for a way to dissociate from them, and began by using some of his $5,000 share to pay off rent arrears he owed Pope's mother. Then he moved to bachelor digs on Market Street. With obvious pleasure he wrote to tell his parents in Port Huron, Michigan, that they could "take it easy after this."[8] Nancy Edison, to whom he owed almost all his inquiring spirit, was bedridden with dementia.

---

* Over the next three years, Edison manufactured and sold 3,600 gold printers at home and abroad.

*Dont do any hard work and get mother anything She desires =
You can draw on me for money—write me and Say how much
money you will need in June and I will Send the amount on the
first of that month = give Love to all the Folks—and write me the
town news—What is Pitt doing. . . .*

Thos A*9

Pitt was his much older brother William Pitt Edison, superintendent of the Port Huron street railway and a perennial loser in speculative local schemes. For the rest of his life Edison was going to have to
deal with "Dear Al" letters from relatives whose number, and financial difficulties, increased in ratio to his own prosperity.[10]

He continued to work on a variety of printing devices through the
summer, at a pace that struck more sedate observers as freakish. An
assistant at F. Brunner, Engraver & Die Sinker, in Manhattan recalled
him slamming bound rolls of typewheel blanks on the counter and
asking "When can I have them—when can I have them?" as if the
future of telegraphy depended on their immediate embossing. Sometimes he was in such a hurry, he was out the door before the assistant
had time to reply.[11]

In the fall, Edison and Pope created an elegant glass-domed,
private-line printer that won first prize in its class at the American
Institute fair in New York. It was "comparatively slow" and simple,
but so were most of the operators it was designed for. Gold & Stock
saw its potential and bought not only the machine but the company
the partners had formed to market it.[12]

At this point Edison's patent attorney, Lemuel Serrell, expressed
concern that his client was not doing enough to protect all the technological ideas that he kept innocently talking about. Edison at once
filled a notebook with sketches, explanatory notes, and dates of inspiration. The fact that most of the entries were made on the spot in
Serrell's office, with no reference to the instruments themselves, testified to the photographic accuracy of his memory.[13]

---

* A characteristic of Edison's orthography was his use of the sign = to signify something
more than a dash and less than a period. His odd distribution of initial capitals (mostly
cursive ones), seems to have come from simple enjoyment of the way they looked as he
inscribed them.

Out of this skein of shapes two major projects emerged: a refined system of automatic transmission, and an evolving series of "universal" printers that promised to be his most important achievement yet.

Automatic telegraphy—codified electric pulses shot down the line mechanically, rather than by a slow human hand tapping a key—was a technology that inventors had been struggling to perfect ever since the first Morse operator developed repetitive motion disorder. It involved the perforation of a paper ribbon with holes that corresponded to dots and dashes, each one permitting an electrical contact to be made as the ribbon whirled through a transmitter. In theory, the speed at which pulses were released was limited only by the pace of the ribbon. In practice, however, electrical induction—current arising in a wire because of charges in the magnetic field around it—caused a problem known as "tailing," with each hole, like a miniature comet, allowing part of its charge to drift behind it and blur the oncoming roundness of the next. At a speed of more than ten words a minute, the blurring erased the distinction between hurtling dots and dashes. Distance compounded the problem, so that often only an unbroken streak printed out at the end of the line.[14]

Earlier in the year, Edison had designed a shunted automatic transmitter that to a certain extent reduced tailing. He returned to that work with a will in October, when an independent backer, George Harrington, helped him open a second factory on Railroad Avenue, the American Telegraph Works, and followed up with the incorporation of the Automatic Telegraph Company a month later, capitalized in the extraordinary amount of $13 million. Harrington was a former assistant secretary of the treasury under Abraham Lincoln, and Edison looked forward to tapping him as a treasury in himself. Another promising source of funds was Harrington's friend Daniel H. Craig, a founding director of the Associated Press and the most ardent promoter of automatic telegraphy in the country. He had heard enough of Edison's skills to write him, "If you should tell me you could make babies by machinery, I shouldn't doubt it."[15]

Encouraging as these commitments were to a young inventor still largely unknown, they were eclipsed by the scarcely credible sight of a draft contract, in Marshall Lefferts's handwriting, offering him "forty thousand dollars" for a universal printer that would deliver text copy as well as numerical information. The sum was so large that it hardly bothered Edison to see that Lefferts, thinking twice, had crossed out the word *forty* and substituted "$30,000"—nor that the dread phrase "payable in stock" followed.[16]

Yet he still sounded not quite grown up when he wrote again to boast to his parents.

> *I am in a position now to Let you have some Cash, so you can write and say how much = I may be home some time this winter = Can't say when exactly for I have a Large amount of business to attend to. I have one shop which Employs 18 men, and am Fitting up another which will Employ over 150 men = I am now—what "you" Democrats call a "Bloated Eastern Manufacturer."*[17]

Edison's latest engineering recruits included Charles Batchelor, John Kruesi, and Sigmund Bergmann. All were recent immigrants, and destined to be among his longest-serving aides. Batchelor, a product of the Lancashire cotton mill industry, was slow, calm, and meticulous, with a deft pair of hands. He could draw like a draftsman and make his own precision tools.[18] Kruesi had many of the same qualities, combined with what passed in Switzerland for charm. Bergmann, newly arrived from Germany, had few words of English, but was a dogged perfectionist at whatever mechanical task he was assigned. "It doesn't matter if his tongue falters," Edison said. "His work speaks."[19]

THE SWING LAMP OF PISA

By January 1871 Edison had so much work on hand that his clients competed for his full attention. George Harrington wanted to know when some marketable devices might be expected from the inventor's new shop, in return for the thousands of dollars he had so far been charged for equipment and supplies. Lefferts was irked to find that he was designing perforators for both Gold & Stock and Automatic Telegraph, and wondered which company was getting the best for its investment in him. "I do see most clearly, that I shall *through you* be a very heavy looser [*sic*]," he wrote.[20] The plaint sounded more cajoling than bitter, because Lefferts had not yet signed their universal printer agreement and could afford to wait until Edison had something to show that justified its execution.*

Edison counseled patience, reminding all that technological breakthroughs could not be hurried. "Galileo discovered the principle of accurate Holology in the swing lamp of Pisa," he wrote Daniel Craig. "It wouldn't be a very sage remark to say—why damn it that lamp aint a clock." He told Harrington that he was working nineteen hours a day on his behalf. "The Machines that I am making now will be made well & Complete, and if they don't perforate more than 80 words a minute then there will be a funeral over here pretty quick."[21]

Harrington, reassured, continued to sign checks on demand, as did Craig, who was grateful for Edison's cheerful serenity. "Your notes, like your confident face, always inspire us with new vim."[22]

So, presumably, did the originality and specificity of his patent applications, such as one that month for a printer designed for the New York Cotton Exchange. It featured two typewheels mounted on a single shaft, one for characters and one for digits, as well as a polarized relay, characteristic of many of Edison's telegraph designs. This "cotton instrument" so impressed Lefferts that he ordered a production run of 150 models. When Edison also turned out a dozen experimental universal private-line printers, Lefferts was emboldened enough to list him as an asset in negotiating a merger between Gold & Stock and the Western Union Telegraph Company, the most powerful conglomerate in America.[23]

Western Union was ruled by William Orton, an executive of commensurate stature. Widely respected, he was forty-five years old, an active Republican and Episcopalian, forceful, incorrupt, choleric, austere, and frail. Unlike many tycoons of the early Gilded Age, Orton was not just a financier. He had trained as a printer, written his graduate thesis on magnetic telegraphy, and risen through teaching, religious publishing, and politicking to become commissioner of internal revenue under President Andrew Johnson. His intellect was, like his

---

* Between October 1870 and May 1871, Edison spent about $11,000 of Harrington's money on experiments alone, and probably twice as much of Lefferts's. Equipping the Ward Street factory cost Harrington a further $16,000.

prose style, clear and cold as glass, and his dignity such that the only nickname employees dared accord him was the single initial O.[24]

When Edison was presented to him on 13 February, he may not have known that a few years before, this affable slouch had been one of his army of gypsy operators, working in Western Union offices as far afield as Memphis and Boston. But Orton was already convinced that Edison was, as he put it, "probably the best electro-mechanician in the country," and saw the value of acquiring his past and future telegraphic patents.[25] Gold & Stock had a prior claim to them, and Lefferts made it a condition of purchase that they would remain within his firm, as a subsidiary of the larger company.

Negotiations toward the merger proceeded. Harrington followed them apprehensively, not wanting to lose the services of the Automatic Telegraph Company's house inventor. On 4 April he took the precaution of reserving unto himself two-thirds of the patent rights to any invention or improvement that Edison might make, "applicable to automatic telegraphy."[26]

A week after signing this conveyance, which included power of attorney, Edison was drawn back to Port Huron by the death of his mother. Nancy Edison was sixty-three. He had been her youngest child, and the one who most benefited from her store of bookish knowledge—if not her religious instruction. Back then she was, in his grateful memory, the only person who did not find him strange. "My mother was the making of me. She understood me; she let me follow my bent."[27]

COOPERATION AND GOOD WILL

Further legal responsibilities piled up on Edison after his bereavement, ending what little remained in him of immaturity. On the first of May he and William Unger leased the third and fourth floors of a building on the corner of Ward Street and Pear Alley in Newark, "with the privilege of four horse steam power," and transferred their thriving telegraph equipment works there. On the tenth, Harrington increased Edison's obligation to the Automatic Telegraph Company by bringing in five more investors to share the burden of financing his other operation at 103 Railroad Avenue. It said much for Harrington's reputation that he could recruit such distinguished partners and trustees as Gen. William Jackson Palmer, Erastus Corning, William P. Mellen, and Josiah C. Reiff, who all seemed eager to contribute more money to an operation that had already cost him $16,000.[28]

Two weeks later Marshall Lefferts and William Orton concluded an agreement for the sale of Gold & Stock to Western Union, with Lefferts retaining the presidency of the former company. Both executives congratulated themselves on securing "the cooperation and good will of Mr. Edison for the future."[29]

On 26 May Edison signed a five-year contract with Lefferts that guaranteed him an annual salary of $3,000 a year if he continued to supply Gold & Stock with marketable inventions. It made particular mention of the universal private-line printer he had just finished working on. He was to receive the title of "Consulting Electrician and Mechanician," and would be rewarded in addition with $35,000 worth of company shares. Assuming those went up rather than down, his total pay package might be much greater than its face value of $50,000.*[30]

Manifestly, Edison was being wooed by potent, grave, and reverend signors. But his new status also brought, in abundance, the jealousy that accompanies professional success. As part of the Gold & Stock/Western Union deal, the American Printing Telegraph Company he had formed with Franklin Pope and James Ashley ceased to exist. Although both partners were generously paid off, their severance did not compare to the size of Edison's windfall, which they felt was partly due to work they had put into the patents concerned. Ashley especially would never forgive him.[31]

VERY HANDSOME EYES

One wet evening that spring three Newark schoolgirls took shelter from the rain in the hallway of Edison's factory on Ward Street. They were invited inside by an employee who knew them, and came upon Edison working on a stock ticker. Mary Stilwell, aged fifteen and a half, was struck by him for two reasons. "First, I thought he had very handsome eyes, and next, because he was so dirty, all covered with machine oil, &c."[32]

---

* Equivalent to $1.06 million in 2018.

The eyes won out, and she was emboldened to ask him about his work. Outside as they talked, the rain grew worse. She and her friends decided they would have to make a dash for home. The man who had brought them in offered to escort two of the girls with his umbrella. It was not big enough for all of them, so by chance or more likely design, Mary found herself left alone with Edison. He pulled an overcoat over his work clothes and escorted her home himself.

> When we got to the house I saw that he was determined to go in and I had to invite him, and when my mother came down she asked who that was. I told her and said that he had brought me home and she went in. I was in mortal terror lest she should ask him to stay, but she did, and then he got up and took off his overcoat and stayed till nine o'clock, and then when he went away he asked permission of my mother and myself to call again. When he got it he availed himself of it almost every evening, and at last after five months . . .[33]

GRAY MORNING SAUNTERS

During that long period of courtship, Edison discovered that the money due to him from his various contracts, though substantial, was not enough to meet the expense of maintaining three shops and a workforce approaching seventy. He also had to bear the necessarily expensive business of experimenting, and such essentials as patent fees, food, and cigars.[34] He developed a severe cash flow problem, and dealt with it by paying his creditors only when their demands turned into threats. At the same time his perfectionism retarded delivery of factory orders and promised prototypes. "When are you agoing to have something to show in the way of the new Perforator and Printer?" Daniel Craig wrote in early June.[35]

When Harrington—always nervous about having to share him with Lefferts and Orton—also complained about an apparent lack of progress on the automatic telegraph, Edison was provoked into a rare outburst of anger. "I cannot stand this worrying much longer. . . . You cannot expect a man to invent & work night and day, and then be worried to a point of exasperation about how to obtain money to pay bills—If I keep on in this way 6 months longer I shall be completely broken down in health and mind."[36]

Harrington responded with a check that met his current payroll problem. Two days later Edison felt flush enough to invest $300 in a Port Huron liquor store. As he joked years later, "I have too sanguine a temperament to keep money in solitary confinement."[37]

He was telling the truth, however, about working night and day. In June he delivered two prototypes of his most important invention to date, a universal stock printer, to Gold & Stock, and on 26 July he executed the first of two patents on improvements in automatic telegraphy to gladden Craig's heart. He shuttled back and forth between his shops and office in Newark and the headquarters of his various clients in New York. Being peripatetic stimulated his creativity: "I have innumerable machines in my Mind." Not wanting to lose any of them, he developed a lifelong habit of carrying pocket notebooks to record every inspiration.[38]

His principal interest at this time (aside from Mary Stilwell) was the electrochemistry of automatic telegraphy. A printing method invented by George Little in 1869 allowed signals to flow from a metal stylus onto paper sensitized with potassium iodide or some other aqueous solution. The marks were "fugitive" if the recording point was platinum, and permanent if it was iron. But recorders based on Little's patent "tailed" terribly, encouraging Edison to come up with a superior method of his own. Harrington let him open a small research laboratory for the purpose, on the top floor of the Automatic Telegraph Company building in downtown Manhattan.[39]

Edward Johnson, the company's superintendent, was deputized to work with him after being warned that Edison was "a genius . . . and a very fiend for work."[40] So began the most important friendship of Johnson's life. "I came in one night and there sat Edison with a pile of chemical books that were five feet high when laid one upon another. He had ordered them from New York, London, and Paris. . . . He ate at his desk and slept in his chair." Within six weeks all the books had been read and reduced to a volume of handwritten abstracts. One result of Edison's consequent erudition was that he was able to concoct a ferric solution for automatic reception that cost only five or six cents a gallon, as opposed to the seventeen dollars a gallon of Little's.[41]

Johnson was awed. Hyperactive himself, wing-mustached and torrentially talkative, he was destined to become an excellent salesman for Edison products, able to prove or pretend their superiority in every particular. But his chief value at this early stage of their association was to help "the Old Man"—as Edison was already quaintly called—deal with the challenges of long-distance automatic telegraphy, a subject in which he was an expert.

They discovered they had a shared gift for penury. Often, Johnson recalled, they worked without dinner through dawn:

> Along about daylight, when weariness and hunger combined to paralyze our mentality . . . we would find ourselves without so much as a nickel wherewith to purchase a bun, not to speak of a bed; and upon such occasion we would combat the weariness and the hunger by taking a brisk walk to Central Park and back, by which time the office boy at least would be on hand to assist us with a meager but grateful breakfast at Coffee Pat's, a well-known penny-lunch establishment on Park Row, whose sole other patrons were the "printer's devils" of the various newspapers in that vicinity.[42]

On one of these "gray morning saunters," when they could not afford even to eat at Pat's, Edison pointed at the decorations in the window and said, "Say, Johnson, we had better quit inventing and hire ourselves out as a pair of Chinese gods. We would be a more brilliant success. At least, we needn't go hungry."[43]

HOW TO TREAT A WOMAN

Edison's wooing of Mary Stilwell became serious around the time of Thanksgiving, when he was taking her home from a walk. By now she was sixteen, and he, at twenty-four, was "Thomas" to her "Mame." In the only interview she ever gave, she recalled how the unspoken issue between them came up.

"Have you ever thought you would like to be married, Mame?"

"Why no—not yet anyhow."

"Well, I have and I would like to, and I would like you for my wife."

"Oh I couldn't."

"Well, and why not? Don't you like me well enough? Think, now, and try not to make a mistake."[44]

Mary stammered something about being too young.

He waved her protest aside. "If you meant no you would say no, so now I'll see your father tomorrow night, and if he says yes we'll be married Tuesday."

What Edison lacked in romance he made up in directness. "I love your daughter and I'll make her a good husband," he told Mr. Stilwell, holding Mary's hand. "I am honest, and I am good, and I know how to treat a woman."

He agreed to wait a week for Mr. Stilwell's answer, which turned out to be positive.

"And so," Mary told Olive Harper in the last year of her life, "we were married."[45]

POPSY WOPSY

The wedding took place on Christmas Day 1871. By then Edison had bought a house at 53 Wright Street in Newark and delivered a lucrative shipment of six hundred universal stock printers to Gold & Stock. Thanks to this and other shop orders and the regularity of his salary as the firm's consulting electrician, he was back in the black, able to lavish $2,000 on his bride for the acquisition of furniture, domestic help, and—important to her—a new wardrobe.[46]

Mary was a pretty picture in fine clothes, especially when she tied a sexy ribbon around her neck and let white ruffles spill out of her loose sleeves and collar. She was a slender, pensive-looking girl with lovely eyes, unused to money (Mr. Stilwell was a sawyer)* but quite willing to spend it, now her husband had it.[47]

They were married by a Methodist Episcopal minister and spent their first night at home—Edison typically asking if he could stop by the factory for an hour or two to solve a consignment problem—before traveling to Niagara Falls and Boston for an abbreviated honeymoon. They were back home on New Year's Day. After that their life together remained private except for two jocular complaints that Edison entered into his notebooks: "My Wife Popsy Wopsy Can't Invent" and

---

* Nicholas Stilwell's occupation was probably the reason Edison included, in his last notebook entry before the wedding, a double-tooth design to prevent bandsaws from running out of line.

Mary Stilwell Edison, circa 1871.

"Mrs Mary Edison My wife Dearly Beloved Cannot invent worth a Damn!"[48]

He certainly could not say that of himself. During the first year of his marriage, and particularly in the months before Mary conceived her first child, he executed thirty-nine successful patents: printing telegraphs, typewheels, perforators, chemical papers, rheotomes, autotel instruments, electromagnetic adjusters, transmitters, unison stops, galvanic batteries, circuits, and signal boxes.[*49] These were amplified by hundreds of notebook notions, some to do with a word-printing idea that had come to him on the eve of his wedding: that of interpos-

---

* At this point, notice might be taken of the remarkable early parallels between Edison's marriages. In both cases he fell in love with a schoolgirl in the spring, courted her assiduously through the summer, made a formal request for her hand in the fall, wed her under the rites of the Methodist Episcopal Church, and overflowed with inventions in the months immediately following. Each wife presented him first with a daughter, then with two sons.

ing a band of silk, or some other resilient fabric, between a metal point and a roll of paper. The silk would be saturated in ink, and electrical impulses coming through the point would impress the paper with dots that formed letters. In an entry dated 18 January ("This is a novel affair"), he drew a batch of such points tapping down on the silk band as it flowed over the paper roll.[50] More elaborately in the same note, he imagined an electrochemical recorder in which the roll became a "platina faced drum" and the points "platina pens" and the interposing band sensitized paper that, as the pens tapped down in response to charges agitating them, spelled out the word BOSTON.

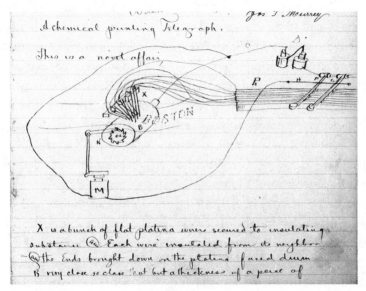

Edison chemical printer spelling out the word "BOSTON,"
January 1872.

Edison was doubtless unconscious that somewhere in the near future, like a shadow of these drawings thrown on an unseen wall, wavered the stylus, the foil, the cylinder, and the stippling vibrations of his most historic invention.[51]

COPPER AND TIME

When he needed a witness to countersign and date ideas that looked patentable, Edison turned more often than not to Joseph Murray, a senior machinist at American Telegraph Works. When in February Edison & Unger began to have trouble handling the volume of orders it attracted from telegraph companies (some of them as far away as England), he decided to make Murray his partner in a supplementary factory, manufacturing equipment primarily for Gold & Stock. On 5 February the new shop opened its doors on Railroad Avenue. This, with the addition of an annex Edison rented for himself on Mechanic Street, brought the number of his facilities in Newark to five, a total he soon found to be unsupportable administratively and financially. Again, in the rags-or-riches toggle that would always characterize his way of doing business, he found himself unable to pay bills on demand.*[52]

In July he worsened his difficulties by buying William Unger out for $17,100 worth of mortgage and notes, and concentrating all his manufacturing activities at Ward Street. The new firm was organized as Edison & Murray, and its payroll grew to an impressive seventy-four.

After the transaction a representative of the R. G. Dun credit rating agency noted that Edison was "talented in the way of inventive genius [but] Is at present Considbly spread out." Although his ultimate success "might in a measure be Consid assured with Careful Management is Yet Consid problematical & cr[edit] shd be extended with a good deal of Caution."[53]

Edison's creative flow subsided only slightly in the second half of the year, while he fought off creditors and tried to apportion equal amounts of time to his competing clients, Gold & Stock and the Automatic Telegraph Company. That meant alternating, or concurrent, development of information-sharing devices for Lefferts and Orton, and automatic transmitters for Harrington.

In one experiment aimed at pleasing the latter, he achieved the almost incredible printout speed of eighteen hundred words a minute by changing his universal private-line printer into an electric typewriter.[54] The device was of little immediate use to the Automatic Telegraph Company—it lacked a sending mechanism—but was a significant advance nonetheless. After working for years with dots and dashes, then with dots alone inscribing the shapes of letters, he was now producing

---

* Threatening letters did not prevent Edison borrowing $3,100 to help his impecunious brother Pitt start up a street railway in Port Huron.

Edison & Murray workforce, Ward Street, Newark, 1873.

neat roman capitals, and punctuation marks besides. "TO MR HAR-
RINGTON," the instrument rattled out, by way of demonstration,
"THIS IS A SPECIMEN DONE ON THE PRINTING MACHINE—DO YOU
THINK IT ANY IMPROVEMENT OVER THE LAST SPECIMEN, IT IS NOT
SO VERY MARKED AS TO KNOCK A MAN DOWN BUT STILL A STEP IN
THE RIGHT DIRECTION."[55]

During the fall and early winter Edison (buoyed by a $4,000 devel-
opment deal from Josiah Reiff) attained a high degree of sophistication in duplex telegraphy.* It was a process, pioneered by Joseph
Stearns and much pondered by himself, of sending simultaneous messages along a single wire. While one stream of signals dot-dashed its
way from *A* to *B,* another stream did the reverse. The idea was to
stagger the streams, so that going pulses would not collide with coming ones, just as pedestrians on a New York sidewalk avoided one
another as they headed downtown or uptown.[56]

Edison's initial duplex designs were sketched rapidly, yet with ele-

---

* Reiff, unaware of Edison's lifelong habit of getting Peter to pay for Paul, was under the
impression he was financing developmental work on the automatic telegraph.

gance and precision. Most of them violated normal telegraphic procedure by counterposing a neutral relay at one end of the line and a polarized relay at the other. He boasted that given enough funding, he could invent any number of such machines. "Very well," William Orton told him, "I'll take all you can make—a dozen or a bushel." Huge as Western Union was, its size meant that it handled the bulk of the nation's message traffic, and it was constantly on the lookout for devices that would speed up transmission. More than that, it was willing to buy patents of any sort that would cramp innovation among its competitors. Edison responded with twenty-one further designs, some penciled while he was waiting in Orton's anteroom, all probing past duplex toward what he called "diplex" messaging. It was a method that dispatched signals the same way in pairs. He intimated that if he were allowed to use Western Union lines to test both kinds of transmission, he might even succeed one day in coordinating duplex and diplex signals to create quadruplex telegraphy, with enormous savings of copper and time.[57]

This idea was so audacious that Orton did not comprehend its import. He gave Edison permission to run night tests on a loop wire between New York and Boston, to occupy experimental space in the Western Union building for the duration, and to use the company's own factory to fabricate necessary equipment. In return, Orton re-

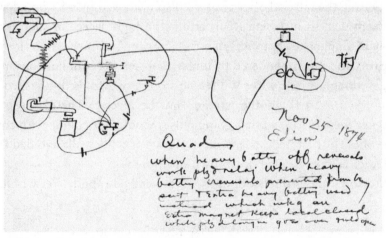

A freehand sketch by Edison of his quadruplex system, 25 November 1875.

served first-refusal rights to any duplex/diplex patents that Edison might come up with.[*][58]

NIGHTS ALONE

The birth of a baby daughter, Marion, on 18 February 1873 enabled Mary Edison to begin a series of duplex transmissions of her own. She saw little of her husband that winter, as he was working at Edison & Murray by day and camping out in the basement of Orton's headquarters most evenings. A Western Union employee recalled a week "during which time he never went to bed or had any regular hours for meals."[59] His model networks on the floor, which he webbed with copper wires unspooled from his pocket, threatened to trip unwary passersby:

> *When he was hungry, he visited a coffee and cake establishment in the neighborhood and absorbed what he was pleased to call the Bohemian Diet, and, returning with an unlighted cigar between his lips, he would begin his experiments anew. After a while, he would throw himself into a chair and doze, sometimes for an hour, and again for shorter or longer periods. He used to say that when he was thus napping, he dreamed out many things that had puzzled him while awake.*[60]

Edison's tests created as many problems as they solved, keeping him away from home for more than a hundred consecutive nights of experiment. At least one system that he sketched for possible patenting was labeled "Four-plex." It laid out a circuit of intimidating complexity with the waggish addendum, "Why not."[61]

He had to postpone any further work on this ne plus ultra of telecommunication in the spring, when Craig and Harrington sent him to England on a delicate assignment. He was charged with demonstrating to skeptical engineers of the British Post Office Telegraph Department that his automatic transmitter-receivers were faster than the so-called rapid printing telegraph of Charles Wheatstone.[62]

---

* Edison's specific mandate was to develop duplex or diplex designs that would amplify but not conflict with the Stearns patent, which Western Union owned.

Edison packed a small satchel, stuffed three boxes with instruments, chemical paper, and preperforated message tapes, and sailed for Liverpool on 23 April, accompanied by an assistant, Jack Wright. The crossing on HMS *Java* was stormy. Having never navigated a body of water bigger than Lake Huron before, he only now discovered that he had an iron stomach. It was not until the green fields of Lancashire hove over the horizon that most of his fellow passengers came on deck for fresh air.[63]

He left Wright with some equipment in Merseyside and reached London at the beginning of May. Most other young Americans visiting the world's largest city for the first time would have wanted to sightsee, but Edison was austerely focused on work.*[64] On the morning after his arrival he set up automatic receiving apparatus in the Post Office headquarters on Telegraph Street. Examiners there wanted to see the quality of sample messages sent to him by Wright in Liverpool. One of them said, encouragingly, "You are not going to have much show. They are going to give you an old Bridgewater Canal wire that is so poor we don't work it." He was also informed that Wright would have to draw his signal charge from "sand batteries"— cells filled with a weak sedimentary electrolyte.[65]

Sensing that the odds were stacked against him, Edison sought the help of Col. George Gouraud, the Automatic Telegraph Company's London representative. Gouraud was a large, majestic, expatriate American who had won a Medal of Honor during the Civil War and made the most of it in building a business career. Glossy of mustache and spit-polished shoes, he was a handsome man, except that his eyes, as a diplomat who knew him observed, "were not quite in tune."[66] Both of them, however, looked out for George Gouraud, and they reflected Thomas Edison as a comer worth cultivating.†

When asked if the company would stand for the purchase of "a powerful battery to send to Liverpool," Gouraud said yes. The only one available turned out to be a monster unit that John Tyndall had

---

* He seemed unaware that his lodging, the famous "Hummums" of Covent Garden, was a favorite haunt of Dickens, Thackeray, and Lewis Carroll and had a lubricious reputation as a hotel/bathhouse for single gentlemen checking in pseudonymously.

† Gouraud became Edison's London agent on 1 June 1878.

used for demonstrations at the Royal Institution. It consisted of a hundred cells and cost a hundred guineas, and Wright had to come to London to get it. But when grounded and connected, it gave messages such momentum that test printouts on Telegraph Street were, in Edison's words, "as perfect as a copper plate."[67]

He had to wait three weeks for the Post Office trial. In the meantime, his stomach proved less able to tolerate English cooking than it had turbulence at sea. After forcing down repeated dinners of roast beef and fried flounder, he complained that his "imagination was getting into a coma" and rejoiced to find a French patisserie in High Holborn that soothed him with carbohydrates.[68]

On 23 May Edison was finally able to demonstrate the superior speed of his automatic telegraphy to that of the Post Office's Wheatstone system. Wright sent him Morse messages at an average rate of five hundred words a minute over the next five days. The chief examiner thought this excellent, but pointed out that such an increased flow of copy would require the hiring of extra transcription clerks. The fact that the flow would, in turn, generate more telegraph income did not seem to outweigh the traditional British preference for muddling through.[*69]

Before returning to America, Edison won permission to conduct self-educating experiments on reels of transoceanic cable stored underwater at Greenwich. He was mystified to find that the signals he had received so sharply along the land line from Liverpool were distorted when sent through two thousand miles of coiled cable. A single dot printed out at the other end as a dash twenty-seven feet long. When after endless adjustments he managed to send whole words, only two or three of them dragged through per minute. Eventually he realized that the coil was the culprit—or rather, that electrical induction caused his signals to bleed from one winding to another.[70]

Those dark hours in Greenwich were as bleak as any he ever spent, the only hostelry open to him being a roach-infested snug for longshoremen. It purveyed molasses cake and coffee that tasted like burnt

---

* Despite the Post Office's decision to pass on Edison's system for this and other reasons, it was considered so promising that two British investors bought the foreign rights to it. They were unsuccessful in their subsequent efforts to introduce it in England.

bread. Gouraud made the mistake of coming down once to breakfast with him there, and felt so ill afterward that he had to be revived with gin. Edison sailed home in mid-June with no particular desire to revisit the Old World.[71]

A PECULIAR EFFORT OF THE MIND

He was greeted in Newark with the disagreeable news that a sheriff had been bothering Mary about debts. Joseph Murray reported that he had lent her $200 to help, but she "did not pay one bill out of it." Murray was a soft-hearted soul and quickly added, "I don't Blame her or find fault with her"—as well he might, since she had a baby to feed and the household accounts were, after all, her husband's responsibility. The larger problem was that the Edison & Murray shop was being bled white by the credit cost of buying out William Unger. And the largest problem of all was that the national economy, like Edison himself, was suffering the consequence of growing too fast and spreading too thin. The railroad-cum-telegraph expansion following the Civil War had petered out for lack of more places to go and a withdrawal of speculative capital. "Business is very dull money worse than ever," Murray wrote his partner. "Believe I have had a hard time of it since you left. I lost 11 lbs in one month but I shall die in the Harness if I ever do die."[72]

For the next year and a half Edison flirted with insolvency while his brain spouted inventions, and payments for them flew in and out of his pocket. A full-scale panic occurred in mid-September 1873, driving scores of banks under and triggering a depression that would last most of the decade. William Orton, conserving Western Union's resources, declined to finance the diplex just as it evolved into a promising prototype model. He considered Edison "a very ingenious man, but erratic," always with his hand out for laboratory funds. Gold & Stock cut back on its orders, compelling Edison & Murray to hustle for manufacturing contracts. The Automatic Telegraph Company, struggling to survive, looked around for a purchaser, even as a group of British investors paid $50,000 in gold for the foreign rights to its patents. Edison had to beg for his one-third share of that money, learning how his creditors felt when he imposed the same indignity on them. He had to beg again, this time to Orton, when a note for

$10,000 payable to Unger became due, along with a threatened lien on his Ward Street shop. Orton gave him only $3,000 on account of unspecified future work. A sympathetic Automatic investor, William Seyfert, came up with $6,600 more to cancel Unger's lien, generating another note that would one day trouble Edison severely.[73]

His rescuer then became his workbench colleague now. Charles Batchelor, who had shone as a precision machinist and mechanic on the factory floor at Ward Street, joined him upstairs in a series of quasi-scientific experiments provoked by Edison's belated discovery of induction in the wet cables of Greenwich.[74] Edison was not alone in being ignorant of the phenomenon, since American electricians had never had much to do with long-distance, undersea signals; they thought instead in terms of relay transmission overland.

He affected contempt for English innovators—"They do not eat enough pie"—but he had met and interviewed enough of them during his off-hours in London to realize that they were far more sophisticated than he in electrical matters.[75] During the winter and spring of 1874 he somehow kept his financial troubles in one compartment of his mind while occupying the rest of it with intense self-education in the arcana of electromagnetic and electrochemical science.* He unembarrassedly began with a primer, John Pepper's *Cylcopaedia of Science Simplified,* then studied and annotated the telegraphic handbooks of Robert Savine and Latimer Clark before absorbing such formidable tomes as William Crookes's *Select Methods of Chemical Analysis* and Charles Bloxam's *Laboratory Teaching.* All these authors were British. Their erudition inspired him to such a degree that he began to write a book of his own, a telegraphic treatise based on his ongoing experiments with Batchelor. Although he never completed it, several chapters appeared in various trade journals, in probable imitation of the first-serial publication method of the great Scottish physicist James Clerk Maxwell.†

Out of this six-month swirl of personal and intellectual turbulence,

---

* "At that time I was very short of funds and needed it more than glory," Edison recalled in later life. "I was paying a sheriff $5 a day to withhold a judgment which had been entered against me in a case which I had paid no attention to."

† Edison's book has been reconstituted from various archival sources. It is available in the digital edition of the *Papers of Thomas A. Edison,* under code NS7402.

like two great birds tossed up by a storm, came Edison's most important contributions yet to the science of communications. The first was
the quadruplex telegraph, which he had meditated on for years but
made practical now through his new understanding of induction. On
8 July 1874 he let Orton observe it in action between New York and
Philadelphia, conscious for the first time that he had an invention
worth millions. "I had my heart trying to climb up around my
oesophagus." The demonstration was a success, and two days later
The New York Times announced it to the world. Although Edison was
mentioned only in passing, he got his first addictive taste of renown.[76]

In its planned form, which he would continue to elaborate and refine for years, the quadruplex was a superbly symmetrical grid that
opposed two terminal stations, each sending and receiving two transmissions. The system allowed them to do so simultaneously—four
signals bypassing one another along the same wire—a feat akin to
playing Beethoven's *Grosse Fuge* on a one-string violin. Edison added
the further counterpoint of "phantom wires" at either end of the main
line, duplicating the variations in its resistance and creating, in theory, another four transmissions. He allowed with some smugness that
the overall concept "required a peculiar effort of the mind, such as the
imagining of eight different things moving simultaneously on a mental plane."*[77]

The dry language of his patent application explained that each pair
of terminals transmitted mismatched signals. Sender A drew constant
power from its source battery, the polarity of which reversed between
every dot-or-dash pulsation of current, with no break in the circuit.
Receiver A was polarized—a feature of most of Edison's telegraph
designs—and unsprung, so that it responded only to what it felt of the
reversals coming its way. Sender B, controlled by a neutral relay with
a retracting spring, was indifferent as to polarity, depending instead
on varying strengths of induced current to keep in touch with receiver
B.[78]

Orton was not sure if the beauty of the quadruplex as an electrical
concept would make it a practical addition to his system. He needed

---

* Later in life Edison remarked that his incandescent light system was "simple" compared to the quadruplex.

to be persuaded of that by seeing it thoroughly tested on a Western Union line. But he did let company shareholders know that the quadruplex was "an invention more wonderful than the duplex," which, coming from an executive famous for circumspection, amounted to a torrent of praise.[79]

NEW FORCE

In August Edison executed a flurry of patents in multiplex and automatic telegraphy, including one for his second great invention of the year, modestly entitled "Improvement in Telegraph Apparatus." It arose out of his earlier discovery that if a positively charged piece of lead was pushed across a slice of damp chalk resting on a negatively charged plate, the current running through made the chalk surface slippery, and caused the lead to skate with such agility it could be impelled in any direction, with no apparent friction or inertia—"as upon ice." Although at first the phenomenon had seemed to be simply a hydrostatic translation of electrical signals into motion, he now saw the lead as a potential pen point, and the chalk as a tabula rasa wanting to be written on. "Hence I term my invention the electric motograph."[80]

Writing—or in telegraphy, printing—consisted of marks imposing themselves on blankness. White paper soaked in an electrosensitive chemical solution and drawn over a drum could substitute for white chalk, Edison reasoned. When a bolt of current vibrated the pen, the point would slide, and the paper beneath darken in reaction, as if inked. (Ferrocyanide of potassium turned Prussian blue.) Conversely, whenever the pen lost its charge, the paper would become less slippery, and there would be a microsecond of drag—the skater digging in— before the next charge, and the next mark. That meant economical printing, with not too much white space separating the characters.[81]

He congratulated himself, with reason, on having discovered a "new force"—so new that he could not for the moment figure how best to utilize it. Nor, since he was used to the ticking of telegraph recorders, did he pay due attention to that of the motograph. Every dash or dot that it registered on the ear as well as the eye; every advance of the paper recorded a downward vibration felt. Perceptively, the editor of *Scientific American* noted on 5 September, "The salient feature of Mr. Edison's present discovery is the production of motion and of sound by the pen, or stylus, without the intervention of a magnet or armature."[82]

THE PROFESSOR OF DUPLICITY

Edison sought to augment the professional respect he earned from inventing the quadruplex and the motograph by assuming the science editorship of a new telegraphic journal, *The Operator*. He contributed a series of articles, including one on the duplex for the 1 October issue that was so densely technical as to dissuade many a young electrician from taking up multiple telegraphy. It ended with a warning: "To be continued." He wrote for other journals too, sometimes posing abstruse problems for readers to solve, and paid Robert Spice, a Brooklyn professor of natural philosophy, to give him a one-month crash course in chemistry.[83]

Although the extra notice accorded him as a result was modest (some of his articles were published anonymously), Edison began to attract those twin concomitants of celebrity, the sycophant and the scourge. George F. Barker of the University of Pennsylvania wrote on 3 November to congratulate him on his "remarkable little instrument," the motograph, and asked if he would be "willing to come on and show it to the highest scientific body in the country, the National Academy of Science." And James Ashley, still fuming over his deal with Gold & Stock three years before, began to mock him in *The Telegrapher* as "the professor of duplicity and quadruplicity."[84]

Edison ignored Ashley and was unable to gratify Barker, perhaps because he could not afford the fare to Philadelphia. He was so desperate for money as the winter came on that he had to sell his house at a loss, and move his wife and daughter into an apartment over a drugstore in downtown Newark. The sale netted him nothing, relieving him only of attendant credit. His private anguish around this time was implicit in one of his periodic doodlings of dream imagery: "*A yellow oasis in hell. . . . the wrestling of shadow, a square chunk of carrion with two green eyes held by threads of gossamer which floats at midnight in bleak old rural graveyards.*"[85]

By the beginning of December he could stand poverty no longer. The quadruplex was now testing superbly on Western Union lines from New York to Boston, Buffalo, and Chicago. Orton saw the system as a crucial asset to the company, worth at least $10 million in the near future and incalculably more beyond. But he had yet to make an offer for the patent rights. Edison, with the ingenuousness that was part of his charm and a large part of his deficiencies as a businessman, wrote him on the sixth: "I need 10, 9, 8, 7, 6, 5, 4, 3, or 2,000 dollars—any one you would like to advance."[86]

Orton met him halfway with $5,000 "in part payment" of a purchase price still to be negotiated, and Edison simultaneously got Harrington to roll over a due note for $3,351 that might otherwise have ruined him.[87] With frigid weather and Christmas approaching, he had more than a hundred workers depending upon him in his various shops, not to mention Mary and Marion. He appealed to Orton for more money, saying that the quadruplex should pay him $25,000, plus annual royalties per circuit. "Edison is almost wild over the Quadruplex," Orton told one of his officers. He was excited about it himself, in his phlegmatic way, but was in no hurry to conclude a deal, for the good reason that patents covering the system had not yet been applied for. Edison had withheld executing them until the last minute of development, in fear of anticipatory imitation.[88] The most Orton would do, pending a formal application to the Patent Office, was to give Edison & Murray a manufacturing order, dated 17 December, for twenty sets of quadruplex instruments. It was worth $15,000 on delivery, but in the meantime Edison had to bear the cost of fabrication. When he appealed yet again, Orton treated him with Scrooge-like coldness (or so it felt to Edison) and departed for Chicago, saying he would be back after the holidays to resume discussions.[89]

At this juncture, like another Dickensian schemer looming out of the darkness, the financier Jay Gould paid a late-night visit to Edison's shop on Ward Street. His ulster drooped shabbily to the ground, and he wore a hard black bowler.[90] A mild, charming man despite his sharklike reputation, Gould owned several railroads plus the Atlantic & Pacific Telegraph Company. Unknown to Orton, he was about to acquire the Automatic Company from Harrington and Reiff, thus gaining control of Edison's printers. He now sensed an opportunity to snatch the quadruplex too—along with its inventor.[91]

Edison was struck by his "far off look" as he listened to an explanation of how the system worked.[92] Gould left without making a proposal. But a few nights later, before Orton's return from Chicago, Edison found himself being escorted through the servants' entrance to a house on Fifth Avenue in Manhattan. There, in a basement office, Gould

> ... started in at once and asked me how much I wanted. I said—make me an offer—then he said—I will give you $30,000—I said I will sell any interest I may have for that money, which was somewhat more than I thought I could get. The next morning I went with Gould to Sherman and Sterling's office and received a check for $30,000, with a remark by Gould that I had got the steamboat Plymouth Rock as he had sold her for $30,000 and had just received the check.[93]

The date was 4 January 1875, and at the flip of a millionaire's deed to a yacht, Edison was on the way to becoming a rich man himself. If not one yet, he was relieved of his current despair and hurt pride. There would be other cash crises in his life, some of them acute, but not till after he was famous and able to rely on the credit that is fame's reward.

Orton returned from his alleged "business" trip to Chicago, only to find that Edison had left town on an alleged "family" trip to Port Huron. They respectively took each other's travels to mean that the one man declined to be hustled, and the other would not be toyed with. At all events Orton was the loser, and Western Union's share value dropped four points when *The New York Tribune* (a Gould-controlled newspaper) announced on 15 January that the expanded Atlantic & Pacific Telegraph Company would shortly be putting one of Edison's automatic systems into operation between New York and Washington, and that he would become the chief electrician of all its telegraph lines.[94]

The announcement was a major blow to William Orton. He tried in vain to "accept" Edison's proposed sale of quadruplex rights for $25,000, then resorted to a Jarndycean lawsuit to claim them. If he had lived until 1913, he would have seen the quadruplex case's final resolution and realized the importance of the great invention he had gambled away.*⁹⁵

TO PRICK AS WE WRITE

Edison celebrated by splurging on scientific books and equipment and helping his father, brother, and in-laws out with loans and settlements. Mary celebrated too, throwing a masquerade party for her husband's twenty-eighth birthday on 11 February and treating herself to a fresh wardrobe.⁹⁶ She would not be able to wear the latest formfitting fashions for long. By the time the Edisons settled into a new house on South Orange Avenue in the early spring, she was again pregnant.

In May, Edison, wanting to free himself from manufacturing responsibilities, dissolved his partnership with Joseph Murray. At the same time he took advantage of Gould's embroilment in disputes over the acquisition of his quadruplex and automatic system patents to fold up his tent as electrician of the A&P Telegraph Company and quietly steal away to the life he had always coveted—that of an independent inventor in a laboratory of his own.⁹⁷

For the time being, it consisted of a few rooms in the Ward Street shop. He hired five extra men to assist him and Batchelor in their experiments. They were the machinists John Kruesi and Charles Wurth, both from Murray's shop floor; James Adams, an old friend from Boston days; his nephew Charley, an excitable boy of fifteen; and the indefatigable Sam Edison, willing to do any job necessary, from carpentry to cleaning up, in return for twenty dollars per week. With occasional extra help from Edward Johnson (still working for A&P) and another former telegraph buddy, Ezra Gilliland, Edison now had the makings of a research and development team.⁹⁸

---

* Orton's bill of complaint, filed 28 January 1875, cited the fact that Edison, the previous summer, had allowed Western Union's chief electrician, George B. Prescott, to assume co-ownership of the quadruplex in return for vital access to the company's wires. Prescott contributed almost nothing to the design of the system but insisted on a half share of its rights bounty. Competing claims and counterclaims disputed the interests of Harrington, Jay Gould, and Edison himself. The quadruplex case, which over time jarndyced into three state and three federal proceedings, as well as others administratively involving successive directors of the Patent Office and secretaries of the interior, is summarized in appendix 3 to volume 2 of the book edition of *The Papers of Thomas A. Edison*.

They began work on 1 June with a list of nineteen experimental projects, including "a copying press that will take 100 copies." The first such press was a messy device that saturated tissue paper with an ink of violet aniline dye and apple pomade. It was fragrant but ran slow, needing frequent blotting. Then on the last day of the month Batchelor noted, "We struck the idea of making a stencil of the paper by pricking with a pen & then rubbing all over with ink."[99]

The "pen" was nothing but a platina point that had to be jabbed manually at the paper while it lay, like a flattened fakir, on a bed of miniature nails. "It is not much good," Batchelor conceded, with English understatement. "Resolved to make a machine to go by clockwork or engine to prick as we write."[100]

The Edison Electric Pen with batteries and press, 1875.

Thus was born Edison's electric pen, a battery-wired stylus with a needle point that flickered in and out faster than a snake's tongue. Held as steeply as possible, to balance the tiny electromagnetic motor on its top, it pricked a sheet of stencil paper in a near-continuous line, allowing the penman to write—or draw—any cursive shapes he pleased.* The resulting perforate was framed, pressed against blank

---

* Edison's electric pen was the first consumer product to use an electric motor. It was also the first mass-copy duplicator and the precursor to A. B. Dick's mimeograph, which

stationery, and ink-rolled to print as many sharp duplicates as required.

At first the electric pen was a cumbersome instrument, for all the mobility of its point, but Edison progressively miniaturized the drive components, making it lighter and less vibrant in the hand. Even so, it required considerable skill to use: an *O*, for example, would drop right out of the paper if inscribed too slowly. He assigned the manufacturing rights to Ezra Gilliland, and it became the showcase product of his Newark laboratory, ubiquitous in businesses and government offices as far away as Russia. Over the next ten years it would sell some sixty thousand units and be remembered as "the grandfather of automatic stencil duplication."* Edison gratefully gave Batchelor and Adams a percentage of the profits.[101]

Helping him develop the pen gave Batchelor an education in electricity, which he had understood imperfectly hitherto. It also introduced him to the charms of nocturnal labor. "We work all night experimenting & sleep till noon in the day," he wrote his brother in England. "We have got 54 things on the carpet. . . . Edison is an indefatigable worker & there is no kind of a failure however disastrous affects him. He stands today the foremost inventor & electrician in this country by far."[102] He showed his respect for the Old Man (who was more than a year his junior) by never addressing him as "Al" or "Tom," as a few old friends were allowed to do. It was always "Edison" or "Mr. Edison," while he in return was "Batch."

TRUE UNKNOWN FORCE

By now Edison had lost interest in the multiplex and automatic aspects of telegraphy. Instead, he became fascinated by the "acoustic" or "harmonic" telegraph, which Elisha Gray and Alexander Graham Bell were separately developing. It involved the transmission of Morse sound signals along a single wire by several differently pitched "reeds," similar to tuning forks. If an identical array of reeds was mounted at the receiving end, each pair would vibrate at their mutual

---

is often misattributed solely to Edison. See Bruce Watson, "A Wizard's Scribe," *Smithsonian*, August 1998.

* One of the electric pen's satisfied customers in 1877 was Lewis Carroll.

frequency, while the others stayed quiet. Ideally, though, all the pairs ought to commune simultaneously, enabling a greater message traffic than even the quadruplex could handle.

The science of sound was a new one for Edison, although it occurred to him that a relay he had invented in 1873, with an oscillating electrode varying the resistance of water or glycerine, could have been adapted to produce audible signals.[103] He began a series of acoustic telegraph experiments for Western Union in November, assisted again by Batchelor and Adams. In less than a week he improved on Gray's acoustic transmitter by employing a delicate balance of electromagnets, resonators, and spring pendulums, each connected to its own battery, responding to its own frequency, and moving in and out of circuit without breaking the flow of current through the whole. "I do not wish to confine myself to any particular form of vibrating pendulum," he wrote in his caveat, "as a tuning fork or string secured at both ends, or wind instrument may be used." He also claimed as unique his idea of keeping current waves distinct as they undulated over long distances (the old telegraphic problem of "tailing") by placing the receiver in a derived circuit and shunting it with a condenser.[104]

At this time the same day the experimental trio were intrigued by a bizarre side effect of magnetizing a vibrator of Stubb's steel. A spark leaped from the core of the magnet, larger and stronger than any that could have been caused by induction. Curiouser and curiouser, when a wire attached to the vibrator was connected to a gas pipe in the laboratory, all the other fixtures in the building sparked in sympathy. Whatever energy they shared followed none of the laws of voltaic or static electricity. It jiggled no galvanometer, lacked polarity, and did not even taste of electrical discharge. Yet a knife stroked across the stove twenty feet away drew a splendid chain of scintilla from the hot metal. "This is simply wonderful," Batchelor jotted in a notebook entry, "and a good proof that the cause of the spark is a *true unknown force*."[105]

For Edison, it was almost as if he had willed the phenomenon to happen. Five months before, in his list of projects for experiment, he had vowed to find a "New force for Telegraphic communication." Already that imponderable looked to be something more marvelous than acoustic transmission. He, Batchelor, and Adams had much subsequent fun by applying the force to vibrators made of twenty-eight different other metals. Only two—boron and selenium—failed to react. Carbon and thallium produced "actinic" sparks; tellurium gave off with "a strong and disgusting smell of garlic"; cadmium outperformed bismuth; and silver produced the best flash of all, a spurt of "magnificent green."[106]

Edison wondered if he could utilize these sparks to send messages on underground and underwater wires without insulation. It did not occur to him at first that he might not need wires at all. Then he discovered that a spark could be made to hover between two lead pencils if they were tilted so that their points almost touched, providing that one was grounded and the other hooked to about a foot of free wire. To view the phenomenon better, he enclosed them in a "dark box" that was penetrated by a pair of brass eyepieces.[107]

The next step toward magic was when he made a "trembling bell" generator of a spark coil inside an evacuated glass tube, and carried the dark box into another building with the free wire still projecting. Incredibly, the sparking persisted, though much enfeebled. He could only assume that his new force was reaching the pencils through the "ether," to use the nineteenth century's favorite word for whatever it was that separated matter from matter. And he would have to wait thirty or more years before his box set's protruding wire could be called a radio antenna.*[108]

INVENTION FACTORY

The birth of Thomas Alva Edison, Jr., on 10 January 1876 saw his father immersed in a serious study of acoustics, and his grandfather constructing a long two-story shed on a hillside in Menlo Park, New Jersey.

Sam was an experienced builder, tough enough at seventy to hoist roof beams with men a third his age. Edison had given him carte

---

* Edison announced his discovery of what he called "etheric force" to newspaper reporters on 28 November 1875. The headlines created considerable popular interest. But he failed to publish his findings in proper academic form, and they were largely mocked by the scientific community. His future industrial rivals Elihu Thomson and Edwin Houston conducted a series of related experiments that proved to their satisfaction that the force was nothing but electrical induction. Edison had, however, discovered high-frequency electromagnetic waves, as confirmed in theory by James Clerk Maxwell and in later practice by Hertz, Lodge, and Marconi.

blanche to find a rural location where he and "the boys" could live and work, far from the worldly distractions of New York City—but not so far as to inhibit business trips and deliveries of supplies. Sam loved a deal as much as he loved whiskey (four slugs of which he tossed back daily, to no apparent effect), and he saved his son many dimes by choosing Menlo Park, a failing, half-finished development planned along a convergence of the Philadelphia turnpike and the Pennsylvania Railroad. Its forty-odd houses and many empty lots overlooked a landscape of corn and fruit farms, with a small lake lying like a mirror in the middle distance and Manhattan, twenty-four miles away, clearly visible in bright weather.[109]

Edison bought two tracts for $5,200. The one smaller and closer to the railroad had a show house on it, big enough to accommodate his family, visiting relatives, and three black servants. The larger field up Christie Street (a steep, muddy boardwalk uninviting to Mary) became by spring a campus for the laboratory of his dreams—an isolated research and development facility, staffed by a team of talented young experimenters and serviced by a parallel team of machinists who would manufacture and sell whatever he invented. Soon to be known as Edison's "Invention Factory," it was a concept new to technology, and for that matter new to science too: communal, democratic, daring in the scope of its ambitions. With William Orton's encouragement, he dedicated the facility first to the new science of telephony.[110]

If anyone that month had a prior claim on the word *telephone* (so far understood to mean the transmission only of pitched sounds, rather than speech), it was neither Elisha Gray nor Alexander Graham Bell but the German telegrapher Philipp Reis, who had invented a diaphragmatic *Telefon* in 1861. Edison had been familiar with its make-and-break circuitry as long ago as 1869. And in the summer just past, he had written the word *speak* above the resonant box of a sketched telephonic instrument, built around a tuning fork whose resistance was varied by mercury.

The idea had not worked, but now, on 14 January, he executed an acoustic telegraph caveat that contained what he afterward called, rather wistfully, the "First Telephone on Record"—a resonating receiver with a membrane vibrating to the incoming oscillations of the line wire, transferred via a tiny electromagnetized coil. Edison thought of it, however, only as a device for measuring wave sounds.[111] Its adaptability to speech reception dawned on him only after Bell's variable resistance telephone design was patented on 7 March.*

All three inventors, plus many others across the country, were preparing to show at the great Centennial Exhibition scheduled to open in Philadelphia in May. Edison secured a space of four hundred square feet, not as much as he wanted but enough to mount imposing displays of his quadruplex and automatic telegraph systems, the motograph and electric pen, and a range of printers, including those that operated chemically and spelled messages out in roman type. He elected not to provoke scientific wrath with a demonstration of "etheric force."[112]

With many items still to perfect, he waited impatiently for Sam to finish the new laboratory. Mary Edison was less excited than he was about their impending move. She was still only twenty years old, and had spent her whole life in Newark. Knowing her husband's eccentric work schedule, she was not looking forward to long nights alone in bed in an unlighted hamlet twelve miles from the nearest police station. Rosanna Batchelor—also a mother of two—had the same fear. The "surprise party" that some of Mary's relatives gave her at home on 16 March was a probable attempt to cheer her up.[113]

Nine days later Edison opened the new laboratory, and life changed drastically for everybody in his immediate orbit.

THE SILENCE OF COUNTRY NIGHTS

For the next year or so, his band of familiars remained small, amounting to a dozen or so veterans of the Ward Street factory.[114] Bergmann quit their ranks and set up an independent shop in New York, where he quickly prospered. Batchelor, Adams, Kruesi, Wurth, John Ott,

---

* In 1880, when Gray's and Bell's rival claims to have invented the telephone were being hotly contested in court, Edison claimed that in July 1875 he had sketched three acoustic telegraph devices with liquid transmitters that permitted the phenomenon of "undulating" or variable resistance current, the fundamental principle of telephony. These remarkable sketches do exist but are undated. There is no surviving evidence that Edison built a model based on one of them around November of that year. He frankly confessed, however, that the model did not work and always gave Bell full credit for his invention.

and Gilliland all moved to Menlo Park—the last on the understanding that he would establish a factory for the electric pen in a shack alongside the railroad.[115] Sam and Charley Edison came too, although the old gentleman was periodically obliged to return to Port Huron, where he had impregnated his housekeeper.[116]

One day he returned with torn clothes and many bruises, having leaped from the New York–Philadelphia express as it steamed past the depot at full speed. He said he had merely imitated what his son used to do with a bundle of newspapers when the evening train came home from Detroit. "I tell you, Tom, I wouldn't do it again for ten dollars!"[117]

Edison was grateful to Sam for finding what he insisted was "the prettiest spot in New Jersey," although Menlo Park's beauty had to be taken on trust in the raw light of early spring. The laboratory was especially stark with its fresh white paint glaring in contrast to the dull yellows and browns of houses elsewhere in the hamlet. The only features that (invisibly) distinguished it from an elongated, double-story schoolhouse were two deep subterranean brick columns, to give it stability. A vibration-free floor was essential to Edison when testing acoustic equipment.[118]

He kept saying that he had come to Menlo Park in search of "peace and quiet," not seeming to realize it was a strange remark for a near-deaf man to make. No less strange was the obsession with sonics that grew upon him now, unless the explanation simply was that the silence of country nights enabled him to measure frequencies that were disturbed by urban noise. He had never had any difficulty with the hard tapping and ticking of telegraph sounders, and he could "read" Morse by ear as easily as he swept his eye across pages of prose. But the much more complex harmonics and changing volumes of Bell's speech telephony—early rumors of which reached him in April—came as a challenging shock. To catch up and then compete with Bell and Gray (even as they competed with each other), he would have to stop thinking of telegraphy as the rapid transfer of signals meant to be decoded as handwriting or print, and adapt to the notion of messages sent and received purely as sound—not even needing, in most cases, to be written down.[119]

The last twist he found hard to accept. Edison was a compulsive, even fanatical recorder of every word, thought, and deed that he deemed to be of practical value. Like another deaf scribbler—Beethoven— he was at a loss without a pencil, and perpetually stuffed his pockets with notebooks and loose memoranda. Missing as much as he did of everyday conversation—gossip at Mary's parties, banter among "the boys"—he drew a distinction between it and communication, which he felt to be his specialty as an inventor. Never far from his mind, as he delved deeper and deeper into the intangibles of acoustics, was the re- cording point of his motograph, the ink wheel of his automatic printer, and the stipple of his electric pen.

It followed that his first five patents at Menlo Park were all de- scribed as "telegraphs," even though three of the designs were really telephonic, thrumming with multiple reeds, electromagnets, bulb and tube resonators, and sounder boxes, all tuned in pairs to different frequencies. He executed the last of them on 9 May, the day before the opening of Centennial in Philadelphia.[120]

His display there attracted less attention than it should have. He mounted it late, and rashly accepted an offer from William Orton to trade his own space for a share of Western Union's. Orton did not want Thomas Edison to look too independent at a time when the company was financing most of his work. The electric pen and auto- matic telegraph, however, were exhibited separately, and both won prizes, as did the quadruplex. Sir William Thomson, the British math- ematical physicist and chairman of the awards committee, praised the pen as "an invention of exquisite ingenuity."[121]

But the quiet sensation of the show was Alexander Graham Bell's demonstration on 25 June of his telephone to a private audience in- cluding Thomson, Elisha Gray, Josiah Reiff, and Edward Johnson. As the son of the creator of an instructional method for the speechless deaf, and a teacher of the deaf himself, Bell had an understanding of phonetics far more sophisticated than that of Edison, who was not present. He modestly described his membrane transmitter and linked, iron box receiver as "an invention in embryo." It was just as well he did, because when he called Sir William in another room, Johnson saw that the judge, ear to the flapping lid of the iron box, was bewildered.* Gray applied his own ear and heard at first only "a very faint, ghostly ringing sort of a sound." Eventually he caught the phrase "Aye, there's the rub," and was able to inform the rest of the party that Bell was quoting Shakespeare.[122]

REEDS, FORKS, BELLS, TAUT STRINGS, TIN TUBES

If Edison attended the Centennial, he did so unobtrusively. He was still so little known in Pennsylvania that a local paper referred to him as "an Englishman named Edison who has detected what is described as a new natural force." A more sharply etched image was necessary if he was to imprint himself and his work on the public mind. Working in Newark and New York, he had at least been in the way of capturing press attention; as far removed as he now was, the best he could hope for was more respect from the trade. *"I'm going to send something within next six weeks to patent ofs,"* he scribbled in a note to an old operator buddy, *"that will make the Teleghers [sic] eyes stick out a little."*[123]

This turned out to be a supersystem of acoustic transfer telegraphy, which he detailed in a lengthy caveat filed on 8 July. It required the synchronization of multiple stations in a frequency range "only limited to the amplitude of vibration which is practicable to give the reeds and the delicacy of the receiving instruments." Edison's novel idea was for acoustic signals of different length to swap intervals of wire time, so rapidly and smoothly that the flow of mainline current was never interrupted, nor would the messages themselves sound broken up at the ends of their respective branch lines. It was a concept of electrical time-sharing or, in future jargon, time-division multiple access.† He supplemented it with thirteen elaborate technical memoranda, entered and indexed in a set of notebooks he had initiated to keep track of experiments at Menlo Park. The record amounted to a retrospective survey of all his telegraph inventions, as well as a

---

* Not to mention another auditor, the Emperor of Brazil.

† TDMA is an essential operating system for mobile phone networks, which have the same synchronization challenges that Edison posed for acoustic transfer telegraphy in 1876.

grounding for the specialized research he now undertook in the field of sound.[124]

While still working on multiple telegraph technology, he conducted a series of experiments in telephone transmission, sure he could improve on the weakness of Bell's short-range signal. By speaking into a magnetized brass diaphragm held under pressure of a damp felt washer, he succeeded in getting a parchment receiver to say "How do you do." But that nonsibilant phrase hardly matched the complexity of "Mr. Watson—come here—I want to see you," a message Bell claimed to have sent coherently four months before. Edison proceeded to stick tiny tacks to diaphragms at various degrees of the curve to gauge where best he might cut in for particular pitches, and explore the acoustic potential of his electromotograph, which he found activated tuning forks as well as electromagnets. If this discovery was still more relevant to telegraphy than telephony, it at least taught him something about Helmholtz's use of forks and magnets to study the mechanics of speech.[125]

Sometimes a thousand twangling instruments—reeds, forks, bells, taut strings, tin tubes—would hum about his ears:

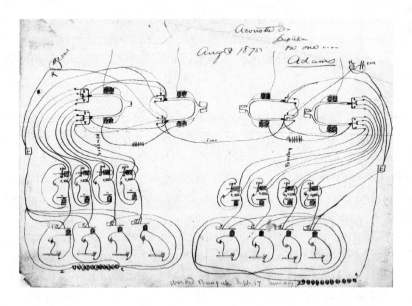

and sometimes voices:

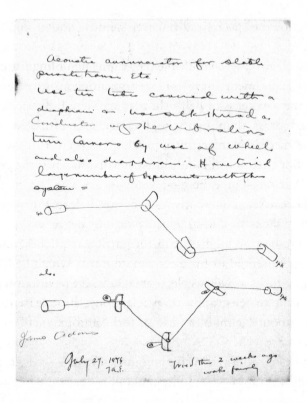

It was not always clear to him which discipline—acoustic telegraphy or telephonic transmission—he was exploring, or even if he was dimly envisaging another, not yet invented. Again and again cylinders and disks invaded his drawings. A cylinder might be an empty drum with a resonant base, or a hand-cranked harmonic receiver, or a roller duplicating lines of perforated print, or a spinning electromagnet,* or a telescopic tube that would enable him to gauge, precisely, the "column of air" necessary to send the "th sh ch s and other hissing sounds" so resistant to electrical dispatch. A disk might be stiff waxed paper revolving under the stylus of a "recorder-repeater," taking a message in dots and dashes that spiraled inward from the perimeter, or an electrochemically coated plate turning in a telephone receiver, or a hard rubber button coated with carbon and touching the vibrant tinfoil face of another disk—a pairing he thought promising, but unaccountably delayed acting on.[126]

---

* Edison typically segued at this point in his notebook jottings to the notion of an artificial rose buttonhole drawing its perfume from a tiny phial of attar.

The most beautiful, and technologically pregnant, instrument to come out of all this speculation was the Edison translating embosser of 3 February 1877. Despite its name, it was not a linguistic device. It merely sped up the distribution, or "translation," of enormous quantities of telegraphic text, such as presidential addresses, down long-distance wires.*[127] Nor was it acoustic in operation. But its design was so sleekly geometric, with twin turntables and twin recorder/reproducer arms tracking volute grooves, that it would look contemporary to audio engineers a century later.

The machine held blanks of oiled paper (Edison found that lard lubricated best) under its circular clamps, pressing them flat against the grooved platen of each turntable. Incoming electromagnetic pulsations caused a lightly sprung embossing point to indent the paper of the first turntable, rotated by an electric motor, while the recording arm, wormed to the platen's degree of spiral, made sure the point stayed on track. A hidden double lever started the second turntable the moment the first was full. Repetition (to use the current term for reproduction) was a simple process of letting the sprung point ride again at high speed over its own indentations, sending the recorded signals on to as many subsidiary stations as could be connected to a sounder in the embosser's circuitry.[128]

MOLECULAR MUSIC

In that same February that saw Edison turn thirty and show his first streaks of silver hair, he and Batchelor began a new series of experiments on what they called, variously, the "telephonic telegraph," the "speaking telegraph," and the "talking telephone." This confusion of names was common in the communications industry, and would last as long as Americans took to adjust to the startling notion that an electrically transmitted message did not necessarily have to be transcribed. It was beyond even Bell's imagination that people might one day use his invention just to chat. As far as Edison was concerned, the telephone was a device to speed up the process of turning words into pulsations of current, then turning the pulsations back into words at

---

* Edison meant it to improve on the already impressive performance of his automatic telegraph, which on 5 December 1876 transmitted President Grant's 12,600-word annual message from Washington to New York in just over an hour.

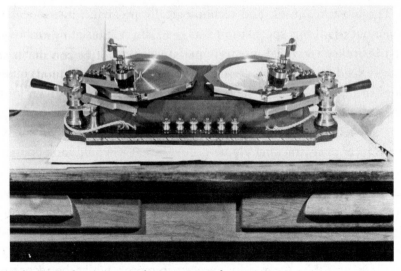

Edison's embossing recorder-repeater, February 1877.

the other end—words intended to be heard only by a receiving operator, who would then (as Edison had done thousands of times as a
youth) copy out the message for delivery. Hence the instrument really
was, for all its crackly noise, telegraphic in function.

Audibility was key, and he had failed so far in his efforts to improve on the wretched Bell magneto transmitter. He thought he might
achieve full vocability through the principle of variable resistance
within a closed circuit, which he held essential to the electrification of
speech.[129] Working often through the night, he and Batchelor made
transmitters out of membranes that shifted rollers or pins along
graphite tracks in circuit, but when they tried to speak clearly through
them, got only "a mumbling sound." Not until they returned to Edison's old idea of a wired button held against a diaphragm, and molded
it out of crushed black lead instead of rubber, did they achieve a dramatic increase in clarity. "With this apparatus," Batchelor recorded
on 12 February, "we have already been able to distinguish clearly
(known) sentences well between New York and Menlo Park."[130]

The excitability of pure carbon under pressure was a major discovery—or rather, rediscovery. Four years before, while constructing a
tubular rheostat, Edison had found that the electrical resistance of
packed, powdered graphite shot up and down, like a rogue barome-
ter, "with every noise, jar or sound." Eerily, when a carbon button

was framed in an iron ring and warmed with the heat of his hand, it gave off creaky, harmonic tones that he called "molecular music."

If so oversensitive then, surely carbon could not be too much so now, when he was looking for a relay that would accommodate the infinite gradations of the human voice—even to the nonvocal breaths, sighs, coughs, and hesitations that punctuated speech?[131]

The question was how compressed the texture of his button should be, or how loose, to give the widest resistance range.* Carbon came in an infinite variety of forms, from softest lampblack to rocklike anthracite. He would have to compound and test most of them for resilience and porosity, with the aural help of Batchelor and Adams. "I am so deaf that I am debarred from hearing all the finer articulations & have to depend on the judgment of others."[132]

UTTERANCES

That May Edison was in the midst of sketching some devices for the capture of sibilants when Rep. Benjamin Butler of Massachusetts challenged him to invent a telephone recorder that would convert sound into text. Edison brooded for a day or two, then came up with the opposite idea.[133] He drew what looked like a xylophone floating in space and scrawled:

Keyboard Talking Telgh,

I propose to have a long shaft with wheels on having breaks (ie electrical) so arranged with a Key board that by depressing say the letters T H I S simultaneously that contact springs will one after another send the proper vibrations over the wire to cause the Emg[†] & diaphragm to speak plainly the word this. . . . No difficulty will be had in obtaining the hissing consonants and as the break wheels & contact springs may be arranged in any form and as many as required used the overtones harmonics of the parts of speech can easily be obtained Turn this over in your mind Mr E & hoop it up.[134]

---

* That of anthracite coal, for example, varied from 300 to 1,700 ohms, yet Edison complained that it was good only for the *o* in *coach* and failed to register "the lisps & hissing parts of speech."

† Electromotograph receiver.

The xylophone bars turned out to be lettered keys, each ending in a tiny metal wheel serrated to make or break signals in the high frequencies. Edison apparently thought he could play the keys—one for each unit of the alphabet—in such legato combinations that *T* would blend into *H,* then into the vowel *I,* which would sharpen into a hiss as the last key was depressed. It was hardly the text recorder Butler had suggested, nor was it workable. Edison soon realized that letters had little to do with phonetics. Instead, he had dreamed up something truly radical: the notion of text transformed digitally into sound.*[135]

Impractical as it was, the keyboard talking telegraph—which he believed could be made to print as well as speak—marked a significant advance in his acoustic understanding. It featured, at least in theory, "tonewheels" rolling out the shapes of sound waves, an acknowledgment that speech consisted of overtones as well as volumes of air pressure,[136] and the double embrace in one instrument of recording and reproducing functions.[137]

Sibilants continued to elude Edison through June. Until he could get a diaphragm to articulate such a word as *scythe,* he felt he could not realize the full potential of his carbon rheostat. In its current experimental form, it took the form of a granular graphite disk about the width of a dime, sometimes sheathed in silk. When laid on the cupped poles of an electromagnet, then compressed beneath a battery-connected armature, the buttons were put in local circuit with them and a sounder. An inflow of main line current lightened the magnetic "weight" of the armature, reducing the resistance of the carbon to a mere few ohms; withdrawal of the current had the reverse effect, increasing resistance to several hundred ohms and again activating the sounder. In an article headed "Edison's Pressure Relay," the *Journal of the Telegraph* remarked, "It is probably the only device yet invented which will allow of the translation of signals of variable strengths, from one circuit into another, by the use of batteries in the ordinary manner."[138]

Despite the ongoing problem of high-frequency registration, Edison was able by midmonth to construct a combination telephone

* Today's text-to-speech computer applications are the realization of Edison's dream of 1877.

transmitter-receiver that tested "far plainer and better than Bell's." The normally phlegmatic Batchelor was so pleased with it that he boasted to his brother, "We have just got our 'speaking telegraph' perfected." That turned out not to be the case, and the pace of around-the-clock sonic experiments increased to the point that the *Operator* reported, "T. A. Edison is gray as a badger, and rapidly growing old."[139]

If so, Edison was not lacking in vitality. He was flush with contractual money from Gold & Stock and Western Union and enjoying the first of the "insomnia" blitzes that would characterize his life as an inventor. Not until 16 July did he feel he had a telephone worth patenting. The application he signed that day specified multiple tympani that "reproduced" vocal inflections and a sibilant-sensitive diaphragm with a layer of platina foil interposed between it and the contact point. But a laboratory visitor (spying for Alexander Graham Bell) found the instrument more powerful than clear, with the word *schism* sounding more like *kim*: "If Edison gets the articulation more perfect, which he is now working at, he can talk in thunder tones any distance."[140]

"We have had terrible hard work on the Speaking telegraph," Batchelor complained to Ezra Gilliland. "This last 5 or six weeks frequently working 2 nights together until we all had to knock off from want of sleep."[141]

Edison's gray look may well have come from the rub-off of carbon dust, platinum black, hyperoxide of lead, graphite, and other sooty conductors that besmirched him as the summer progressed. After one of these adjournments Mary entered the spare bedroom of her house and found an apparent chimney sweep lying dead to the world "on my nice white counterpanes and pillow shams."[142]

VOICES

"Just tried experiment with a diaphragm having an embossing point & held against parafin paper moving rapidly," Edison wrote on 18 July. "The spkg vibrations are indented nicely & theres no doubt that I shall be able to store up and reproduce automatically at any future time the human voice perfectly."[143]

When exactly, that summer, did the sum total of all Edison's past work on the sending and reception of sound coalesce into his greatest invention? As a discovery, it was so sensational that legends began to accrete around it almost at once, and his own memories of the moment swam confusedly. Perhaps it was when, listening to the hum of his elegant translating embosser—not an acoustic instrument, yet strangely melodious when its disks whirled at high speed—he thought he heard voices, "apparently talking in a language which could not be understood." Again, it was the faint sound of his own voice reciting the alphabet, when he retraced some diagrammatic scratches he had made on a strip of paraffined paper. Again, it was when he shouted "Halloo! Halloo!" into the mouthpiece and, pulling the paper through a second time, heard as from the far side of a valley, ^{"Halloo! Halloo!"} Yet again, it might have been the behavior of a voice-activated toy he made of a little man sawing wood to the loud recitation of any nursery rhyme. Or it was the sight, rather than the sound, of a needle attached to a live diaphragm, punching out oscillations as he said into it, "A—A—A." It may even have been when, absent-mindedly caressing such a needle as it vibrated, he felt a prick on his thumb—a sonic wave inscribing itself in his own flesh.*144

"Kruesi—make this," he recalled saying to his master machinist, giving him a drawing of a mounted, foil-wrapped cylinder, with a handle on one side to turn it, and a vibrant mouthpiece projecting a stylus that just touched the surface of the wrap.

> I told him I was going to record talking, and then have the machine talk back. He thought it absurd. However, it was finished, the foil was put on; I then shouted Mary had a little lamb, etc. I adjusted the reproducer, and the machine reproduced it perfectly. I never

---

* When talking on the telephone, Edison started using the greeting "Hello" rather than the old-fashioned "Halloo" and Bell's preferred "Ahoy!" In 1987 the audio historian Allen Koenigsberg established, with the agreement of editors of the *Oxford English Dictionary*, that the word *hello* was indeed coined by Edison. He wrote it for the first time on 15 August 1877, in a note boasting that his latest telephone receiver did not need to ring, "as Hello! can be heard 10 to 20 feet away." By 7 September 1880, delegates to the National Convention of Telephone Companies were wearing HELLO buttons on their lapels. See Allen Koenigsberg, "The First 'Hello!': Thomas Edison, the Phonograph and the Telephone," *Antique Phonograph Monthly* 8, no. 6 (1987).

was so taken aback in my life. Everybody was astonished. I was always afraid of things that worked for the first time. . . . But here was something that there was no doubt of.[145]

The "reproducer" was simply the mouthpiece assembly going back over its tracks and throwing back into the air the same waves of nursery rhyme that, a moment before, had gone into it. What awed Edison beyond any other thought was that the moment did not have to be a moment; it could be a century, if the foil and the stylus were preserved; and then in 1977, if some unborn person turned this same handle, the voice of a man long dead would speak to him. No wonder Kruesi, listening with incredulity to the thing he had made talking with Edison's voice, exclaimed, *Mein Gott im Himmel!*"[146]

All those who heard the miraculous machine in the ensuing months, from the president of the United States on down, reacted with equal disbelief. Since the dawn of humanity, religions had asserted without proof that the human soul would live on after the body rotted away. The human voice was a thing almost as insubstantial as the soul, but it was a product of the body and therefore must die too—in fact, did die, evaporating like breath the moment each word, each phoneme was sounded. For that matter, even the notes of inanimate things—the tree falling in the wood, thunder rumbling, ice cracking—sounded once only, except if they were duplicated in echoes that themselves rapidly faded.

But here now were echoes made hard, resounding as often as anyone wanted to hear them again. Breath had been turned into metal, metal was convertible back into air. It was a form of resurrection harder to credit (since faith was unnecessary) than that of Jesus Christ, which may have been why the most eloquent of the talking machine's early witnesses was an Anglican priest, the Rev. Horatio N. Powers. He not only wrote but spoke into an Edison cylinder the first poem ever written for acoustic preservation. It was entitled "The Phonograph's Salutation."

I seize the palpitating air. I hoard
Music and speech. All lips that breathe are mine.

I speak, and the inviolable word
Authenticates its origin and sign. . . .

In me are souls embalmed. I am an ear
Flawless as truth, and truth's own tongue am I.
I am a resurrection; men may hear
The quick and the dead converse, as I reply.[147]

IT WOULD GIVE US THE SPEECH

When poetry and myth are correlated with more factual records of Edison's invention of the phonograph (a name he gave it himself, based on the Greek particles for "sound" and "inscription"), they are not wholly disproved. Something extraordinary happened between his application for a new telephone patent on 16 July and his confident prediction, forty-eight hours later, that he would soon be able "to store up and reproduce" the human voice at will.[148]

Around daybreak on the seventeenth, amid a swirl of acoustic drawings and consonant-laden phrases—*Hemidemisemiquaver, Protochloride, The majestical myth which Physicists seek*—he wrote in his notebook, "*Glorious = Telephone perfected this morning 5 am = articulation perfect got ¼ column newspaper every word. had ricketty transmitter at that.*" Clearly an epiphany of sorts had occurred. His note said nothing about playing back the sounds he had heard.[149] But on a fragment of the same date he sketched both his telephone and his translating embosser and wrote the scattered words *reproduced, indenting,* and *needle.* The embosser was shown as a spiral platen with two tonearms, and the telephone had a strange device attached that might be a wheel, but on the other hand might not: if a wheel, why were two points stroking its circumference—one from the receiver and the other from what was clearly a reproducing diaphragm? The note's purpose became even clearer if the stroked outline was seen as the side view of a cylinder. Here was Edison (signing his name above, with Batchelor and Adams below as witnesses) thinking in terms of sending sound, receiving sound, inscribing sound, and playing sound back—all in one connected sequence.[150]

Supplementary testimony as to his moment of epiphany was supplied later by Charles Batchelor, a matter-of-fact man not given to Edisonian flights of fancy.

> The first experiment, as I remember it, was made in this way: Mr. Edison had a telephone diaphragm mounted in a mouthpiece of rubber in his hand, and he was sounding notes in front of it and feeling the vibration of the center of the diaphragm with his finger. After amusing himself with this for some time, he turned round to me and said, "Batch, if we had a point on this, we could make a record on some material which we could afterwards pull under the point, and it would give us the speech back." I said, "Well, we can try it in a very few minutes," and I had a point put on the diaphragm in the center. . . . We got some of the old automatic telegraph paper, coated it over with wax, and I pulled it through the groove, while Mr. Edison talked to it. On pulling the paper through the second time, we both of us recognized we had recorded the speech. We made quite a number of modifications of this the same night, and Mr. Edison immediately designed a machine which would be better adapted for giving us better talking.[151]

Just how "immediately" that design ensued, neither man could be sure. For ninety-one years another signed document was taken as proof that Thomas Edison saw the phonograph whole, in three dimensions, on 12 August 1877. It was his order to Kruesi to "make this":[152]

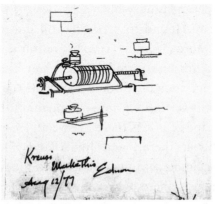

Edison's sketch of his first phonograph, circa November 1877.

Not until 1968 was it discovered that while the drawing may have been original, the inscription was of much later date, scrawled by Edison to please a publicist when he was too old to recall, or care, just when the model was built. But he did file a provisional British patent specification on 30 July, to confirm as discreetly as possible that he had been able "to make a record of the atmospheric sound waves" of human speech. He might have followed up at once with construction of a model in proof—except that the articulation he was getting from handheld diaphragms and strips of paraffin paper did not yet suggest any practical mechanics. Before there was a model, there had to be a design. It would take all that summer and much of the fall for Edison's mental pairings of waves and grooves, disks and buttons, drums and cylinders, point and pen, and script and sound to merge into an instrument that talked just as he talked, and remembered better.[153]

THE ILLUSION OF REAL PRESENCE

The prototype phonograph he received from Kruesi at the beginning of December was as simple and solidly built as a railway coupler. It consisted of a small brass cylinder, spirally engrooved, with an axle and turning handle cut at the same pitch, so that the advance of the needle (recording or reproducing) from left to right was identical with that of the cylinder. A sleeve of tough, yet indentable tinfoil was clamped on for every fresh recording, and the diaphragm units were toggled so only one vibrated at a time.

After much experimenting with needle design, a rounded rather than chisel point was found to press the foil into the groove more gently, and hence more faithfully, responsive to the vibrations of the diaphragm above. Edison's original stylus had been so sharp that when he recited "Mary had a little lamb" into the mouthpiece, all Batchelor heard was *ary ad ell am*—"Something that was not fine talking, but the shape of it was there . . . we all let out a yell of satisfaction and a 'Golly it's there!!' and shook hands all round."[154]

Now, with the cylinder turning steadily and the reproducer riding smoothly in its groove, there was no mistaking Mary or the dimensions of her lamb, and every sibilant in "its fleece was white as snow" sounded clear. By the end of November, Edison was ready to demonstrate his "talking machine" to the world.

Thanks to the evangelism of Edward Johnson, who had appointed himself a roving huckster for Menlo Park products, word had gotten around that a young engineer in New Jersey had invented a "recording telephone."[155] But the difficulty most professionals had in believing such a story had prevented it from becoming news. Edison decided to let the machine announce itself.

On 8 December *Scientific American* went to press with the biggest scoop in its history.

THE TALKING PHONOGRAPH

Mr. Thomas A. Edison recently came into this office, placed a little machine on our desk, turned a crank, and the machine enquired as to our health, asked how we liked the phonograph, informed us that it was very well, and bid us a cordial good night. These remarks were not only perfectly audible to ourselves, but to a dozen or more persons gathered around, and they were produced by the aid of no other mechanism than the simple little contrivance explained and illustrated below.[156]

There followed a technical drawing that needed only four indicative letters to show *A,* the comfortably curving rubber mouthpiece, *B,* the cylinder on its shaft, *C,* the crank handle, and *D,* the reproducing speaker. Indentations in the foil wrap could be seen. "There is no doubt," the magazine continued, "that by practice, with the aid of a magnifier, it would be possible to read phonetically Mr. Edison's record of dots and dashes,* but he saves us the trouble by literally making it read itself. The distinction is the same as if, instead of perusing a book ourselves, we drop it into a machine, set the latter in motion, and behold! the voice of the author is heard repeating his own composition."[157]

Scientific American needed more than fifteen hundred words to describe the phonograph's deceptively intricate operation. "No matter how familiar a person may be with modern machinery and its wonderful performances, or how clear in mind the principle underlying this strange device may be, it is impossible to listen to the mechanical

---

* [*Sic*]. This reference to "dots and dashes," rather than the hill-and-dale continuity of Edison's grooves, shows how hard it was even for a scientific journalist to adjust to the newness of the phonograph in 1877.

speech without his experiencing the idea that his senses are deceiving him." And the imagination also boggled at the uses its technology could be put to. Great singers would continue to sing, in their prime, long after they had lost their voices and died. Witnesses in court would have their testimony recorded down to the last stammer of denial. The children of the rich would hear proof of Papa's determination and soundness of mind when deeding all to his mistress. And if one day, *mirabile dictu,* some other Edison were "to throw stereoscopic photographs of people on screens in full view of an audience [and] add the talking phonograph to counterfeit their voices . . . it will be difficult to carry the illusion of real presence much further."[158]

TINFOIL COULD TALK

Edison did not wake up, in clichéd fashion, to find himself famous after the *Scientific American* article. Professional and popular plaudits were inhibited at first by the almost occult nature of his invention. When he applied for a patent on Christmas Eve, its originality so stunned examiners at the Patent Office that they issued one without question, not having any precedent to judge the instrument against. Then in an article published on the last day of the year, the *Daily Cincinnati Enquirer* referred to him as "Professor Edison," and soon the honorific was applied routinely.[159] Joseph Henry, secretary of the Smithsonian Institution, called him "the most ingenious inventor in this country," adding after a pause, "or in any other." To Sir William Thomson, he was the "first electrician of the age." The French Académie des sciences allowed Tivador Puskás, Edison's hastily appointed European agent for phonograph sales, to exhibit the instrument to its members. They greeted it with a reported "storm of applause," even though many of them were aware that Charles Cros, a Parisian amateur engineer, had filed specifications for something similar, when Edison was still working on his motograph telephone receiver.* From London to Milan to San Francisco, the pho-

---

* On 18 April 1877, Cros, too poor to apply for a patent, filed a letter with the Académie describing his idea of a *paléophone* that would reproduce sound by combining the "phonautograph" voice-sketching method of Scott de Martinville with duplicative photoengraving, an almost prohibitively difficult process. Cros never built a working model. Despite conspiratorial theories to the contrary, he and Edison do not appear to have been aware of each other before December 1877.

nograph was the subject of scholarly lecture-demonstrations and celebrated as the greatest acoustical phenomenon of the century. Its inventor
was compared to Franklin and Faraday, and became the subject of
schoolgirl essays, religious editorials, and cartoon caricature. By March
the "fire of genius" could always be seen burning in his "keen gray
eyes," hitherto blue. Then on 10 April *The Daily Graphic* hailed Edison
as "The Wizard of Menlo Park," an appellation that stuck even after
Menlo Park was no more.[160]

His laboratory lost its status as a secluded retreat. "Every day a
dozen of the heavy lights of literature and science come here," Edison
complained to Benjamin Butler. Journalists and sightseers came up the
boardwalk in such hordes he talked of "taking to the woods."[161] In fact
he loved publicity and cultivated more of it, going so far as to thank *The
Daily Graphic* for its coverage of the phonograph in an exquisitely calligraphed letter. His signature at the end was anything but modest —

Edison's letter of thanks to The Daily Graphic, *16 May 1878.*

—and in due course was adopted as his trademark.[162]

Celebrity became eminence on 18 April, when he accepted an invitation to present the phonograph to the National Academy of Sciences during its spring assembly at the Smithsonian Institution in Washington. Before he was introduced at the afternoon session, George Barker of the University of Pennsylvania arranged a comparative hookup of Bell, Phelps, Gray, and Edison telephones. The first three systems, all magnetic, were plagued by weak signals and interference along a line to Philadelphia, whereas the last, boosted by its carbon button transmitter, sounded sharp and clear.[163]

Edison declined to appear onstage and instead held court behind a desk in the president's adjoining office. There was such a press of academicians wanting to see him that the doors to the room had to be taken off their hinges. Meanwhile he sat with the phonograph before him, nervously twisting a rubber band between his fingers. Never having been mobbed before and inhibited by his deafness, he proved a shy, awkward public figure and let Charles Batchelor do most of the demonstrating.[164]

The machine—longer than the one he had shown to *Scientific American,* its rotation steadied by a flywheel—performed well, if not loudly, dropping some sibilants but responding with fidelity to some of the shouts, songs, whistles, and laughter projected at it by scientists forgetful of their dignity.[165] They were uniformly amazed that tinfoil could talk. When the astrophysicist Henry Draper sought to resume the formal proceedings with a lecture on spectrography of the sun, he was hard put to make his own voice sound as thrilling.

Edison relaxed as the day wore on. An interviewer dispatched by the *Washington Evening Star* noticed how animated he became in describing the latest products of his laboratory. He said he had just invented a device to measure the heat of stars. As for acoustic instruments, he was developing a disk phonograph "three or four times more powerful" than the cylinder model, and a "sort of improved ear trumpet," with an air chamber inside, which should help him listen to far-off sound "with perfect distinctness."[166]

Quizzed about his deafness, he said it did not bother him. If he needed to hear the output of his acoustic devices clearly, he simply put a stick into his teeth and jammed it against the speaker diaphragm.

That way, he explained, "I can hear more plainly than through the external ear."

For the rest of that night until two A.M., and through an equally long day following, he and Batchelor wore through yards of foil for the entertainment of Washington's elite, from hundreds of lawmakers in the Capitol to a private demonstration for the commanding general of the U.S. Army, William Tecumseh Sherman. By way of a climax, an invitation came for Edison to do the same for President Rutherford B. Hayes—who would not let him leave the Executive Mansion until after midnight.[167]

He sought relief in the Smithsonian museum from incessant gasps of astonishment and requests for replays of "Mary Had a Little Lamb." There he discovered, to his evident surprise, that in 1860 Scott de Martinville had used a needle-bristling membrane to trace speech patterns laterally on glass blackened with soot.* One of Scott's "phonautographs" was on display and proved to be primarily a visual device, meant to show the unique waves of every voice. As such it was a recorder only, incapable of sonic reproduction. Edison remarked that Scott would have been the father of the phonograph, had he been smart enough to inscribe in tinfoil instead of on glass.[168]

Before heading home for Easter, he stopped by Mathew Brady's photograph studio on Pennsylvania Avenue, sat in a chair that may once have been warmed by Lincoln, and posed with a hand on his lustrous invention, looking too tired to crank it one more time.[169]

WHISPERS

Surprisingly to some, Edison seemed to want to free himself of the phonograph after inventing it, patenting it, publicizing it, and licensing it out to the Edison Speaking Phonograph Company.[170] This venture, organized on 24 April, listed Edward Johnson as its general agent and made the risky decision not to sell phonographs, but to exhibit them to paying customers. While Johnson took a machine on the road to regale lecture audiences with recorded "Recitations, Con-

---

* Modern computer programmers at Lawrence Berkeley National Laboratories have dramatically translated some of Scott's visual patterns into actual audio.

versations, Songs (with words), Cornet Solos, Animal Mimicry, Laughter, Coughing etc.," Edison made side deals for the marketing of talking toys and clocks, demanding a 20 percent royalty on every item sold and offering no help with the mechanics involved.[171]

Before moving on to a project he considered more to his serious purpose as an inventor, he executed the first of two British phonograph patent applications that amounted to an astonishingly prescient survey of all the improvements and alternative designs he and other acoustic engineers would essay over the next quarter-century—disk records as well as cylinders; wax grooves instead of tinfoil ones; electromagnetic recording and reproduction; mass duplication by electroplating and the use of printing presses; even compressed-air amplification. For no inferable reason other than haste, he neglected to file for protection of these ideas in the United States and thereby committed himself to years of bitter later litigation.[172]

As soon as he could, he resumed work on the telephone, seeing that it needed his carbon button to realize its prodigious business potential. The Bell Telephone Company evidently felt the same, and tried in vain to buy the device from him. But Edison turned instead to Western Union, demanding $100,000 for a virtually solid-state transmitter that combined an induction coil with a disk of pure lampblack, tamped right against the diaphragm, doing away even with a needle.[173]

In offering this breakthrough device to William Orton, he was sure it would be judged fairly. Orton had always been a hard bargainer—too hard, in one instance—but had never held Edison's defection to the A&P Telegraph Company against him, nor hesitated to give him more assignments, out of frank respect for his inventiveness. Edison privately admitted to loving the man, despite their past squabbles over money, and was cheered that Orton did not flinch at his demand, stipulating only that the transmitter be tested first. It functioned perfectly on Western Union's wires, picking up whispers from three feet away and sending them without interference for seventy miles. A Bell transmitter, in contrast, failed to carry a shouted call from New York to Newark.[174]

The last thing Edison anticipated at this moment of triumph was that Orton, who was only fifty-two, would die on him. But no sooner had they agreed on his own specified terms of sale—"six thousand dol-

lars per annum for seventeen years payable in monthly installments"—
than Western Union announced that its president had been felled by a
stroke.[175]

"His last words to me were of you," Grosvenor Lowrey wrote Edison.[176]

The company's directors could well have renegotiated the pending
deal. But they chose to honor it—not surprisingly, since Edison had as
usual undersold himself. His carbon transmitter would remain a staple of American telephones for another century.[177]

At the time, he thought he had behaved like a canny businessman.
"I knew that I would soon spend this money experimenting if I got it
all at once," he reminisced when he was older and not much wiser. "I
fixed it so I couldn't."[178]

CLEAR STEALING

Notwithstanding Edison's attempt to dam one income stream, others
flowed pleasantly into his personal account. They included a $10,000
development grant from the Speaking Phonograph Company, which
also guaranteed him 20 percent of all exhibition receipts, carbon button purchase orders (he retained the manufacturing rights to that
item), and royalties on the sale of his telephone sets, five hundred of
which went to Chicago alone. Wanting to share his good fortune, he
awarded Charles Batchelor a 10 percent share of phonograph royalties, offered his father a tour of Europe, and bought Mary an elegant
team of horses for her spring outings.[179]

He also hired a personal secretary, Stockton Griffin, to handle the
sacks of mail and interview requests that came with his new fame.
This relieved him of bureaucratic paperwork just as his output of
laboratory notes swelled enormously. He perfected the astronomical
device he had mentioned in his Washington interview, calling it at first
a "carbon electro-thermometer," then a "tasimeter."[180] It was based
on his discovery that a rod of hard rubber was so supersensitive to
heat as to register even that of a star when aimed correctly. Its expansion altered the resistance of an adjoining carbon button, which could
then be calibrated electrically on a galvanometer. He also adapted the
principle of variable resistance to a plethora of "phone" products—an
aerophone that used compressed air to blast speech at large crowds or

wandering children; a megaphone that served the same purpose; a telephonoscope that reportedly picked up conversation from two miles away; an auriphone that did the same, in miniature, for deaf persons; and a phonomotor that converted sound waves into rotary mechanical action, allowing compulsive talkers to bore in more ways than one.*[181]

In mid-May he was upset by the public claim of David Edward Hughes, a British experimental physicist, to have discovered variable resistance in packed semiconductors long before him. The Royal Society was reportedly in receipt of a "Hughes Telephone" that featured a solid-state carbon transmitter like his own. "Evidently Mr H don't read the papers," Edison wrote William Preece, the newly appointed electrician to the British Post Office. "That is nothing but my carbon telephone . . . I'll bet £100 on it."[182]

Pique turned to anger when Preece, in reply, not only denied this but supported a new claim by Hughes to have invented a "microphone" featuring the carbon button. Edison had regarded Preece as a friend since his visit to London in 1873, and relied on him to act as his advocate before the local scientific establishment. Less than a year before, he had welcomed the Englishman to Menlo Park, and shown him all his sonic devices featuring pressure relays. Now Preece wrote with apparent relish, "The recent discoveries of Professor Hughes have thrown your telephone completely in the shade."[183]

There was a degree of truth to these words, because Edison had recently sent some demonstration telephones to Britain, in hope of breaking into the local market, and they had proved so vulnerable to interference as to be unworkable. He had forgotten that most lines in England were laid underground, unlike the high wires that gave him clear transmission in America. "You were on the very threshold of a great discovery," Preece lectured him, ". . . had it not been for the phonograph distracting your attention."[184]

The interference problem Edison was sure he could solve. But he was enraged by what he saw as Hughes's "piracy" and Preece's dis-

---

* Edison's need for instrumental names was so great in the spring of 1878 that he ordered his bookseller to get him a copy of Jacob Boyce's *Etymological Glossary* of Greek-derived words. *Papers*, 4.247. Charles Batchelor teasingly signed off a letter to him with "Yours phonographicarbontelephonically, Batch!!!"

loyalty, and decided to make a public issue of them. He brushed aside the argument of Sir William Thomson that while he was the real inventor of the microphone, Hughes had developed one independently. "It is not coinvention," Edison wrote an English acquaintance, "because after a thing is known all over the world for two years its sudden reinvention is clear stealing."[185]

A prolonged battle of claim and counterclaim ensued on both sides of the Atlantic, waged mainly in newspapers and technical journals.*[186] It embittered Edison's relations with Preece and slowed his development of an improved phonograph for exhibition purposes. Physically spent, he spent some days in bed toward the end of June. When Professors Barker and Draper invited him to join a scientific expedition to Rawlins, Wyoming, to observe the total eclipse of the sun on 29 July he accepted, seeing it as a chance to try out his tasimeter almost seven thousand feet above sea level.[187]

Mary was five months pregnant with her third child and not happy to be left alone. He was barely a week from home when Griffin wired him to ask, "How long are you going to stay there Mrs E wants to know."[188]

SOMETHING LIKE NIGHT

It was clear from Mary's querulous inquiry that Edison had said nothing to her about his intent to continue west after the eclipse, until the Pacific Ocean stopped him. "This is the first vacation I've had in a long time, and I mean to enjoy it," he told a reporter, saying he wanted to see Yosemite and San Francisco. But first he was determined to measure the heat of the sun's corona at the moment when the moon blocked out all the photosphere.[189]

The tiny town of Rawlins, which he reached on 18 July, was hard put to accommodate a trainload of scholarly strangers hauling almost a ton of astronomical and photographic equipment. It consisted of little more than a long street of bars and bawdy houses that some-

---

* Although at least one British journal, *Engineering*, deliberately suppressed data that supported Edison's case, his reputation in Britain suffered as a result of bringing it. Late in the year Thomson criticized him for failing to acknowledge that he had overreacted: "There is no doubt he is an exceedingly ingenious inventor, and I should have thought he had it in him to rise above . . . the kind of puffing of which there has been so much."

times echoed, at night, with the sound of gunfire settling local disputes. There was a hotel of sorts that found space for Edison only by
doubling him up with Edwin Fox, a reporter for *The New York Herald*. Their sleep was disturbed that night when a drunk frontiersman
barged into their room and said he wanted to see the famous inventor
he had read about in the newspapers. He introduced himself as "Texas
Jack" and demonstrated his skill with a gun by shooting through the
window at a weather vane. When Edison, who did not sleep well for
the rest of the night, inquired about him downstairs, he was assured
that "Jack was a pretty good fellow" and not one of the "badmen"
who frequented the town.[190]

He looked around for a suitable site to set up his tasimeter and
found that the scientists had bagged all the most sheltered places for
their telescopes and Draper's big wet-plate camera. Rawlins was on
the cusp of the Continental Divide, exposed to the atmospheric turbulence that a total eclipse can cause. Edison had no choice but to establish himself in a henhouse, temporarily displacing its current residents,
and pray for calm weather.[191]

The activities of the astronomical delegation caused much local
rubbernecking. Edison and his fellow observers gave notice that they
should be left undisturbed during the short "totality" of the eclipse,
when their instruments would have to be kept perfectly focused. Law
enforcement authorities in Rawlins were sympathetic and gave them
permission to shoot any unqualified intruder "on the spot."[192]

On Sunday 28 July there was a dawn-to-midnight rehearsal of all
the telescopic, spectroscopic, and other procedures that would have
to be coordinated when the cosmic moment arrived. Edison sacrificed
sleep altogether that weekend, trying to ensure that nothing but solar
rays would strike the infrared sensor of his invention. The tasimeter
was a camera-like, slit-visored box built around a lampblack button
that was pressed between two battery-connected disks of platinum
and backed up behind an expansive disk of vulcanite. An adjacent
mirror galvanometer, playing a spot of light along a graduated scale,
registered degrees of heat as minuscule as one-millionth of a degree
Fahrenheit. Any stray source of warmth—even his little finger moving
past the visor five feet away—edged the light spot sideways. For that
reason the box had to be aligned with a roof telescope precisely aimed

at the target pulsator. He practiced by focusing on two bright stars, Arcturus and Vega. Rawlins was scheduled to rotate through the lunar shadow at three-fifteen on Monday afternoon, and he would have less than two and a half minutes to register the corona at full flare.[193]

Encouragingly, the day began with pristine weather. "Not a cloud obscured the heavens," the *Laramie Daily Sentinel* reported, "and the air had that clear, deep blue which is found nowhere else but in the mountain region."[194] Later on a mass of cumulus drifted toward the sun and thickened, casting gloom over the landscape as well as on the astronomers. They were cheered when it passed away around noon, but then a wind began to blow, as if agitated by the darkness fast approaching from the northwest. It grew to gale force, buffeting through Edison's henhouse. Feathers and thistledown filled the air. He tried in vain to balance his telescope as the moon notched across the sun and the light faded. At five minutes past three only one-eighth of the sun still shone. The citizens of Rawlins watched through pieces of smoked glass as the crescent pared down. Something like night descended. Cattle along the ranges stopped grazing. The eclipse was total at three-fifteen, but Edison's rig was still unsteady. Then with only one minute of totality left, the wind dropped, he got a fix on the corona, and was rewarded with a rightward sweep of his graduating light.[195]

But he discovered that the tasimeter was ten times too sensitive for the rays it was receiving. It was thrown off scale, and he had no time to adjust it for further measurements before daylight returned, and puzzled cocks began to crow.[196]

STREAMERS OF THE SUN

Many years later, after the tasimeter had been shelved and forgotten, a western legend grew up that Thomas Edison "invented the electric light" while stargazing, or sunscoping, in Rawlins. Another story had him accidentally dropping a bamboo rod into a campfire and seeing it glow in the flames.[197]

The yarns were of course fanciful, since various forms of incandescent light had been invented (or at least attempted) before, from Jacques Thénard's briefly luminous platinum wire of 1801 to Stanislas Konn's self-destructing carbon rod lamp of 1873.[198] Edison had already experimented with some makeshift lamps at Menlo Park, firing them up from batteries and concluding, like so many electricians before him, that there was no way an incandescent element could shine for long without suffering a total, and permanent, eclipse of its own.[199]

If he did not experience an epiphany during his stay in Rawlins, he definitely mused about ways of harnessing electric power for work and light after visiting San Francisco and Yosemite in the first week of August. Recrossing the Sierras and Rockies, with their tumbling rivers, he wondered why hydroelectricity was not being used to drill mines and detect ores. The harvest-ready flatness of Iowa's cornfields, with overloaded wagons crawling toward distant elevators, cried out for electric trains—even automatic trains—speeding along lines that matched the geometry of the landscape.[200]

Throughout his career so far, he had thought of electricity solely in terms of telegraphy and telephony, the tasimeter being little more than a by-product of his experiments in variable resistance. It had, however, interested him in astronomy. (On reaching home, he meant to use it to sweep the sky for undiscovered stars.) And working with Henry Draper had also made him curious about the new science of spectroscopy. Even more "illuminating," perhaps, was the fact that he had just experienced, at close hand, a cosmic event withheld from the sight of most human beings. Edison would have been less than flesh if he did not experience some Kantian emotions when he connected with the light of Arcturus and the streamers of the sun, through the open roof of a henhouse in Wyoming.[201]

Before returning east via Chicago, Edison was invited to present a report entitled "On the Use of the Tasimeter for Measuring the Heat of the Stars and of the Sun's Corona" to the American Association for the Advancement of Science in St. Louis on 23 August. He had a terror of public speaking, and had an excuse not to break his journey after receiving a disturbing letter from his secretary. "Mrs. E's health is not of the best," Stockton Griffin wrote. "She is extremely nervous and frets a great deal about you, and about everything. I take it to be nervous prostration—She was so frightened yesterday for fear the children would get on the [railroad] track that she fainted." Griffin

The total eclipse seen from Creston, Wyoming, 29 July 1878.
Astronomical drawing by E. L Trouvelot.

had summoned the family physician, Dr. Leslie Ward, and Mary was better again, but "needs a change and right away."[202]

There was a slight tone of reproach in the last words, since newspapers were saying how tanned and healthy Edison looked after his sojourn in the mountains. He felt obliged, however, to accept the AAAS's invitation, if only because Barker and Draper wanted to induct him as a member. No sooner had he done so than Griffin wired him to "return at once," as Mary had had a relapse. Edison left St. Louis immediately after delivering his address.[203] He reached Menlo Park on the twenty-sixth to find that the cure she needed was to have him back.

Later that day he was seen in a rockaway buggy cresting the hill above the hamlet, with Mary driving and his children nestled behind him.[204]

TO BE ABLE TO SUBDIVIDE

Within twenty-four hours of returning home Edison had sketched something he labeled "Electric Light," although it looked more like a battery-operated thumbscrew. On close examination, the pressure points of the rack were shown to be platinum, and the small element between them was either boron or silicon. Current passing through was supposed to produce an arc of light for as long as the element separating the points remained in place.[205]

It was not the first time he had doodled a scintillant or luminous device. During his early experiments with variable resistance, he had noticed that a piece of metallic silicium held between two live carbons took on a steady glow that might well empower "a Common Electric Light," if only the problem of burnout—oxidation and fusing of the incandescent element—could be solved. He had experimented with wicks of carbonized paper, electrifying them in weakly evacuated glass chambers, but gotten little more than a brief radiance, then smoke. In the spring he had talked of "subdividing" electric light into a multiplicity of lamps after hearing that the veteran inventor Moses Farmer had helped William Wallace, a wire manufacturer in Ansonia, Connecticut, build an eight-horsepower dynamo. But telephone work had prevented him going north to see it, despite the plea of one of Wallace's engineers that "it will be a blessing to the world, to be able to subdivide."[206]

While out west he had discussed the dynamo with George Barker, who knew Wallace and suggested they visit Ansonia together sometime soon. This suggestion rekindled Edison's dormant interest in illumination technology. "He came home full of projects for producing light in large quantities and distributing it in small units as is done by gas," Batchelor wrote, in a reminiscence of his boss's high state of excitement after the eclipse. The two men sat up several nights "figuring out stations that could deliver current to houses which could be used equally for light, or for small powers such as pumps, sewing machines, printing presses & all sorts of manufacturing—All these could be turned on or off at will without affecting any other."[207]

Barker and Batchelor accompanied Edison to Ansonia on the afternoon of 7 September. A fellow passenger on the train was amused to see Edison, wearing a frayed straw hat and long linen duster, squeezed into the amen corner of the parlor car, with the professor's considerable bulk crowding him.[208]

The same duster ballooned back from his legs next day as he stood facing the rotary whir of Wallace's little machine, ruminatively chewing tobacco and figuring out how it worked. He did not hear the other three men talking and joking around him. When Batchelor interpreted, he laughed but was soon absorbed again. The dynamo ignited eight dazzling carbon arc lamps via a single thick copper wire. Edison visibly gloated in its promise of long-range distribution. In the words of a reporter present, "He sprawled over a table with the simplicity of a child, and made all kinds of calculations. He calculated the power of the instrument and of the lights, the probable loss of power in transmission, the amount of coal the instrument would save in a day, a week, a month, a year, and the result of such saving on manufacturing."[209]

Eventually he told Wallace that a machine that lit only one lamp per horsepower was not what the world was looking for. "I believe I can beat you in the search," he said.[210]

A BIG BONANZA

Edison did not mean to be critical of the dynamo itself, which he arranged to buy on the spot. Here was a single machine generating more power than all the batteries he had ever handled, illuminating a whole foundry as it did so. But the steadily eroding carbons, as well as the red heat and quarter-inch thickness of the copper conductor linking them, made him optimistic that he could "beat" Wallace, and other experimenters too, in attaining the double chimera of incandescence and subdivision. "It was all before me," he said afterward. "I saw that the thing had not gone so far but that I had a chance. . . . The intense light had not been subdivided so that it could be brought into private houses."[211]

That same night in Menlo Park he doodled some electric arc and kerosene lamps, in the apparent hope of creating a fire in his mind. One drawing in his notebook showed a pair of spiral wires around metal poles sitting on switches. He wrote beneath, *It may be possible that one regulator at the Central Station may be made to do it for all the main current being regulated by the heat of a large spiral.* Then it occurred to him that an incandescent lamp could be made to regulate itself in such a way that its wire never melted. Over the next few days he and Batchelor sketched forty-five variations of this idea and filed a caveat to protect them. On 13 September Edison telegraphed Wallace, HURRY UP THE MACHINE. I HAVE STRUCK A BIG BONANZA.[212]

As always when diverging into a new course of experiment, he saw himself at the glorious end of it, rather than the fraught beginning. He forgot that he was now so famous that he could no longer afford to boast an invention without being sure that it would work. "I have it now!" he said to one of the newspapermen who had found that Edison loved to give interviews. "When ten lights have been produced by a single electric machine, it has been thought to produce a great triumph of scientific skill."[213] With weird precision, he described what he was going to achieve in the immediate future:

With the process I have just discovered, I can produce a thousand—aye, ten thousand from one machine. Indeed, the number may be said to be infinite. When the brilliancy and cheapness of the lights are made known to the public—which will be in a few weeks, or just as soon as I can thoroughly protect the process—illumination by carburetted hydrogen gas will be discarded. With fifteen or twenty of these dynamo-electric machines perfected by Mr. Wallace I can light the entire lower part of New York City, using a 500 horsepower engine. I propose to establish one of these light centers in Nassau Street, whence wires can be run uptown as far as the Cooper Institute, down to the Battery, and across to both rivers. These wires must be insulated, and laid in the ground in the same manner as gas pipes. I also propose to utilize the gas burners and chandeliers now in use. In each house I can place a light meter, whence these wires will pass through the house, tapping small metallic contrivances that may be placed over each burner. Then housekeepers may turn off their gas, and send the meters back to the companies whence they came. Whenever it is desired to light a jet, it will only be necessary to touch a little spring near it. No matches are required.

Again, the same wire that brings the light to you will also bring power and heat. With the power you can run an elevator, a sewing machine, or any other mechanical contrivance that requires a motor, and by means of the heat you may cook your food. To utilize the heat, it will only be necessary to have the ovens or stoves properly arranged for its reception. This can be done at trifling cost.[214]

Virtually every electrical device Edison described had yet to be invented, and he would soon start building his own dynamos too. The words *volts, amperes,* and *ohms* were not yet in parlance. When, on

5 October, he executed his first lighting patent, it only presupposed with the miraculous subdivision process he claimed to have devised. With deceptive modesty, he claimed an "Improvement in Electric Lights," exemplified by a design plucked more or less at random from his regulatory caveat. Nevertheless it was the first incandescent lamp he ever invented, and as time passed, its originality became more apparent. A spiral of platinum, or any wire with a high fusing point, hung in a glass cylinder and wrapped itself loosely around a vertical zinc rod. The rod lengthened as the wire incandesced and depressed a lever, just as the spiral shone with maximum brilliance and was about to melt. The lever shut off current to the spiral, allowing it to cool while still glowing. Meanwhile the rod contracted and another surge of current flowed into the cylinder. The make-and-break cycle recurred with such rapidity that the human eye was not much aware of fluctuations in the light. But in practice the rod kept bending and the constant vibration of metal against metal caused the lamp to literally die of fatigue. Edison experimented with some other thermostatic devices, but they all failed.[215]

He had to accept that it would be a while yet before he strung ten, let alone ten thousand lights across "the entire lower part of New York City." So great was his reputation, however, that the mere fact he was confident of doing so prompted the avid interest of Wall Street. While gas stocks slumped on both sides of the Atlantic, Grosvenor Lowrey drew him into negotiations with a group of financiers associated with Drexel, Morgan, & Co. The bank graciously offered to relieve Edison of all financial distractions in exchange for title to his lighting patents, present and future. J. P. Morgan sought to market those rights in Britain and Europe, seeing vast imperial revenues once Edison managed (as surely he would) to subdivide the light. "Impossible overestimate result if such success attained," Morgan cabled his London partners.[216]

Edison hesitated. He had already promised to let George Gouraud and Tivador Puskás handle his foreign patent sales, but their influence did not compare with Morgan's. He absolved himself of guilt in the matter by handing over power of attorney to Lowrey, telling him, "All I want is to be provided with funds to push the electric light rapidly." The result was the hasty formation on 16 October of the Edison Electric Light Company by a board of blue-chip incorporators representing the interests of Western Union as well as Drexel, Morgan. They awarded Edison $250,000 in stock, an experimental budget of $130,000, a guaranteed minimum share of annual royalties, and other allowances for a total of $395,000.*[217]

"With the English patents," Lowrey told him, "I think we can get money enough to set you up forever."[218]

TWO YEARS, OR MORE

Edison's sudden semi-deification as the modern Prometheus, without yet having brought light to anybody, made him realize, with more private apprehension than Lowrey could imagine, the consequences of too much braggadocio. If he failed to deliver what he had so airily promised, he would be humiliated and likely ruined. That prospect loomed almost at once, when trustees of the Light Company pressed for a demonstration of the system they were investing in. He dared not risk showing them the few fallible models he had so far managed to construct. Burner after platinum burner was fizzling out or fracturing in his hands. Nor could he prove his claim to have "just discovered" the secret of subdivision. He began to show signs of stress and closed the doors of his laboratory to visitors.[219]

One of the few who still managed to get in was, inevitably, another reporter. He bribed his way upstairs with a cigar and found "the Professor" much more subdued than normal. With rain falling outside and blue smoke curling over his head, Edison admitted that he might need two years, or more, to make all the improvements necessary to his electric lamp. He connected one workbench model to the Wallace dynamo, just delivered, and its platinum strip glowed an intense, cold white before he prudently turned the current off. "Now, old man, get out and let me go to work."[220]

The interview was published on 20 October, and Edison felt renewed pressure to "show and tell" his new invention. Sleepless and half-starved, with bloodshot eyes and a week's worth of beard, he strove to construct a light that would last as long, or even half as long, as a candle. After three days of further failure he was felled by a slashing attack of facial neuralgia that kept him bedridden for the rest of

---

* Equivalent to $10.3 million in modern money.

the month.* Mary, suffering hardly less pain, gave birth to a twelve-pound son, William Leslie, on the twenty-sixth.[221]

Edison had not fully recovered when he heard that the Light Company believed a rival lamp designer, William E. Sawyer, might have anticipated his recent patent. "I was astonished at the way Mr. E received the information," Stockton Griffin wrote Lowrey. "He was visibly agitated and said it was the old story—i.e., lack of confidence—The same experience which he had had with the telephone, and in fact with all his successful inventions. . . . He said it was to be expected that everyone who had been working in this direction . . . would immediately set up their claims upon ascertaining that his system was likely to be perfect."[222]

Actually, Sawyer, an impoverished and unstable alcoholic living in New York, thought the opposite of Edison's chances. He knew from experience that the self-regulating platinum lamp would never work, being an impossible balance of cost and inefficiency. With his partner, Albon P. Man, he now claimed to have invented "a means of making carbon incandescent without consuming it." Edison tried to downplay this disturbing news, telling Griffin that the line he was developing was "entirely original and out of the rut."[223]

SUCH A DREARY PLACE

When in December the directors and bankers of the Light Company were finally allowed to visit Menlo Park, their image of Edison as a solitary, inspirational genius was, in Lowrey's tactful phrase, "somewhat tempered." The much-publicized white laboratory on a green hillside overlooking New York was now the center of a muddy construction site. Bricklayers and carpenters were racing the onset of winter, getting ready a new office-library building in the front yard and a massive machine shop at the back. The laboratory itself was in a state of apparently chaotic expansion, as Griffin's clerks and file cabinets and Kreusi's artisans and heavy equipment moved out of the ground floor to their new accommodations. The space vacated was already filling up with new experimenters and researchers, many of

---

* Trigeminal neuralgia, or *tic douloureux,* is one of the most painful afflictions known to medicine. It is a spasmodic condition often brought on by stress.

them with university or polytechnic degrees. With wads of Morgan money in his pocket, Edison was recruiting at a rate that would more than triple the size of his workforce over the course of the next year.[224]

"Such a dreary place," a young Princetonian, Francis Upton, wrote his father. "The work of course keeps my mind full."[225]

The directors were concerned at his profligacy and dismayed to find little going on in the lighting department. Edison seemed to be concentrating all his intellectual effort on the design of a generator that looked like nothing more than a monstrous tuning fork. He explained that it was a "magneto-electric machine" which, if successful, would give him as much electricity as twenty or thirty Wallace generators. Before there could be a successful lamp, there had to be a steady flow of power, and since power could derive only from mechanics—something Edison understood better than any electrician in the world—he had invested heavily in two big new engines and boilers, fixed to deep foundations at the end of the machine shop.[226]

Lowrey, shuttling between the separate planets that Edison and his backers lived on, begged him not to resist corporate scrutiny, while persuading the board of the Light Company that its funds were being wisely invested. Both sides were in any case pretty sure, as the year came to an end, that no other inventors had the capital and creativity vital to subdivision of the light—not Sawyer and Man with their brittle carbons and cracking glass tubes, nor Hiram Maxim with his graphite rod glimmering in a globe of hydrocarbon vapor, nor the Englishman St. George Lane Fox-Pitt with his iridium loops trying to keep aglow in nitrogen.[227] None of these money-strapped men had the benefit of the unique research, developmental, and manufacturing facilities of Menlo Park. Lowrey assured Edison that his backers trusted him and wanted only to feel they were partners in his endeavor:

All they, or I, shall ask from you is to give confidence for confidence. Express yourself, when you come to a difficulty, freely. You naturally, having an experience of difficulties and of the overcoming of them, in your line (which none of the rest of us can have), may feel that it would be prejudicial, sometimes, to let us see how great your difficulties are, lest we, being without your experience in succeeding, might lose courage at the wrong time.[228]

THE BUSIEST MAN IN AMERICA

"He is an untiring genius," the R. G. Dun & Co. credit assessor wrote in his latest report on Edison, "apt to run from one effort at invention to another without fully completing the work he is on."

That was less true in 1879 than in previous years, for his pursuit of universal electric light—what William Preece in Britain publicly called "an absolute ignis fatuus"—preoccupied him to a degree that soon became obsessive. "I think there is no doubt I am the busiest man in America," he informed a more sympathetic English friend, the otologist Clarence Blake. "The phonograph gets very little consideration from me nowadays."[229]

Yet his fascination with sound persisted, especially after Blake gave a lecture on the telephone in London and closed with a tribute to Alexander Graham Bell without mentioning Edison's carbon button transmitter. George Gouraud, who was desperate to open a telephone company in England before Bell attained a full monopoly there, kept beseeching Edison to finish and send over a receiver he had invented some months before, in the hope that it would circumvent Bell's British patents. It was a startlingly loud device based on the motograph principle of a reproducing point traveling over an electrosensitive surface—in this case, a thimble-size cylinder of hard chalk, slicked with water and rotated by hand. If it was spun fast enough, a person calling in normal tones from New York could have his voice amplified in Menlo Park to a field outside the laboratory.[230]

Edison had handed the receiver over to his nephew Charley to develop. Now, with his competitive instincts aroused by Gouraud's pleas, he ordered Sigmund Bergmann to make him two wall-mounted telephones with the new instrument boxed inside. The handle protruded on the right, a central lever pressed a wet roller up against the chalk, and a mouthlike orifice, complete with what looked like lips, emitted the vibrations of a hidden diaphragm. An erectile transmitter tube curved up from below the box for outgoing calls. It was the ugliest instrument Edison produced in a life generally unchastened by aesthetics. But its speaking volume and almost stereoscopic fidelity put the phonograph to shame. No less a British authority than John Tyndall postponed a lecture he was due to give on Edison acoustic devices at the Royal Institution, in order to add it to his program.[231] The two sets were ready by the end of February and sent to London under Charley Edison's care.

Tyndall was delighted with them. He demonstrated the chalk receiver during his lecture, getting Charley to call from Piccadilly Circus so he could beam the young man's voice distinctly to scientists in the audience. "I congratulate you with all my heart on this beautiful achievement and realization of all your promises," Gouraud wrote Edison. "After this people will doubt you less concerning the Electric Light."[232]

THE DARKNESS OF IGNORANCE

If by *people* Gouraud meant the discerning sort who lived in England and read the London *Times,* he was too sanguine. On 22 March their newspaper of choice reported that "Mr. Edison has failed in his experiments." Fourteen of his sixteen claims to have advanced the technology of light had been rejected by the U.S. Patent Office. "The most he has ever yet accomplished has been to maintain 400 coiled iron wires in a state of partial incandescence with a 16-horsepower steam-engine." So much for his promise to ignite twenty thousand lights from one central station. The attempts of "this impulsive man" to make a self-regulating lamp with a platinum burner had all been unsuccessful, leading to "great discouragement at Menlo Park." Platinum had to be heated to 2,700 degrees Fahrenheit before it shed any appreciable light. It melted so quickly thereafter that his vaunted switch-off rod could not expand in time to stop the runoff. That was why Edison had not mounted a single public exhibition of his work so far.

> A favored few who have been admitted to his laboratory at Menlo Park have beheld it—a single lamp, enclosed in a glass globe, beautiful as the light of the morning star. But he has refused to let anyone inspect it closely, and has never allowed the exhibition of it privately to last long. He has never been able to depend on its durability. His apparatus is as far from perfection as it ever was, and, in fact, well-informed electricians in New York do not now believe that Mr. Edison is even on the right line of experiment.[233]

Edison reacted both defensively and humorously, telling the *The Daily Graphic* that he had "never before read a statement containing so many lies." However, the *Times* had done him a favor: its false report had thinned out the crowd of visitors who constantly encroached on his time at Menlo Park: "I have prayed for an earthquake or something of the sort to keep some of them away." Far from being bothered by abuse from overseas, he said, "I rather like it, and it wouldn't bother me a particle if they kept up the cry—at least until I am ready to show what I have accomplished."[234]

He also insisted that his employees were "as happy as clams," but that was not altogether true. Francis Upton, for one, was losing heart. "The light does not yet shine as bright as I wish it might." After working long nights all through the winter at Edison's side, he foresaw no imminent success, if indeed subdivision of the light could be achieved at all. What the *Times* had to say about the nondurability of platinum lamps was true, and his boss seemed to be the only man in the lab (apart from the inscrutable Batchelor) who refused to accept that.[235]

Upton was a brilliant young man of mathematical and statistical bent, and because of those qualities, he was slow to comprehend the way Edison's mind worked. To him, four months of failed experiments on one intractable thing meant that the thing was no good. To Edison, failure itself was good. It was the fascinating obverse of success. If studied long enough, like a tintype image tilted this way and that, it would eventually display a positive picture.

He almost blinded himself by peering through a microscope at the incandescence of platinum, iridium, and nickel burners, observing—as if he were still focusing on the sun's corona—that they mysteriously cracked and popped just before melting. After seven hours his eyes began to throb "with the pains of hell," but he was able to confirm the Russian physicist Alexander Lodygin's discovery that certain gases, including oxygen, seeped out of fusible metals at white heat. This made the maintenance of any kind of vacuum in a lightbulb impossible after sealing.[236]

Edison understood from the start of his experiments that oxygen in any appreciable quantity decomposed a wire even as it incandesced. But having only a hand pump in the laboratory, he could never suck more than a token amount of air out of his experimental lamps. Gaseous occlusion at *le moment critique* tantalizingly shortened the lovely mellow glow he got from a platinum spiral. Around the same time he noticed that the finer the wire, and the tighter it was wound, the greater its luminosity. Or to put it another way, the higher the resistance of the filament, the more efficient the lamp. This phenomenon, simplified, led him to formulate Edison's Electric Light Law: "The amount of heat lost by a body is in proportion to the radiating surface of that body."[237]

This was a key insight—a tilt of the tintype—that stimulated another, just as radical. If his lamps as *resistors* increased the dissipation of energy as heat and light, they would commensurately decrease the size of the *conductors* needed to feed them with current. Edison thus reversed the consensus among illumination engineers that an extended network of subdivided lamps would offer as little resistance as possible to the circulation of current was ludicrously wrong, calling for prohibitive amounts of copper. He likened the flow of unresisted electrical power to that of city water rushing through overlarge pipes, losing pressure at the same time as it drained the reservoir upstate. The central station's "reach" could be infinitely extended if only a minuscule amount of current was allowed to trickle into each burner in the circuit.

He further insisted that electrical conductors should be looped in multiple arc—not in series, like telegraph relays—so that if any number of lamps were switched off, the rest would continue to shine.*[238] Upton, his studies at the University of Heidelberg still rigidly in mind, could not adjust to these arguments when Edison first advanced them. They conflicted with orthodox opinion and must therefore be wrong. As he ruefully conceded in 1918, "The one great impression of my years in Menlo Park [was] how impenetrable the veil of the future seems to be when new problems are to be solved, and how simple the result often is when the darkness of ignorance is lighted by the genius of one man."[239]

LINES OF FORCE

In April, Upton as mathematician, Batchelor as technician, and Edison as designer achieved a major breakthrough in dynamo design,

---

* Multiple-arc circuitry is now generally known as parallel wiring.

obsolescing that of any other generator on the market. Having already tried two of William Wallace's dynamos and found them wanting, Edison was convinced—again contrary to general belief—that those of Zénobe Gramme and Werner von Siemens did little more than consume their own energy. Over the course of the winter he had tested the European machines with dynamometers, bearing always in mind that they had to be powered with coal and steam before they could pass on any power of their own in the form of electricity. Batchelor drew many graceful graphs to plot the ideal curvature of armature windings, and Upton, probably the only man in Menlo Park who understood the theories of James Clerk Maxwell, translated the graphs into electromagnetic algebra and converted the energy output of each generator into foot-pounds. At basis was the team's uncertainty whether the new dynamo should adopt the "ring" winding pattern favored by Gramme, with wires coiled in series around a revolving wheel, or the "drum" pattern of Siemens, which had a continuous wrap of wire encircling a fat cylinder. Edison decided on the latter configuration.

In a more seminal decision, guided by instinct rather than theory or even experimental evidence, he specified two unusually massive field electromagnets and an armature of extremely low resistance. The latter was designed to conserve as much energy as possible within the dynamo and maximize its efficiency—again, a notion counter to standard practice, which was to maximize output instead. When integrated into a phalanx of duplicate dynamos and connected to a complete illumination system (such as Edison intended to set up and exhibit in Menlo Park soon), the resultant strength of field would be regulated to supply only as much electricity as was called for in the mains—each machine sharing an equal amount of load, whatever the demand for power from outside.[240]

The prototype bipolar dynamo looked so lankily strange, as it stood on its armature like the bottom half of Paul Bunyan, as to evoke the hilarity of engineers more used to squat generators. John Tyndall, who had lavished such praise on Edison's loudspeaker telephone, mocked it as "wholly new" and wholly misguided. Writing in the *Journal of Gas Lighting*, he harrumphed, "It is difficult adequately to express the ludicrous inefficiency of the arrangement; but one thing

is abundantly certain, and that is that the person who seriously proposed it was wholly destitute of a scientific knowledge of either electricity or the science of energy."[241]

All Edison knew in his American ignorance was that when he put the
dynamo through its first paces, "it developed so much power that the
coil on the bobbin [armature] was torn to pieces and I had to stop." He
made no apologies, then or later, for the slenderness, and minimal
winding, of its four-foot iron poles. As far as he was concerned, their
diameter was simply based on the resistance of the magnet, or the number of lines of force it could furnish, rather than the length or space
through which the lines of force were propagated. He placed greater
emphasis on their length, which governed how far the lines extended
into space. It was that dimension, and the sheer intensity of the magnetic field around the armature, that had wrecked the spinning bobbin.
"This fact seems never to have been brought out by any person in connection with dynamo machines but it is of the greatest importance. It
explains the reason of my employing long magnets."[242]

Countless small modifications were necessary before the machine's
performance was smoothed out, but its eccentric design made ultimate sense, and by July Upton could justifiably boast, "We have now
the best generator of electricity ever made."*[243]

THE MOMENT WHEN

That summer, the last of the decade, brought a general relieved sense
among Americans that the dragging depression of 1873 had at last
run its course. Edison received a $24,500 advance royalty payment
from the backers of his chalk telephone in Britain, and gave Mary a
thousand to spend on herself.[244] He splurged on five hundred books
and periodicals for his brick library and hired some new assistants
with a view to making a final, all-out blitz on light development in the
fall. The young men were given quarters in a boardinghouse on Christie Street operated by "Aunt Sally" Jordan, Mary's stepsister.

---

* Late in 1879 Francis Hopkinson of Great Britain showed that the efficiency of the bipolar dynamo could be further improved by simple dimensional changes. In doing so, he
legitimized Edison's radical invention.

The most important of these recruits was Ludwig Böhm, a glass-blower trained in the celebrated Bonn workshop of Heinrich Geissler. He played the zither, sported the red student cap of an elite German university, and liked to recite his many social distinctions, humor not being one of them. As a result he was hazed so unmercifully that Edison took pity on him and let him stay in the attic of a little glass shop adjacent to the laboratory. When not puffing hundreds of globes and tubes for the lamp team, he would retire to his room and yodel Alpine songs until silenced by pebbles hurled against the pitched roof. The only person who liked him was six-year-old Marion, for whom he made many colored glass animals.[245]

There was no question of his skill with a long pipe. He effortlessly blew flasks and tubes of flamingo-like delicacy, some of them, designed for mercury pumps, with an internal bore of only an eighth of an inch. They lightened the labor of Francis Jehl, a stocky eighteen-year-old who wanted to be an electrician but was assigned most of the time to the exhaustion of blank bulbs, exhausting himself in the process. Until the laboratory acquired its first Geissler and Sprengel evacuators, which operated automatically, Jehl had to bear down with both arms and shoulders on a stiff piston pump, seesawing it until the gauge told him he was within a few millimeters of a perfect vacuum. The difficulty of maintaining that state in a bulb, once the base burner unit was introduced, sealed, and wired up, was extreme. At first incandescence, the platinum curl would give off its occluded gases, lessening the vacuum unless they were at once pumped out. If any gas remained once the bulb was "necked off" with an oxyhydrogen flame, they would reenter the wire and weaken its structure.[246]

Edison's infatuation with platinum, protracted by his delusion that somewhere in the world a vast lode of the precious metal could be found and mined to bring its cost down, lasted through the summer. He used an electric pen to duplicate fifteen hundred querulous letters to local authorities as far away as St. Petersburg, Russia—"Dear Sir: Would you be so kind as to inform me if the metal platinum occurs in your neighborhood?"—and sent a prospector all over Canada and the American West in the hope of striking lucky. Although the search yielded him nothing, he developed an interest in mining and mineralogy that would profoundly affect the future course of his life.*[247]

By the end of August, when he went with Mary to Saratoga Springs to attend the annual meeting of the American Association for the Advancement of Science, Edison was persuaded that he at last had the makings of a workable electric light. He had several bipolar dynamos built or nearly finished, bulbs evacuating to a degree of 0.00001 atmospheres, and test filaments of various metals glowing for as long as four hours before they immolated themselves. In two days at the resort he wrote a triumphant paper, "On the Phenomenon of Heating Metals in Vacuo by Means of an Electric Current," and got Upton to read it for him.[†] He claimed to have produced an unoccluded platinum that was the best of all elements for the production of domestic electric light—"a metal in a state hitherto unknown, a metal which is absolutely stable at [a] temperature where nearly all substances melt or are disintegrated, a metal which is as homogenous as glass, as hard as steel wire, in the form of a spiral . . . as springy and elastic when dazzling incandescent as when cold."[248]

But on returning to light experiments in September, Edison had to acknowledge that platinum had other liabilities besides its cost. One was the tightness with which it had to be wound to produce the radiance, and resistance, he wanted. Some of Batchelor's spirals were so fine they could be straightened to a length of thirty inches. What was more, a superfine coating of "pyro-insulator"was needed to prevent them from short-circuiting on the curl.[249] This delicacy, plus an obstinate tendency to oxidize even in Jehl's best vacuums, forced him to return, almost in despair, to carbon as a potential source of *lux aeterna*.

As with his invention of the phonograph two years before, the moment when he discovered his first viable filament (or when it discovered him) became myth so quickly that he could never be sure how

---

* Seven of the five hundred volumes Edison ordered for his library in 1879 were studies of mineralogy and mining.

† Edison overcame his stage fright enough to read another paper, describing his invention of the chalk cylinder telephone receiver while Alexander Graham Bell sat in the audience. As far is is known, this was the last time he delivered a speech in public, until his shaky appearance at Light's Golden Jubilee fifty years later.

and when the miracle happened. It must have been after Böhm blew together an amalgam of Geissler and Sprengel mercury pumps and permitted a bulb vacuum of nearly one million atmospheres, in the first week of October. It must have been after Charles Batchelor began to carbonize soft spirals of lampblack, scraped from the funnels of smoky oil lamps, in the second week of October. It could not have been when Edison had to deal with a crisis involving the chalk telephone in Britain, and his nephew Charley's mysterious death in Paris,* in the third week of October.

Most probably—even certainly, according to the compulsion of all concerned to fix a momentous event in time—it was at the beginning of the fourth week and on the night of Tuesday 21 October blending into the small hours of Wednesday, that a length of carbonized thread, or a twist of carbonized paper, or a carbonized fishing line, or some other carbonized fiber began to glow in vacuo with a light that would not go out. The delight of watching that one filament shine and shine was so great that Edison could be excused for saying, later on, that it incandesced for "over forty hours."[250]

According to Batchelor's contemporary notes, the light lasted no longer than thirteen and a half hours, but that was more than enough to signal that the Old Man was destined, in spite of all doubts, to make it shine as long as he chose.

BRIGHTENING OF THE HUMAN OWL

On the eve of New Year's Eve, when all the excitement was over, and Edison's "Eureka" moment (he actually wrote the word in his laboratory logbook) had been headlined around the world, and fifty-nine reliable lamps were strung up around Menlo Park, ready for the grand public exhibition he had so long promised, "the boys" gathered in the laboratory for an anticipatory celebration. Edwin Fox of *The New York Herald* was there to record the occasion.[251]

At first, he wrote in his account of the evening, Edison was nowhere to be seen. Batchelor prevailed upon Ludwig Böhm to bring up

---

* Charley Edison died in October in the midst of what appears to have been a homosexual entanglement with an English friend. His uncle had to pay for the expatriation of his remains and subsequent funeral in Port Huron.

*Charles Batchelor in the Menlo Park laboratory. The first
photograph ever taken by incandescent light, 22 December 1879.*

his zither from the glass shop. "Play us something with those shake
notes in it. They go right down my back." Böhm obliged with an exquisite melody.[252]

> During the playing a man with a crumpled felt hat, a white silk hand-
> kerchief at his throat, his coat hanging carelessly and his vest half
> buttoned, came silently in, and, with his hand to his ear, sat close by
> the glass blower, who, wrapped up in his music, was back perhaps in
> his native Thuringia again.
>
> "That's nice," said he, looking around. It was Edison.
>
> The glassblower played on, and the scene was curious. Standing
> by a blazing gas furnace he had lighted, Van Cleve,* with bare folded
> arms, listened or else shifted the hot irons [of the filament furnace]
> with his pincers, but he did it gently. Edison sat bent forward. The
> others who had taken up one tool or another moved them slowly. Far
> back through the half-darkened shop young Jehl might be seen lifting
> the heavy bottles of gleaming quicksilver at the vacuum pumps, and
> the soft music was delicately thrilling through it all. It was the wed-
> ding of spirit and matter, and impressed me strangely.

* Cornelius Van Cleve, a carbonizer married to Mary Edison's half-sister.

"Can you play The Heart Bowed Down?" said Edison, suddenly.
"No, I cannot."
"Here—whistle it, some of you."[253]

Five or six obeyed, but Böhm shook his head. Edison lost interest in the music. He took a pad and pencil out of his pocket, sketched a glass implement, and held it out to Böhm.

"Can you blow that?"

"Yes," said the youth, and hurried back to his shop. It was ten-thirty P.M., and Edison was clearly ready to begin one of his nocturnal workbench sessions.

During the hours that followed, Fox was struck by the acuteness and force of Edison's remarks as he attacked theoretical scientists. "Take a whole pile of them that I can name and you will find uncertainty if not imposition in half of what they state as scientific truth. . . . Say, Van Cleve, bring me the *Dictionary of Solubilities*."[254]

He scornfully pointed out an entry stating that platinum was infusible, except in the heat of an oxy-hydrogen flame. "Come here; I'll melt some in that gas jet. . . . Look in here now. You see along the magnified wire a number of little globules? That is where the platinum has fused."

Next he turned to the subject of electrical illumination. "The peculiar moonlight color in the voltaic arc light is due to the impurities in the carbon, magnesium among the rest. What's the matter with you, Francis?"[255]

JEHL: I'm hungry.
EDISON: Where's the lunch?
JEHL (*Despondently*): There was none ordered. We didn't think
you were coming back to work all night, and now we're here and
there's nothing.
EDISON: Get something to eat. (*To FOX*) You see, the carbon used
is made out of a powder, held together by various substances. . . .
George, get me a stick of carbon and a filament.[256]

He proceeded to demonstrate, explain, and propound some of his discoveries, reeling off chemical and metallurgical names with what Fox described as "the peculiar nocturnal brightening of the human

Edison's "New Year's Eve Lamp," 1879.

owl." Midnight came and went. Unstoppably garrulous, Edison reverted to his idée fixe about the superiority of the empirical over the academic scientist. "Professor This or That will controvert you out of the books, and prove out of the books that it can't be so, though you have it right in the hollow of your hand and could break his spectacles with it."[257]

When, at length, "lunch" arrived, all Jehl had managed to buy at the depot was a brown paper bag of smoked herring, and another of crackers. Van Cleve found some lager to wash the repast down, but Edison chose a tin mug of water.

By the time he stopped talking it was four A.M., and everybody but the reporter had fallen asleep (Jehl with his head on the *Dictionary of Solubilities*). Only then did Edison take off his coat and look around for a bench to nap on. Fox went out into the night feeling both inspired and bilious. "I shall place those smoked herrings, biscuit and cold water on a high shelf, a very high shelf, in my memory; my stomach may never forget them."[258]

The night was cloudy, and snow was forecast for the morning.[259] Menlo Park's two hundred other residents were asleep. In about twelve hours their peace, and the isolation of the hamlet from the rest of the world, would be disturbed by pilgrims to the Festival of Light.

Telegraphy

1860–1869

AT THIRTEEN, EDISON drove a freight train for the first time, down forty-seven and a half miles of track—"all alone," he boasted afterward, although the engineer's eye must surely have been on him.[1] Solo or not, it was an ecstatic moment for a boy on the edge of adolescence, with the engine roaring loud enough to make him forget that he could no longer hear birds sing.[2]

The Chicago, Detroit & Grand Trunk Junction Railroad ran a daily round trip between his hometown of Port Huron and Detroit, Michigan. Every morning at eight, young "Al"* boarded the baggage car of the southbound local, well supplied with candy, hickory nuts, and popcorn balls to sell during his three-hour journey to the big city.[3] Lake Huron receded behind him, and the St. Clair River slid by on his left, with Canada—his family's ancestral country—spreading out beyond. Then for a while both river and frontier disappeared, as the train edged inland to pass through Smith's Creek, Ridgeway, New Haven, New Baltimore, Mount Clemens, and Utica—small towns that half a year earlier had seemed unattainably remote but were now familiar, almost neighborly parts of his expanding environs. When the river reappeared, it had swelled into Lake St. Clair, and Detroit, opening up ahead, offered him almost four hours of urban freedom.

Al was not yet old enough to qualify for membership of the Young Men's Society Reading Room, but the city's stores were open for browsing and for the purchase of chemical and electrical supplies to aid his experiments at night. Useful bits of brass and iron and, with

* At this stage of his life Edison was called "Alva" by his parents and "Al" by friends. His earliest surviving letter, dated 10 August 1862, is signed "Alva."

luck, the occasional old battery cell could be filched from dumps around the Grand Trunk's machine shop and engine houses on Michigan Avenue. Every afternoon on his way back to Michigan Central Depot, he picked up a hundred copies of the *Detroit Free Press,* to supplement his candy sales on the return journey home. If any papers were left unsold by the time the train, slowing, approached Port Huron, he would jump off with them onto a sandbank and walk the last quarter-mile into town, hawking his last stock en route.[4]

Otherwise he would stay on the train until its run ended at Fort Gratiot, a sleepy old military reserve guarding the confluence of the river and Lake Huron.* Sam Edison's big white house and observation tower stood within the stockade. So Al did not have far to walk, past the hospital and graveyard (not his favorite locality, on dark winter nights) before reaching the comforts of home, and dinner, and his reeking basement laboratory.[5]

NOW WE HAVE RODE

Free time, free throwaways, the *Free Press,* and even free railroad freight privileges, as his business expanded to include fruit and grocery sales—Al was his own man now, no longer subject to Nancy Edison's disciplined schooling. Earning an excellent income of forty to fifty dollars a week, he conscientiously paid her a dollar a day for his keep and invested the rest of his fortune in chemical and electrical equipment. At the same time he voluntarily continued his liberal education, reading the works of Thomas Paine at the behest of his father, a lifelong libertarian who espoused the rights of the southern states to secede from the Union.[6]

That issue became fraught on 18 May, when the *Free Press* gloomily reported the nomination in Chicago of the "black Republican," Abraham Lincoln, for president of the United States. The newspapers Al sold that summer covered the election campaign with apocalyptic "by Magnetic Telegraph" bulletins prophesying rebellion in the South and "irrepressible conflict" if Lincoln was elected. When, around midnight on 6 November, the first dispatches confirming his victory

---

* Now the Thomas Edison Depot Museum in Port Huron.

came down an accessible wire into Port Huron, Al was able to "read" some of the results to fellow urchins by putting his tongue to it and tasting the tiny shocks of each dot and dash. From that moment on, war between the states was inevitable. Fort Gratiot awoke from its slumbers, and recruits began to drill on the parade ground.[7]

Al Edison, newsboy, circa 1860.

He was fourteen by the time Fort Sumter fell, and for most of 1861 no more aware than any Michigan youth his age of the catastrophe unfolding in the South and East. The state sent regiment after regiment to the distant battlefields, but was otherwise peaceful as ever.[8] Al saw more immigrant Norwegians—daily trainloads of them heading for Iowa and Minnesota—than he did men in uniform. Meanwhile his grocery and news businesses prospered to such an extent that he began to employ other boys. One sold bread, tobacco, and stick candy aboard the immigrant "special." Another loaded baskets of market vegetables onto the morning express to Port Huron, where a German lad collected them and sold them on commission downtown. Al himself continued his lucrative commute, buying butter and, in season, immense quantities of blackberries from farmers en route, and purveying them at either end of the line.

Early in 1862 he bought three hundred pounds of old type slugs from a junk dealer, and a small secondhand press to accommodate them. It occurred to him that since the front part of the baggage car, a small, unventilated compartment, was never used, he could turn it into a mobile print shop, teach himself how to set type, and produce his own onboard newspaper. The first issue of *The Weekly Herald* ("Published by A. Edison") appeared on 3 February as a double-sided, copy-rich broadsheet, elegantly laid out and even decorated verso with a woodcut of a puffing locomotive. Only the orthography left something to be desired.[9] Under the headline LOCAL INTELE-GENCE, the paper's chief correspondent reminded the Grand Trunk Railway that it had a policy of rewarding meritorious service.

> *Now we have rode with Mr. E. L. Northrop, one of their Engineers, and we do not believe you could fall in with another Engineer, more careful, or attentive to his Engine, being the most steady Engineer that we have ever rode behind (and we consider ourselves some judge having been Railway riding for over two years constantly,) always kind, and obligeing, and ever at his post.* *[10]

Elsewhere Al reported that "Gen. Cassius M. Clay, will enter the army on his return home," announced the forthcoming thousandth birthday of the Empire of Russia, listed the latest per-pound price of dressed hogs and turkeys, allowed himself a pause for philosophical reflection ("Reason Justice and Equity, never had weight enough on the face of the earth, to govern the councils of men"), and even found space at the foot of column four for a joke: " 'Let me collect myself,' as the man said when he was blown up by a powder mill."[11]

Eager readers were offered a subscription to the *Herald* of eight cents monthly, and were promised that their names would be grate-fully published in future issues. These inducements, coupled perhaps with curiosity as to whether anywhere else in the world a newspaper was being printed and sold aboard a moving train, led to a rapid rise in circulation to more than four hundred copies a week.[12]

---

* The suspicion arises that Mr. E. L. Northrup was the "kind, and obligeing" engineer who allowed Edison to drive his freight train for 60½ miles.

Gradually appropriating unto himself more and more of the baggage car, Al set up a traveling laboratory where he could consult Fresenius's *Qualitative Analysis* and mix chemicals at only moderate risk of blowing out the windows. Since the time it took to put the *Herald* to bed each Saturday interfered with both his experiments and his candy sales, he soon subcontracted the latter chore to a Port Huron schoolboy. Forty-five years later Barney Maisonville recalled their partnership:

> Al was very quiet and preoccupied in disposition. . . . Most boys like to have money, but he never seemed to care for it himself. The receipts of his sales, when I sold for him, were from eight to ten dollars the day, of which about one-half was profit. But when I handed the money to him, he would simply take it and put it in his pocket. One day I asked him to count it, but he said: "Oh, never mind, I guess it's all right." . . .
>
> He was always studying out something, and usually had a book dealing with some scientific subject in his pocket. If you spoke to him he would answer intelligently enough, but you could always see that he was thinking of something else when he was talking. Even when playing checkers he would move the pieces about carelessly as if he did it only to keep company, and not for any love of the game. His conversation was deliberate, and he was slow in his actions and carriage.
>
> Still, he showed sometimes that he knew how money could be made.[13]

25 CENTS APIECE, GENTLEMEN

This was evident on Wednesday 9 April, when Al arrived in Detroit on his usual midmorning train and found crowds milling anxiously around the great bulletin boards that city newspapers posted outside their headquarters for breaking news. According to headlines being chalked up by editors with telegrams in their free hands, an epochal clash of arms had taken place on Sunday at Shiloh, on the Tennessee River, and the first accounts were only just coming in. In twelve hours of conflict, starting with a surprise dawn attack on General Grant's Union Army, more blood had been shed than ever before in American history. Some pints of it had fatally filled the boots of the commanding Confederate general, Albert S. Johnston, who was now succeeded by Gen. P.G.T. Beauregard. Grant had repelled the onslaught, but only just, and had failed to pursue the enemy as rain and darkness descended. Monday's fighting had been almost as savage, with one late wire reporting that sixty thousand may have been killed or wounded by the time Beauregard was beaten.[14]

Al was already enough of a journalist to realize that there would be a phenomenal demand for the afternoon edition of the *Detroit Free Press* at every stop on his return trip home. Thinking and moving much faster than Barney gave him credit for, he copied the main headlines on display and hurried to give them to the Grand Trunk telegraph operator at Michigan Central. In exchange for a healthy bribe—three months' worth of complimentary magazines—the operator sent an alert to the railroad's upstate stationmasters, instructing them to post the headlines locally and announce that a major delivery of newspapers was coming north on the four P.M. train.[15]

Al's next stop was the *Free Press* office, where the afternoon edition was already thumping off the presses.[16]

He demanded one thousand copies "on trust." Henry N. Walker, the paper's editor, was touched by his sass and authorized the order.[17] Enlisting another boy's help, Al got the papers onto the train and had them folded by the time it reached Utica, where he usually sold only two copies.

THE GREAT BATTLE ON THE TENNESSEE.

The Fight Lasted Two Full Days.

ALBERT SIDNEY JOHNSTON KILLED.

BEAUREGARD'S ARM SHOT OFF

Gen. Prentiss, of Illinois, Taken Prisoner.

GEN. W. H. WALLACE KILLED.

GEN. W. T. SHERMAN WOUNDED.

The Detroit Free Press *reports the Battle of Shiloh, 10 April 1862.*

I saw a crowd ahead on the platform [and] thought it was some excursion, but the moment I landed there was a rush for me; then I realized that the telegraph was a great invention. I sold 35 papers; the next station, Mt. Clemens, [was] a place of about 1000. I usually sold 6 to 8 papers. I decided that if I found a corresponding crowd there that the only thing to do to correct my lack of judgment in not getting more papers was to raise the price from 5 cents to 10. The crowd was there and I raised the price; at the various towns there were corresponding crowds. It had been my practice at Port Huron to jump from the train at a point about 1/4 mile from the station where the train generally slackened speed. I had drawn several loads of sand at this point and become very expert. The little German boy with the horse met me at this point; when the wagon approached the outskirts of the town I was met by a large crowd. I then yelled 25 cents apiece, gentlemen, I haven't got enough to go round. I sold all out and made what to me then was an immense amount of money. I started the next day to learn telegraphy.[18]

LO! YOU HAVE O

Al's studies for a new career were given impetus by a Grand Trunk conductor who became exasperated by his use of the train for personal business.[19] Stacks of newsprint and crates of groceries, with fruit flies traveling free, were bad enough, but chemistry experiments posed the additional threat of an onboard fire. Sure enough, late that summer one of his phosphorus sticks fell to the floor, nearly immolating the baggage car. Al found himself ejected—instruments, printing press, and all—onto the platform at Mount Clemens, a newsboy no longer.

As luck would have it, the local stationmaster, James Mackenzie, was a skilled telegraph operator and glad to teach him the trade. He was indebted to Al for having pulled his infant son from the tracks one August morning when cars were shunting.* For about four months he taught the former newsboy how to send and receive (or in telegraphic parlance, "write" and "read") Morse code.[20] Memorizing mnemonic rhymes helped:

---

* According to the account given in Edison's authorized biography, Al was loitering on the platform when an unbraked boxcar, pushed out of a siding, bore down on Mackenzie's son, who was playing on the main track: "Edison dropped his papers and his glazed cap, and made a dash for the child, whom he picked up and lifted to safety without a second to spare, as the wheel of the car struck his heel, and both were cut about the face and hands by the gravel ballast on which they fell."

•	*One dot stands for E, for enterprise sure,*
••	*And two stands for I, for self ever pure,*
• •	*Yet divide them a trifle, and lo! you have O,*
— —	*Or space them a bit, and M is the go.*[21]

Al studied eighteen hours a day for most of the fall and early winter of 1862. He spent his mornings and afternoons with Mackenzie, returning to Fort Gratiot on the evening train for solitary practice at night. His tapping technique—the ability to "pound brass" in a fragmented yet flowing rhythm—gradually improved, although it would take years before he attained real wrist freedom, and could hear a ticking stream of dots and dashes as if it were ordinary speech.[22]

He still took frequent trains to the city, and used the facilities of a local gun shop to make his own set of instruments. The mechanics of telegraphy was still simple, although the demands of war communications and reportage for greater speed and message volume would soon make them less so. All Al needed at first was a brass transmitter with a sprung key and, at the other end of a length of stovepipe wire, an electromagnetic relay that either printed dots and dashes on tape or ticked them out on a sounder. Grove batteries, each consisting of a clutch of sour-smelling, open-top cells, provided the energy to send and receive. When he added a subsidiary station to his experimental network, he found that a less powerful Daniell battery was enough to activate the local circuit.[23]

Over and above technology, Al had a normal adolescent's longing for information. He persuaded the Detroit Young Men's Society that he was almost grown up and hence qualified to hang out in its well-stocked reading room.* He became a member well in advance of his sixteenth birthday, and proceeded to devour *Les Misérables* and several other novels of Victor Hugo, along with Thomas Burton's *Anatomy of Melancholy* and many volumes of the *Penny Cyclopædia*. If he also, as he later claimed, read Isaac Newton's *Principia Mathematica*, it was surely with incomprehension.[24]

At some point in the course of the winter of 1862–63 he worked for a while as a "plug," or trainee operator, in Miciah Walker's jewelry-

---

* Later the Detroit Public Library.

cum-bookstore in Port Huron. There was a small telegraph office on the premises that received press dispatches overnight, as a service to the local newspaper. He found, in the course of taking copy there until three every morning, that his temperament was nocturnal, and that solitude suited him. If he felt drowsy during the day, there was always a chair somewhere for him to catnap in, his muffled hearing providing all the peace he needed.[25]

By spring he was proficient enough in Morse to apply for a night operator's job at Stratford Junction, on the Grand Trunk Railway just across the Canadian border. The work involved monitoring the movements of trains by means of telegraph messages exchanged with other stations along the line. Mr. Walker tried to keep him in Port Huron as an apprentice at twenty dollars monthly, but Al was not drawn to the jewelry business, and his father, who would have to agree to his indenture, was not tempted by the terms.[26]

Al in any case had good reason to get out of town. Fort Gratiot was no longer a congenial place to live in. Sam Edison's libertarian views and history of radical opposition to any authority grated on the pro-Union nativism of the fort keeper, Henry Hartsuff. "He is from Canada—reported to be a very dishonest man," Hartsuff wrote the War Department. "His sympathies are intensely secession[ist] which renders his presence here very odious."[27]

Sam, who delighted in confrontation, could be relied on to resist eviction. Al in contrast wanted nothing so much as a paid post not too far from home, with lots of spare time to continue his self-education in chemistry and electricity. Stratford Junction promised all this. By pleasant coincidence, it was located in the same general area of Ontario where his parents had spent their early married life. His ninety-six-year-old grandfather, Samuel Ogden Edison, Sr., still lived in Elgin County, along with sundry other paternal and maternal relatives. So when Al's job application was accepted, Nancy need not feel her youngest son had entirely left the bosom of the family.[28]

A BOY FREE FROM FEAR

Although he persuaded himself that working as a night operator for twenty-five dollars a week would give him the whole day for experimenting, Al was still a teenager who needed more sleep than he would in adulthood. The Stratford depot was a quiet one between dusk and dawn. Were it not for a standard security procedure that required him to flash *dah-di-di-di-dit* down the line every hour, he could have enjoyed long periods of repose. This nuisance precipitated his first invention, an automatic sender. It consisted of a notched wheel driven by his office clock and attached to the transmitter in such a way that on the hour, a wooden hammer would rise and fall on the key in the precise rhythm required. The device worked well until a dispatcher came to investigate why "Stratford" was sometimes impossible to "raise," despite the regularity of its signal.[29]

Al incurred further disapproval one night when he almost succeeded in getting two freight trains to collide head on. He responded affirmatively to a call asking if he would stop one of the trains, so as to let the other one through. But by the time he descended from the telegraph office to the yard and looked around for the signalman, a rush and a roar told him that the wrong train had gone through. He dashed back upstairs and wired ahead that he had been unable to "hold it." The reply from the next station was "Hell."[30]

Fortunately the track was straight, and the converging trains saw each other in time to brake to a halt. Next morning the general superintendent of the Grand Trunk Railroad ordered the Stratford agent to bring Al to his office in Toronto and explain why a sixteen-year-old had been permitted to hold such a responsible position after dark.

> He took me in hand and stated that I could be sent to Kingston States Prison, etc. Just at this point, three English swells came into the office. There was a great shaking of hands and joy all around; feeling this was a good time to be neglected I silently made for the door, down the stairs to the lower freight station, got into the caboose going to the next freight . . . and kept secluded until I landed a boy free from fear in the U. S. of America.[31]

Al was also a boy rich in platinum. During his brief stay in Canada he had heard about a stash of old Grove battery cells at the Grand Trunk depot in Goderich, and slyly asked the agent if he might strip them of their "tin" cathodes. Some of those precious metal strips, amounting to several ounces of reworked scrap, would be used in his laboratory experiments forty years later.[32]

NEVER HEARD YOU ON HERE

He now became a "tramp" telegraph operator, in common with hundreds of other youths, skilled or semiskilled, who rode the nation's rails in the latter days of the Civil War. They were in such demand, as the wires thrummed with news from the front, that train conductors often let them travel free to whichever town they liked.

Although Morse code had ended the historic dependency of communication upon transportation, the two fields were still intertwined, in that telegraph companies needed rights of way for their wires, and railroads needed wire stations to control the movements of their trains—ideally staffed by personnel who could be relied on to stay awake.[33]

Not a few of the young "sparkers" or "lightning slingers," as they called themselves, were seeking to evade the draft. In the late summer of 1863 Al was still a year and a half from that dread fate. Meanwhile the wandering life suited him, as did the glamour of working in an industry at the forefront of modern technology. Every issue of *The Telegrapher* magazine carried the front-page motto "Is it not a feat sublime? Intellect hath conquered time." There were so many job opportunities across the country that any qualified candidate stood a chance of being hired, for decent money, the moment he stepped off the train.* Often as not his immediate predecessor would be stepping on, in search of a town that had prettier girls or served better corn whiskey.[34]

All but the smallest of small-town branches had a number of operators working together, sharing digs and dirty jokes (many of which would go on the lines in "blue" Morse code) with the exaggerated camaraderie of theater folk during the course of a limited production.[35] It was unlikely that any pair of temporary buddies would meet again, after deploying in different directions from a six-month part-

---

* An estimated fifteen hundred skilled telegraph operators had been siphoned off for war duty, thus creating an urgent need for help in the civilian sector.

nership in Carson City, Nevada, or Cleveland, Ohio—although for years they might weirdly recognize each other's "fist" in interchanged, unsigned signals, gibberish to the outside world.[36]

> *BOSTON:* Your next number is 1.
> *ST. LOUIS:* Thank you. Number 1, New York 9th to—.
> *BOSTON:* Please sqe.
> *ST. LOUIS:* I sign &.
> *BOSTON:* Never heard you on here before. Where did they dig you out? That's a hot sig. Ha! Ha![37]

Should one of them, against all probability, go on to become a famous inventor, he would have to endure countless reminders of past intimacy from lonely old tappers with unremembered names. By the same token, there was always a chance that the vagaries of later life would lead to surprise reunions that might be sweet, or turn sour.

Al's next roost was Adrian, Michigan, where he bonded with a local boy named Ezra Gilliland.[38] He was again employed as a night operator, this time by the Lake Shore & Southern Railroad, at seventy-five dollars a month. His stay there lasted long enough for him to establish a little workshop of his own. But when he made the mistake, one evening, of "breaking in" with an urgent message on a wire occupied by the superintendent, he was obliged to move on to a day job in Fort Wayne, Indiana. The schedule there did not suit him, so in the fall of 1864 he became a "second-class operator" for the Western Union Company at Union Station, Indianapolis.[39]

WORDS PER MINUTE

Here Al turned eighteen and became at least the beginnings of an inventor. His clockwork sender in Stratford had been little more than a toy, but he now devised an instrument of serious purpose that prefigured two of his most important mature achievements, the translating embosser and the phonograph.

It was a recorder-repeater that satisfied his aesthetic desire to transcribe "press report"—inflowing, often enormously long journalistic dispatches—in script that was both clear and beautiful, while at the same time dealing with the prohibitive speed at which most dispatches came over the wire. He was, for all the elegance of his handwriting, by no means slow as a copyist. But even the most agile taker found it hard to keep up with the forty words per minute characteristic of most transmissions. (Virtuoso senders in New York and Washington took a sadistic delight in forcing provincial operators to beg for the occasional *ritardando*.) It was an open secret, as a result, that takers often rephrased sentences, or even omitted whole paragraphs, especially late at night when their wrists became tired. Editors complained about patchy press reports but had to make do with what came through.[40]

Al lined up two pendulum-driven Morse registers and geared the one to receive fast copy, indenting every dot and dash on a cylinder of paper tape. The tape unspooled into a bin, whence it was drawn up by the other register, and played back through a sounder at whatever speed—usually twenty-five words per minute—a transcriber found comfortable. He invented the machine primarily as a practice instrument to improve his own taking, but when he won permission to use it officially, the high quality of his copy embarrassed the station's top press man, and he was encouraged to seek employment elsewhere.[41]

The waning days of the war found him at the Western Union branch on Fourth Street in Cincinnati, ambitious now to become a top press man himself.* An office that could increase the speed and volume of its message handling was an office unlikely to fire any proficient operator, at least at the rate to which Al was becoming accustomed. Putting aside experiments for the moment, he practiced so hard that he acquired the automaton-like trance typical of takers, whereby they heard code, and wrote words, without absorbing the meaning of either.[42]

This was chillingly apparent on the night of 14 April 1865, when he and his fellow operators became aware that a huge crowd was gathering half a block away, outside the headquarters of *The Cincinnati Enquirer*. They sent a boy to inquire about the excitement, and he came back shouting, "Lincoln's shot."[43]

---

* How Edison evaded, or avoided, being drafted in 1865 is not known. He may have paid a $300 commutation fee or simply kept one step ahead of summons by moving from Indianapolis to Cincinnati.

The newspaper must have gotten its story from a Western Union telegram, which meant that someone in the room—who?—had received, automatically transcribed, and messengered the century's biggest news down the street without paying attention to its contents. "Look over your files," the office manager said. After a short search the scribbled report was found and held up. Al, reminiscing forty-four years later, chose not to identify its author.[44]

His current salary was eighty dollars a month, only five dollars more than he had earned in Adrian and not enough to support much recreation in a large, expensive city. He economized by sharing digs with a pair of actors and a pair of office friends—Ezra Gilliland, who had followed him to Cincinnati, and Milton Adams, a sophisticated dandy uncomplimentary about Al's hickish demeanor. "The boys did not take to him cheerfully, and he was *lonesome*."[45]

Influenced as much, perhaps, by the dramatic circumstances of Lincoln's death as by the profession of his roommates, Al developed a taste for plays and playacting and showed occasional signs of wanting to tread the boards himself. He attended performances at the National Theater in Cincinnati whenever he could afford to, and memorized chunks of Shakespeare to quote aloud, notably the opening soliloquy in *Richard III* (performed with a limp and hunched back). For the rest of his life he would write *Now is the winter of our discontent* when he wanted to show off his calligraphy.[46]

He was justifiably proud of the telegraph taker's script he developed this year—line by line perfectly straight across unruled paper, with no time-wasting flourishes, and only the occasional character misshapen as he kept trying to increase his speed.* When he stayed away from a union meeting and took the press report alone into the small hours (writing with an agate pen on five layers of oiled tissue

---

* Edison delighted in all kinds of pen play. He became an expert forger, effortlessly imitating the handwriting of Washington, Jefferson, Napoleon, and amusing himself—if not his dupes—by presenting notes for large sums of money owed: "It's your signature, isn't it?" The publisher Edward Bok recalled his drawing a circle round a dime, then filling it with a minutely inscribed copy of the Lord's Prayer that included all commas and periods. Marshall, *Recollections of Edison*, 94; N. N. Craig, "Thrills," autobiographical ms. ca. 1930, Biographical Collection, 41, TENHP; *Providence* (RI) *Journal*, 13 Aug. 1927.

and interleaved carbons), he was rewarded with a visit from the day manager, J. F. Stevens. "Young man, I want you to work the Louisville wire nights. Your salary will be $125."[47]

The appointment was worth more than extra money. Al was now officially a first-class operator, entitled to special respect wherever he wandered next on the "linked lightning" network.[48]

NO COUNTRY LIKE THE US

Three months in Nashville. Another three in Memphis. Four more in Louisville. . . .[49] Peacetime and the erasure of the Mason-Dixon Line brought a sense of exultant reopening of the American South, to a Northern youth just turned nineteen, full of wanderlust and curious to explore that beaten land. "I have growed Considerably," he advised his parents on a message form stamped by the South-Western Telegraph Company. "I dont look much like a Boy Now."[50]

When he wrote again on a similar blank, not bothering to date it or say where he was, he seemed to have wider travel in mind, and was studying foreign languages in books: "Spanish very good now before I Come home I will be able to Speak Spanish & Read & write it as fast as any Spaniard. I can also Read French too but Cant Speak it." In early August 1866 he was in New Orleans and conspiring with two fellow sparkers to take a steamer to Brazil, possibly unaware that the preferred language there was Portuguese. They had heard that the imperial government was spending many milreis on an extension of the national telegraph system and assumed that with their technical skills they could partake of this flow of gold. But a race riot broke out in the city, the steamer they planned to take to Rio de Janeiro was commandeered for the use of federal troops, and an Ancient Mariner who had lived in South America shook a skinny hand in Al's face, advising him that "there was no country like the US" for a young man who sought to better himself.[51]

Chastened, Al told his friends he was going back home and set off on the long trip to Port Huron.* The railroad north through the depressed landscape of Alabama was so rotten—"scrap iron laid on

---

* Afterward Edison heard that both had died of yellow fever in Vera Cruz.

wooden stringers"—that the train had to chug along at little more than a walking pace. He was able to lean out his car window and pick peaches off passing trees.[52]

Like many another returning prodigal, he found that "home" was not the happy place it had once been. During the late years of the war in Fort Gratiot, Henry Hartsuff's animus against Sam Edison ("a villainous and malignant Copperhead rejoicing over rebel victories and abusing our government and public men") had increased to the point of paranoia. As caretaker of the fort, Hartsuff had a reputation for hard drinking and official corruption, but he enjoyed the protection of a son who was a general in the army. Using that influence, he got the chief quartermaster in Detroit to requisition the big white house in the grove, by right of eminent domain in a military reservation. In consequence he at last succeeded in evicting Sam from a property bought in good faith twelve years before, paying him only $500 of its appraised $2,300 value.[53]

The elderly Edisons were now living in a dark little cottage just outside the fort. Sam was as ornery and antigovernment as ever,* pursuing a furious lawsuit against the army that anyone could see was doomed. Nancy was ailing and losing her mind. Al stood their misery for no more than a month, then returned to the South-Western Telegraph Company office in Louisville.[54]

EXPERIMENTAL RESEARCHES

His second spell in Kentucky lasted until the following July. By then he was well into his twenty-first year, able to copy an average of eight to fifteen columns of Associated Press report every day in a highly individual scrip: "I found that the vertical style, with each letter separate and without any flourishes, was the most rapid, and that the smaller the letter the greater the rapidity." He could write between fifteen hundred and two thousand words without tiring, in a minute hand as readable as diamond type.[55]

This virtuosity was perfected, moreover, in conditions that strained his faculties to the limit. He worked on the second floor of a dilapidated downtown building, in a room that was never cleaned and only

---

* At sixty-four years of age, Sam outjumped 250 men in Fort Gratiot.

sporadically heated in the winter. The ceiling was half stripped of its plaster, and ornamented with dried splats of chewing tobacco. The copper wires that connected his telegraph set to the switchboard were eroded with blue crystals, and the board's brass leads were black with the smoke of lightning strikes, "which seemed to be particularly partial to Louisville." Every now and again in stormy weather there would be an explosive crack from the wall, not good for the health of operators with heart problems. Al's principal feed came via the "blind" side of a repeater in Cincinnati, meaning that he was unable to interrupt a transmission and request the repetition of a dropped word or sentence. The wire also tended to leak when it crossed the Ohio River at Covington, so he got violent changes of current when receiving. He smoothed out some of the fluctuations by playing the signal through a quartet of relays and sounders. "The clatter was bad, but I could read it with fair ease. When, in addition to this infernal leak, the wires north to Cleveland worked badly, it required a large amount of imagination to get the sense of what was being sent . . . and as the stuff was coming in at the rate of thirty-five to forty words a minute, it was very difficult to write down what was coming and imagine what wasn't coming."[56]

His champion feats as a taker were nighttime transcriptions of President Andrew Johnson's 7,126-word Second Annual Message to Congress in December, and Johnson's subsequent 6,111-word veto of the District of Columbia franchise bill early in the new year of 1867. In memory, he conflated both messages into one avalanche of verbiage, with section after section being scissored off as he wrote and rushed to the *Courier-Journal*'s office for typesetting. "I was fifteen hours in the chair on this occasion without a moment's intermission for food." Actually the articles appeared a month apart, but each involved Herculean effort, and justified his reputation as one of the fastest takers in the country.[57]

By midsummer Al was back at the Western Union office in Cincinnati, for a three-month stint that saw him use his mastery of telegraph technology as a key to open up wider fields of science, such as magnetism, metallurgy, and conductivity. He began to keep a notebook of experimental ideas (some his own, some copied for instruction): self-adjusting and polarized relays, long-distance electromechanical repeaters, duplex transmitters, a secret signaling method for the army, a private-line telegraph system for Procter & Gamble. He drew them in a free-flowing, two-dimensional style that showed a palpable pleasure in the way each design spilled out of his pen, without a single short-circuit of electricity or ink. He compiled lists of scholarly books to study. Among them was the first volume of Faraday's *Experimental Researches in Electricity*, borrowed from the Free Library around the same time the great man's death was announced in London. It and its three companion volumes became his lifelong bible, and Faraday of all scientists the one he most revered. He also satisfied a growing appetite for current and cultural affairs by devouring twenty volumes of *The North American Review*, which he had bought for two dollars in Louisville and could not be parted from thereafter.[58]

The more Al brooded over his books and clapped together strange-looking models and forgot to eat and declined booze and wore, according to one incredulous observer, the same suit season after season, the more eccentric he seemed to some of his less cerebral colleagues. They called him "Luny Edison" or "Victor Hugo" because of his fondness for French novels, and wondered why he never seemed to have any money, although he had been earning an excellent salary for two years. "He was always broke," one of them complained. "The day after pay day he'd come to me to borrow a dollar."[59]

Ezra Gilliland, J. F. Stevens, and others in Cincinnati who understood him better knew that Al was spending all he had on technical equipment and supplies.[60] It was a compulsion, the instinctive behavior of a man—boy no longer, a technician already mutating into a thinker—who could not help doing what he was born to do.

THE JAY FROM THE WOOLLY WEST

In January 1868 James Ashley, the editor of *The Telegrapher*, received an article submission from an aspiring inventor in Port Huron, Michigan. It illustrated a double transmitter of exquisite symmetry, and the accompanying text guaranteed that "by means of this ingenious arrangement, two communications may be transmitted in opposite directions at the same time on a single wire."[61]

Ashley thought the manuscript, which was long and technical, interesting enough to publish on the front page of his journal, but by the time he got around to doing so in the spring, the author had moved from Port Huron and was identified as "Mr. Thomas A. Edison, of the Western Union Telegraph Office, Boston, Mass."[62]

Edison's arrival at that major branch of the nation's biggest corporation marked the formal beginning of his inventive career, although it would be six months yet before he applied for his first patent, and four more before he felt confident enough to make the ultimate transition to New York.[63]

He certainly did not look like a person of substance when he arrived at the office in Boston, being half-starved and more than usually shabby after a blizzard-slowed train journey. But he came recommended as an "A1 man" by Milt Adams, who had befriended him during his first spell in Cincinnati and was now working for the Franklin Telegraph Company in Boston. He was assigned to begin taking copy at five-thirty that evening. His fellow night operators lost no time in setting up what they thought was a trap for "the jay from the woolly West." They gave him a cheap pen and put him onto the number-one wire from New York, saying that a fifteen-hundred-word dispatch from the *Boston Herald* was about to come through. By prearrangement, they had lined up one of the fastest tappers in Manhattan to send it, starting *moderato assai* and then accelerating to his best *presto*.[64]

Edison effortlessly kept up with the transmission, writing very small, conscious that he had a surprised audience watching over his shoulder. Taker's instinct reassured him he could do four or five words per minute more than the New York man could send. His correspondent began to tire, slurring rhythms and sticking the signals, but as Edison remarked afterward, "I had been used to this style of telegraphy in taking report, and was not in the least discomfited." When he sensed his invisible tormentor had tired, he opened the line and tapped a message of his own: "Suppose you send a little while with your other foot."*[65]

---

* In 1898 Fred Catlin, a veteran operator, wrote Edison, "While you were not in your day among the stars as a sender, yet, as a receiver I do not think you had a peer. Thirty years ago I was considered as fast as the fastest, and I recall the pleasant hours spent in tapping off press matter and messages to you. . . . It was a pleasure because I could sail along indefinitely without interruption."

From then on, the number-one wire was his privileged conduit. He was by his own admission a poor sender, but now became nationally known for his "chirography." Ashley used that imposing word as a headline in *The Telegrapher*, claiming below that Thomas Edison was "about the finest writer we know of." He said he had seen one of the young man's press report cards, measuring "five by eight inches, and there are 647 words upon it . . . the whole plain as print."[66]

Milt Adams fell on hard times as soon as he had helped Edison to good. Franklin Telegraph could not stand the competition of Western Union and fired him. Edison compassionately put him up in the "hall bedroom" of his apartment on Bullfinch Street. Neither of them had any money because Adams's finances were reduced to "absolute zero centigrade," and Edison's, as usual, were invested in experimental equipment. Poverty drew them together, and they became insepara-ble, eating together in an "emaciator" boardinghouse where the por-tions were so small as to be almost affordable.[67]

It followed that they had no spare cash for social life. One night the headmistress of a girls' private school dropped by the Western Union office and asked Edison if he would stage a telegraph demonstration for her students. The invitation led to his first recorded experience of an electromagnetic shock not induced by a battery.

A few days before I carried the apparatus and with Adams's assis-tance, set it up in the school, which was in a double private house near the public library. The apparatus was set up when school was out. I was then very busy building private telegraph lines and equip-ping them with instruments which I had invented, and forgot all about the appointment and was only reminded of it by Adams who had been trying to find me and had at last located me on top of Jordan, Marsh & Company's store, putting up a wire. He said, we must be there in 15 minutes and I must hurry. I had on working clothes and I didn't realize that my face needed washing. However, I thought they were only children and wouldn't notice it. On arriv-ing at the place, we were met by the lady of the house and I told her I had forgotten about the appointment and hadn't time to change my clothes. She said that didn't make the slightest difference. Ad-ams's clothes were not of the best because of his long estrangement from money. On opening the main parlor door, I never was so paralyzed in my life. I was speechless, there were over 40 young ladies from 17 to 22 years, from the best families. I managed to say that I would work the apparatus and Mr. Adams would make the explanations. Adams was so embarrassed that he fell over an ottoman, the girls tittered and this increased his embarrassment, until he couldn't say a word. The situation was so desperate that for a reason I could never explain, I started in myself and talked and explained better than I ever did before or since.[68]

It was a matter of smug satisfaction to him afterward, when strolling around town with colleagues from the office, that any chance encounter with the girls would attract smiles and nods in his direction.[69]

HOW TO USE A JACKKNIFE

Another epiphany occurred when Edison, who loved to browse the secondhand bookstores on Cornhill in Boston, bought all three volumes of Faraday's *Experimental Researches*. Owning and studying the complete work affected him much more than his exposure to a borrowed volume in Cincinnati. He tried in various makeshift laboratories to perform all the procedures of the "Master Experimenter" himself, rejoicing in the simplicity of his prose and the spare use of mathematics. Adams recalled him coming home from Western Union at four A.M. and studying Faraday instead of going to bed. "I am now twenty-one," Edison burst out over breakfast. "I may live to be fifty. Can I get as much done as he did? I have got so much to do and life is so short, I am going to hustle."[70]

By now Ashley had published his article on the double transmitter and praised it in *The Telegrapher* as an "interesting and ingenious" device, if not particularly original. The editor pointed out that double transmission had been used for many years in Germany. "But Mr. Edison has simplified the process by which it is effective." This set Edison on the road to improving, and ultimately amplifying, duplex technology. His hopes of an early patent were frustrated when Joseph B. Stearns, a fire alarm specialist in Boston, was awarded protection for a similar system in June. Refusing to be discouraged, Edison listed fourteen ways in which he thought his transmitter was superior, and arranged with a local machinist to build three complete sets on spec, advertising them at the lofty prices of $400, $450, and $500. He got no orders.[71]

Ashley's editorial goodwill also encouraged him to contribute more articles to *The Telegrapher*. On 9 May "Edison's Combination Repeater" described another derivative invention—the original in this case being George Phelps's seminal combination printing telegraph of 1859. Edison argued that the Phelps machine, which was operated by a keyboard much like a piano's, rat-tatted roman characters onto paper so rapidly that "repeaters in general use on the Morse lines" were unable to duplicate them accurately. His repeater (which he illustrated with one of his mirror-like binary designs) was built on a new principle, and employed magnets "of a peculiar construction," in such a way that it could keep up with vibrations of any speed and, what was more, operate on "a current so feeble that its action would not be perceptible upon a Morse relay."[72]

As the summer and early fall progressed, he continued with unflagging energy to experiment and publish—his most ambitious effort being an eighteen-hundred-word survey of Boston's major manufacturers of electrical and telegraphic equipment in the 15 August issue of *The Telegrapher*. The article showed how thoroughly he had familiarized himself with the technological resources of the city in the four and a half months of his residence there. Among them was the prestigious shop of Charles Williams, Jr., where Moses Farmer had a little laboratory crammed with apparatus. Edison mentioned this fact without saying that he had taken space in the same building. But for some reason—possibly Farmer's own austere, devoutly religious self-effacement—he made no attempt to ingratiate himself with the American Faraday.[73]

Working nights at Western Union, and by day literally under Williams's roof in a third-floor attic, Edison invented and made half a dozen devices, including a stock ticker, a fire alarm, and a facsimile telegraph printer ("which I intend to use for Transmitting Chinese Characters").[74] He executed his first successful patent application on 13 October for an electrochemical vote recorder, whittling the submission model himself from pieces of hardwood. "To become a good inventor, you must first know how to use a jackknife."[75]

It was a clever device—too clever to be commercial, as he soon found out. Designed to speed up the laborious process of vote counting in legislative bodies, it took signals of "aye" or "nay" from electric switches on every desk and imprinted them on a roll of chemically prepared paper, in each case identifying the signal with the legislator's name. At the same time it separately tabulated the votes on an indicator dial. Edison's dream of seeing his "recordograph" clicking and spinning in the chambers of Congress dissolved when he heard that speedy voting was the last thing politicos wanted in the passage of bills. They needed time to lobby one another in medias res. Edison resolved that hereafter he would invent only things that people wanted to use.[76]

Patenting, he learned, was an expensive procedure for a young inventor, with heavy fees for each application and approval, payable to agents, attorneys, draftsmen, and the Patent Office itself—not to mention the often enormous costs inherent in defending letters patent against charges of infringement, or prosecuting his own charges against plagiarists. Throughout Edison's career from now on, a large part of his income, whether it totaled four or six figures, was absorbed into patent litigation like water through sand.

Another disagreeable aspect of professional inventing was the necessity to cultivate financial backers and influential businessmen. Edison happened to be good at it. Once money men got used to his uncouth appearance, they were beguiled by his humor and swayed by his utter self-confidence. But that did not make it any less humiliating to have to ask, and ask, and ask, and subsequently tolerate the maddening desire of investors to involve themselves in the creative process or, worse still, venture ideas of their own.

He started modestly, biting the ear of a fellow operator, DeWitt C. Roberts, who agreed to "furnish or cause to be furnished sufficient money to patent and manufacture one or more . . . Stockbroker Printing Instruments," in return for a one-third interest in their potential sale. Roberts subsequently sold part of that interest to another investor—a warning to Edison that he could not always choose his pecuniary company. But Roberts also financed the vote recorder, and various Boston brokers, merchants, and telegraph company directors, most notably E. Baker Welch, lined up to support the young inventor as his reputation spread around the city.[77]

LABORATORY OVER THE GOLD-ROOM

On 30 January 1869 an announcement appeared in the "Personal" section of *The Telegrapher*: "Mr. T. A. Edison has resigned his situation in the Western Union office, Boston, Mass., and will devote his time to bringing out his inventions."[78]

Eighteen days later Edison executed his second patent application, for an elegantly ratcheted black-and-gold "Stockbroker Printing Instrument," almost as complex as a clock.[79] It was aimed at the booming new market for alphanumerical tickers that reported fluctuations in prices on the New York gold and stock exchanges to subscribing brokers. The technology had been pioneered the year before by Edward Calahan, whose Gold & Stock Telegraph Company now sought to expand to Boston. So did the almost synonymous Gold & Stock Reporting Telegraph Company of his rival Samuel Laws.

Edison saucily chose to open a stock quotation service of his own at 9 Wilson Lane, in the same building Laws planned to put on line. He took two rooms on the floors above them, one for price monitoring and posting, the other for an experimental workshop he called his "Laboratory over the Gold-room." Knowing little about the trading business, he relied on one of his backers, the broker Samuel W. Ropes, Jr., for market expertise. His first customer could not be bluer of chip or more Back Bay Brahmin: the banking and brokerage house of Kidder, Peabody. In time, Edison added twenty-five subscribers.[80]

If he expected Laws to buy his printer, he was disappointed. It was superior to Calahan's original, having a single typewheel instead of two, and operating with a single drive wire rather than three. For a couple of months Edison and a new operator buddy, Frank Hanaford,* eked out an impecunious living, installing cheap, private-line dial telegraphs in the Boston-Cambridge area. They had to shell out considerable sums for the purchase of such supplies as forty-seven glass and porcelain insulators for pole wiring, 1¾ pounds of blue vitriol electrolyte powder, and a variety of tars, oils, and sulfur to proof their wires against the city's corrosive coal smog.[81]

By spring it was evident that Boston was either too small or too

---

* Milt Adams by now had succumbed to wanderlust and gone west.

conservative for Edison's restless ambition. Welch declined to provide funds for his "magnetograph" printer, which would have relieved customers of the need to maintain messy, accident-prone batteries.* Nor would Western Union allow him to test an improved version of his double transmitter on any of its busy long-distance wires. He had high hopes for this instrument, and accordingly arranged to use a line belonging to the small Atlantic & Pacific Telegraph Company, in Rochester, New York.[82]

A STROKE OF LUCK

With an advance from Welch of only forty dollars in his pocket, Edison arrived in Rochester on 10 April. He had to wait four money-draining days before the A&P gave him access to its New York line, late at night in an office off the Reynolds Arcade. The four-hundred-mile wire turned out to be so badly insulated, and his apparatus so complex in operation, that the taker he had hired in Manhattan became confused and the test failed. He nevertheless posted an announcement in *The Telegrapher* that his transmission had been a "complete success," returned briefly to Boston to settle up with various creditors there, then spent all the cash he had left on a one-way steamship ticket to New York.[83]

His excuse to Welch for this sortie was that he needed to get the double transmitter locally altered to handle the problem of long-distance induction—electromagnetic interference slowing and blurring the passage of signals. That work would take time, and the A&P would also have to fix its line before he could resume testing. But the real reason he gravitated to New York was that, like countless aspiring youths before him, he sniffed the city's air and caught—or thought that he caught—the intoxicating perfume of success: "People here come and buy without your soliciting."[84]

If so, they were not much in evidence on his first seventy-two hours there. He had to walk the streets all night with only enough money to buy a cup of coffee and a plateful of apple dumplings at Smith & McNell's restaurant on Washington Street. For the rest of his life he

---

* Edison gave himself an unwanted winter tan when he splashed himself with nitric battery acid in his laboratory. "My face and back were streaked with yellow; the skin was thoroughly oxidized."

would talk about the deliciousness of those dumplings. They kept him going until, on his third day in town, he stopped by Samuel Laws's Gold & Stock Reporting Telegraph Company on Broadway—just in time to find the office going into a panic over its jammed general transmitter. He studied the machine's workings and told Dr. Laws that a contact spring had snapped off and fallen between two gear wheels.[85]

"Fix it! Fix it!" Laws cried, losing revenue by the minute. "Be quick!"[86]

Edison removed the spring and set the contact wheel at zero, while employees scattered through the financial district to reset branch indicators. In two hours the system was working again. Dr. Laws gratefully hired him on the spot as an operator-mechanic at one hundred dollars a month.[87] Although this was less than he had earned during his best years as a tramp operator, it meant that he would no longer have to sleep on shavings and could subsist on more than stodge and sugar. He proceeded to work with frenzied energy, sometimes up to twenty hours a day: "I'll never give up for I may have a stroke of Luck before I die." At the beginning of August, Laws appointed him superintendent of the company's entire operating plant and tripled his salary.[88]

He justified his promotion by reconfiguring and improving a stock printer that Laws had designed but could not patent, because it replicated several features of an instrument designed by Edward Calahan. Edison produced what was in effect a new stock printer of his own, radically simpler, smaller, and smoother in operation. In the process, he lost the lucrative position he had only just won. Laws sold out to the competition on 27 August, creating a virtually monopolistic Gold & Stock Telegraph Company. The expanded firm came with its own superintendent, so by the end of the month Edison was back on the street.[89]

This time, however, he was already known in New York as the patentee of four useful inventions—the vote recorder, the private dial telegraph, the double transmitter, and the stock printer—as well as the author of several informative technical articles. A new book by Franklin Pope, *Modern Practice of the Electric Telegraph,* had a section that illustrated and praised "Edison's Button Repeater" (one of the long-distance relays he had invented in Cincinnati) as "a very simple and ingenious arrangement of connections . . . which has been found to work well in practice."[90]

Pope had preceded Edison as superintendent of the Laws company. Now that they both found themselves at large, with an abundance of skills to share, they decided to announce their own merger, and formed "a Bureau of Electrical and Telegraphic Engineering in this city."[91] Before it could be formally constituted, Edison had an experience downtown that ensured he would never regard money men as pillars of probity and responsibility.

On Friday 24 September he was on transmitter duty in the balcony of the "Gold Room" on New Street, a suffocatingly nicotinous parlor where speculators in gold stocks traded bullion under the stare of a water-spitting, gold-leaf dolphin. During the course of the day, soon to be known as "Black Friday," Jay Gould launched a stealth assault

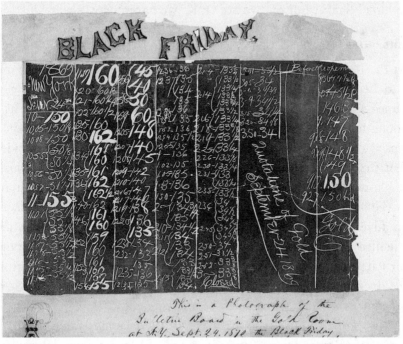

Black Friday 1869. Gold price postings annotated by future president James A. Garfield.

on the gold market, attempting to "corner" a majority of it for himself. The room's price indicator, which had drooped to as low as 144¼ the previous evening, surged to 155 in just six minutes. Tumult broke out on the floor. Edison, fascinated, climbed on top of the Western Union telegraph booth to watch sober-suited men behaving like a pack of howling coyotes. The financier Albert Speyers, who had bought $6 million worth of gold the day before, gave every appearance of going insane as he bid the price up to 160. At the climax of the hysteria, just before noon, the indicator reached 162½. Then news came that President Ulysses S. Grant had authorized the Treasury Department to sell $4 million in government gold. This turned Gould's attack into a rout and caused a concurrent panic on the Stock Exchange. After closing, so many fortunes had been lost that the prospect of homicides, or suicides, on Wall Street was serious. A company of militia was posted to keep order.

About the only observers who stayed calm through the afternoon were Edison, the dolphin, and the Western Union operator, who said to him, "Shake, Edison, we are O. K. We haven't got a cent."[92]

MRS. POPE'S HOUSE

At the beginning of October the firm of Pope, Edison & Co. came into being, and opened an office downtown at 78–80 Broadway. James Ashley signed on as third partner, touting Edison in print as a young man who, in addition to being a master of electrical science, was also "of the highest order of mechanical talent." The trio decided that their inaugural specialty was to be Edison's latest invention, a single-wire "Financial and Commercial Instrument" that both received and printed Morse signals without a local battery. They placed an advertisement in *The Telegrapher* to announce, "We possess unequalled facilities for preparing Claims, Drawings, and Specifications for Patents." All custom instruments would be made to order across the river in New Jersey, where Pope lived with his mother—and where Edison, too, would thenceforth be a paying guest.[93]

Mrs. Pope's house was in Elizabeth, two Pennsylvania Railroad stops away from Jersey City, where Edison found some laboratory space. As the days shortened toward December, he fell into a commuter's routine of rising every morning at six to catch the eastbound train there, working his customary eighteen-hour day, then waiting in frigid dark for the one A.M. local to take him back to Elizabeth.[94]

The cold he could stand with multiple layers of underwear, and the darkness he would one day do something about.

Natural Philosophy

1847–1859

IN HIS THIRD year, Alva Edison's memory began to retain and correlate the fragmentary impressions of the world that comprise the blur of infancy.[1] The most primary of these vignettes was of a broad-faced, dark-haired woman holding him and instructing him, and—more distant yet lustily present—a big man with a chinful of graying beard.[2]

Although primary, they were not his very first memories, which dated back to the summer of 1849. He was crawling across the floor toward a Mexican silver dollar, thrown down by the young man who came to court his big sister. He was watching a fleet of prairie schooners load up to join the westward gold rush, one of them with his uncle Snow Edison aboard. He was again in Nancy Edison's arms, but this time she held him up to see Marion, a vision in white at twenty, being given away to the same young man.[3]

Somewhere near at hand, that December day, would have been his brother Pitt, eighteen, and sister Tannie, sixteen. And somewhere infinitely farther off, in the Baptist heaven his mother kept talking about, were three children dead before he was born.[4]

That made Alva the youngest of seven and the only child in his father's house—a vacant-eyed little boy with a huge head. The security that surrounded him was physical (Sam Edison, prospering, had laid every brick and nailed every shingle of the seven-room house overlooking the canal basin) as well as emotional: he had the bulk of his mother's attention, composed equally of love and discipline. Both parents were, by the standard of the day, well into middle age. Sam, or Samuel Ogden Edison, Jr., in the family Bible, was forty-five at the time of Marion's wedding, and Nancy forty-one. Since she was un-

Edison's birthplace in Milan, Ohio.

likely to have any more children, whatever she had left of her genteel, minister's-daughter culture would be shared exclusively with Alva.[5]

He was unique among his siblings in another respect, being the only one who was not Canadian born. The Edison family was remotely American in origin, but because of their pro-British sympathies, they had fled north after the Revolution. Sam grew up in Vienna, Ontario, a farming village that had also attracted Nancy's family, the Elliotts, from upstate New York. He and she had married in 1828 and would probably have continued to live and work in Vienna—Sam as an innkeeper, Nancy as a schoolteacher—had it not been for the "contrary" streak in Sam that had him rebelling against any prevailing sentiment, even that of his own parents. Just as they had rejected the Declaration of Independence in 1776 and rooted for Great Britain in the War of 1812, he reacted in 1837 against the conservative, clerical regime that had begun to repress free thought in Ontario. Sam participated in an antigovernmental revolt that year, and after it failed, he made the best use of his long legs in skipping across the border to Detroit, with provincial troops in hot pursuit. He was indicted for high treason in Canada and deemed it prudent to start a new life as a carpenter in Milan, Ohio. By 1839 his family had settled with him

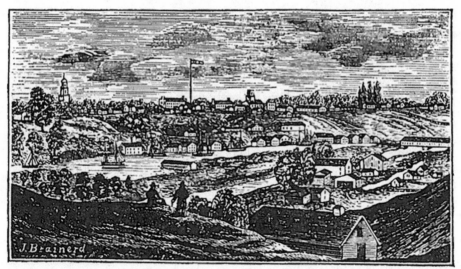

"Milan from near the Sandusky City Road," by J. Brainerd, 1847.

there. Sam did well enough from various lumbering and auctioneer-
ing enterprises to provide a solid bourgeois environment for Thomas
Alva Edison to be born in, on the snowy early morning[6] of 11 February 1847.

NOT PLAYING MUCH

Seen from the street, Sam's house presented its redbrick, white-
trimmed facade to the rising sun. Inside and to the west, it com-
manded a panorama of Milan's canal basin. He had built it at the
highest point of a bluff overlooking the Huron River Valley. Any
small boy standing at one of its parlor windows had to be thrilled by
the sight, far below, of dozens of lake schooners, laden with wheat,
leaving the harbor via the canal and gliding north down the river.[7]

Milan in the late 1840s was simultaneously a small town of about
fifteen hundred inhabitants, and one of the largest primary grain mar-
kets in the world. Often a line of five hundred or more four-horse
farm wagons could be seen creaking down the hillside into town,
laden with wheat. Elsewhere on the slope, a complex system of chutes,
hoists, and trolleys offered Alva elementary visual lessons in physics.
There was a shipyard adjoining the harbor, and crowds would collect
around the Edison house to watch whenever a new schooner was
launched below.[8]

When Alva was old enough to imitate his father's way with a saw (shingle maker Sam, sizing, splitting, and shaving three-foot "bolts" from Canada), he nailed together plank roads from mill debris, and hung about the yards at the docks memorizing the songs of lumber gangs and canal men. He briefly attended a little red schoolhouse, but there was something about him, with his domed forehead and peculiar habits, such as laboriously drawing copies of the signs on storefronts, that mystified his teachers and repelled many children. "I often run acrost him in town," a schoolmate wrote many years later, "with just as dirty nose & face as the other boys But he seemed to Be thinking of something all the time & not Playing much . . . well some of the Boys called him E-dison fool you know how Boys are. But he was far from a fool in some of his remarks his cousin [Lizzie Wadsworth] told me it Puseled her sometimes to get what he meant."[9]

Another old Milaner, racking her memory after the fool had become famous, dimly recalled "a child that was always doing funny things different from other children, loved to be by himself."[10] Sam Edison maintained that "Thomas Alva never had any boyhood days." He was more interested in "steam engines and mechanical forces" than in elementary lessons learned at school.

> Was he a remarkable smart boy? Why, no. Some folks thought he was a little addled, I believe. Teacher told us to keep him in the streets, for he would never make a scholar. All he ate went to support his brain, and he was puny. He was forever asking me questions, and when I would tell him I didn't know, he would say, "Why don't you know?"*[11]

One reason the word *addled* was to recur often in descriptions of young Alva may have been the strange equableness that kept him calm in crises or catastrophes. He seemed unable to understand that burning

---

* The word *addled* recurred so often in Edisonian lore that Alva himself claimed to have heard his teacher use it in conversation with an inspector, saying that there was no point in keeping the boy at school: "I was so hurt by this . . . that I burst out crying and went home and told my mother about it." In his telling, Nancy waylaid the teacher and told him "that I had more brains than he himself."

his father's barn down, "just to see what it would do," would enrage Sam and result in a public flogging. More seriously, he lost a friend, George Lockwood, while swimming in one of the many streams that fed the canal and river south of Milan. "After playing in the water a while, the boy with me disappeared in the creek," Edison recalled, still matter-of-fact about the tragedy fifty-odd years later. "I waited around for him to come up but as it was getting dark I concluded to wait no longer and went home. Some time in the night I was awakened and asked about the boy. It seems the whole town was out with lanterns and had heard that I was last seen with him. I told them how I had waited and waited, etc. They went to the creek and pulled out his body."[12]

Alva's public education in Milan did not last much beyond the beginning of second grade, if indeed it lasted till then: "I used never to be able to get along at school. I don't know now what it was, but I was always at the foot of the class. . . . My father thought I was stupid, and at last I almost decided I must really be a dunce."[13]

SHINGLES, STAVES, SPARS

At Detroit in the spring of 1854 Sam Edison loaded Nancy, Pitt, Tannie, and Alva aboard a little paddleship, the *Ruby*. They were bound for Port Huron, forty miles north down the St. Clair River. Marion remained behind with her husband Homer Page, to make what life they could in Milan. The town had gone into precipitous decline, killed by the advent of railroads. In just a few years its canal would silt up and disappear, along with the harbor and shipyard. Sam was smart to move when he did, from an inland port lapsing into decay to a lake port twice its size and thrumming with sawmills, annually shipping 93 million board-feet of pine lath, shingles, staves, and spars, along with fragrant cedar poles and bales of tanning bark.[14] Having profited substantially from grain trading in Milan, as well as from his lumber business, he was poised to do even better in Port Huron.

He spent $2,800 on a double-story white frame house in the defunct military reserve of Fort Gratiot. It measured well over thirty-six hundred square feet inside—not counting a spacious cellar that would become Alva's first laboratory. The ceilings were high and the windows large, opening to a surrounding ten-acre grove—also Sam's property—with virgin woodland rising to the west and the St. Clair River entering Lake Huron in the east.[15]

Fort Gratiot's most prominent architectural feature was its seventy-four-foot white brick lighthouse, signaling the junction of river and lake. It was fitted with smoky whale oil lamps that cast an oddly greenish glow through glass that daily had to be wiped clean of soot. Sam, ever the entrepreneur, was inspired by its height, and lack of accessibility to tourists, to erect an "observation" tower twenty-six feet higher on his own ground and open it to the public. For the first two or three years of the tower's existence, he earned a supplementary income from this attraction. It afforded a magnificent, if rickety, panorama of lake, river, and the apposition of two nations, both of which he could in a sense call home.[16]

Sam during this brief period was at the crest of his modest success in business, dealing not only in lumber and grain and seventy-five-cent observatory tickets but also selling apples and pears from his orchard and vegetables from his small farm. Next he got into real estate, buying and selling local properties, and by the fall of 1856 he was worth an estimated $6,000.* But his libertarian ethics were ill matched to the principles of property law, which required among other things that titles should be searched before a sale, and taxes paid after it.[17]

On 24 August 1857, just when such nuisances were beginning to affect Sam's cash flow, the collapse of the Ohio Life Insurance Company, depressing railroad stocks, caused the nation's first panic spread by telegram. Michigan was especially hurt by the subsequent recession. Detroit's economy crashed; banks and businesses closed; bread lines formed; child beggars roamed the streets. Port Huron suffered similarly. In December Sam was indicted for real estate fraud. Rumor had it that he escaped bankruptcy, or worse, by putting his holdings in Nancy's name. Credit reports issued during the winter of 1857–58 stated that he had "totally failed" as a businessman. Were it not for a livery business operated by Pitt Edison, and child farm labor eagerly supplied by Alva, Sam might have been forced to sell his white house in the grove.[18]

---

* About $183,000 in today's money.

THE ONLY TEACHER

Alva's boyhood in Port Huron is so little documented that it is impossible to say just when he went to school there. In his sixties he claimed to have had only three months of formal education. This was almost certainly an exaggeration, since he attended a private primary establishment run by the Rev. George Engle and, after a gap of several years, went to the town's Union School. However short a stint he passed in each, it seems clear that he left the first for the same reason he was sent home from the little red schoolhouse in Milan—his "addled" strangeness—and that he quit the second at age twelve to begin his career as a railroad newsboy. It is also clear, in view of his learning difficulties in class, that the only teacher who understood him and fertilized his brain until then was Nancy Edison.[19]

"My mother was the making of me," he once allowed, in a rare moment of self-revelation. "She was so true, so sure of me; and I felt that I had someone to live for, someone I must not disappoint." Short, heavyset, gray-eyed, dark-browed, and austere, Nancy was the sister of two Baptist ministers and the granddaughter of a Quaker. She was a disciplinarian and kept, in Alva's vivid memory, "a switch behind the old Seth Thomas clock that had the bark worn off." Grim, however, Nancy was not: she had an ironic twinkle and perpetual hint of a smile about the lips, both of which she passed on to her son. There is no evidence she was disturbed by her failure to instill any piety in him. In all other respects her influence upon his intellectual development was profound.[20]

A family friend described her as "industrious, capable, literary and ambitious," the significant adjective being the third. Another recalled how Alva and Nancy were left alone much of the time after Tannie married and moved out of the house in June 1855: "I well remember the old homestead, surrounded by the orchard, and frequently saw Mrs. Edison and her son conversing. Sometimes I noticed that she was instructing him in his lessons and I often wondered why he never went to school." Yet another wrote him in old age, "One thing I can very distinctly remember is being in front of your house one day and seeing your mother standing at the door calling you to come in to your lessons."[21]

Nancy Elliott Edison, circa 1854.

The first books of any consequence they read together appear to have been a history of the Reformation and—more to Alva's taste—Robert Sears's *The Family Instructor, or, Digest of General Knowledge Embracing the Various Divisions of History, Biography, Literature, Geography, Natural History, and the Other Sciences.* Three and a half inches thick and profusely illustrated, it prepared him for some of the more formidable titles he and Nancy set themselves over the next four years, including Burton's *Anatomy of Melancholy,* Hume's six-volume *History of England,* and Gibbons's *Decline and Fall of the Roman Empire.* By the time they reached Richard Green Parker's *Natural Philosophy,* the book that most influenced Alva in youth, his fundamental area of curiosity—the workings of nature in a godless world—was established, and Nancy effectively could teach him no more.[22]

THE WHOLE CIRCLE OF THE SCIENCES

Late in life Edison told Henry Ford that Parker's encyclopedic tome was "the first book on science I read when a boy nine years old." If he indeed studied it that early, he was well prepared to do so again when he entered the Union School at age eleven.[23] Its formal title was *A School Compendium of Natural and Experimental Philosophy,* and the 1854 edition featured an ancillary subtitle: *Containing Also a Description of the Steam and Locomotive Engines, and of the Electro-Magnetic Telegraph.**[24]

"The whole circle of the sciences," Alva read in the introduction, "consists of principles deduced from the discoveries of different individuals, in different ages, thrown into common stock. The whole, then, is common property, and belongs exclusively to no one."[25] Before he thought of venturing into the charmed circle himself, he was thus cautioned that countless others had preceded him, and even the greatest individuals among them—Copernicus, Newton, Franklin, Faraday—had based their work on the accumulating wisdom of mankind. There was no originality, except that of God the Creator. All inventions were but rediscoveries and rearrangements of primal matter.

Nancy Edison could have had no objection to that notion. But Parker made clear that his book would have nothing to do with the spiritual, moral, and intellectual branches of philosophy, namely, theology, ethics, and metaphysics. He was concerned only with the science of natural philosophy—"that which treats of the material world"—and within that science, "only the general properties of unorganized matter."[26]

During the course of Alva's lifetime, the word *physics* would replace *natural philosophy* as a scientific term. But Parker in 1856 insisted that the former discipline was more general than the latter. Physics comprised both natural history and natural philosophy—

---

* An elaborate frontispiece showed "Electrical Telegraphic Communication" in action, with a frock-coated gentleman sending a message along wires that traversed hill, dale, and coastline to another gentleman checking it in printout and looking by no means pleased at the news it contained.

which was to say, organized as well as unorganized matter. He excluded chemistry from his book, on the grounds that it introduced human agency into the conditions and relations of bodies.[27] Having thus restricted himself to an unbiological view of the universe, he proceeded to enumerate and describe every field that Alva would one day explore, with the exception of X-rays and botany—as if in some occult way he was remembering the boy's future.

In a preliminary section, *Of Matter and Its Properties,* Parker listed the sixty-one known elements, exemplified in the elementary composition of rocks like granite. Forty-nine of the basic ingredients of nature were metallic, and although a proper consideration of them should be left to chemists, he wrote enough about the "malleability" and "ductility" of gold and platinum to make metallurgy sound like the most rewarding of sciences. An ounce of gold could be attenuated more than fifty miles in length, and "platinum can be drawn even to a finer wire than this."[28]

Under *Mechanics,* Alva learned that there were six fundamental instruments: the lever, the pulley, the wedge, the screw, the inclined plane, and the wheel. If the last was built with a heavy rim and long, light spokes, force had to be applied to make it spin, but once put into motion, it would become a "fly-wheel," and inertia would steady the revolutions of any machine that it drove. Those parts of a wheel or disk "furthest from the center of motion move with the greatest velocity; and the velocity of all the parts diminishes as their distance from the action of motion diminishes."[29]

Parker defined *Friction* (in words guaranteed to convulse any schoolroom) as "the resistance which bodies meet with in rubbing against each other." A *Vacuum* was "unoccupied space; that is, a space which contains absolutely nothing." He showed a drawing of a hand-pump laboriously evacuating the air from a "glass vessel or bulbed receiver" and explained that air was essential to combustion. "Place a lighted taper, cigar, or any other substance that will produce smoke, under the receiver, and exhaust the air; the light will be extinguished, and the smoke will fall." Under "Pneumatics," Alva was informed that air was so elastic, total evacuation was almost impossible. However, when a tall, closed-top tube full of heavy mercury was dipped into a bowl of more mercury, gravity would partially pull the column down, leaving behind a "Torricellian vacuum . . . the most perfect that has been discovered."[30]

The heading "Steam-Engine" rated a fourteen-page discussion, with explanatory diagrams and a beautiful woodcut of a paddle-wheeler thrashing along through undulant waves. Another illustration showed a pineapple-stacked locomotive, evocative of the one Alva would soon drive, solus, down forty-seven and a half miles of track.[31]

Under "What is Electricity?" he read the most important (if vague) definition he would ever have to memorize: "Electricity is the name given to an imponderable* agent which pervades the material world, and which is visible only in its effects." Apparently the term derived from the Greek word *elektron* for "amber"—a material that, when rubbed, became "excited," drawing toward itself "pieces of paper, thread, cork, straw, feathers, or fragments of gold leaf." There were many "non-electric" substances impervious to this force, and the "attraction and repulsion" between those that responded and those that did not was the fundamental, empowering conflict of nature.[32]

Some theorists thought of electricity as a fluid. Others denied its materiality and deemed it to be "a mere property of matter," like magnetism. Parker believed it to be two fluids of opposite qualities,

---

* Parker explained elsewhere that "imponderable" meant "weightless." With its opposite *ponderable*, it made up the two great classes of natural philosophy.

eternally seeking a counterbalance between "positive" and "nega-tive," and said that a modern consensus of scientific opinion had de-veloped along those lines. Professor Faraday, he wrote (imprinting the name of the great experimentalist on Alva's mind for the first time), had proposed a nomenclature of electricity that included the words *anode* for the positive element and *cathode* for the negative, in any flow of power through a conductor, or wire.[33]

Parker's references to fluidity made Alva think, early on, of electric-ity in hydraulic terms. For the rest of his life he would see it pumping back and forth like water, but silently and weightless, palpable only in its power to shock or even kill. Possibly it might even be used, one day, to revive a dying person.[34] There was an interesting account under "Galvanism" of an electrical test recently performed on a hanged murderer in Glasgow:

> *The galvanic battery employed consisted of 270 pairs of four-inch plates. On the application of the battery to different parts of the body, every muscle was thrown into violent agitation; the leg was thrown out with great violence, breathing commenced, the face ex-hibited extraordinary grimaces, and the finger seemed to point out the spectators. Many persons were obliged to leave the room from terror or sickness; one gentleman fainted, and some thought that the body had really come to life.*[35]

Alva was able to study all the basic types of batteries, from voltaic piles to zinc-platinum Grove cells—powerful units, apparently, but inefficient, requiring constant replenishment of their sulfuric and ni-tric acid levels: a dangerous, cough-inducing chore.

Detailed discussions of "Magnetism" and "Electro-Magnetism" followed. Parker temptingly wrote that if "fine black sand," or mag-netite, was spilled on a sheet of paper with a horseshoe magnet be-neath it, "the particles will be disposed to arrange themselves, in a regular order, and in the direction of the curve lines." In a footnote, he printed Faraday's thought-provoking speculation, on the basis of recent studies of sunspot activity, that there might be an electrical relationship between "astral and terrestrial magnetism."[36]

Acoustics, Alva read, was "the science which treats of the nature and laws of sound. It includes the theory of musical concord or harmony." Sound waves needed air to travel through, and the more humid the air the better: "A bell can be more distinctly heard just before a rain." Conversely, if a bell was rung in an exhausted receiver, its chime could not be heard. Geometry and acoustics were intimately related: "The smooth and polished surface of interior parts of certain kinds of shells, particularly if they are spiral or undulating, fit them to collect and reflect the various sounds which are taking place in the vicinity."[37]

Parker compared sound to light, in that it could be collected at one point after reflecting off several surfaces, for greater intensity and focus. Under "Optics"—"the science that treats of light, of colors, and of vision"—he cautioned that nobody, not even Sir Isaac Newton, really knew what light was. Newton had supposed it to consist of minute particles streaming from source to viewer—the "corpuscular" theory. But the sound analog, if valid, implied that light also traveled in waves. Hence "the opinions of philosophers of the present day are inclining to the undulatory theory." Whichever notion was true, light moved so fast that it effortlessly delineated any movement perceptible to the human eye. This posed the counterquestion, *When is motion imperceptible?* Parker's answer was "When the velocity of a moving body [a distant freighter on Lake Huron, say] does not exceed twenty degrees in an hour."[38]

Oddly, he did not refer to the phenomenon of persistence of vision, although few boys Alva's age were unfamiliar with the way a rapid series of still images drawn on flipped cards, or the drum of a zoetrope, seemed to move of their own accord. But there was a description on pages 245–46 of the magic lantern, with an illustration showing how it magnified and threw a glass slide's colored image onto "a white surface prepared to receive it."[39]

All optical media, Parker wrote, were luminous, transparent, or translucent—and the dioptrical subcategory, exemplified by lenses, bent light beams to their will according to whether they were convex or concave. This led him to an extended discussion of the physiology of the eye, including accounts of its sensitivity to abrupt changes of light and darkness—useful to any young reader who might one day be asked by the U.S. Navy to improve the vision of "splash observers" in wartime.[40]

The final sections of the book were devoted to "The Electro-Magnetic Telegraph" (with a useful Morse table that rendered Alva's surname as • — •• •• ••• • • — •) and "Astronomy," warning him that if he ever wanted to observe a total eclipse of the sun, it would last "little more than three minutes."[41]

TWO HUNDRED BOTTLES

Before mastering some, or all, of the sciences Parker described, Alva at age eleven became a greengrocer—"marketing garden truck," in the unglamorous phrase of the day. Sam Edison's financial difficulties in the summer of 1857 were such that he had to grow and sell as much produce as the orchard and field around the house would supply. Thanks to the richness of Fort Gratiot's soil, their bounty was copious.

Sam Edison tilling his field at Fort Gratiot, date unknown.

Alva took to the work with a will, plowing and planting eight acres of sweet corn, radishes, onions, parsnips, and beets: "I was very ambitious about this garden and worked very hard. My father had an old horse and wagon and with this we carried the vegetables to the town which was 1½ miles distant and sold them from door to door." When, later in the season, pears and apples ripened in the orchard, Alva's takings increased, and he was smart enough to hand them to his mother. In time he learned the fine art of packing figs in boxes with reinforced bottoms, suggesting more fruit than materialized in actual delivery.[42]

He persevered in greengrocery through the following two summers, at one point doing so well that he was able to give Nancy $600.[43] Whatever pocket money he kept for himself, he invested in a chemical laboratory in the cellar of the Edison house.

Alva may have been motivated to equip this facility by Parker's provocative insistence that chemistry had no part in natural philosophy, which was the study of the material world "as is." Beyond physics, chemistry was the science that creatively investigated and "altered the natural arrangement of elements to bring about some condition that we desire." In other words, it was an active rather than reactive application of mind to matter, in that respect more humanistic and challenging. He loved it from the start and was always happiest when he could retreat from the noise and fallibility of machines and privately mix powders. He once said that he wondered "how it was he did not become an analytical chemist instead of concentrating on electricity, for which he had at first no great inclination."[44]

His juvenile laboratory's principal feature was an array of two hundred bottles, all labeled poison in a deliberate, and effective, ploy to dissuade anyone else from tampering with them. Aided sometimes by a friend, Joseph Clancy, he concocted many volatile compounds that Nancy worried might explode. He failed to blow out the cellar windows, but he did succeed in wrecking a corner of the old telegraph office downtown, burning several other boys in the process.[45]

At some point in 1858 Alva hitched together enough lengths of stovepipe wire, along with old jars, nails, zinc, insulated copper coil, a Grove cell, and sprung brass keys, to put his laboratory in Morse communication with the Clancy house, one and a half miles away.

With puberty looming ("Ma, I'm a bushel of wheat, I weigh eighty pounds"), it was his last childish indulgence before two sobering experiences transformed him into the beginnings of a man. One was a return to school—the Union School, in downtown Port Huron—possibly for only one eleven-week term. It flagged him as no more suited to the classroom than he had been twice before. The other was the mysterious trauma, whether medical, accidental, or in some other aspect pathological, that shut him off forever from ambient sound. No official record survives of what it was, if indeed any records were kept; by the time he was old enough to speak of his deafness without embarrassment, he did so without ever giving a convincing account of its cause. The casual admission he wrote at twenty-nine—"I haven't heard a bird sing since I was twelve years old"—must stand in its simple poignancy as all that will ever be known of the matter.[46]

If his dating was precise, Alva lost three-quarters of his hearing several months before the advent of puffing, whistling locomotives and clanking cars broke Fort Gratiot's harmonious web of natural sounds. At any rate he was a child no longer when, in the late fall of 1859, he persuaded Nancy to let him ride the Grand Trunk Railway wherever its steam would take him, down the track to all the years ahead.

EPILOGUE

1931

S EVENTY-TWO YEARS LATER, as Edison lay dying, it was suggested to President Hoover that the entire electrical system of the United States should be shut off for one minute on the night of his interment. But Hoover realized that such a gesture would immobilize the nation and quite possibly kill countless people. Nor would he countenance an alternative idea, that he order the extinction of all public lights at that moment. It was not only inconceivable, it was impossible that America could recapture, even for sixty seconds, the dark that had prevailed in 1847, when Thomas Alva Edison was born.[1]

The president emphasized this in a statement dated 20 October. "The dependence of the country on electrical current for its life and health is itself a monument to Mr. Edison's genius." Acknowledging, however, that there was "a universal desire" to pay personal respect to the old man as a benefactor of humanity, he called on all private individuals and organizations to put out their lights from 10:00 to 10:01 P.M., Eastern time, the following evening—which turned out to be, appropriately, the anniversary of the night in 1879 when Edison achieved his first viable lamp.[2]

He was buried at dusk in Rosedale Cemetery, Montclair, New Jersey. The sun went down behind Eagle Rock just as his coffin was lowered into the grave. Across the river in Manhattan, an immense crowd began to assemble along Broadway between Forty-second and Forty-third streets. At two minutes before ten, the CBS and NBC radio networks broadcast an advance reminder of Hoover's call, some stations playing Haydn's setting of the words of Genesis, *darkness was upon the face of the deep.*[3] On the hour in Milan, Ohio, the town clock struck, and Edison's birthplace went dark as the chimes continued to toll at six-second intervals. All lights in the White House were doused, including the big globes surrounding Executive Park, and large areas of the national capital and its suburbs followed suit. In New York Harbor, the torch held by the Statue of Liberty flickered out. Simultaneously the billboards and marquees of "the Great White Way" faded, and a hush descended on the crowd, which by now extended north into the Fifties. The absence of sound was more remarkable than that of light, because some small stores continued to glow. Not a vehicle moved in the entire theater district.[4]

Midway through the fight program at the American Legion arena in Ybor City, Florida, gloves dropped and the audience stood in darkness as the gong sounded taps. In the movie houses of Reading, Pennsylvania, talkies stopped talking and the picture faded from the screen until nothing could be seen but the dim red glow of exit lights. The small city of Franklin, at the opposite end of the state, attempted a total blackout, but was foiled by a wash of autumn moonlight. Chicago skyscrapers lost their sparkle, and several high beacons turned off, posing a momentary threat to air traffic. Farms and villages in occluded parts of the country vanished like crystals dissolving in ink. The Pacific Gas & Electric Company doused all its lights in northern California. A strange hush obtained over urban areas on the West Coast. Pedestrians came to a halt as streets darkened. Men took off their hats, and women bowed their heads.[5]

EDISON'S DEATH LEFT behind a legend so potent that it quickly grew to the dimensions of myth. For a quarter of a century he was deified in adulatory biographies and movies, to the mingled satisfaction and perplexity of his wife and children, none of whom could escape the stretch of his shadow. They tried to adjust to its inhibiting chill with varying degrees of success.[6]

Mina Edison married again in 1935 to Edward Everett Hughes, a rich old businessman who persuaded her, before his own death in 1940, to enjoy the pleasures of cocktails (frowned on in Chautauqua) and world travel. She took his surname but was quick to jettison it upon resuming widowhood. For the last seven years of her life, which she divided between Glenmont and Seminole Lodge, she was again, imposingly, Mrs. Thomas Alva Edison.

Tom died cuckolded and alone in a Massachusetts hotel room in 1935, allegedly of heart failure. William retained his coarse vitality to the end, patenting five radio devices and signal systems before his death in Wilmington, Delaware, two years later. Marion never remarried. She survived in Norwalk, Connecticut, mourning Tom and consoling herself with opera, until 1965. Charles ran the huge but atrophying conglomerate of Thomas A. Edison, Inc., until it was absorbed by the McGraw Electric Company in 1957. In wordly terms the most successful of Edison's sons, he was appointed assistant secretary of the navy by Franklin D. Roosevelt in 1937, and was promoted to secretary before resigning in 1940 to campaign for the governorship of New Jersey. He served as governor for only one term, then returned to business and became, in wealthy old age, a crotchety red-baiter. Childless, like all his siblings except Madeleine, he died in 1969. She followed him ten years later, having produced four sons by John Sloane. Theodore was the last of the primary family to die, a scrupulously principled intellectual, conservationist, and—in old age—opponent of the Vietnam War. After his death in 1992 the name of Edison lingered only among descendants of the Sloane family. Of old Sam Edison's lusty blood, no patrilineal trace remains.[7]

ONE OF THE imponderables of a dying inventor's coma is that watchers around the bed can never be sure what dreams of fact or imagination may be playing inside his motionless, white-haired skull. When he is, on top of everything, stone deaf, that makes his last consciousness even more private.[8] But if Edison's aural memory in October 1931 was capable of reaching back beyond his mysterious inner-ear ailment at age twelve, who knows but what he heard again the harmonious noises that made Fort Gratiot such a haven of natural sounds before the loud arrival of Grand Trunk Railroad trains: the tooting of bugles on the parade ground; and before that the spring chorus of skylark and blackbird and quail around the house in the grove; and before that the humming of Port Huron's seven sawmills; and before that the crunch and thump of logs in the St. Clair River; and before that his mother's voice calling "Alva" as she summoned him to his lessons; and earlier still, among school bells and church bells, the songs of shipyard workers he had memorized in Milan—his first recordings!—and farther, even subliminally back, whatever outside sounds penetrated the encompassing dark of his first nine months of life.

ACKNOWLEDGMENTS

My principal debt for expert assistance with the research and writing of this biography is to Leonard DeGraaf, the chief archivist at Thomas Edison National Historical Park. I am also indebted to Paul Israel and Thomas E. Jeffrey, respectively editor and senior editor of the *Papers of Thomas A. Edison* project at Rutgers University. Aside from their general help, Messrs. DeGraaf and Jeffrey gave my manuscript a thorough scholarly review, as did my wife and fellow biographer Sylvia Jukes Morris. The scientific, technological, and medical portions of the text were scrutinized by Louis Carlat, associate editor of the *Papers* project, Dr. Carl M. Horner, laboratory consultant at Edison-Ford Winter Estates in Fort Myers, Florida, and Dr. Karen Chapel of Northville, Michigan. Dr. David Edison Sloane gave me privileged access to his collection of Edison family papers. I am profoundly grateful to all these generous and patient people, as well as to my past and present editors, Robert Loomis and Andy Ward, and to the other kindly souls listed below.

Michele Albion; Marie Arana; David Ball; Pierson Ball; Konstantin Batygin; Antony Beaumont; Pamela A. Brunfelt; Sam Brylawski; Karen Chapel; Ned Comstock; Mike Cosden; Lee A. Craig; Anthony Davidowitz; Judy Davidowitz; Charles DeFanti; Dino Everett; ; Marc Greuther; Tom Griffith; George Herrick; Chris Hunter; Dodie Kazanjian; Georgianne Ensign Kent; Ginny Kilander; Clifford Laube; David Levesque; Richard Lindsey; Charles Macpherson; Stephen Morgan; John Novogrod; Harry Pennington; Kate Armour Reed; Alexandra Rimer; Benjamin M. Rosen; Donna Perrett Rosen; David Seubert;

Walter Suskie; Rachel Weissenberger; George Willeman; Hiram P. Williams Jr.; and Lois Wolf.

And finally, my gratitude to Scott Moyers, who first suggested I write a biography of Thomas Alva Edison.

—E.M.

SELECT BIBLIOGRAPHY

The main published primary source for Edison studies is Rutgers University's ongoing "The Papers of Thomas A. Edison" project (see Notes). There have been only two indispensable biographies, issued at opposite ends of the twentieth century: *Edison: His Life and Inventions* by Frank Dyer and T. C. Martin (1910) and *Edison: A Life of Invention* by Paul Israel (1998). The former, though reverential in tone, is fact-filled and contains many of Edison's dictated reminiscences. The latter is scholarly, objective, and dense with technological and business detail, benefiting from the author's long service as editor of the Edison Papers. Neither volume offers much coverage of the last decades of Edison's life.

The bibliography below confines itself to book sources of especial relevance to the text of this biography. All other sources, including individual manuscripts, oral histories, and periodicals, are cited *passim* in the Notes.

ARCHIVES

CHC Charles Hummel Collection, Wayne, NJ
COL Columbia Center for Oral History, Thomas Edison Project, Columbia University, NY
DSP David E. E. Sloane Papers, Hamden, CT (private collection)
EFW Edison-Ford Winter Estates, Fort Myers, FL
FSP Frank J. Sprague Papers, New York Public Library, NY
HFM Henry Ford Museum, Dearborn, MI
JDP Josephus Daniels Papers, Wilson Library, University of North Carolina at Chapel Hill, NC

PTAE   The Papers of Thomas A. Edison (online edition offering digital access to collections of Edisonia in more than 140 public and private repositories)
TENHP Thomas Edison National Historical Park, NJ

BOOKS AND MONOGRAPHS CONCERNING EDISON

The Papers of Thomas A. Edison (Johns Hopkins University Press, Baltimore, 1989–)

Vol. 1: *The Making of an Inventor, February 1847–June 1873.* Edited by Reese V. Jenkins et al. 1989.
Vol. 2: *From Workshop to Laboratory, June 1873–March 1876.* Edited by Robert A. Rosenberg et al. 1991.
Vol. 3: *Menlo Park: The Early Years, April 1876–December 1877.* Edited by Paul Israel et al. 1994.
Vol. 4: *The Wizard of Menlo Park, 1878.* Edited by Paul Israel et al. 1998.
Vol. 5: *Research and Development at Menlo Park, 1879–March 1881.* Edited by Paul Israel et al. 2004.
Vol. 6: *Electrifying New York and Abroad, April 1881–March 1883.* Edited by Paul Israel et al. 2007.
Vol. 7: *Losses and Loyalties, April 1883–December 1884.* Edited by Paul Israel et al. 2011.
Vol. 8: *New Beginnings, January 1885–December 1887.* Edited by Paul Israel et al. 2016.
Vol. 9: *Competing Interests: January 1888–December 1889.* Edited by Paul Israel et al. Forthcoming.

Albion, Michele W. *The Florida Life of Thomas Edison.* Gainesville, FL, 2008.
———. *The Quotable Edison.* Gainesville, FL, 2011.
Ballentine, Caroline Farrand. "The True Story of Edison's Childhood and Boyhood." *Michigan History Magazine* 4 (1920).
Bryan, George S. *Edison: The Man and His Work.* New York, 1926.
Carlson, W. Bernard. "Edison in the Mountains: The Magnetic Ore Separation Venture, 1879–1900." In Norman Smith, ed., *History of Technology*, vol. 8. New York, 1983.
Collins, Theresa M., and Lisa Gitelman. *Thomas Edison and Modern America: A Brief History with Documents.* New York, 2002.
DeGraaf, Leonard. *Edison and the Rise of Innovation.* New York, 2013.
———. *Historic Photos of Thomas Edison.* Nashville, TN, 2008.

Dennis, Paul M. "The Edison Questionnaire," *Journal of the History of Behavioral Sciences* 20, no. 1 (1984).

Dickson, William Kennedy. "A Brief History of the Kinetograph, the Kinetoscope, and the Kineto-Phonograph." *Journal of the Society of Motion Picture Engineers* 21, no. 6 (Dec. 1933).

Dickson, William Kennedy, and Antonia Dickson. *The Life and Inventions of Thomas Alva Edison.* New York, 1894.

Dyer, Frank Lewis, and Thomas Commerford Martin. *Edison, His Life and Inventions,* 2 vols. New York, 1910.

Edison: The Invention of the Movies: 1891–1918. Museum of Modern Art / Library of Congress DVD set, 2005.

Feaster, Patrick. "Speech Acoustics and the Keyboard Telephone: Rethinking Edison's Discovery of the Phonograph Principle." *ARSC Journal* 38, no. 1 (2007).

Friedel, Robert, and Paul Israel. *Edison's Electric Light: The Art of Invention.* Baltimore, 2010.

Fritz, Florence. *Bamboo and Sailing Ships: The Story of Thomas A. Edison and Fort Myers, Florida.* Fort Myers, FL, 1949.

Gall, Michael J. "Thomas A. Edison: Managing Menlo Park, 1876–1882." MA thesis, Monmouth University, 2004.

Hammer, W. J. "Edison and His Inventions," 12-part memoir, as told to Willis J. Ballinger. NEA Press Service, Oct.–Nov. 1931.

Hendricks, Gordon. *Origins of the American Film.* A reprint compilation of this author's *The Edison Motion Picture Myth* (1961), *Beginnings of the Biograph: The Story of the Invention of the Mutoscope and the Biograph and Their Supplying Camera* (1964); and *The Kinetoscope: America's First Commercially Successful Motion Picture Exhibitor* (1966). New York, 1971.

Hounshell, David. "Edison and the Pure Science Ideal in Nineteenth-Century America." *Science* 207 (1980).

Israel, Paul. *Edison: A Life of Invention.* New York, 1998.

Jeffrey, Thomas E. "'Commodore' Edison Joins the Navy: Thomas Alva Edison and the Naval Consulting Board." *Journal of Military History* 80 (April 2016).

———. *From Phonographs to U-Boats: Edison and His "Insomnia Squad" in Peace and War, 1911–1919.* Bethesda, MD, 2008.

———. "Tom and Beatrice Edison." Unpublished biographical essay, 2018, privately held.

———. "When the Cat Is Away the Mice Will Work," *New Jersey History* 125, no. 2 (2010).

Jehl, Francis. *Menlo Park Reminiscences,* 3 vols. Dearborn, MI, 1937, 1938, 1941.

Jones, Francis Arthur. *Thomas Alva Edison: Sixty Years of an Inventor's Life.* New York, 1908.

Josephson, Matthew. *Edison: A Biography,* 1959; New York, 2003.

Lathrop, George P. "Talks with Edison." *Harper's Magazine* 80 (Feb. 1890).

Martin, Thomas C. *Forty Years of Edison Service, 1882–1922: Outlining the Growth and Development of the Edison System in New York.* New York, 1922.

McClure, James B., ed. *Edison and His Inventions, Including the Many Incidents, Anecdotes . . .* 1879.

McPartland, Donald Scott. "Almost Edison: How William Sawyer and Others Lost the Race to Electrification." PhD diss., City University of New York, 2006.

Meadowcroft, William H. *Boys' Life of Edison*. New York, 1921.

Millard, André. *Edison and the Business of Invention*. Baltimore, 1990.

Nerney, Mary Childs. *Thomas A. Edison, A Modern Olympian*. New York, 1934.

Öser, Marion Edison. "The Wizard of Menlo Park, by His Daughter," March 1956, TENHP.

Pretzer, William S., ed. *Working at Inventing: Thomas A. Edison and the Menlo Park Experience*. Dearborn, MI, 1989.

Tate, Alfred O. *Edison's Open Door: The Life Story of Thomas Alva Edison, a Great Individualist*. New York, 1938.

Vanderbilt, Byron. *Thomas Edison, Chemist*. Washington, DC, 1971.

Warren, Waldo P. "Edison on Invention and Inventors." *Century Magazine,* July 1911.

Wile, Raymond S. "Edison and Growing Hostilities." *ARSC Journal* 22, no. 1 (Spring 1991).

———. "The Edison Invention of the Phonograph." *ARSC Journal* 14 (1982).

———. "The Rise and Fall of the Edison Speaking Phonograph Company, 1877–1880." *ARSC Journal* 8, no. 3 (1976).

BOOKS (GENERAL)

Allerhand, Adam. *An Illustrated History of Electric Lighting*. Bloomington, IN, 2016.

Association of Edison Illuminating Companies. *"Edisonia": A Brief History of the Early Edison Electric Light System*. New York, 1904.

Bowers, Brian. *Lengthening the Day: A History of Lighting Technology*. New York, 1998.

Bowser, Eileen. *The Transformation of Cinema, 1907–1915*. New York 1990.

Bright, Arthur A., Jr. *The Electric Lamp Industry: Technological Change and Economic Development from 1800 to 1947*. New York, 1949.

Carlson, W. Bernard. *Tesla: Inventor of the Electrical Age*. Princeton, 2013.

Copeland, George A., and Michael W. Sherman. *Collector's Guide to Edison Records* (Monarch Record Enterprises, 2012).

Craig, Lee A. *Josephus Daniels: His Life and Times*. Chapel Hill, 2013.

De Borchgrave, Alexandra Villard, and John Cullen. *Villard: The Life and Times of an American Titan*. New York, 2001.

Dickson, W. K. L., and Antonia Dickson, *History of the Kinetograph, Kinetoscope and Kineto-Phonograph 1895*; New York, 2000.

Fahie, J. J. *A History of Wireless Telegraphy*. New York, 1901.

Finlay, Mark R. *Growing American Rubber: Strategic Plants and the Politics of National Security*. New Brunswick, NJ, 2009.

Geduld, Harry M. *The Birth of the Talkies: From Edison to Jolson*. Bloomington, IL, 1975.

Gelatt, Roland. *The Fabulous Phonograph, 1877–1977*, 2nd rev. ed. London, 1977.

Gleick, James. *The Information: A History, a Theory, a Flood.* New York, 2011.

Gitelman, Lisa. *Scripts, Grooves, and Writing Machines: Representing Technology in the Edison Era.* Stanford, CA, 1999.

Grant, James. *The Forgotten Depression: 1921: The Crash That Cured Itself.* New York, 2015.

Hammond, John Winthrop. *Men and Volts: The Story of General Electric.* Philadelphia, 1941.

Harvith, John, and Susan Harvith, eds. *Edison, Musicians, and the Phonograph: A Century in Retrospect.* New York, 1987.

Helmholtz, Hermann. *On the Sensations of Tone as a Physiological Basis for the Theory of Music.* Translated by Alexander J. Ellis. London, 1875.

Hughes, Thomas P. *Networks of Power: Electrification in Western Society, 1880–1930.* Baltimore, 1983.

Johnson, Rodney P. *Thomas Edison's "Ogden Baby": The New Jersey & Pennsylvania Concentrating Works.* Highland Lakes, NJ, 2004.

Lief, Alfred. *Harvey Firestone: Free Man of Enterprise.* New York, 1951.

Marshall, David T. *Recollections of Edison.* Boston, 1931.

McDonald, Forrest. *Insull: The Rise and Fall of a Billionaire Utility Tycoon.* Washington, DC, 1952.

Millard, André. *America on Record: A History of Recorded Sound,* 2nd ed. New York, 2005.

Musser, Charles. *Before the Nickelodeon: Edwin S. Porter and the Edison Manufacturing Company.* Berkeley, CA, 1991.

———. *The Emergence of Cinema: The American Screen to 1907.* New York, 1990.

National Park Service. *Historic Furnishings Report: Edison Laboratory.* Harpers Ferry, WV, 1995.

New York Edison Company. *Thirty Years of New York, 1882–1912: Being a History of Electric Development in Manhattan and the Bronx.* New York, 1913.

Newton, James. *Uncommon Friends: Life with Thomas Edison, Henry Ford, Harvey Firestone, Alexis Carrel, and Charles Lindbergh.* New York, 1987.

Parker, Richard G. *School Compendium of Natural and Experimental Philosophy: Embracing the Elementary Principles of Mechanics, Hydrostatics, Hydraulics, Pneumatics, Acoustics, Pyronomics, Optics, Electricity, Galvanism, Magnetism, Electro-Magnetism, Magneto-Electricity, and Astronomy, Containing Also a Description of the Steam and Locomotive Engines and of the Electro-Magnetic Telegraph.* New York, 1856.

Prescott, George. *The Speaking Telegraph, Electric Light, and Other Recent Inventions.* New York 1879.

———. *Electricity and the Electric Telegraph,* 2 vols. New York, 1888.

Rowsome, Frank Jr. *The Birth of Electric Traction: The Extraordinary Life and Times of Frank Julian Sprague.* North Charleston, SC, 2013.

Scott, Lloyd N. *Naval Consulting Board of the United States.* Washington, DC, 1920.

Simil, Vaclav. *Creating the Twentieth Century: Technical Innovations of 1867–1914 and Their Lasting Impact.* New York, 2005.

Slosson, Edwin S. *Creative Chemistry.* London, 1921.

Smoot, Tom. *The Edisons of Fort Myers.* Sarasota, FL, 2004.

Spehr, Paul C. *The Man Who Made Movies: W. K. L. Dickson.* New Barnet, Herts., UK, 2008.

Stamps, Richard, Bruce Hawkins, and Nancy Wright. *Search for the House in the Grove: Archeological Excavation of the Boyhood Homesite of Thomas A. Edison in Port Huron, Michigan 1976–1994.* Rochester, MI, 1994.

Sward, Keith. *The Legend of Henry Ford.* New York, 1948.

Thirty Years of New York: Being a History of Electric Development in Manhattan. NY Edison Co., 1913.

Taylor, Jocelyn Pierson. *Mr. Edison's Lawyer: A Biographical Sketch of the Founder of the Edison Electric Light Company, Grosvenor Porter Lowrey.* Privately printed, 1978.

Venable, John D. *Out of the Shadow: The Story of Charles Edison.* East Orange, NJ, 1978.

Welch, Walter L., and Leah B. S. Burt. *From Tinfoil to Stereo: The Acoustic Years of the Recording Industry, 1877–1929.* Gainesville, FL, 1994.

White, Wallace D. *Milan Township and Village: One Hundred and Fifty Years.* Milan, OH, 1959.

Wile, Raymond E., and Ronald Dethlefson, eds. *Edison Artists and Records, 1910–1929,* 2nd ed. New York, 2012.

Williams, Samuel Crane. *Historical Sketch of the Growth and Development of the Town of West Orange, NJ, 1862–1937.* West Orange, NJ, 1937.

NOTES

The main archive of Edison documents at Thomas Edison National Historical Park (TENHP) in West Orange, New Jersey, comprises some five million pages. Although much of that gigantic collection remains unexplored by scholars, a Rutgers University project, The Papers of Thomas A. Edison, is in the process of editing and publishing the records that relate most closely to Edison and his work. Most of these are housed at TENHP, but the Papers project also includes documents from a wide variety of other repositories. Core material from whatever source is being issued in three related yet dissimilar forms: a multivolume book edition of selected and annotated documents; a larger digital edition accessible online; and a very large microfilm edition available in research libraries. At the time this biography was written, the book edition totaled eight volumes out of a planned fifteen and covered the period 1847–87, roughly half of Edison's eighty-three-year lifespan. (A ninth volume, *Competing Interests: January 1888–December 1889,* is scheduled for publication in 2020.) The digital edition extends the coverage of TENHP documents to 1898. The microfilm edition, structurally similar, extends it to 1919, but this extension is not digitized. The period 1920–31 is still unselected.

Consequently, much of this biography was researched among original documents held at Thomas Edison National Historical Park. But advantage has been taken of the digital edition's supplementary scans (unrestricted as to time period) of virtually all other collections of Edisonia in the United States.

It will be seen from the above-described complexity that referring directly to specific editions of *The Papers of Thomas A. Edison,* let alone to outside repositories embraced by the project, would complicate citations to the point of algebra. For that reason, references in the endnotes relate to

the online index to *The Papers of Thomas A. Edison* at http://edison
.rutgers.edu/, abbreviated henceforth as PTAE. This index is forbiddingly
sophisticated but once mastered directs scholars to the exact document
sought, often in downloadable image form.

An exception to the general citation of PTAE as the gateway to *The
Papers of Thomas A. Edison* is when the book edition offers scholarly editorial commentary unavailable elsewhere. In such instances, it will be simply cited as *Papers*, with volume and page number.

The abbreviation TENHP refers to original documents in the archive at
West Orange. Other source abbreviations are given in the Select Bibliography.

One major collection of Edison papers will remain closed until 2025.
The author is grateful to Edison's great-grandson, Dr. David Edward Edison Sloane, for privileged access to it, and for permission to publish certain
quotations. A minor but historically important collection of love letters
from Edison to his wife, Mina, has for years been withheld without explanation by the board of the Charles Edison Fund of Newark, New Jersey.

Thomas Edison is identified as TE in the notes, and Mina Miller Edison
as MME. Other family members retain their full names.

PROLOGUE (1931)

1. *Key West Citizen,* 11 Feb. 1931; Albion, *Florida Life of Edison,* 169–70; *New
   Castle* (PA) *News,* 11, 12 Oct. 1931.
2. TE to Dr. C. Ward Crompton, 21 Dec. 1921, and to George W. Barton, 10
   Jan. 1923, TENHP; Hammer, "Edison and His Inventions"; TE to Josephus
   Daniels, quoted in New York *Sun,* 11 Oct. 1914; MME to Theodore Edison,
   23 Mar. 1925, PTAE. Asked by a Yale hygienist in 1930 about his tobacco
   habit, TE responded, "Chew constantly + 2 to 3 cigars a day," and said he had
   developed his taste for plug at age ten. TE to Irving Fisher, 12 Oct. 1930,
   TENHP.
3. John Coakley Oral History, 36, Biographical Collection, TENHP; Israel, *Edison,* 10; Dr. Frederick M. Allen to TE, 28 June 1931, TENHP; *Pittsburgh
   Press,* 19 Oct. 1931. In a letter dated 9 Jan. 1931, TE calibrated his milk intake at 375 cubic centimeters, rather less than a pint, every two hours. Just
   when he adopted this all-milk diet is uncertain. He told his correspondent he
   had been on it "for over 8 years," whereas his personal physician stated in a
   posthumous report that TE had given up on solid food "three years ago."
   Bishop William F. Anderson noticed on 23 June 1929 that TE ate nothing at
   dinner, saying that "he had already dined" on milk. In the fall of 1930 TE
   reported that he took "only milk and orange juice: 6 times a day." On 1 Dec.
   1930 he told a former employee, "I live on milk now." On 5 February 1931,

the famed Battle Creek surgeon/dietitian John Harvey Kellogg was unable to persuade him to break this liquid diet. Three months later TE was living, or rather dying, on milk alone. TE to Clifton S. Wady, 9 Jan. 1931, TENHP; William F. Anderson, "A Sunday in the Home of Mr. Thomas A. Edison," ts., ca. 1931, PTAE; TE to Irving Fisher, 12 Oct. 1930, TENHP; Marshall, *Recollections of Edison*, 100; *San Antonio Express,* 19 Oct. 1931.

4. *Fort Myers News-Press,* 11 June 1931.

5. *Piqua* (OH) *Daily Call,* 19 Oct. 1931; Joseph Lewis, "A Visit With Thomas Alva Edison," in *Atheism and Other Addresses* (New York, 1938).

6. Albion, *Florida Life of Thomas Edison, 174.*

7. Narney, *Edison, Modern Olympian,* 16; Paul Kasakove, "Reminiscences of My Association with Thomas Alva Edison," 8–9, TENHP; *Pittsburgh Press,* 2 Aug. 1931.

8. TE's courtship of personal publicity throughout his career had as much to do with his iconic stature as with the things he invented. Three critical studies of TE the self-promoter are Wyn Wachhorst, *Thomas Alva Edison: An American Myth* (Cambridge, MA, 1981); Gordon Hendricks, *The Edison Motion Picture Myth* (1961), in Hendricks, *Origins of American Film*; and David Nye, *The Invented Self: An Anti-Biography from the Documents of Thomas A. Edison* (Odense, Denmark, 1983). While necessarily corrective, these books advance myths of their own.

9. TE in 1925, quoted in Dagobert Runes, ed., *The Diary and Sundry Observations of Thomas Alva Edison* (New York, 1948) 50.

10. TE notebook N-88-01-03.2, PTAE; Israel, *Edison,* 409.

11. TE Patent 1,908,830, approved 16 May 1933.

12. Israel, *Edison,* 461; MME to Theodore Edison, 12 July 1931, PTAE; *Pittsburgh Press,* 3 Aug. 1931.

13. Israel, *Edison,* 461; Finlay, *Growing American Rubber,* 21, 107–8; *Chicago Tribune,* 22 Oct. 1931.

14. *Chicago Tribune,* 22 Oct. 1931; Albion, *Florida Life of Edison,* 176–77; *Jefferson City* (MO) *Tribune,* 21 Oct. 1931.

15. William Edison to TE, 18 Feb. 1931, TENHP.

16. Madeleine Edison Sloane to MME, 28 May 1928, TENHP. The last of TE's grandsons, Michael Edison Sloane, was born on 8 Jan. 1931.

17. This selection is partial. For a full listing, see "Edison Companies" in the Thomas A. Edison Papers website, Edison.rutgers.edu.

18. For TE's merchandising of his own name, see Gitelman, *Scripts, Grooves,* 163–64 and *passim.*

19. Medical bulletins, 2 Aug.–18 Oct. 1931, TENHP; *Oakland Tribune,* 19 Oct. 1931; Newton, *Uncommon Friends,* 6.

20. Albion, *Florida Life of Edison,* 176; Associated Press releases, 4 and 8 Oct. 1931; *Chester* (PA) *Times,* 9 Oct. 1931.

21. Tate, *Edison's Open Door,* 146; William Pretzer, "Edison's Last Breath," *Technology and Culture* 45 (2004).

22. L. W. McChesney, "A Light Is Extinguished," privately printed booklet (approved by Mina and Charles Edison), 4–6, TENHP; *Oakland Tribune,* 20

Oct. 1931. The following account of the ceremonies attending TE's death is based on the official schedule of events, ts., 18 Oct. 1931, Funeral File, TENHP; Charles Edison to Henry Ford, telephone transcript, 19 Oct. 1931, HFM; Nerney, *Edison, Modern Olympian,* 296–305; and Associated Press, United Press, and International News Service reports in syndicated newspapers, 18–22 Oct. 1931 (henceforth AP, UP, and INS).

23. Quoted in *Galveston* (TX) *Daily News,* 20 Oct. 1931.
24. The original aphorism can be found in Helen Zimmern, ed., *Sir Joshua Reynolds' Discourses* (London, 1887), 194.
25. *Pittsburgh Press,* 19 Oct. 1931.
26. Einstein and Ford quoted in AP report, 19 Oct. 1931; *Public Papers of the Presidents of the United States: Herbert Hoover* (1931), 3.362.
27. TE quoted by Lucile Erskine in *St. Louis Post-Dispatch,* 10 Mar. 1912.

PART ONE · BOTANY (1920-1929)

1. TE to William Ores, 24 Jan. 1921, TENHP; TE marginalia in his copy of Helmholtz, *On the Sensations of Tone,* 13–14, TENHP. These notes appear to have been made ca. 1911.
2. In 1862 Kelvin, then Sir William Thomson, defined the second law of thermodynamics as: "Although mechanical energy is *indestructible,* there is a universal tendency to its dissipation, which produces . . . diffusion of heat, cessation of motion, and exhaustion of potential energy through the material universe. The result of this would be a state of universal rest and death." Quoted in Gleick, *Information,* 271.
3. McPartland, "Almost Edison," 214. Confirmation of Einstein's theory by observations of bent starlight during a solar eclipse had been announced by the Royal Society on 9 November 1919.
4. TE's copy of Albert Einstein, *Relativity: The Special and General Theory,* ed. Robert Lawson (New York 1920), 4, TENHP. De Bothezat was an aeronautical scientist in the employ of the U.S. government and an expert on fluid dynamics. His paper, which he sent to TE in typescript (TE General file, 1920, TENHP), appears to be an early draft of a lecture on relativity that he later delivered before the Indiana Association of MIT. *Technology Review* 24 (1922).
5. F. A. Christie, e.g., on John Wesley Powell's *Truth and Error, or the Science of Intellection* in *Unity,* 9 Mar. 1899; "Edison's Views on Life and Death," *Scientific American,* 30 Oct. 1920; TE quoted in "Edison Working on How to Communicate with the Next World," *American Magazine,* Oct. 1920. TE's "units of life" or "entities" perpetuating certain evolutionary characteristics seems to have derived from Darwin's theory of primitive cells built up of "gemmules" that subsequently transubtantiated into organs with specific functions. TE marginal notes on pp. 355–69 of his copy of Darwin's *The Variation of Animals and Plants Under Domestication* (New York, 1899), TENHP. In another such note, TE wrote, "Entities explain everything" over a discussion of the genetic patterns latent in a fertilized egg. T. Brailsford Rob-

ertson, *The Chemical Basis of Growth and Senescence* (Philadelphia, 1923), 197, TENHP. In his copy of Sherwood Eddy's *New Challenges to Faith* (New York, 1926), p. 4, TE defined intelligent behavior as the result of "previous stimuli stored in the organ of memory." For an analysis of TE's deterministic theory of "director" particles, see Anthony Enns, "From Poe to Edison," in Martin Willis and Catherine Wynne, eds., *Victorian Literary Mesmerism* (New York, 2006), 65ff.

6. Richard Outcault interview, *GE Monogram*, Nov. 1928.

7. New York *World*, 17 Nov. 1889; *Papers*, 8.206; Marion Edison Öser, "The Wizard of Menlo Park, by His Daughter," Mar. 1956, TENHP; Dyer and Martin, *Edison*, 773; *The American Magazine*, 78.5 (Nov. 1914).

8. For the reorganization of Thomas A. Edison, Inc., in 1915, see Jeffrey, *Phonographs to U-Boats*, 62 ff.

9. Charles Edison, *Flotsam and Jetsam* (privately printed, New York 1967), introduction; "Charles Edison: From Bohemia to the Boardroom" in Jeffrey, *Phonographs to U-Boats*, 104. See also "Ex-Jersey Governor Was Once a 'Village' Poet," *New York Times*, 27 Sept. 1967.

10. Jeffrey, *Phonographs to U-Boats*, 107–8; Venable, *Out of the Shadow*, 79; "Business Activities of Mark Jones," memo, ca. 1925, 8, TENHP.

11. Jeffrey, *Phonographs to U-Boats*, 107–8; Venable, *Out of the Shadow*, 79–80; Charles Edison to MME, 6 Jan. 1930, TENHP.

12. Charles Edison to MME, 18 Mar. 1920, TENHP.

13. MME to Theodore Edison, 23 Mar. 1920, PTAE.

14. Ibid.; Jeffrey, " 'Commodore' Edison," 33–34.

15. TE interviewed by Edward Marshall, *National Labor Digest*, July 1919; U.S. Senate, 66th Cong., 2nd sess., *Awarding of Medals in the Naval Service: Hearing Before a Subcommittee on Naval Affairs* (Washington, DC, 1920), 546; Josephson, *Edison*, 454; William H. Meadowcroft to Frank Baker, 21 Aug. 1920, TENHP; TE superscript on Henry Lanahan to TE, 3 May 1920, Legal File, TENHP.

16. MME to Theodore Edison, 23 Mar. 1920 and 9 May 1929, PTAE; Lynn Given interview, 20 Mar. 1990, TENHP; TE to George M. Wise, Bombay, ca. July 1920, and TE superscript on J. K. Small to TE, 6 Oct. 1920, TENHP; TE to J. F. Menge, 24 May 1920, TENHP.

17. The following account is taken from an unidentified newsclip, TE General File 1920, TENHP.

18. Ibid.

19. Grant, *Forgotten Depression*, loc. 1425, 1995; Sullivan, *Our Times*, 165.

20. Sullivan, *Our Times*, 176ff.; *New York Times*, 17 Sept. 2003; Grant, *Forgotten Depression*, loc. 1370; McDonald, *Insull*, 209.

21. Israel, *Edison*, 454–55; MME to Theodore Edison, 16 Oct. 1920, PTAE.

22. MME to TE, n.d., ca. Oct. 1920, EFW. The executives fired by TE were John Constable, Frank Fagan, Stephen Mambert, and William Maxwell.

23. MME to Theodore Edison, 18 Oct. 1920, PTAE. TE was still berating his son at the beginning of a family visit in November. "If Papa keeps at Charles as he did at dinner tonight I would not ask him to stay for such a week of torture—

Charles is too sensitive and finely put together to stand it." MME to Theodore Edison, 1 Nov. 1921, PTAE.

24. MME to Theodore Edison, 5 Nov. 1921, PTAE; Grant, *Forgotten Depression,* loc. 1390.

25. TE to C. S. Williams, 5 Jan. 1921, TENHP; Nerney, *Edison, Modern Olympian,* 196; TE to Sherwood Moore, 5 Jan. 1921, TENHP; Jeffrey, *Phonographs to U-Boats,* 109–13, 126; Israel, *Edison,* 455; TE pocket notebook 20-08-04, TENHP. In one, seemingly arbitrary case, TE fired a young college-educated employee, the friend of his electrochemical assistant Paul B. Kasakove. "He's lost the lustre in his eye. You mark my words, he's going to be a very sick man." Kasakove scoffed at this opinion, but several months later the dismissed youth suffered a manic collapse. Kasakove, "Reminiscences," TENHP.

26. Marion Edison Öser to TE, 23 Apr. 1920, TENHP.

27. TE superscript on A. Holt to TE, 5 Mar. 1920, "Family" folder, TENHP; MME to Theodore Edison, 6 Apr. 1920, PTAE; Jeffrey, "Tom and Beatrice."

28. R. W. Kellow to William Edison, 9 Jan. 1920, and William Edison's reply, same date, "Family" folder, TENHP.

29. TE memo, 3 Jan. 1920, and TE superscript on William Edison to R. W. Kellow, 9 Jan. 1920, "Family" folder, TENHP.

30. Miller Reese Hutchison Diary, 1 Jan. 1921, TENHP.

31. TE superscript on Roland Collins to TE, 29 Dec. 1920, TENHP; MME to Theodore Edison, 6, 7, and 12 Jan. 1921, PTAE.

32. Sward, *Legend of Henry Ford,* 114.

33. Charles Edison, *Flotsam and Jetsam,* quoted in Venable, *Out of the Shadow,* 57.

34. Venable, *Out of the Shadow,* 81; unidentified newsclip, datelined New York, 7 May 1921, TENHP.

35. Thomas A. Edison, Inc. (hereafter TAE Inc.) had made a record profit of $2.9 million in 1919. In 1920, this figure fell to $820,000; in 1921, the company lost $1.3 million. The effect of the depression was particularly evident in sales of Edison disk phonographs, which plummeted from 141,907 units to 34,326 in the same period. Millard, *Edison and Business,* 292, 294; Grant, *Forgotten Depression,* loc. 5796, 2948, 2982; Gelatt, *Fabulous Phonograph,* 210.

36. Grant, *Forgotten Depression,* loc. 2790; TE quoted by Charles Edison in Venable, *Out of the Shadow,* 81. For TE's convenient deafness, see, e.g., his interview in New York *Sun,* 27 Aug. 1884.

37. Charles Edison to TE, 12 Sept. 1921, TENHP; Venable, *Out of the Shadow,* 81.

38. One new intellectual venture for TE at this time was a Bryanesque monetary plan, extensively researched, for a replacement of the gold standard by a commodity-based currency. Although it was not taken seriously by contemporary economists, his *Proposed Amendment to the Federal Reserve Banking System* (West Orange, NJ, 1922) has recently been praised as imaginative and anti-inflationary by David L. Hammes in *Harvesting Gold: Thomas Edison's Experiment to Re-Invest American Money* (Silver City, NM, 2012). See also

"Says Edison Beat Bryan Money Plan," *New York Times,* 24 Nov. 1922; Nerney, *Edison, Modern Olympian,* 197–212; and Israel, *Edison,* 446, 527.

39. TE pocket notebook 20-10-15, TENHP. This recording, played on a "William and Mary"-style Edison console, may be heard at https://www.youtube.com/watch?v=ewe-bnrx1kA.

40. TE pocket notebook 23-12-23, TENHP; TE superscript on an article ms. by Francis A. Grant of *Musical America,* ca. Dec. 1920, TENHP; TE memos to Sherwood Moore, 1920–21, TENHP; Kasakove, "Reminiscences," 145; TE pocket notebook 20-08-000; TE patent 1,492,023; Theodore Edison Oral History 1, 118–19, TENHP.

41. Meadowcroft, *Boys' Life of Edison,* 241; MME to Theodore Edison, 8 Apr. 1921, PTAE.

42. Warren G. Harding Inaugural, 4 Mar. 1921, *American Presidency Project,* http://presidency.ucsb.edu.

43. *New York Times,* 5 Aug. 1921; unidentified newsclip, datelined 7 May 1921, TENHP; Venable, *Out of the Shadow,* 81.

44. Israel, *Edison,* 455; *New York Times,* 6 May 1921. See also Dennis, "Edison Questionnaire."

45. T. C. Martin, "Edison a Student at Seventy-Four," *World's News,* 18 June 1921; "Edison Answers Some of His Critics," *New York Times,* 23 Oct. 1921; Newton, *Uncommon Friends,* 7. According to his publicity assistant, John Coakley, TE's papers of choice were the *Times* and staunchly Republican *New York Herald Tribune.* John Coakley Oral History, 36, TENHP.

46. Oliver Lodge folder, Biographical File, TENHP; Sherwood Eddy, *New Passages to Faith* (New York, 1926), 119, copy in TENHP; TE to Myron Herrick, 28 June 1921, TENHP; TE interviewed in *Detroit News,* 26 Oct. 1921; *Electrical Review and Western Electrician,* 10 Nov. 1914.

47. *New York Evening Mail,* 10 May 1921; *New York Tribune* 11, 12 May 1921.

48. *New York Globe,* 12 May 1921; *New York Times,* 13 May 1921; Lancaster (PA) *Intelligencer,* quoted in Dennis, "Edison Questionnaire."

49. *New York Tribune,* 12 May 1921; *Boston Globe,* 15 May 1921; Thomas Edison National Historic Park, "150 Questions" list, Nps.gov; Dennis, "Edison Questionnaire."

50. *New York Times,* 18 May 1921; Einstein folder. Biographical Collection, TENHP. In a 1924 letter, recently discovered, Einstein wrote his sister, "Scientifically I haven't achieved much recently—the brain gradually goes off with age." *Guardian,* 14 Mar. 2018.

51. H. Winfield Secor, "An Interview with Nicola Tesla," *Science and Invention,* Feb. 1922; Edwin R. Chamberlain to TE, 17 May 1921, TENHP.

52. TE superscript on Edwin R. Chamberlain to TE, 17 May 1921, TENHP; Dennis, "Edison Questionnaire." See also "What Do You Know? The Edison Questionnaire," *Scientific American,* Nov. 1921; TE quoted in "Mr Edison's Brain Meter," *Literary Digest,* 28 May 1921.

53. G. W. Plusch to TE, 12 May 1921, TENHP; "Mr Edison's Brain Meter," *Literary Digest,* 28 May 1921; *Washington Times,* 23 May 1921; *Arizona Republic,* 22 May 1921; "Diogenes Looking for a Man that Can Answer a Few

Simple Questions," https:lccn.loc.gov/2016678705; *Boston Globe,* 15 May 1921.

54. TE superscript on H. C. Stratton, 11 May 1921, TENHP; TE quoted in *Newark Evening News,* 14 May 1921. For a comprehensive survey of reaction to what the *San Francisco Examiner* called "Thos. Edison's Mental Teasers," see Dennis, "Edison Questionnaire." See also "The Edison Questionnaire—Its Aim, Its Results, and Its Collateral Significance," *Scientific America* 125 (Nov. 1921).

55. Dennis, "Edison Questionnaire"; *Literary Digest,* 28 May 1921. TE issued two more, equally forbidding questionnaires that summer. See *Chicago Tribune,* 12 May and 30 July 1921.

56. MME to Theodore Edison, 6 July 1921, PTAE.

57. Leland Crabbe, "The International Gold Standard and U.S. Monetary Policy from World War I to the New Deal," *Federal Reserve Bulletin,* June 1989; *Olean* (NY) *Evening Herald,* 12 July 1921; John W. Dean, *Warren G. Harding* (New York, 2004), loc. 1653. Harding was successful in retarding passage of the Bonus Bill in 1921 and successful again in vetoing it the following year.

58. MME to Theodore Edison, 18 July 1921, PTAE. Harding was also a regular visitor to Chautauqua, Mina's family resort. But according to a UP report, this was to be TE's "first meeting" with the president. *Minneapolis Tribune,* 25 July 1921.

59. Lief, *Harvey Firestone,* 208; *Akron Times,* 23 July 1921. Except where otherwise noted, this account of the TE/Harding weekend is derived from Harvey S. Firestone with Samuel Crowther, *Men and Rubber: The Story of Business* (New York, 1926), 228ff., plus eyewitness reports in the July 1921 Clippings File, TENHP. (Harding brought a large press party with him to the campsite.) See also MME to Theodore Edison, 23 July 1921, PTAE, and photographs in the Library of Congress Prints and Photography Division online catalogue, Nos. LC-H27, A3138-3151 and F81-15260.

60. Firestone, *Men and Rubber,* 230.

61. *Canton* (OH) *Daily News,* 24 July 1921.

62. Quoted in Lief, *Harvey Firestone,* 209.

63. Francis Champ Chambrun, "Famous Travelers: Edison, Ford, Firestone," dnr .maryland.gov. Harding's visit with the Vagabonds received enormous publicity, due to the presence of fifteen White House reporters, photographers, and movie cameramen. Although Mina cannot have been gratified by headlines such as "EDISON SLEEPS AS BISHOP PREACHES," many reports noted the president's fascination with TE. Clippings Folder, July 1921, TENHP.

64. Grant, *Forgotten Depression,* chap. 18, *passim.* TE memo, ca. 31 Dec. 1921, pocket notebook 20-08-00, TENHP; Israel, *Edison,* 409, 521; *Edisonian* 5, no. 1 (Winter 2009); *New York Times,* 5 Aug. 1921; MME to Theodore Edison, 20 Nov. 1921, PTAE.

65. Charles Edison to TE, 21 Oct. 1921, TENHP.

66. MME to Theodore Edison, 19 Feb. 1921, PTAE.

67. Charles confirmed this in later life. "If father hadn't stepped in, we would have been completely broke." Venable, *Out of the Shadow,* 81. A report on

the cost of operating Edison Industries at this time put the total at slightly over $1 million per annum, of which $298,832 went to labor and help. Meanwhile there was only $28,000 worth of work in progress. John V. Miller to TE, 30 June 1922, TENHP.

68. H. F. Miller to TE, 31 Dec. 1921, TENHP; Beatrice Edison to MME, 22 June 1922, TENHP; TE to Dr. Gaunt, n.d., 1922, TENHP; John V. Miller to TE, 23 Jan. 1922, TENHP.

69. Sources differ as to the precise extent of TE's 1920–21 purge. The figure given here is that of Charles Edison, who calculated a reduction from 10,000 to 3,000 by 28 Feb. 1922 (Israel, *Edison*, 455.) As early as 5 Aug. 1921, *The New York Times* reported it as 8,000 to just over 1,000. A typescript, "Business Activities of Mark Jones," ca. 1925 (TENHP), gives it as 11,000 to 11,500.

70. TE interviewed in *Detroit News*, 26 Oct. 1921.

71. Edmund Morris, *Dutch: A Memoir of Ronald Reagan* (New York, 1999), 37, 693.

72. "Radio Currents," *Radio Broadcast* 1, no. 1 (May 1922).

73. Millard, *America on Record*, 137–38; TE superscript on L. R. Garretson to TE, 1 Feb. 1922, TENHP.

74. Memo, ca. early 1920s, William Benney folder, Biographical Collection, TENHP.

75. MME to Theodore Edison, 9 Feb. 1922, PTAE.

76. MME to Theodore Edison, 10 May 1922, PTAE.

77. TE to William J. Curtis, 7 May 1920, TENHP.

78. Millard, *America on Record*, 132; DeGraaf, *Edison and Innovation*, 112; TE Patents 1,369,272 and 1,411,425; Harold Anderson Oral History, 7–8, COL; Wile and Dethlefson, *Edison Artists*, 148–50.

79. TE superscript on William D. Johnstone, Jr., to TE, 15 July 1922, TENHP. TE's pocket notebooks, 1920–23, concentrate mainly on recording improvements.

80. *Rubber Age*, 25 Oct. 1922; *Western Canner and Packer*, May 1922; Finlay, *Growing American Rubber*, 53–54, 34, 47; Lief, *Harvey Firestone*, 239. *Rubber Age*, 25 Oct. 1922, "conservatively" estimated current U.S. rubber consumption at 270,000 tons, "eighty percent of which will go into tubes and tires." World production was put at 330,000 tons.

81. *Rubber Age*, 25 Oct. and 10 Nov. 1922; Kendrick A. Clements, *The Life of Herbert Hoover: Imperfect Visionary, 1918–1928* (New York, 2010), x; Silvano A. Wueschner, "Herbert Hoover, Great Britain, and the Rubber Crisis, 1923–1926" (Ebhsoc.org, 2000); Finlay, *Growing American Rubber*, 55–57.

82. *New York Times*, 23 July 1922.

83. Harvey S. Firestone to TE, 9 Jan. 1923, TENHP; Harvey S. Firestone and Samuel Crowther, *Rubber: Its History and Development* (Akron, OH, 1922); TE to Firestone, 16 Jan. 1923, TENHP.

84. "I was astounded at the knowledge of rubber that he had on hand. . . . He told me more than I knew and more than I think our chemists knew." Firestone, *Men and Rubber*, 226–27.

85. TE Patent 60,646; see Part Seven; *Papers*, 2.651, 2.670, and 3.71; "An Hour with Edison," *Scientific American*, 13 July 1878.

86. TE to Firestone, 16 Jan. 1923, TENHP. TE went through the book (preserved in TENHP) in less than a week and added none of his usual marginalia. There was one world map that might have cautioned him, showing that rubber plants flourished nowhere outside the tropics, except for a thin strip of natural guayule in north-central Mexico.

87. TE wrote to ask the British chemist S. J. Peachey's advice when conducting these experiments. Peachey was delighted to cooperate with "the greatest inventor of the age." TE pocket notebook 22-09-30 and 22-12-22; Peachey to TE, 19 Sept. 1921, TENHP; Joseph P. Burke, "Chlorinated Rubber," ts. memo, 24 July 1922, found on the reopening of TE's desk on 23 Aug. 2016, TENHP; Francis S. Schimerka to TE, 24 Nov. 1922, TENHP.

88. Loren G. Polhamus, *Plants Collected and Tested by Thomas A. Edison as Possible Sources of Domestic Rubber* (USDA Agricultural Research Service, ARS 34–74, July 1967), 7, 190; Harvey M. Hall and Frances L. Long, *Rubber Content of American Plants* (Washington, DC, 1921), 60.

89. TE pocket notebook 22-09-30, TENHP (the preceding entry is dated 22 Dec. 1922); Stephen S. Anderson, "The Story of Edison's Goldenrod Rubber: Constructed from the Records of the Original Researchers" (1952), TENHP.

90. According to Karl Ehricke, "we finally got the [sound quality] grading up to almost perfect, 98 percent." Karl Ehricke Oral History 1 (1973) 13, TENHP.

91. Quoted in Bryan, *Edison: The Man*, 102.

92. Except where otherwise indicated, the information in this section derives from "Thomas Edison's Attic," a radio documentary by Gerald Fabris, sound archivist at TENHP, 31 May 2005, https://wfmu.org/playlists/shows/15231; and Jack Stanley, "The Edison 125-foot Horn," a two-part YouTube presentation, https://www.youtube.com/watch?v=yjPzfTAuZNo and https://www.youtube.com/watch?v=2Q4vDBA_38I. Both sources feature contributions from Theodore Edison and Ernest L. Stevens, TE's music director in the 1920s. See also Ernest L. Stevens Oral History (1973), COL.

93. TE quoted in Fabris, "Thomas Edison's Attic." See also Stevens Oral History, 9–12, COL.

94. https://www.nps.gov/edis/learn/photosmultimedia/upload/EDIS-SRP-0198-09.mp3.

95. Quoted in Stanley, "Edison 125-foot Horn." "Father didn't understand mathematics at all." Theodore Edison Oral History 2, 26.

96. Theodore Edison to MME, 25 Mar. 1923, PTAE.

97. MME to Theodore Edison, 4 and 15 Apr. 1923, PTAE.

98. MME to Theodore Edison, 1 May 1923, PTAE.

99. TE pocket notebook 20-08-00, TENHP; MME to Theodore Edison, 1 May 1923, PTAE; TE to Ernest G. Liebold, 16 July 1923, TENHP; *New York Times*, 12 June 1923. For a sample Edisonian "negro joke," see Nerney, *Edison, Modern Olympian*, 244.

100. Theodore Edison to TE, 16 June 1923, TENHP.

101. Ibid.

102. MME to Grace M. Hitchcock, n.d., ca. May 1923, PTAE.

103. Gelatt, *Fabulous Phonograph*, 220, 223; Millard, *America on Record,* 142.

104. At the Edison Association Convention in White Sulphur Springs, 12 Oct. 1922, a representative of the Radio Corporation of America duplicated some of the sparking experiments in Edison's 1875 notebooks, connecting them to a detector-amplifier wireless set and Western Electric loudspeaker. "The signals came through with most gratifying clearness." Edwin W. Hammer to TE, 17 Oct. 1922, TENHP.

105. See Gelatt, *Fabulous Phonograph,* 223–24.

106. TE to Walter H. Miller, 14 Jan. 1926, TENHP ("I do not want to touch this scheme at present. . . . They cannot record without distortion"); *Victor Record Sales Statistics (1901–1941),* mainspringpress.com; Gelatt, *Fabulous Phonograph,* 223–24; TE to Harvey Firestone, 19 Dec. 1923, TENHP.

107. J. V. Miller memorandum, "Laboratory of Thomas A. Edison: Work in Progress," 30 Sept. 1923, TENHP.

108. TE pocket notebook 22-12-22, TENHP.

109. Henry Watts, *Dictionary of Chemistry and the Allied Branches of the Other Sciences* (London, 1883); Vanderbilt, *Edison, Chemist,* 277. Volumes 1 through 6 of TE's copy of Watts (TENHP) are all annotated for "plant extraction" information.

110. TE to Henry Ford, 16 Sept. 1923. The correct per-acre rubber yield using TE's figures would have been 661.39 pounds. He gave no reason for adjusting it upward.

111. Lief, *Harvey Firestone,* 234; Finlay, *Growing American Rubber,* 77; TE to Henry Ford, 16 Sept. 1923.

112. MME to Theodore Edison, 5 Aug. 1923; *Norwalk (OH) Reflector-Herald,* 13 Aug. 1923. TE's personal Lincoln, a dark green convertible, was a gift from Ford for this trip and is preserved at HFM.

113. TE quoted in Henry Fairfield Osborn letter to *New York Times,* 9 Oct. 1931 ("Osborn, you are always in the past in billions of years. . . . I am always thinking of the future"); TE in *T.P.'s Weekly,* 29 Nov. 1907.

114. TE used the phrase "the beautiful hills of Milan" in a letter to his elder sister Marion Edison Page, a longtime resident of the town. She turned it into what was locally regarded as a poem. See "The Birthplace of Edison Dreams of Her Fallen Greatness," *Firelands Pioneer* 13 (1900), 716. The following account of TE's visit to Milan is based on reports in *Norwalk (OH) Reflector-Herald,* 13 Aug. 1923; *Sandusky (OH) Register,* 12 Aug. 1923; and *Sandusky Star-Journal,* 13 Aug. 1923, Image AL00737, ohiohistory.org.

115. Dialogue from *Sandusky (OH) Register,* 12 Aug. 1923.

116. White, *Milan Township and Village,* 16–19.

117. DeGraaf, *Edison and Innovation,* 229; Sandusky (OH) *Star-Journal,* 13 Aug. 1923. On TE's insistence, the house was electrified the following winter. *Norwalk (OH) Reflector-Herald,* 3 Feb. 1924.

118. *Sandusky (OH) Register,* 12 Aug. 1923.

119. Ibid.

120. Harvey S. Firestone, Jr., reminiscence on the death of TE, 19 Oct. 1931, HFM. See note 5 ("units of life") above.

121. Anderson, "Sunday in the Home," PTAE.

122. TE, introduction to William Van der Weyde, ed., *The Life and Works of Thomas Paine* (New Rochelle, NY, 1925).

123. Ibid.

124. Albion, *Quotable Edison,* 108. Albion's volume is notable, in an unreliable field, for its documentation of original sources—in this case, a TE interview with Edward Marshall in *Forum,* Nov. 1927, entitled "Has Man an Immortal Soul?"

125. TE, introduction to Weyde, *Life and Works of Thomas Paine.*

126. TE quoted by John F. O'Hagan, undated clip in 1931 scrapbook, "Edison's Philosophy of Life," TENHP. TE marked a passage in Darwin's *Variation of Animals and Plants,* 230, noting that the Dutch florist Peter Voorhelm had no difficulty distinguishing among twelve hundred varieties of the hyacinth.

127. "Marginalia Project of Books in the Thomas Edison Historical Park Library," ts., TENHP. See also Vanderbilt, *Edison, Chemist,* 283–84.

128. Nerney, *Edison, Modern Olympian,* 15. See, e.g., TE's marginalia in his copy of Darwin's *Variations of Animals and Plants,* 356, 358; TE pocket notebooks, 1924–25 *passim,* TENHP. See also "Some Ideas Sent by Mr. Edison . . . to Captain Coulter, U.S.A.," ca. 22 June 1922, TENHP.

129. TE to Henry Ford, 15 Sept. 1923, TENHP.

130. Vanderbilt, *Edison, Chemist,* 275; Finlay, *Growing American Rubber,* 75.

131. MME to Theodore Edison, 2 July and 12 June 1924, PTAE; Ann Osterhout to MME, 11 Nov. 1924, PTAE; Theodore Edison to Ann Osterhout, 30 Oct. 1924, PTAE. Ann's given name was Anna, but she shortened it shortly after becoming Theodore's fiancée.

132. Winthrop J. Van Leuven Osterhout (1871–1964) was "probably the leading man in the country on plant physiology." Theodore Edison to MME, 14 Mar. 1924, PTAE.

133. MME to Theodore Edison, 26 May 1924, PTAE; A. E. Johnson and Karl Ehricke Oral Histories, 29 Mar. 1971, 26, TENHP. After marrying Theodore, Ann was "at the laboratory almost every day," engaged on colloidal lead research. Ann Edison to MME, 23 Feb. 1926, PTAE. The substance was then believed to be a promising cure for cancer. Wilhelm Stenström and Melvin Reinhard, "Some Experiences with the Production of Colloidal Lead," *Journal of Biological Chemistry* 69 (Aug. 1926).

134. Marion Edison Öser to TE, 24 Sept. 1924 and 6 June 1923, PTAE.

135. Marion Edison Öser to MME, 11 Feb. 1925, PTAE.

136. MME to Theodore Edison, 23 Mar. 1925, PTAE.

137. MME to Theodore Edison, 1 Jan. 1924, PTAE.

138. MME to Theodore Edison, 13 Mar. 1925 and 1 Jan. 1924, PTAE. Mina told another reporter around this time that TE was "the most even man I ever saw. He is happy in his home. . . . But he does not want to be bothered. Nor does

he enter into what is going on around him." *Norwalk* (OH) *Reflector-Herald,* 11 Feb. 1927.

139. "The Most Difficult Man in America," *Collier's Magazine,* 18 July 1925.

140. TE Last Will and Testament, 1 Feb. 1926, TENHP. The will was slightly amended on 30 July 1931, with a codicil redefining the terms by which Theodore and Charles inherited the troubled assets of the Edison Cement Company. TE's disposition of his assets (estimated at $12 million in 1931) favored his two sons by Mina, to the anger of Madeleine, mother of his only grandchildren, while reinforcing the feelings of Marion, Tom, and William that they rated low in his esteem. A predictable intrafamily legal squabble ensued. The current (2018) value of TE's estate would be about $198 million.

141. Charles Edison to MME, 27 Mar. 1927, PTAE ("Our 1926 balance sheet . . . is the best one we ever had, with a ratio of current assets of 11 to 1 & in a very liquid condition"); Newton, *Uncommon Friends,* 7. In effect, TE and Charles swapped titles, TE now becoming chairman of the board of his eponymous company (on 2 Aug. 1926).

142. For the origin of professorial prejudice against TE in 1880, see McPartland, "Almost Edison," 201–4. Ian Wills, "Edison, Science and Artefacts," PhilSci Archive 2007, dates it even earlier, to 1875. http://philsci-archive.pitt.edu /3541/.

143. Raymond C. Cochrane, *The National Academy of Sciences: The First 100 Years, 1863–1963* (Washington, DC, 1978), 284.

144. Notwithstanding his often tongue-in-cheek condemnation of scientists, TE in 1924 compiled a list of those he most admired in history. They were, in order, Faraday, Tyndall, Ampère, Galileo, Newton, Whitney, da Vinci, Becquerel, Bertholet, Darwin, Pasteur, Humboldt, Helmholtz, and Siemens. He included, then deleted, Volta, Ohm, Morse, and Bell, and he made no mention of Hertz or Einstein. TE to Edsel Ford, ca. 26 May 1924, TENHP.

145. TE quoted in Israel, *Edison,* 307; P. B. McDonald to TE, 12 Oct. 1923; TE memorandum, mid-Oct. 1923, TENHP.

146. Michael I. Pupin to P. B. McDonald, 6 Nov. 1923, TENHP.

147. TE to P. B. McDonald, 30 Nov. 1923, TENHP. *Electrical World* never published its proposed article on TE the scientist, and after TE's death Norman Speiden, director of the Edison archives in West Orange, discouraged publication of the McDonald/Edison/Pupin correspondence for fear of "an unsavory argument regarding priority." In 1937, however, Speiden presented the letters, plus a selection of TE's scientific papers, to Harvey Cushing for inclusion in a proposed library of the history of science. "There can be no doubt of their importance," Cushing replied. "I shall treasure them most highly." Cushing to Speiden, 27 Mar. 1937, TENHP.

148. Gelatt, *Fabulous Phonograph,* 221–28; TE introduction to George E. Tewksbury, *A Complete Manual of the Edison Phonograph* (Newark, NJ, 1897), 14.

149. MME to Theodore Edison, 4 May and 29 June 1924, PTAE; Israel, *Edison,* 457. The most detailed account of TAE Inc.'s long and disastrous attempt to

enter the radio business is given in Theodore Edison Oral History 1, 60–76, TENHP.

150. Josephson, *Edison,* 471.

151. American imports of foreign crude rose to 413,338 long tons in 1926, while 77,300 long tons of fresh rubber were registered as "afloat" in December alone. Both figures broke records. U.S. Bureau of the Census, *Record Book of Business Statistics, 1927* (Washington, DC, 1927–29), 48–50.

152. Secretary, Edison Pioneers, to Robert Treat Hotel, 9 Feb. 1927, TENHP; *Edison Monthly,* Mar. 1927.

153. "Edison, Deified but Lonely," *Sandusky* (OH) *Star-Journal,* 11 Feb. 1927.

154. Edison Pioneers 10th Annual Luncheon seating list, 11 Feb. 1927, TENHP. Ford at this time was reportedly the richest man in the world, eclipsing John D. Rockefeller. Gerald Leinwand, *1927: High Tide of the Twenties* (New York 2001), 35.

155. *Portsmouth* (OH) *Daily Times,* 14 Feb. 1924; Frank Lewis Dyer and Thomas Commerford Martin, with the collaboration of William Meadowcroft, *Edison, His Life and Inventions,* rev. ed. (New York, 1929), 813.

156. TE to Henry Ford, 15 Feb. 1927, TENHP.

157. Albion, *Florida Life of Edison,* 116; Newton, *Uncommon Friends,* 10.

158. MME to Theodore Edison, 16 May 1920 and 23 July 1921, PTAE. TE's one-time personal secretary Alfred O. Tate remarked that what really drew TE to Ford "was the sheer bulk of the man's fortune." Sward, *Legend of Henry Ford,* 115.

159. *The International Jew* (Dearborn, MI, 1920), 141 ff.; *Dearborn Independent,* 6 Aug. 1921; TE superscript on Harry A. Harrison to TE, 14 June 1921, TENHP. TE's attitude to Jews was common among many middle-class Christian businessmen during his lifetime. In his view, more casual than bitter, Jews dominated the music, banking, and media industries, and were always after a bargain. Though unquestionably biased by modern standards, he had none of the paranoidal xenophobia displayed by, say, Henry Adams or Henry Ford. His fullest statement on the subject was made to Isaac Markens, author of *The Hebrews in America,* on 15 Nov. 1911: "The Jews are certainly a remarkable people, as strange to me [in] their isolation from all the rest of mankind, as those mysterious people called gypsies—While there are some 'terrible examples' in mercantile pursuits, the moment they get into art, music, & science & literature the jew [sic] is fine. . . . The trouble with him is that he has been persecuted for centuries by ignorant malignant bigots & forced into his present characteristics and he has acquired a 6th sense which gives him an almost unerring judgement in trade affairs—Having this natural advantage & got himself disliked by many as I saw in Europe, I believe that in America where he is free that in time he will cease to be so alarmist & not carry to such extremes his natural advantages. I write you this as I can see from the tone of your book that you are trying to uphold the honor of the Jewish race." Edison Personal folder, 1911, TENHP.

160. *New York Times,* 27 Feb. 1927; Finlay, *Growing American Rubber,* 78, 80–81; *Marion* (OH) *Star* and *Xenia* (OH) *Evening Gazette,* 28 Feb. 1927; 1927 Clippings File, TENHP.

161. *Xenia* (OH) *Evening Gazette,* 28 Feb. 1927; Albion, *Florida Life of Edison,* 122; *Fort Myers Press,* 2 Sept. 1927; TE, "Notes on Rubber Plants and Their Care," June 1927 notebook, TENHP.

162. Finlay, *Growing American Rubber,* 81; TE pocket notebook 27-05-26, TENHP.

163. Gano Dunn, attendee at the 1927 NAS annual meeting, in *News of the Edison Pioneers,* no. 2 (1946). TE's nominator was unidentified.

164. MME to Chautauqua Bird and Tree Club, summer 1928, PTAE; *New York American,* 12 June 1927; Finlay, *Growing American Rubber,* 81–82. See also Lisa Vargues, "In Search of Thomas Edison's Botanical Treasures," New York Botanical Garden, 30 Dec. 2013, http://blogs.nybg.org/science-talk/2013.

165. Josephson, *Edison,* 470; Israel, *Edison,* 457; Finlay, *Growing American Rubber,* 82–84, 87, 262; Vanderbilt, *Edison, Chemist,* 291, 287; TE quoted in MME to Theodore Edison, 21 Aug. 1927, PTAE.

166. MME to Theodore Edison, 21 Aug. 1927, PTAE.

167. Finlay, *Growing American Rubber,* 81. For an evocation of the acrid conflict at the laboratory between TE's "Old Guard" staff and younger executives unable to prevail against them, see Nerney, *Edison, Modern Olympian,* 248–50.

168. Marion Edison Öser, "Wizard of Menlo Park"; Tate, *Edison's Open Door,* 298; *Popular Science,* Dec. 1927.

169. Vanderbilt, *Edison, Chemist,* 286, 291; "Edison Hunting for Rubber in Weeds," *Literary Digest,* 22 Nov. 1927.

170. TE Patent 1,740,079, granted 17 Dec. 1929.

171. Ibid.

172. Ibid.

173. Ibid.

174. Venable, *Out of the Shadow,* 82.

175. MME to Theodore Edison, 13 Jan. 1928, PTAE.

176. *Fort Myers Press,* 13 Jan. 1928; Jerome Osborn reminiscences, Biographical Collection, TENHP; Finlay, *Growing American Rubber,* 21, 92–93.

177. MME to Theodore Edison, 2 Mar. 1928, PTAE; Newton, *Uncommon Friends,* 10–11; Jehl, *Menlo Park Reminiscences,* 1134–35. Ford's purchase of Menlo Park was announced in *New York Times* on 16 Feb. 1928.

178. In Oct. 1922, at the New York Electrical and Industrial Exposition, the Pioneers displayed a "Museum of Edisonia" curated by Frank Wardlaw and Francis Jehl. It featured, along with Hammer's bulb collection, one of the "Jumbo" dynamos from the Pearl Street project, TE's first electric locomotive, and an "original phonograph with its tinfoil records." *Edison Monthly,* Nov. 1922.

179. TE quoted in *Hagerstown* (MD) *Globe,* 2 June 1928.

180. Emil Ludwig, "Edison: The Greatest American of the Century," *American Magazine,* Dec. 1931.

181. TE on 12 June 1928, quoted in L. M. Roberston, "'Inspirations': The Plant Studies of Thomas Alva Edison," ts., TENHP; Finlay, *Growing American Rubber,* 101.

182. Finlay, *Growing American Rubber,* 101; MME to Theodore Edison, 6 June 1928, PTAE.

183. MME to Theodore Edison, 26 Mar. 1928, PTAE; Thomas Jeffrey, Edison family historian, to author, 19 Mar. 2017 ("Carolyn's birth year advanced forward with each decennial census, so that by 1930 she was exactly the same age as Charles"); Ann Edison to MME, 26 Mar. 1927, PTAE; Emil Ludwig 1928 notebook, 20 Feb. (transcribed from German shorthand by Gordon Ludwig), Schweizerisches Literaturarchiv, Bern, Switzerland.

184. MME to Theodore Edison, 26 Mar. 1928, PTAE; Thomas Jeffrey, Edison family historian, to author, 19 Mar. 2017 ("Carolyn's birth year advanced forward with each decennial census, so that by 1930 she was exactly the same age as Charles"); Ann Edison to MME, 26 Mar. 1927, PTAE.

185. TE quoted in *Superior* (WI) *Telegram,* 9 July 1923. The author has, with permission, copied the last thirteen words of this sentence from Jerrard Tickell's *Odette* (London, 1949).

186. During this stay in Fort Myers, TE personally collected 553 specimens and tested a total of 1,756 wild plants for latex. None satisfied him. Anderson, "Story of Edison's Goldenrod Rubber, 10–11, TENHP.

187. *Fort Myers Press,* 13 June 1928; Josephson, *Edison,* 475. On TE the autocratic employer, see, e.g., Frank L. Dyer Diary. 9 and 13 Nov. 1906, TENHP; Hammer, "Edison and His Inventions," II; Tate, *Edison's Open Door,* 294; Jeffrey, *Phonographs to U-Boats,* 10–13; Israel, *Edison,* 454–55.

188. Finlay, *Growing American Rubber,* 86; Nerney, *Edison, Modern Olympian,* 283; William Meadowcroft to Everett Holt, 18 July 1927, TENHP.

189. Millard, *Edison and Business,* 310–11; Jeffrey, "Tom and Beatrice," 6; "Conference re Edison Museum," 20 Aug. 1928, TENHP.

190. Ibid. Ford ultimately spent $30 million on the Edison Institute/Greenfield Village complex, including $3 million on Edison artifacts alone. Douglas Brinkley, *Wheels for the World: Henry Ford, His Company, and a Century of Progress, 1903–2003* (New York, 2003), 377.

191. Nerney, *Edison, Modern Olympian,* 240.

192. TE's copy of Adolf Heil's *The Manufacture of Rubber Goods* (1923) is heavily marked up on pages to do with "the inferiority and inclination to tackiness of certain rubbers," TENHP.

193. *New York Times,* 23 Sept. 1928; Israel, *Edison,* 376–77; Vanderbilt, *Edison, Chemist,* 337.

194. *Brooklyn Daily Eagle, Chicago Tribune,* and *Louisville* (KY) *Courier-Journal,* 21 Oct. 1928.

195. MME to Theodore Edison, 5 Nov. 1928, PTAE.

196. A medical examination of TE by Dr. Frank Sladen, physician-in-chief of the Henry Ford Hospital, Detroit, showed him to be "unusually alert and active mentally," but troubled by acute spasmodic cramps, possibly related to an ulcer or diabetes. Sladen to MME, 14 Aug. 1928, HFM. He sent a complete report to TE the same day.

197. Nerney, born ca. 1880, was a professional librarian and editor who achieved some fame in 1915 when, working as secretary of the NAACP, she chastised the leadership for not reacting strongly enough to the racism of D. W. Griffith's *Birth of a Nation,* 6 Dec. 1915, NAACP Papers, Library of Congress. Her

archival work at TAE Inc. lasted two years. At first Charles Edison intended it to be the basis of an official life of his father, but evidently he found her insufficiently worshipful. After leaving the company's employ, she published her charming if dated *Edison, Modern Olympian.*

198. Nerney, *Edison, Modern Olympian,* 17–18; Ernest L. Stevens Oral History, 15 ("He knew every cuss word in the English language").

199. Bryan, *Edison: The Man,* 272; Nerney, *Edison, Modern Olympian,* 228–229, 237; Walter S. Mallory, "Edison Could Take it," ts. memoir, ca. 1931, TENHP; New York *World,* 6 May 1894; Dyer and Martin, *Edison,* 780; Newton, *Uncommon Friends,* 9.

200. Vanderbilt, *TE, Chemist,* 297.

201. Finlay, *Growing American Rubber,* 95.

202. Laboratory notebook 29-02-01, TENHP; Finlay, *Growing American Rubber,* 94.

203. Samuel Crowther, "Thomas Edison: A Great National Asset," *Saturday Evening Post,* 5 December 1929.

204. The Book of Job, 28:12; Mary C. Nerney Notebook N-28-11-01, 14 Jan. 1929, TENHP.

205. MME to Theodore Edison, 1 Mar. 1929, PTAE; Hubert S. Howe quoted in *Lowell* (MA) *Sun,* 16 Oct. 1931: "Edison suffered from diabetes for forty years [and] never took insulin." When TE's laboratory desk was reopened on 23 Aug. 2016, the author found a pair of unused Luer syringes nestling in a pigeonhole.

206. TE, "Best in My Index," reproduced in Finlay, *Growing American Rubber,* 96–98; TE pocket notebook 29-01-25 and rubber notebook 29-02-01, TENHP. Mina was amused by TE's new habit of talking in botanical Latin. "Our common speech has no place with him. When I think that three years ago he did not know a rose from a turnip, I am astounded. Of course it has to have the promise of rubber in it or he is not interested but when it comes to weeds—he is an authority." To Theodore Edison. 15 May 1930, PTAE.

207. *Los Angeles Times,* 12 Feb. 1929. TE's dyspeptic remark aroused widespread criticism. See, e.g., *Brooklyn Daily Eagle,* 12 Feb. 1929.

208. *Los Angeles Times,* 12 Feb. 1929.

209. Ibid.; *Electrical World,* 16 Feb. 1929.

210. Laboratory notebooks 29-02-00 ("Solidago") and 29-02-01, TENHP.

211. Ibid.; Anderson, "Story of Edison's Goldenrod," 15, TENHP.

212. Fritz, *Bamboo and Sailing Ships,* 35.

213. Anderson, "Story of Edison's Goldenrod," 17–18, TENHP; TE pocket notebook, ca. 1929, TENHP.

214. Ibid. See also TE rubber notebook, 29-02-01, TENHP. During the previous month, TE had succeeded in vulcanizing the rubber of an unidentified plant. "He is as happy and proud of it as a king," MME wrote. "It is the softest sheet rubber I ever felt . . . the first rubber ever produced from weeds." MME to Theodore Edison, 15 Apr. 1929, PTAE.

215. MME to Theodore Edison, 25 Apr. 1929, PTAE; Nerney, *Edison, Modern Olympian,* 283; Fritz, *Bamboo and Sailing Ships,* 28.

216. MME to Theodore Edison, 6 June 1929, TENHP.

217. Allan Sutton, *Edison Blue Amberol Cylinders* (Denver, 2009), xiv; Charles Edison to MME, 22 Feb. 1929, PTAE. For a discussion of the innovative but overdelicate Edison microgroove LP, see Wile and Dethlefson, *Edison Artists,* 122Bff.

218. MME to Theodore Edison, 6 June 1929, TENHP.

219. Ben H. Tongue, "Some Information on the Edison Company's Two-Year-Run in the Radio Business," http://www.bentongue.com/edison/edison.html; MME to Theodore Edison, ca. mid-Dec. 1929, PTAE; Charles Sumner Williams, Jr., to TE, "12.2.29," TENHP. Charles's venture into the radio business, involving the purchase of the Splitdorf Radio Company on 14 Jan. 1929, was a disaster. See below, note 253.

220. Williams to TE, "12.2.29," TENHP. The four drafts, one in Charles's hand, one in Williams's, and two typed, were retained by Mina, who recorded the reactions of Theodore and Ann Edison. Copies supplied to the author by Thomas E. Jeffrey, who notes the possibility that the letter was never seen by TE.

221. Ibid.

222. Dr. Frank Sladen to Frank Campsall, 25 Aug. 1929, HFM; Campsall to Sladen, 26 and 28 Aug. 1929, HFM. TE was still ailing weeks after the Jubilee, "not yet able to put in his usual long hours in the laboratory." Mary Nerney to Charles S. Palmer, 15 Nov. 1929, TENHP.

223. Unless otherwise indicated, the following account of TE's attendance at Light's Golden Jubilee is based on reports in the *Detroit Free Press* and *Chicago Tribune,* 20–22 Oct 1929, and "Miss Nichols Describes Dearborn Light Jubilee," *Manitou Springs* (CO) *Journal,* 14 Nov 1929. Marian Nichols was a member of the Edison family party. Extra visual details come from photographs of the event in HFM and TENHP.

224. Harrisburg (PA) *Evening News,* 19 Oct. 1931.

225. Theodore Edison Oral History, 35–36, TENHP; *Detroit Free Press,* 21 Oct. 1929.

226. Ibid.; Jehl, *Menlo Park Reminiscences,* 1141.

227. *Detroit Free Press,* 21 Oct. 1929.

228. Ibid.

229. Ibid.; *Chicago Tribune,* 20 Oct. 1929; Garet Garrett, "The World Henry Ford Made," unpublished article, Ford biographical file, TENHP; Jehl, *Menlo Park Reminiscences,* 340–41.

230. Jehl, *Menlo Park Reminiscences,* 342.

231. *Chicago Tribune,* 20 Oct. 1929; *Detroit Free Press,* 21 Oct. 1929.

232. *Chicago Tribune,* 20 Oct. 1929; Golden Jubilee box, 1929, TENHP; *Battle Creek* (MI) *Enquirer,* 21 Oct. 1929; *St. Louis Post-Dispatch,* 22 Oct. 1929.

233. Tate, *Edison's Open Door,* 304; Jehl, *Menlo Park Reminiscences,* 1139.

234. *Chicago Tribune,* 22 Oct. 1929.

235. *Detroit Free Press,* 22 Oct. 1929.

236. The following dialogue is taken from the soundtrack of a film record of the

evacuation ceremony, uploaded by the Henry Ford Museum at https://www
.youtube.com/watch?v=ARqyM9nvWuw. Note: the video is wrongly described as having been filmed after the banquet that evening.

237. See Larry Tye, *The Father of Spin: Edward L. Bernays and the Birth of Public
Relations* (New York, 2002), 63–68. While acknowledging Bernays's work in
staging the Golden Jubilee, Tye emphasizes the more important promotional
role played by Henry Ford.

238. Audio quotations from "Edison Golden Jubilee Radio Broadcast," aircheck
fragment, 21 Oct. 1929, https://www.youtube.com/watch?v=tFCX4OybnlY.

239. *Minneapolis Star*, 22 Oct. 1929.

240. *Detroit Free Press*, 22 Oct. 1929.

241. *Chicago Tribune*, 22 Oct. 1929.

242. *Philadelphia Inquirer* and *Cincinnati Inquirer*, 22 Oct. 1929.

243. Madame Curie was shocked by how "very old" TE looked, and by the effort
it cost him to speak. *Marie Curie et ses filles: Lettres* (Paris, 2011), 317.

244. The following transcript of TE's speech at the Jubilee dinner is taken from a
pallophotophone recording resurrected by the Museum of Science in Schenectady, New York, and kindly provided to the author by Chris Hunter. Most
of it can be heard online at https://www.youtube.com/watch?v=G4SbydoXWLg.

245. "Miss Nichols Describes Dearborn Light Jubilee," *Manitou Springs* (CO)
Journal, 14 Nov. 1929.

246. *St. Louis Post-Dispatch*, *Los Angeles Times*, and *Philadelphia Inquirer*, 22
Oct. 1929.

247. *Battle Creek Enquirer*, 22 Oct. 1929.

248. Wilmington (DE) *News Journal*, 22 Oct. 1929.

249. Plainfield (NJ) *Courier-News*, 24 Oct. 1929; Josephson, *Edison*, 481. For a
poignant eyewitness description of TE's return to his desk in the West Orange
laboratory, see Nerney, *Edison, Modern Olympian*, 292.

250. Arthur Walsh circular to the trade, 29 Oct. 1929, facsimile in Wile and Dethlefson, *Edison Artists*, 171. Sales of Edison disk phonographs had declined
from 141,907 in 1921 to 15,320 in 1928, respectively. During the same period, Edison disk records sales fell from 7,721,080 to 495,500. Phonograph
folder, 1929 General File, TENHP.

251. *Los Angeles Times*, 12 Feb. 1929. See MacDonald, *Insull*, 244ff.

252. Sutton, *Edison Blue Amberol Cylinders*, xiv; Israel, *Edison*, 456.

253. Charles's "suicide" threat at this time (see above) may have been aggravated
by TE's criticism of his "extravagance" in building himself a magnificent
house, just as the stock market was collapsing, and along with it TAE Inc.'s
radio venture. MME to Theodore Edison, "Dec. 1929," PTAE. In a telling
anecdote, Ernest L. Stevens recalled TE's ritual arrival at the laboratory on
winter mornings, huddled in his open-sided electric car, followed half an hour
later by "Charlie Edison in a big fat limousine . . . all dolled up in a fur coat
and all." Stevens Oral History, 43, COL.

254. TE, Dec. 1929, quoted in Millard, *Edison and Business*, 322; Polhamus,
Plants Collected, 34–37.

255. *Time*, 16 Dec. 1929. By now TE had identified 13,344 plants representing 2,222 species. TE remained in Fort Myers from 5 Dec. 1929 until 11 June, 1930, concentrating on goldenrod development and giving crucial support to the passage of the Plant Patent Act of 23 May, which for the first time in history recognized plant breeders as inventors. Glenn E. Bugos, "Plants As Intellectual Property: American Practice, Law, and Policy in World Context," Caltech Working Paper 144, May/Oct. 1991. TE continued with rubber research for the rest of the year in West Orange, keeping in fair health and good spirits. By the fall of 1930 he was extracting over 8 percent of rubber from a four-foot goldenrod, while at Fort Myers, the EBRC produced a variety of *Solidago leavenworthii* that grew more than twelve feet tall and gave 12.5 percent leaf rubber on a dry weight basis. TE pocket notebook 30-07-03; Vanderbilt, *Edison, Chemist,* 295. In October, Charles and Theodore Edison accepted the failure of their $2 million attempt to create a viable Edison radio division, blaming it on the deepening depression and insuperable competition. Charles abandoned it on 31 Dec. 1930. In later life he gratefully noted that TE never said "I told you so." Charles Edison to TE, 16 Oct. 1930, TENHP; Theodore Edison Oral History 1, 60–76, TENHP; Venable, *Out of the Shadow,* 83.

PART TWO · DEFENSE (1910–1919)

1. Dyer and Martin, *Edison,* 705, 671–81. For a detailed description of Edison Industries in 1910, including an analysis of its organizational problems, see Millard, *Edison and Business,* 186ff. The following paragraphs are based on these two sources.
2. Nerney, *Edison, Modern Olympian,* 184; Paul S. Lavery interview, 1963, Biographical File, TENHP.
3. By 1910, TE's "business was in the front rank of the larger organizations formed in the era of consolidation in American industry." Millard, *Edison and Business,* 189.
4. Tate, *Edison's Open Door,* 126.
5. Ibid., 128; Josephson, *Edison,* 89.
6. Ralph H. Beach to Francis Jehl, 20 Dec. 1937, HFM.
7. Edward Pleydell-Bouverie, MP, quoted in *Papers,* 6.391.
8. Josephson, *Edison,* 401–3; Tate, *Edison's Open Door.*
9. Francis Jehl to Walter H. Johnson, 16 May 1921; TE to William Meadowcroft, ca. 15. Dec. 1921, TENHP. Jehl got into acute financial difficulties as an émigré electrical engineer in Budapest during World War I. "He appears very selfish," TE remarked when Jehl appealed for assistance in returning home. "After abandoning his country for 30 years he wants her aid."
10. Millard, *Edison and Business,* 202–3, 190–91.
11. Charles D. Lanier, "Thomas Alva Edison, Greatest of Inventors," *Review of Reviews* 8 (July 1893).
12. Miller Reese Hutchison, "My Ten Years with Edison," a collection of dated diary extracts, TENHP (not to be confused with Hutchison's diary proper,

cited as Hutchison Diary); Millard, *Edison and Business,* 218; Robert Traynor, "The Road to the First Electric Portable Hearing Aid . . . [*sic*] and Beyond" (2015), Hearinghealthmatters.org.

13. According to Hutchison, TE successfully tested his Acousticon hearing aid in 1901 but declined to wear it and asked that the test to be kept secret "because his deafness was his chief asset." See, e.g., Hutchison Diary, entries for 28 May 1909, 5 and 8 July 1910; Hutchison Diary extracts for "My Ten Years," TENHP (hereafter Hutchison Extracts). For a detailed account of Hutchison's courtship of TE, see Jeffrey, *Phonographs to U-Boats,* 76–78.

14. Hutchison Extracts 1907–1931, *passim,* TENHP; "Early American Automobiles," http://www.earlyamericanautomobiles.com/massautos.htm.

15. Miller Reese Hutchison, "Transcontinental Address to Thomas A. Edison" (1915), audio file, www.gutenberg.org.

16. From a list of jokes found in TE's desk by National Park Service staff ca. 1980s, superscribed: "WM [Meadowcroft]—Show this to the Old Man, let him have a good laugh over it, and then destroy it."

17. Jones, *Edison: Sixty Years,* 192ff.; *Scientific American,* 13 July 1889; *Engineer,* 18 Dec. 1891; Dyer and Martin, *Edison,* 587–88; Lawrence Goldstone, *Going Deep: John Philip Holland and the Invention of the Attack Submarine* (New York, 2017), loc. 5021–68.

18. Goldstone, *Going Deep,* loc. 2696, 2402.

19. Miller Reese Hutchison, *The Submarine Boat Type of Edison Storage Battery* (Orange, NJ, 1915), 3; "Hutchison Electrical Tachometer," *Railway Master Mechanic,* February 1910.

20. Hutchison, *Submarine Boat Type,* 3.

21. Hutchison, "My Ten Years," TENHP; Hutchison, *Submarine Boat Type,* 3–4.

22. Hutchison, "My Ten Years," entry for 17 July 1910, TENHP.

23. "[Difficulties] appear to give him a high form of intellectual pleasure." Meadowcroft, *Boys' Life of Edison,* 235.

24. Hutchison to TE, 23 Sept. 1911, TENHP.

25. Ibid.

26. TE to Emil Rathenau, 25 May 1911, TENHP.

27. Hutchison, "My Ten Years," entry for 24–26 Aug. 1910, TENHP.

28. Ibid.; *New York Times,* 27 Aug. 1910.

29. Edmund Morris, *Colonel Roosevelt* (New York 2010), 153–57.

30. *New York Times,* 2 Oct. 1910. The following dialogue is quoted entirely from this interview.

31. For a Kantian analysis of the debate between faith and reason in the context of scientific materialism around this time, see William Barrett, *Death of the Soul: From Descartes to the Computer* (New York, 1986), 56–58 and *passim.*

32. TE anticipated the findings of modern neuroscientists that purportedly "random" selections are in fact deliberate choices made by the brain, often hundreds of milliseconds before it persuades itself to the contrary. See P. Haggard, "Human Volition: Towards a Neuroscience of Will," *National Review of Neuroscience* 9, no. 12 (December 2008).

33. In the course of this long interview, TE also discussed vagaries of memory in

the Broca's fold of the brain, Hertzian waves, Brownian motion, and the revelations of the ultra-microscope. "We may, eventually, be enabled to see the inner structure of matter"; Edward Marshall, " 'No Immortality of the Soul' Says Thomas A. Edison," *New York Times,* 2 Oct. 1910. The article may be seen in facsimile at http://query.nytimes.com/mem/archive-free/pdf?res=9903 EEDC1F39E333A25751C0A9669D946196D6CF&mcubz=1.

34. *New York Times,* 13 Oct. 1910. See also "Edison's Views on Immortality Criticized," *Current Literature,* Dec. 1910.

35. *New York Times,* 9 Oct. 1910.

36. Ibid., 17 and 4 Oct. 1910.

37. *New York Times Book Review,* 31 Dec. 1910.

38. MME to Charles Edison, 6 Mar. 1911, TENHP, CEF.

39. John Sloane to Madeleine Edison, 7 July 1911, DSP; Charles Edison to MME, 28 Aug. [1910], CEF.

40. Madeleine Edison to John Sloane, 10 Aug. 1910, DSP; Madeleine Edison superscript on letter to her from William H. Allen, 17 Apr. 1912, DSP.

41. John Sloane to Madeleine Edison, ca. 6 Mar. 1914, DSP.

42. Miller Reese Hutchison to Frank Dyer, 13 Jan. 1911, TENHP; Hutchison, "My Ten Years," entry for 21 Dec. 1910, TENHP.

43. *Boston Herald,* 13 Apr. 1915; Hutchison to TE, 10 Jan. 1911, TENHP.

44. TE to Leonid Mundingo, 19 Apr. 1911, TENHP ("I am very deaf myself and consider it a great advantage as the modern world is so noisy."); Hutchison to John Monnot, 1 Sept. 1911, TENHP.

45. Hutchison to TE, 10 Jan. 1911, TENHP; *American Year Book 1911,* 372; Hutchison, "My Ten Years," entry for 27 Jan. 1911, TENHP; TE to Carlo Pfister, 27 Jan. 1911, TENHP. The German newspaper *Staats-Zeitung* published slightly variant numbers on 1 Aug. 1911, indicating a European (with Russia) total of 169 boats—nineteen more than Hutchison had estimated in January.

46. Hutchison, "My Ten Years," entries for 27 Jan. and 9 Feb. 1911, TENHP. The Italian and Brazilian embassies also responded.

47. Hutchison, "My Ten Years," entry for 9 Feb. 1911, TENHP.

48. Ibid., entry for 21 Jan. 1911: Hutchison to Nixon & Mannock Ltd., 25 Mar. 1911, and to Richard L. Dyer, 13 Jan. 1911, TENHP.

49. Israel, *Edison,* 426; Jeffrey, *Phonographs to U-Boats,* 61.

50. By 1910 the drain on National Phonograph had reached $100,000 a month. Millard, *Edison and Business,* 196.

51. Ibid., 197, 195, 212. Engineers aware of TE's prejudice against disks had to develop an Edison prototype in secret before he was persuaded to take it over himself. Frank Dyer to TE, 9 Nov. 1912, TENHP.

52. Israel, *Edison,* 428; Millard, *Edison and Business,* 197, 203–5.

53. The treasurer of TAE Inc. calculated that its constituent companies were worth $10,329,036 at the time of absorption. TE's personal share of that total was $379,097. Ernest Berggreen to R. G. Dun, 24 Feb. 1911, TENHP. See also Millard, *Edison and Business,* 197–99; Israel, *Edison,* 427–28; Jeffrey, *Phonographs to U-Boats,* 61–62.

54. MME to Charles Edison, 26 Feb. and 2 Mar. 1911, TENHP.

55. Ibid., 7 Oct. 1910, CEF.

56. Welch and Burt, *Tinfoil to Stereo,* 143, 110; Gelatt, *Fabulous Phonograph,* 166–67; MME AP interview, 10 Jan. 1947, TENHP.

57. Welch and Burt, *Tinfoil to Stereo,* 82–83; TE quoted in Jeffrey, *Phonographs to U-Boats,* 40; TE Patent 1,110,428.

58. TE Patents 1,002,505 and 1,119,142.

59. A rival phenolic product, Bakelite, appeared at the same time and involved both Edison and Aylsworth in a prolonged patent infringement suit that the General Bakelite Company eventually lost. See Vanderbilt, *Edison, Chemist,* 240–45.

60. MME to Charles Edison, 7 Oct. 1910 and 6 Mar. 1911, CEF.

61. Georgianne Ensign Kent, *Vartanoosh: My Grandmother's Story* (New York, 2006), 105–16.

62. Ibid., 105.

63. Hutchison, "My Ten Years," entry for 26 Aug. 1910, TENHP; *New York Times,* 27 Aug. 1910. For the complexity of the synchronization problem in the production of early sound movies, see Geduld, *Birth of the Talkies,* 43ff.

64. Hutchison, "My Ten Years," entries for 13 Mar. and 22 Apr. 1911, TENHP; Hutchison Extracts 23 Jan., 6 and 19 Mar., 4 and 26 May 1911, TENHP.

65. Hutchison, "My Ten Years," entries for 28 July, 1 Apr., and 1 May 1911, TENHP.

66. Items 958–994, "Edison's Patents, 1910931," http://edison.rutgers.edu/pat ente6.htm; "T. A. Edison's Color Pictures," *Nickelodeon,* 1910; TE to Johann S. Bergmann, 28 June 1911, TENHP.

67. TE Patent 1,016,875; Hutchison Extracts, 21 July 1911, TENHP.

68. *Sandusky* (OH) *Star-Journal,* 3 Aug. 1911; 1911 Clippings File, TENHP.

69. MME to Charles Edison, Oct. 1912 ("that horrible tobacco smoke that you are compelled to breathe all the time"), TENHP; Venable, *Out of the Shadow,* 13. Charles soon acquired the nickname "Smoke." Jeffrey, *Phonographs to U-Boats,* 98.

70. Years later the diplomat Myron Herrick, a fellow passenger on the *Mauretania,* reminded TE "how, in your innocence, you did not know that your fame had grown and that all the European world was at your feet." Herrick to TE, 28 June 1921, TENHP.

71. Leon Edel, ed., *Henry James Letters,* vol. 4, *1895–1916* (Cambridge, MA, 1984), 579; Venable, *Out of the Shadow,* 12; TE quoted in *People,* 27 Aug. 1911. Peggy James was a Bryn Mawr classmate of Madeleine Edison.

72. *Manchester Guardian* and *Times,* 9 Aug. 1911; "Mr. Edison's Impressions of Europe," ts., 1911, TENHP; Venable, *Out of the Shadow,* 14.

73. Unidentified newsclip, London, 10 Aug. 1911, TENHP.

74. New York *World,* 10 Aug. 1911. TE was presented with a copy of the Parliament Bill signed by Asquith, Lloyd George, and other senior government members.

75. *Huntington* (IN) *Herald,* 10 Aug. 1911.

76. Hutchison to John Monnot, 8 June 1911, TENHP. Except where otherwise

indicated, the following account of TE's European tour is based on the letters of Madeleine Edison to John Sloane, Aug.–Sept. 1911, DSP.

77. 1911 Clippings File, TENHP; Madeleine Edison to John Sloane, 21 Aug. 1911, DSP.

78. New York *World,* 28 and 31 Aug., 1911; *New York Times,* 3 Sept. 1911; Madeleine Edison to John Sloane, 21 and 29 Aug. and 2 Sept. 1911, DSP; Charles Edison's detailed account of this incident in Venable, *Out of the Shadow,* 16–17.

79. MME to Charles Edison, 15 Sept. 1911, TENHP; Madeleine Edison to John Sloane, 15 Sept. 1911, DSP.

80. Francis Jehl to William J. Hammer, 8 Oct. and 1 Nov. 1911, TENHP.

81. Madeleine Edison to John Sloane, 4 Sept. 1911, DSP; "Edison in Hungary and Moravia," *Electrical World,* 7 Oct. 1911.

82. Madeleine to John Sloane, 4 Sept. 1911, DSP.

83. "Inventor Edison's Daughter," *Wilkes-Barre Times Leader,* 28 Nov. 1894; Francis Jehl to William J. Hammer, 8 Mar. 1912, TENHP ("The old man don't seem to assist her a bit in a pecuniary sense"); Josephson, *Edison,* 301.

84. Quoted in *Schenectady* (NY) *Gazette,* 27 Nov. 1931.

85. Theodore Edison Oral History 1, 17–18, TENHP; Madeleine Edison to John Sloane, 19 Sept. 1911, DSP.

86. Madeleine Edison to John Sloane, 19 Sept. 1911, DSP. Charles returned home before the rest of the family to attend the fall semester at MIT.

87. Madeleine Edison to John Sloane, 23 Aug. 1911, DSP.

88. Marion Edison Öser to TE, ca. 24 Feb. 1912, TENHP; Hutchison, "My Ten Years," entry for 7 Oct. 1911, TENHP.

89. MME to Charles Edison, 6 Mar. 1911, TENHP; *Cleveland Plain Dealer,* 8 Oct. 1911.

90. *New Bedford* (MA) *Mercury,* 31 Aug. 1911; "Mr. Edison's Impressions of Europe," 1911, TENHP; *Ogden* (UT) *Standard,* 26 Sept. 1911; *St. Paul* (MN) *Dispatch,* 29 Sept. 1911.

91. *Syracuse* (NY) *Herald,* 1 Oct. 1911; Sigmund Bergmann to TE, 5 and 17 Oct. 1911, TENHP. The article, translated for TE to read, informed him that he was "more inventive than smart." TE's reply to Bergmann has not survived.

92. *New Bedford* (MA) *Mercury,* 31 Aug. 1911; "Edison's Impressions of European Industries," *Scientific American,* 18 Nov. 1911.

93. *Cleveland Plain Dealer,* 8 Oct. 1911; Madeleine Edison to John Sloane, 2 Aug. 1911, DSP; *Pittsburgh Telegraph,* 19 Aug. 1911; *Detroit News,* 20 Aug. 1911.

94. *Cleveland Plain Dealer,* 8 Oct. 1911; Madeleine Edison to John Sloane, 2 Aug. 1911, DSP; *Pittsburgh Telegraph,* 19 Aug. 1911; *Detroit News,* 20 Aug. 1911.

95. Hutchison, "My Ten Years," entry for 7 Oct. 1911, TENHP; *Seattle Star* and *Washington Times,* 18 Oct. 1911. The most detailed account of TE's reaction ("What's the use talking about it until I get it?") is in *St. Louis Post-Dispatch,* 19 Oct. 1911.

96. Israel, *Edison,* 468. TE received only three votes of support from the NAS. At

about the same time and for a more anatomical reason, Marie Curie was rejected for membership of the Institute of France.

97. *The Merchant of Venice,* IV.1; *Edison Monthly,* June 1911; Dyer and Martin, *Edison,* 742.

98. MME to Charles Edison, 27 Oct. 1911, TENHP; Israel, *Edison,* 429.

99. Hutchison, "My Ten Years," entry for 11 Nov. 1911, TENHP; Jeffrey, *Phonographs to U-Boats,* 76–77.

100. Hutchison to TE, entries for 12 Dec. and 2 Nov. 1911, TENHP. According to Paul Kennedy, *The Rise and Fall of the Great Powers* (New York, 1989), 203, this perception in 1911 may have been slightly exaggerated. Kennedy rates the U.S. Navy as third in the world on the eve of World War I.

101. Hutchison, "My Ten Years," entry for 2 Nov. 1911, TENHP; New York *Sun, Washington Post,* and *New York Times,* 3 Nov. 1911.

102. Hutchison, "My Ten Years," entries for 26 and 27 Nov. 1911, TENHP; *Cincinnati Enquirer* and *Marion* (OH) *Star,* 28 Nov. 1911; *Sandusky* (OH) *Star-Journal* and *Houston Post,* 7 Dec. 1911.

103. Hutchison, "My Ten Years," entry for 31 Dec. 1911. TENHP.

104. Ibid., entry for 9 Jan. 1912; Henry Ford, *Edison as I Know Him* (New York, 1930), 13. The adjective *nouveau* is MME's (to Theodore Edison, 16 May 1920, PTAE). Ford's snapshots are preserved in HFM and can be viewed online. The best analysis of the Edison-Ford relationship is in Sward, *Legend of Henry Ford,* 110–15.

105. W. J. Bee to Henry Ford, 6 Apr. 1911, and Bee to W. C. Anderson (Ford's companion on the plant tour), same date, TENHP. Despite an assertion by Matthew Josephson that Ford called upon TE "unannounced" in 1909 (*Edison,* 456), it is clear from these letters that the two men had not met since their initial encounter in 1896.

106. The loan was at 5 percent and understood to be repayable in storage battery sales to Ford. Contract between Ford and TE, 29 Nov. 1912. Original copy plus draft termination agreement, July 1925, R. W. Kellow File, TENHP.

107. TE to Henry Ford, 29 Oct. 1912, HFM.

108. Hutchison Extracts, 14 Feb. 1912, TENHP. See also Jeffrey, *Phonographs to U-Boats,* 82.

109. MME to Charles Edison, 9 Feb. 1912, PTAE. By now, Hutchison was earning 20 percent on all battery contracts.

110. Hutchison to Charles Edison, 15 Apr. 1912. This letter, one of several similar in TENHP, totals six and a half closely typed pages.

111. The most detailed account of the sessions of the Insomnia Squad is Jeffrey, "When the Cat Is Away." See also O. Simmons, "Edison and His Insomnia Squad," *Munsey's Magazine,* Sept. 1916; Jeffrey, *Phonographs to U-Boats,* 5–60.

112. Jeffrey, "When the Cat Is Away," 11; *Edison Phonograph Monthly,* Feb. 1914.

113. TE to Edward H. Johnson, 30 Oct. 1912, TENHP; Carl Wilson to Peter Weber, 9 July and 5 Sept. 1912, TENHP.

114. Carl Wilson to Peter Weber, 5 Sept. 1912, TENHP.

115. The original of this photograph, preserved at TENHP, has conflicting dates of

11 Sept. and 18 Oct. 1912 inscribed on the back. Thomas Jeffrey believes the latter date (added in the 1960s by a Park Service archivist) to be correct, on the assumption that the supper was a celebration of five weeks' successful work. If so, TE and his team look anything but triumphant. The author credits the earlier date, which Norman R. Speiden, the first curator of the Edison Papers, took from an old print "sent down from Glenmont by Mrs. Edison." Jeffrey also doubts the photo was taken at night. However, the sharpness of the shadows, and the increasing radiance of the light emanating from somewhere beyond TE's left elbow, suggest it was indeed artificially lit.

116. Jeffrey, "When the Cat Is Away," 17.

117. TE Patent 1,197,723. See L. I. Schiff, "Motion of a Gyroscope According to Einstein's Theory of Gravitation," *Proceedings of the National Academy of Sciences* 46, no. 6 (1960); TE Patent 1,234,451. The lateness of TE's record-improvement blitz caused a postponement of the introduction of his disk phonograph system. In January 1913 *Edison Phonograph Monthly* ran a cover picture of TE auditing a record and defensively reported that he had undertaken "a long, hard grind" toward perfection. "But those who have heard the sample records . . . agree that it has been more than worth while."

118. The following quotations are taken from Will Irwin, "Why Edison Is a Progressive," *Californian Outlook*, 12 Oct. 1912. The interview, conducted on 17 September, was widely reprinted and made use of in Progressive Party literature. TE declined to speak at meetings in support of Roosevelt, but he contributed three times to his campaign fund. Because of the Republican/Progressive split, Woodrow Wilson was elected. Roosevelt easily defeated Taft in the final vote.

119. See Morris, *Colonel Roosevelt*, 117–19.

120. TE quoted in Irwin, "Why Edison Is a Progressive"; MME to Charles Edison, 19 Jan. 1913, ("Women will become as shrewd and wordly as the men. Everybody following his & her career and home forgotten entirely"), PTAE.

121. The following interview quotations are taken from Lucile Erskine, "Women Will Not Be Men's Equals for 3000 Years," *St. Louis Post-Dispatch*, 10 Mar. 1912.

122. TE to Edward H. Johnson, 30 Oct. 1912, TENHP.

123. Frank Dyer to TE, 9 Nov. 1912, TENHP.

124. Ibid.

125. Announcement typescript, 18 Nov. 1912, TENHP. See also Israel, *Edison*, 433–34.

126. Hutchison Extracts, entry for 1 Jan. 1913, TENHP.

127. Hutchison to Charles Edison, 10 Jan. 1913, TENHP.

128. Except where otherwise indicated, the following summary of the early history of sound pictures is based on Musser, *Emergence of Cinema*, 178–79, 438–39, and Geduld, *Birth of the Talkies*, 31–39.

129. See Bowser, *Transformation of Cinema*, 73.

130. TE patent 1,286,259, filed 6 Mar. 1913. The basic U.S. patent of the Kinetophone was number 1,054,203, "Combination Phonograph and Moving Picture Apparatus," held by TE's employee Daniel Higham. A perfectly preserved

Kinetophone system is on exhibit at the Henry Ford Museum in Dearborn (Image ID: THF 36593).

131. TE superscript on George Harrold to TE, 10 Mar. 1913, TENHP. The cylinder operated a toothed wheel that, acting as an electrode, repeatedly opened and closed a telegraph-like circuit to the synchronizer. An electromagnetic governor on that device transferred the pulsations to the projector. *Buffalo News,* 23 Feb. 1913.

132. There is a detailed description of the Kinetophone projection process (which TE kept secret for fear of patent infringement) in Hutchison to Charles Edison, 13 Feb. 1913, TENHP.

133. "Lecture," footage restored by the Library of Congress from TE's introductory talking picture *The Edison Kinetophone* (1913).

134. *Dunkirk* (NY) *Evening Observer,* 4 Jan. 1913; New York *Sun,* 4 Jan. 1913.

135. *Dunkirk* (NY) *Evening Observer,* 4 Jan. 1913.

136. *New York Times* and *Paterson* (NJ) *Guardian,* 4 Jan. 1913; *Salt Lake City News,* 16 Jan. 1913; *Canonsburg* (PA) *Daily Notes,* 7 Jan. 1913. For an account of another Kinetophone preview that month, held by TE at the Orange Country Club, see Isaac Marcosson, "The Coming of the Talking Picture," *Munsey's Magazine* 48 (Mar. 1913).

137. "Movies Are 'Talkies' Too," New York *World,* 4 Jan. 1913; "Edison Says 'Talkies' Will Replace Movies," syndicated feature in multiple newspapers, 8–9 Jan. 1913. William Edison bid $15,000 on behalf of some "businessmen of high standing" for restaurant rights at talkie venues in New York. "This is no bull con but a straight out and out proposition," he wrote his father. TE informed him that all rights had been disposed of already. William Edison to TE, 16 Jan. and ca. 27 Feb. 1913, TENHP.

138. P. J. Brady quoted in *New York Times* and New York *World,* 4 Feb. 1913; eyewitness account of Ralph H. Beach to Francis Jehl, 20 Dec. 1937, HFM ("[Edison] needed the money, but did not take it"); Hutchison to Charles Edison, 13 Feb. 1913, TENHP; C. H. Wilson to Charles Wetzel, 25 Jan. 1913, TENHP; Hutchison to TE, 16 Jan. 1913, TENHP.

139. In 1921 TE told an interviewer he had been "all on fire to spread this means of education broadcast over our land." *New York Herald,* 15 May 1921.

140. Israel, *Edison,* 420–21; Jeffrey, *Phonographs to U-Boats,* 82ff.

141. Hutchison Extracts, entry for 24 Jan. 1913, TENHP; MME to Charles Edison, 14 Jan. 1913, PTAE; Hutchison to Charles Edison, 13 Feb. 1913, TENHP.

142. MME to Charles Edison, 26 Jan. 1913, PTAE. According to P. J. Brady, the lawyer who represented Dos Passos, the deal included "large royalties" and would ultimately have been worth "several millions" to TE. New York *World,* 4 Feb. 1913.

143. *Newark* (NJ) *Call,* 8 Feb. 1913.

144. Hutchison Diary, entry for 17 Feb. 1913, TENHP; New York *World,* *New York Evening Telegraph,* and *New York American,* 18 Feb. 1913. The *World* reported that the ovation lasted ten minutes. Hutchison, who stayed behind at the Colonial after TE left, timed it at fifteen. "They say it is the greatest hit they have ever had in their theater." Hutchison to TE, 17 Feb. 1913, TENHP.

145. *New York Times,* 18 Feb. 1913. The Library of Congress restoration of this film gives an approximate idea of how Edison talkie technology worked at its best. Although the program is cornball by modern standards, it includes a beautiful performance of "Silver Threads Among the Gold" by the tenor George W. Ballard. There is an unavoidable lapse in synchronism at the end, when "God Save the King" plays in audio while the minstrels mouth the words of "The Star Spangled Banner" in video. This is because the soundtrack derives from an alternative take, filmed for British release.

146. "Instructions for Installing and Operating the Edison Kinetophone Telephone System," ts., 17 Mar. 1913, TENHP.

147. Hutchison to Charles Edison, 13 Feb. 1913, TENHP.

148. Ibid. A diagram of this complex headset-handset apparatus, complete with buzzers, buttons, "horn transmitter," and multicolored line wires, is preserved in TENHP (Motion Picture folder, 19 Feb. 1913).

149. *New York Evening Telegraph,* 18 Feb. 1913; *New York Tribune,* 13 Jan. 1913.

150. *Philadelphia Inquirer,* 22 Feb. 1913; *Ottawa (KS) Evening Herald,* 26 Sept. 1913; *Philadelphia Item,* 23 Feb. 1913; *World Magazine,* 16 Feb. 1913; *Comanche Chief,* 19 Sept. 1913; *New York Evening Telegraph,* 18 Feb. 1913; *Pine Bluff Daily Graphic,* 31 Dec. 1913.

151. TE Patent 1,286,259.

152. *New York Times,* 4 Jan. 1913; *New York Evening Sun,* 18 Feb. 1913. See also "Edison Gets War Scenes for Talking Pictures," *New York Telegram,* 21 Jan. 1913.

153. *New York Tribune* and *Washington Post,* 28 Nov. 1911. TE also planned to make a talking picture of the inauguration of Woodrow Wilson on 4 Mar. 1913. A week later he was reported as confirming that he had shot parts of Wilson's speech (whether live or in a Kinetophone studio was not clear) and would be exhibiting the film in due course. There is no record of him doing so. *Boston Globe,* 24 Feb. 1913; *Pittsburgh Post-Dispatch,* 10 Mar. 1913.

154. TE quoted in *The New York Times,* 12 May 1912; Hutchison to W. H. Ives, 18 Feb. 1913, TENHP. See also "Discovery Communications Realizes Edison's Vision," http://edison.rutgers.edu/connect.htm#5. TE's educational movies project was a natural sequel to a series of semidocumentary, reformist shorts put out by his studio between 1910 and 1913. They covered such subjects as slumlords, tuberculosis, and child labor. Bowser, *Transformation of Cinema,* 45. Hutchison told Ives that TE wanted to "personally direct" scientific and industrial tutorials that would be detailed enough to amount to "self contained textbooks." For the brief rise and consumer-assisted fall of educational/moral movies before World War I, see Bowser, *Transformation of Cinema,* 37ff.

155. Winthrop D. Lane in "Edison Versus Euclid," *Survey,* 6 Sept. 1913; TE to L. H. Putney, 24 Dec. 12, 1912, 1913 Motion Picture File, TENHP.

156. L. H. Putney to TE, 29 Jan. 1913, TENHP (quoting $80 per projector plus $1.50 a day for film rentals); W. H. Ives to TE, 22 Jan. 1913, TENHP. The films in the sample program were not listed, but probably included some of an

entomological series TE had prepared as a test run. Scenario subjects included "The House Fly," "The Various Ways in Which Insects Build," "The Apple Tree Tent Caterpillar," and "Microscopic Plant Life." Hutchison to TE, 11 Mar. 1913, TENHP.

157. L. H. Putney to TE, 29 Jan. 1913, TENHP.

158. Frederick J. Smith, "Looking into the Future with Thomas A. Edison" *New York Dramatic Mirror*, 9 July 1913.

159. "Edison Versus Euclid," *Survey*, 6 Sept. 1913.

160. Ibid.

161. Ibid.; TE memo to Harry F. Miller, ca. 5 May 1913, TENHP. The loan (modern equivalent, $1.3 million) was repayable in four months and secured by $63,000 in Edison Phonograph Works stock. Hutchison Extracts, entry for 6 May 1913, TENHP. Hutchison seems to have expected 6 percent, but a docket in TENHP indicates that the interest was reduced by one point.

162. Minutes of Kinetoscope and Kinetophone Committee eighth meeting, 8 May 1913, TENHP; Smith, "Looking into the Future with Edison"; TE Patent 1,138,360, "Method of Presenting the Illusion of Scenes in Colors," 16 June 1913; Hutchison Extracts, entry for 24 Aug. 1913, TENHP.

163. TE to Geo. F. Morrison, 18 Aug. 1913, TENHP; Elbert Hubbard, *Selected Writings of Elbert Hubbard* (New York, 1922), 106; MME to Charles Edison, 14 Jan. 1913, PTAE.

164. Hubbard, *Selected Writings*, 107; Madeleine Edison to John Sloane, 2 Sept. 1913, DSP; MME to Charles Edison, 10 Sept. 1913, PTAE.

165. MME to Charles Edison, 10 Sept. 1913, PTAE.

166. John H. Greusel, *Thomas Edison: The Man, His Work, and His Mind* (Los Angeles, 1913), 69–70, 12.

167. Diamond Disc retail advertisement, 1913. New York *Sun*, 23 Dec. 1913. Although demonstration models of the disk phonograph had been playing sample records in selected showrooms for more than a year, the Edison Diamond Disc Phonograph and the Diamond Discs themselves were not officially marketed until Dec. 1913. *Edison Phonograph Monthly*, Oct. 1913; *New York Times*, 17 Dec. 1913.

168. Vanderbilt, *Edison, Chemist*, 132; TE superscript on Thomas Wardell to TE, 26 Mar. 1913, TENHP.

169. *Edison Phonograph Monthly*, Oct. 1913; Phonograph folders, *passim*, TE General Files, TENHP.

170. Gelatt, *Fabulous Phonograph*, 166–67. "Blue Amberols . . . when played with an Edison Diamond Reproducer . . . outperformed any other medium of reproduced music then available. The ears in Edison's recording studios were attuned with extraordinary sensitivity to the elements of good sound reproduction." See also Millard, *Edison and Business*, 212.

171. Frank J. Essig to TE, ca. Dec. 1912, TENHP; S. Willard Cutting to TE, 25 Feb. 1913, TENHP.

172. Allan L. Benson, "Edison's Dream of New Music," *Cosmopolitan* 54, no. 5 (May 1913): 797–800.

173. Jones, *Edison: Sixty Years*, 282; TE interview in *Etude* magazine, Apr. 1917.

TE cited his favorite operatic composers as "Bellini, Rossini, Donizetti, and Verdi."

174. TE interview in *Etude*; Benson, "Edison's Dream of New Music"; TE quoted in *Current Literature* 54, no. 4 (Apr. 1913).

175. See, e.g., Thomas Wardell to TE, 26 Mar. 1913, TENHP.

176. TE draft of a stock letter to the phonograph trade, 17 June 1913, TENHP.

177. Samuel Gardner Oral History, 3–4, TENHP.

178. Theodore Edison Oral History, 58–59, TENHP. "His ear curve was such—we had measurements on that—that he lost his upper register almost entirely, so he lost the diction that depends on the hissing consonants very much for being able to understand things."

179. Samuel Gardner Oral History, 6, TENHP.

180. Ibid.

181. Ibid.; Gardner interviewed in Harvith and Harvith, *Edison, Musicians,* 49.

182. Benson, "Edison's Dream of New Music."

183. Ibid.; Madeleine Edison Sloane Oral History, 12, COL. In 1920 TE wrote to the mother of a deaf daughter, "If she rests the upper teeth on the edge of the piano frame she can probably hear the piano." TE superscript on Victor Robb to TE, 29 Mar. 1920, TENHP. Ludwig van Beethoven used a similar technique to hear high frequencies, biting into sticks that he pressed against the keyboard in order to send sound vibrations directly into his brain. Krystian Zimerman, BBC-3 interview, 21 Apr. 2013.

184. Benson, "Edison's Dream of New Music."

185. When TE studied Hermann Helmholtz, *On the Sensations of Tone,* he made exhaustive marginalia on the pages devoted to piano sound. Most listeners, he wrote, "don't hear the *Thump* being intent on the music. I hear all these thumps ½ an octave lower than middle C & hear only the thump sound of last 14 keys it is as loud as a xylophone & yet normal Ears only hear the Musical vibrations and not the Thump. I cannot hear a musical sound above a certain pitch hence the Thump sounds are noise of vibrations of low pitch." TE's copy, 503, TENHP.

186. Benson, "Edison's Dream of New Music"; Meadowcroft, *Boys' Life of Edison,* 247; Meadowcroft to M. Kline, 29 Dec. 1920, TENHP; Ernest L. Stevens Oral History, 21, COL.

187. TE interview in *Etude,* 1 Apr. 1917; TE in *Musician,* May 1916; Vanderbilt, *Edison, Chemist,* 130.

188. Stevens Oral History, 17, COL. See also Ernest L. Stevens folder, Biographical Collection, TENHP. TE eventually sanctioned the release of eight Diamond Disc recordings by Rachmaninoff, and paid him a "special fee" of $147. Although the pianist went on to become a star of RCA/Victor's classical list, he had a special fondness for his early Edison acoustic recordings, and kept a framed copy of the Liszt disk. Edison Records studio logbook, 1919, TENHP. See also Max Harrison, *Rachmaninoff: Life, Works, Recordings* (New York, 2002), 224–25.

189. Harvith and Harvith, *Edison, Musicians,* 44.

190. Welch and Burt, *Tinfoil to Stereo,* 146–49; Venable, *Out of the Shadow,* 25–44.

191. CE to MME, 1 Oct. 1913, PTAE; Venable, *Out of the Shadow,* 25–44; Hutchison Extracts, entry for 1 Feb. 1914, TENHP.

192. TE quoted in *Detroit News,* 26 Oct. 1921; Venable, *Out of the Shadow,* 76.

193. Madeleine Edison to John Sloane, 23 Feb. 1914, DSP; *Fort Myers Press,* 23 Feb. 1914; *New York Times,* 12 Feb. 1914.

194. Madeleine Edison to John Sloane, 27 Feb. and 1 Mar. 1914, DSP.

195. Josephson, *Edison,* 457; Sward, *Legend of Henry Ford,* 50–58.

196. Josephson, *Edison,* 457; TE to Edward N. Hurley, 1 Jan. 1918, HFM.

197. TE to H. E. Heitman, quoted in *Fort Myers Press,* 25 Mar. 1914; Madeleine Edison to John Sloane, 22 Mar. 1914, DSP.

198. Charles Edison to Hutchison, 23 Mar. 1914, TENHP; John Burroughs quoted in *Fort Myers Press,* 9 Apr. 1914; Madeleine Edison to John Sloane, 25 Mar. 1914, DSP.

199. *Fort Myers Press,* 28 Mar. 1914; Madeleine Edison to John Sloane, 30 Mar. 1914, DSP; TE to Hutchison, 14 Jan. 1914, TENHP.

200. Slogan emblazoned an Edison lobby display triptych, photograph image 23.500/14, TENHP; TE quoted in "Talking Singing Whistling Movies," *St. Louis Post-Dispatch,* 23 Feb. 1913.

201. Madeleine Edison to John Sloane, 29 Mar. and 8 Apr. 1914, DSP; MME to Madeleine Edison, ca. 17 June 1914, DSP.

202. Erskine, "Women Will Not Be Men's Equals"; INS Special, 8 Oct. 1911, Clippings File, TENHP.

203. New York *World,* 1 Oct. 1911.

204. TE marginalia in his copy of Allan H. Powles's translation of Bernhardi (New York, 1914), 18 and 34, TENHP. For a fuller statement of his views on the outbreak of the war, which he blamed on a nervous "overreadiness" of Germany's part, see Edward Marshall, "Edison's Plan for Preparedness," *New York Times,* 30 May 1915.

205. "Edison in Wartime," *American Magazine,* Nov. 1914; Diarmuid Jeffreys, *Aspirin: The Remarkable Story of a Wonder Drug* (New York, 2005), loc. 1612.

206. "Edison in Wartime," *American Magazine,* Nov. 1914.

207. Nerney, *Edison, Modern Olympian,* 229.

208. Except where otherwise indicated, the following paragraphs are based on the chapter, "Organic Chemicals and Naval Research," in Vanderbilt, *Edison, Chemist,* 234–58.

209. TE quoted in *Edison Diamond Points,* Feb. 1917; Vanderbilt, *Edison, Chemist,* 250–55; Thomas H. Norton, "A Census of the Artificial Dyestuffs Used in the United States," *Chemical Engineer and Manufacturer* 24, no. 5 (Nov. 1916); *Edison Diamond Points,* Feb. 1917. See also De Graaf, *Edison and Innovation,* 194–96.

210. *New York Times,* 25 Oct. 1914. In the summer of 1915, the U.S. Secret Service investigated a briefcase mistakenly left on a Manhattan-bound train and found evidence of a $100,000 contract to buy and resell Edison phenol to

German-American firms by means of a fraudulent "Chemical Exchange Association." The funds involved came from an espionage account at the German embassy. Edison was embarrassed when the New York *World* broke the story, although he had already committed the rest of his phenol surplus to the U.S. military. Jeffreys, *Aspirin,* loc. 1843–54; U.S. Navy Bureau of Supplies order 23233, 15 Apr. 1915, TENHP.

211. Goldstone, *Going Deep,* loc. 92; House Naval Affairs Committee, *Hearings on Estimates Submitted by Secretary of the Navy, 1914* (Washington, DC, 1914), 634; Jeffrey, "'Commodore' Edison," 13–14.

212. Jeffrey, "'Commodore' Edison," 15–16.

213. Patrick Coffey, *American Arsenal: A Century of Waging War* (New York, 2014), 17; Craig, *Josephus Daniels, passim.*

214. Josephus Daniels, 10 Oct. 1914, Internet Archive audio dub, https://archive.org/details/EDIS-SRP-0191-06.

215. *Washington Times, Brooklyn Daily Eagle,* and New York *Sun,* 11 Oct. 1914.

216. Hutchison, *Submarine Boat Type,* 5; *Washington Times,* 11 Oct. 1914.

217. Ibid.

218. Except where otherwise indicated, the following paragraphs are based on "Report of the Committee on Edison Fire," *Journal of the American Concrete Institute* 11 (Apr. 1915).

219. New York *Sun,* 10 Dec. 1914; Josephson, *Edison,* 430.

220. New York *Sun* and *Plainfield* (NJ) *Courier-News,* 10 Dec. 1914.

221. "Report of the Committee on Edison Fire," 587.

222. Exchange quoted in James Carson, "Anecdotes of my Association with Thomas A. Edison," ts., 1936, TENHP.

223. Jeffrey, *Phonographs to U-Boats,* 63; Charles Edison in Venable, *Out of the Shadow,* 76.

224. *Hearings of the Secretary of the Navy Before the House Naval Committee* (Washington, DC, 1914), 632.

225. DeGraaf, *Edison and Innovation,* 195; "Report of the Committee on Edison Fire," 647.

226. Hutchison to Alfred DuPont, 1 Feb. 1915, TENHP; illustrations in "Report of the Committee on Edison Fire," 609, 616, 618.

227. Israel, *Edison,* 432–33; Venable, *Out of the Shadow,* 75; Josephson, *Edison,* 431; *New York Times,* 10 Dec. 1914; Josephson, *Edison,* 431.

228. For a detailed account of the reorganization of Thomas A. Edison, Inc., after the fire of 1914, see Millard, *Edison and Business,* chap. 11. See also "The Empire of Stephen Babcock Mambert" in Jeffrey, *Phonographs to U-Boats,* 62–67.

229. Edward Marshall, "Edison's Plan for Preparedness," *New York Times,* 30 May 1915; "New Edison-LaFrance Searchlight," *Safety Engineering* 33, no. 5 (May 1917); *Popular Science Monthly* 92, no. 2 (Feb. 1918); TE interviewed in *Film Daily,* 4 Mar. 1927.

230. Jeffrey, *Phonographs to U-Boats,* 80, 88; TE to Josephus Daniels, 11 Feb. 1915, quoted in Jeffrey, "'Commodore' Edison," 18. See ibid., 18ff. for Hutchison's massive publicity blitz in behalf of the S-type battery.

231. Marshall, "Edison's Plan for Preparedness."

232. Ibid.

233. Daniels to TE, 7 July 1915 (combined draft), http://edison.rutgers.edu
/7JulyLetter.pdf. For a full account of the composition of this letter, begun on
31 May, see Jeffrey, " 'Commodore' Edison," 4–9.

234. Daniels to TE, 7 July 1915, original copy, TENHP.

235. Hutchison to Daniels, 12 Sept. 1935, TENHP. See Jeffrey, " 'Commodore'
Edison," 5–6, for the complicated story of the letter's authorship.

236. Scott, *Naval Consulting Board,* 12–13.

237. Israel, *Edison,* 447–48; Jeffrey, " 'Commodore' Edison," 10.

238. Israel, *Edison,* 448; Jeffrey, " 'Commodore' Edison," 10; *American Review of
Reviews,* Sept. 1915.

239. Photograph copy in TENHP; Hutchison to Daniels, 28 Aug. 1915, TENHP
("Dr. B. and Mr. E. are not very congenial"); Vanderbilt, *Edison, Chemist,*
238–89; TE to Frank J. Sprague, 26 May 1878, New York Public Library;
T. C. Martin, "Frank Julian Sprague," *Scientific American,* 21 Oct. 1911; William D. Middleton and William D. Middleton III, *Frank Julian Sprague: Electrical Inventor and Engineer* (Bloomington, IN, 2009), loc. 434, 2614, and
passim; Frank J. Sprague to Leo Baekeland, 16 Dec. 1915, FSP.

240. *Washington Post,* 7 Oct. 1915.

241. Ibid.

242. *Washington Evening Star,* 7 Oct. 1915.

243. The following account is taken from Naval Consulting Board Minutes, 7 Oct.
1915, TENHP.

244. "Remarks of the Secretary of the Navy," Naval Consulting Board Minutes, 7
Oct. 1915, 6–8.

245. *New York Times,* 16 Oct. 1915.

246. Ibid.

247. Ibid.

248. Copy in TENHP; Jeffrey, " 'Commodore' Edison," 13.

249. *New York Times,* 7 Apr. 1915; Hutchison to TE, ca. 18 Jan. 1916, and to
H. B. Brougham, 20 Jan. 1916, TENHP. "The navies of the world," TE announced after the F-4 disaster, ". . . must expect catastrophes so long as they
continue to use sulphuric acid in those vessels." *New York Times,* 27 Mar.
1915.

250. TE to Leo Baekeland, 16 Oct. 2015, TENHP.

251. *San Francisco Chronicle,* 21 Oct. 1915.

252. Ibid.

253. William J. Hammer to Hutchison, 18 Oct. 1915, TENHP.

254. *San Francisco Chronicle,* 21 Oct. 1915; Hutchison, Edison Day script, ts., 16
Oct. 1915, TENHP. The speech can be heard at https://www.nps.gov/edis
/learn/photosmultimedia/upload/EDIS-SRP-0206-01.mp3.

255. Hutchison, Edison Day script; *San Francisco Chronicle,* 21 Oct. 1915.

256. *Los Angeles Times,* 22 Oct. 1915.

257. See Luther Burbank, *The Training of the Human Plant* (New York, 1907).

258. Firestone, *Men and Rubber,* 190–91. A letter from Firestone to TE, 27 Apr.

1915, TENHP, indicates that the two men were hitherto strangers, although they had business relations, and MME's family had known the Firestones in Akron and Chautauqua for many years. Firestone came to the Pacific-Panama Exhibition specifically to "honor" Edison, and join him in a trip to a trade show in San Diego later that month. *San Francisco Chronicle,* 21 Oct. 1915; *Los Angeles Times,* 30 Oct. 1915.

259. TE to E. G. Liebold, 16 Dec. 1927, HFM.

260. *Los Angeles Times,* 28 Oct. 1915; District Court E.D. (PA), *United States v. Motion Pictures Patents Co.,* 1. Oct. 1915, *Federal Reporter* 225, 811.

261. Israel, *Edison,* 454; Bowser, *Transformation of Cinema,* 224.

262. *New York Times,* 7 Nov. 1915.

263. *Los Angeles Times,* 30 Oct. 1915; undated Denver newsclip describing TE on 4 Nov. 1915, Edison 1931 scrapbook, PTAE; TE interviewed in *Los Angeles Times,* 29 Oct. 1915.

264. TE to Leo Baekland, 17 Nov. 1915, FSP; "Hydrogen Given off by Edison Storage Battery," USN Bureau of Steam Engineering memo, 15 Oct. 1915; Hutchison to Daniels, 16 Dec. 1915; both in legal department records, TENHP.

265. Naval Consulting Board Minutes, 27 Dec. 1915, 18, TENHP.

266. Ibid., 9 Feb. 1916, 98, TENHP.

267. Hutchison Extracts, entry for 31 Dec. 1915, TENHP.

268. Robert M. Lloyd to Admiral Nathaniel Usher, 26 Jan. 1916, TENHP.

269. New York *Sun,* 16 Jan. 1916.

270. "Record of Proceedings of a Court of Inquiry . . . into an accident which occurred on board the U. S. submarine at the navy yard, New York, on January 15, 1916," ts., Edison Storage Battery Company Litigation File, 11–17, TENHP, hereafter "Record of Proceedings." There is an annotated plan of the explosion in box 10 of this file.

271. *New York Times,* 16 Jan. 1916; *Wall Street Journal,* 21 Jan. 1916; *Brooklyn Daily Eagle,* 16 Jan. 1916.

272. *New York Times,* 16 Jan. 1916; Hutchison to TE, Jan. 1916, TENHP; *New York Evening World,* 19 Jan. 1916.

273. *New York Times,* 17 Jan. 1916.

274. New York *Sun,* 17 Jan. 1916.

275. *New York Times,* 19 Jan. 1916; "Record of Proceedings," 75–76, 469, 471–72; Hutchison to Adm. R. Griffin, 14 Jan. 1916, TENHP. This long letter, written the day before the disaster, makes plain Hutchison's nervousness about the inadequacy of E-2's current battery ventilation.

276. "Record of Proceedings," 106–7; *New York Evening World* and *Brooklyn Daily Eagle,* 20 Jan. 1916.

277. Ibid.; E. A. Logan, "Chemistry of Primary Galvanic Cells and General Discussion of Storage Cells," *Proceedings of the US Naval Institute* 41 (Sept.–Oct. 1916); Jeffrey, " 'Commodore' Edison," 33. Hutchison told Daniels that a detective in his employ observed "frequent consultations" during the course of the trial between a member of the court and representatives of Edison's principal competitor, the Electric Storage Battery Company.

278. "Record of Proceedings," 439; *New York Times,* 11 Feb. 1916.

279. Ibid., 257.

280. Ibid., 258.

281. Ibid., 282.

282. New York *Sun* and *New York Times,* 25 Feb. 1916. The exchange was stricken from the record.

283. TE to unnamed recipients, 23 Feb. 1916, TENHP.

284. Hutchison to Adm. R. Griffin, 14 Jan. 1916, TENHP.

285. *Brooklyn Daily Eagle,* 15 Feb. 1916; Daniels to Admiral George Burd, 7 Mar. 1916, TENHP; *New York Times,* 17 Jan. 1916.

286. TE to William Meadowcroft, ca. 7 Feb. 1916, TENHP.

287. *New York Times,* 16 Mar. 1916; *Washington Times,* 15 Mar. 1916.

288. *Hearings Before the Committee on Naval Affairs . . . On Estimates Submitted by the Secretary of the Navy, 1916* (Washington, DC, 1916), 3.3344. Hereafter *Hearings, 1916.*

289. *Hearings, 1916,* 3.3354.

290. *Mansfield* (OH) *News-Journal,* 15 Mar. 1916; *Hearings, 1916,* 3.3351.

291. *Washington Post,* 16 Mar. 1916; *Hearings, 1916,* 3.3355.

292. *Hearings, 1916,* 3.3355.

293. *Washington Times,* 15 Mar. 1916.

294. "His happiness over that little grandson is very great." MME to Madeleine Edison Sloane, 26 Mar. 1916, DSP. Madeleine's news coincided less agreeably with another of Beatrice Edison's avowed pregnancies. Her confinement was "expected" around the end of June, but thereafter she and Tom remained childless. Beatrice Edison to MME, 19 June 1916, PTAE; Madeleine Edison Sloane to MME, ca. late Aug. 1916, DSP.

295. TE to Guy Emerson of the Roosevelt Non-Partisan League, 10 May 1916, FSP; *New York Times,* 13 May 1916; Theodore Roosevelt to TE, 13 May 1916, in *The Letters of Theodore Roosevelt* (Cambridge, MA, 1958), 8.1041–42.

296. Morris, *Colonel Roosevelt,* 456–57.

297. Thomas Robins, "Friends in a Lifetime," ts. memoir, 1944, 18, TENHP; MME to Theodore Edison, 9 Apr. 1916, PTAE; DeGraaf, *Edison and Innovation,* 145; TE to Benjamin Tillman, ca. June 1916, TENHP.

298. Jeffrey, *Phonographs to U-Boats,* 91–92.

299. *New York Times,* 4 Sept. 1916; Minutes of the Naval Consulting Board, 19 Sept. 1916, TENHP. TE remarked of Hughes, "His capacity for hindsight, as we learn from his speeches, is highly developed, but as to his foresight, we are not equally well informed."

300. Navy Bill, 1916, copy in TENHP.

301. Naval Consulting Board to Josephus Daniels, 20 July 1916; Daniels to Lemuel Padgett, 21 July 1916, TENHP.

302. Minutes of the Naval Consulting Board, 19 Sept. 1916, TENHP; Naval Consulting Board to Daniels, 20 July 1916; Daniels to Lemuel Padgett, 21 July 1916, TENHP.

303. Josephus Daniels, *The Wilson Era,* vol. 1, *Years of Peace, 1910-1917* (Chapel Hill, NC, 1944), 464.

304. Ibid., 465–66; Morris, *Colonel Roosevelt*, 472.

305. TE Patents 1,297,294, 1,300,708, and 1,300,709; TE Ship Equipment Notes, 20 Oct. 1916, TENHP.

306. TE to Committee on Sites, 4 Oct. 1916, TENHP.

307. Leo Baekeland to Thomas Robins, 13 Oct. 1916, TENHP; TE superscript on Leo Baekeland to Thomas Robins, 13 Oct. 1916, TENHP. TE had earlier prospected Fort Wadsworth and Governors Island in New York Bay and even prospected the Hudson Valley. "On account of the ice, I did not go beyond Tarrytown." TE to Committee on Sites, 4 Oct. 1916, TENHP.

308. "Proposed Report of Committee on Sites," 7 Dec. 1916, TENHP.

309. TE superscript on Hudson Maxim to TE, 10 Dec. 1916.

310. Frank Sprague to TE, 13 Dec. 1916; TE to Daniels, 15 Dec. 1916, ms. draft, both TENHP.

311. Elmer Sperry to TE, 19 Dec. 1916, TENHP.

312. Jeffrey, " 'Commodore' Edison," 28; Hutchison to Louis Howe, 23 Dec. 1916, TENHP.

313. Daniels to TE, 20 Dec. 1916, CHC.

314. TE to Daniels, 22 Dec. 1916, TENHP.

315. TE superscript on Hudson Maxim to TE, 27 Jan. 1917, CHC; *Bridgewater* (NJ) *Courier-News* and *Oakland Tribune*, 4 Jan. 1917; Hutchison Extracts, entry for 31 Dec. 1916, TENHP.

316. Maxim to TE, 6 Feb. 1917, CHC.

317. Rodney Carlisle, "The Attacks on US Shipping that Precipitated American Entry into World War I," 46, www.cnrs.org; Daniels to TE, 3 Feb. 1917, CHC.

318. TE to Daniels, 10 Feb. 1917, CHC; Joseph Fagan, *Eagle Rock Reservation* (Charleston, SC, 2002), 67ff.

319. TE to Dr. Robert Reese, 6 Feb. 1917, CHC.

320. TE memo to self, 14 Feb. 1917, CHC; TE to Daniels, 17 Mar. 1917, CHC; TE to Sir Eric Geddes, n.d., quoted in William Meadowcroft memo, 23 Jan. 1918, TENHP; Scott, *Naval Consulting Board*, 185, 177, 175.

321. TE to Newton D. Baker, 6 Apr. 1917, CHC; Jeffrey, *Phonographs to U-Boats*, 27.

322. Karl T. Compton, "Edison's Laboratory in Wartime," *Science* 75 (1933); Scott, *Naval Consulting Board*, 183.

323. Compton, "Edison's Laboratory," 75.

324. Ibid.; TE to Daniels, 30 Apr. 1917, CHC.

325. Scott, *Naval Consulting Board*, 162.

326. TE to Daniels, 30 Apr. 1917, CHC.

327. War career summary in TE pocket notebook, 27 Jan. 1920, TENHP.

328. TE to Daniels, 28 Mar., 16, 17, and 26 Feb., 14 May, and 9 and 26 July 1917, CHC.

329. TE to Daniels, 6 Mar. 1917, CHC. When one series of experiments became costly, Daniels had to encourage Edison to "go ahead and spend as much money as will be necessary" to complete them. Hutchison to TE, 11 May 1917, TENHP.

330. William Meadowcroft to H. Gernsback, 16 Feb. 1917, TENHP; TE to Dan-

iels, 23 July 1917, CHC; TE superscript in Lucius Beers to TE, 21 Aug. 1917, CHC; Hutchison to TE, 16 Aug. 1917, TENHP.

331. Hutchison to TE, ca. 10 Aug. 1917, TENHP.

332. MME to Madeleine Edison Sloane, 24 Sept. 1917, DSP; Charles B. Harford to Meadowcroft, 13 Mar. 1918, TENHP.

333. Hutchison to TE, 16 Aug. 1917, TENHP; MME to Madeleine Edison Sloane, 3 September 1917, DSP; John Sloane to Madeleine Edison Sloane, 27 Aug. 1917, DSP; Jeffrey, *Phonographs to U-Boats*, 3 and 119; Venable, *Out of the Shadow*, 64–65.

334. William L. Saunders to TE, 17 and 18 Aug. 1917, CHC; Woodrow Wilson to William L. Saunders, 24 Aug. 1917, CHC; TE quoted by Daniels in *News of the Edison Pioneers*, no. 1 (1946).

335. MME to Theodore Edison, 27 Aug. 1917, PTAE.

336. MME to Madeleine Edison Sloane, 24 Sept. 1917, DSP; Charles B. Hanford to William Meadowcroft, 13 Mar. 1918, CHC.

337. TE to Gen. William Crozier, 20 Aug. 1917, TENHP; Newton D. Baker to TE, 22 Aug. 1917, TENHP; TE to Daniels, 19 July 1917, CHC; Theodore Edison to TE, 15 Nov. 1917, TENHP; "Report on Trench Wheel—Experiments," 9 Dec. 1918, TENHP. Theodore also boasted that the wheel would make an excellent dispenser of poison gas. "Report on Trench Wheel." His long illustrated letter explaining this device to his father shows that he was a born inventor. Theodore Edison to TE, 15 Nov. 1817, TENHP.

338. MME to Madeleine Edison Sloane, 24 Sept. 1917, DSP; MME to Theodore Edison, 22 Sept. 1917, PTAE.

339. See, e.g., TE to William L. Saunders, 1 Sept. 1917, TENHP, and Special Board on Naval Ordnance report on TE's turbine-headed shell, 23 Feb. 1917, CHC; Jeffrey, *Phonographs to U-Boats*, 28.

340. Meadowcroft to Reginald Fessenden, 8 Dec. 1917, TENHP; and to Madeleine Edison Sloane, 17 Dec. 19, 17 Oct. 1917, DSP; Daniels to TE, 13 Oct. 1917, CHC; House Resolution no. 4961, copy in TENHP. For a detailed account of Charles Edison's campaign to oust Hutchison from Thomas A. Edison, Inc., see Jeffrey, *Phonographs to U-Boats*, 93–97.

341. Scott, *Naval Consulting Board*, 166.

342. Thomas Robins, "Friends in a Lifetime," ts. memoir, 1944, 13, TENHP. TE's copy of the Admiralty British Isles sea chart for 1913, with soundings marked, is preserved at TENHP.

343. TE to Eric Geddes, 21 Nov. 1917, in Scott, *Naval Consulting Board*, 167–70.

344. TE to Daniels, 21 Nov. 1917, CHC; Madeleine Edison Sloane Oral History, 22, COL.

345. MME to Theodore Edison, 21 Jan. 1918, PTAE.

346. Frank J. Sprague to TE, 30 Jan. 1918, FSP.

347. Ibid.

348. William Meadowcroft quoted in Jeffrey, *Phonographs to U-Boats*, 29.

349. TE quoted in Israel, *Edison*, 450.

350. MME to Charles Edison, ca. 3 Feb. 1918, and to Theodore Edison, 4 Feb. 1918, PTAE.

351. Jeffrey, *Phonographs to U-Boats,* 30; MME to Theodore Edison, 13 May 1918, PTAE.

352. Charles Edison to Madeleine Edison Sloane, 2 Apr. 1918, DSP. There is much discussion of Carolyn Hawkins Edison in the closed papers of the Sloane family, DSP. She was considerably senior to her husband. See Jeffrey, *Phonographs to U-Boats,* 119–20.

353. Venable, *Out of the Shadow,* 66; Charles Edison to Madeleine Edison Sloane, 5 Apr. 1918, DSP.

354. Venable, *Out of the Shadow,* 67.

355. Madeleine Edison Sloane to John Sloane, 7 Aug. 1918, DSP.

356. Joseph F. McCoy Reminiscences, Biographical Collection, TENHP, 31–32.

357. Jeffrey, *Phonographs to U-Boats,* 97. Hutchison prospered briefly, then became a victim of the postwar depression. By the end of 1925 he was down to his last $275. He lived on until 1944, clinging to his title of "Doctor," and never ceasing to bask in the memory of having once moved among with the great. "I spent the happiest days of my life with Edison. I knew him as did no other man." Hutchison Extracts, entry for 31 Dec. 1925 and *passim,* TENHP; *Journal of the Patent Office Society* 19, no. 3 (1937); "Edison's Right Hand," *Kappa Alpha Journal,* Fall 1998; "The Rise and Fall of Miller Reese Hutchison," in Jeffrey, *Phonographs to U-Boats,* 76–97.

358. Jeffrey, *Phonographs to U-Boats,* 28; TE to Daniels, 30 July 1918, JDP; Madeleine Edison Sloane to John Sloane, 7 Aug. 1918, DSP.

359. John Burroughs, "A Strenuous Holiday," in *The Works of John Burroughs* (Cambridge, MA, 1921), 22.119–20. For detailed accounts of the trip, see ibid., 109–26, and Firestone, *Men and Rubber,* 202ff.

360. Hutchison to C. English, 4 Dec. 1918, TENHP; Jeffrey, "'Commodore' Edison," 33–34.

361. Daniels to TE, 6 Nov. 1918, TENHP; TE to Daniels, 14 Nov. 1918, TENHP.

362. William Edison to MME, 25 Nov. 1918, PTAE.

363. TE to Franklin D. Roosevelt, 21 May 1918, TENHP; Franklin D. Roosevelt to TE, 10 Sept. 1918, TENHP.

364. TE to Franklin D. Roosevelt (draft), 10 Sept. 1918, TENHP; Daniels to TE, 6 Nov. 1919, TENHP; TE in New York *World,* 13 Feb. 1923. Many years later Thomas Robins wrote, "Some of his marginal comments on naval letters which I sent him were of such hair-curling nature that I did not care to take the risk of keeping them in the files of the Naval Consulting Board which were sent to Washington after the war." Robins, "Friends of a Lifetime," 17, PTAE.

365. *New York Times,* 14 Oct. 1919; Hutchison press release, ca. Oct. 1919, TENHP.

PART THREE · CHEMISTRY (1900–1909)

1. William Gilmore to W. S. Logue, 6 Feb. 1900, TENHP; William Edison to TE, 6 Feb, 1900, and ca. Sept. 1899, PTAE.

2. Blanche Travers Edison (1879–1946). TE superscript on C. E. Baker to TE, 2 May 1900, TENHP.

3. Charles Edison to MME, ca. May 1913, PTAE.

4. TE Diary, entries for 15 and 17 July 1885, PTAE; Glenmont curator Beth Miller to L. DeGraaf, 11 Jan. 1918.

5. Marie Louise Toobey Edison (1880–1906). *Chautauquan* 20 (Apr.–Sept. 1899).

6. Baltimore *Sun*, 18 Feb. 1899; William Edison to TE, ca. Sept. 1899, TENHP.

7. William Edison to MME, 8 July 1900, TENHP; *New York Evening World*, 10 Feb. 1900.

8. *Boston Globe*, 3 July 1899; Carl Leibinger to TE, 9 May 1900, TENHP.

9. A. A. Friedenstein to TE, 8 May 1900, TENHP.

10. New York *Sun*, 3 Feb. 1903; advertisement by "The Edison Electric Belt Company" in *Des Moines Register*, 29 July 1900; Dyer and Martin, *Edison*, 512; TE Patent 759,356. L. Barton Case advised Edison on 14 May that Tom regretted signing a partnership agreement with Friedenstein, and had no desire to proceed with him. TENHP.

11. Miller Reese Hutchison, *The Edison Storage Battery: A Series of Non-Technical Letters* (Orange, NJ, 1912), 8; Dyer and Martin, *Edison*, 116–17; Walter S. Mallory reminiscences, Biographical Collection, TENHP.

12. Hutchison, *Edison Storage Battery*, 10–11.

13. TE pocket notebook 10-00-00.2, TENHP. See also "How the Edison Battery Started," *Grid*, Jan. 1920, copy in TENHP.

14. Dyer and Martin, *Edison*, 928; https://www.fhwa.dot.gov/ohim/summary95/mv200.pdf. In 1900 New York City's horses dropped twelve hundred metric tons of manure a day.

15. TE pocket notebook, TENHP; Hutchison, *Edison Storage Battery*, 10; "How the Edison Battery Started"; Vanderbilt, *Edison, Chemist*, 235ff. Gleick, *Information*, notes how frequently the word *imponderable* was used by writers on electricity in the nineteenth century (127).

16. TE quoted in Hutchison, *Edison Storage Battery*, 8. See, e.g., "Storage Battery Problems," 30 Mar. 1900: "Nothing but the lead sulphuric acid cell is at all practicable [for automobile traction], and this has been examined physically, chemically, electrically, and mechanically by a great number of leading physicists, chemists, electricians, and engineers."

17. *Rochester Democrat and Chronicle*, 25 Aug. 1901; Vanderbilt, *Edison, Chemist*, 203.

18. Dyer and Martin, *Edison*, 555; TE interviewed in *Grid*,, Jan. 1920.

19. Meadowcroft, *Boys' Life of Edison*, 86; Vanderbilt, *Edison, Chemist*, 206. The title of Albert Einstein's first relativity essay, published in 1905, directly addressed the paradox of mass-energy equivalence: "Does the Inertia of a Body Depend upon Its Energy Content?" *Annalen der Physik*, 21 Nov. 1905.

20. Israel, *Edison*, 297; TE interviewed in *Grid*, Jan. 1920; Vanderbilt, *Edison, Chemist*, 206–7; Walter Holland, "The Edison Storage Battery," *Electrical World* 55, no. 17 (28 Apr. 1910).

21. William Edison to MME, 8 July 1900, PTAE.

22. William Edison to Walter Mallory, 4 Sept. 1900, PTAE.

23. Ibid.

24. TE to Blanche Edison, n.d., ca. Oct. 1900, TENHP.

25. TE Patents 684,204 and 692,507; Ralph D. Pray, "Edison's Folly," www .mining-engineer.com.

26. TE to T. Cushing Daniel, 14 Dec. 1901, TENHP.

27. TE Patent 684,204; Israel, *Edison*, 412–13;Vanderbilt, *Edison, Chemist*, 206–7.

28. Israel, *Edison*, 412–13; Kevin Desmond, *Innovators in Battery Technology: Profiles of 93 Influential Electrochemists* (Jefferson, NC, 2016), 115; M. V. Schoop, "The Jungner Nickel-Iron Accumulator," *Scientific American Supplement* 17 Sept. 1904; TE Patent 670,024 (amplified 12 May 1901, amended 21 Aug. 1902), certified copy in TENHP.

29. *Elektrochemische Zeitschrift* 7 (Aug. 1900) cited by George S. Maynard, *List of References on Storage Batteries, 1900–1915* (New York Public Library guide), 4; TE Journal Subscription List, TENHP.

30. The following description is taken from the text of TE Patent 871,214.

31. Josephson, *Edison*, 408; TE Patent 871,214.

32. TE Patent 871,214.

33. TE Patents 684,204, and 871,214; Vanderbilt, *Edison, Chemist*, 207; Israel, *Edison*, 413. The current price of cadmium was $1.20 per pound, as opposed to four cents a pound for lead.

34. William Edison to TE, 24 Nov. 1900, TENHP.

35. Israel, *Edison*, 390.

36. Dyer and Martin, *Edison*, 508; Walter Mallory quoted in Dyer and Martin, *Edison*, 512–13. See also 921–25.

37. See Israel, *Edison*, 414, for details of this incorporation.

38. Louis E. Bomeisler to TE, 18 Feb. 1901, TENHP.

39. TE superscript on ibid.

40. TE superscript on C. C. Hickock to TE, 22 Feb. 1901, TENHP.

41. D. Van Nostrand & Co. to TE, 5 and 12 Dec. 1900, TENHP; TE to George Iles, 21 Feb. 1901, TENHP.

42. Albion, *Florida Life of Edison*, 42–53; Smoot, *Edisons of Fort Myers*, 52–56.

43. Smoot, *Edisons of Fort Myers*, 59; Albion, *Florida Life of Edison*, 54–55.

44. *St. Louis Republican*, 14 Apr. 1900. See also *The Edisonian*, Vol. 11 at http:// edison.rutgers.edu/newsletter11.htm#7 for an essay on the relationship of TE and Tesla.

45. Ralph H. Beach to Francis Jehl, 20 Dec. 1937, HFM; Marc Raboy, *Marconi: The Man Who Networked the World* (New York, 2016), 155–56.

46. A. Frederick Collins in *Western Electrician*, 24 Aug. 1901.

47. Except where otherwise indicated, this section is based on Arthur E. Kennelly, "The New Edison Storage Battery," *Transactions of the American Institute of Electrical Engineers* 18 (1901), 219ff., and TE Patent 701,804.

48. Vanderbilt, *Edison, Chemist*, 232.

49. Walter Mallory to William Shelmerdine, 1 May 1901, TENHP. The date of TE's decision to drastically modify the New Village plant is left vague ("its installation was nearing completion") in Dyer and Martin, *Edison*, 518. But the sequence of letters covering the period of construction, filed as the Walter

Mallory Papers in TENHP, show that the modification could only have occurred at this juncture. See TE to Harlan Page, below, and also *Cement and Engineering News,* June 1901.

50. Dyer and Martin, *Edison,* 518–19; TE to Harlan Page, 6 May 1901, TENHP. The typed copy of this letter mistakenly adds a zero to each production figure.

51. TE to Harlan Page, 6 May 1901, TENHP.

52. Walter Mallory quoted in Dyer and Martin, *Edison,* 512–14.

53. W. N. Stewart to TE, 5 July 1901, TENHP.

54. TE to W. N. Stewart, 15 July 1901, TENHP.

55. For a complete list of Edison's 147 battery patents, see Edison.rutgers.edu /battpats.htm.

56. Musser, *Before the Nickelodeon,* 176–77; Howard W. Hayes to William Gilmore, 15 July 1901, TENHP.

57. MME to Madeleine Edison, 17 and 18 Aug. 1901, DSP. MME soon returned home from Sudbury, then immediately returned home, leaving TE there in the male company of her brother John V. Miller and two mining associates. A long letter from TE to MME, ca. 25 Aug., TENHP describes their subsequent camping and prospecting adventures. *What Happened on Twenty-third Street* was shot on 23 Aug. 1901.

58. For some documents covering TE's stay in Sudbury, see "Thomas Edison," Sudburymuseum.ca.

59. TE to MME, "Sunday," ca. 25 Aug. 1901, TENHP; *Edison: Invention of the Movies,* DVD 1. The name of the turn-of-the-century Marilyn Monroe was Florence Georgie.

60. "Thomas Edison," Sudburymuseum.ca; Israel, *Edison,* 524; G. A. Aufrecht to TE, 15 Nov. 1901; W. E. Davenport to Edison attorney Howard Hayes, 12 Sept. 1901, TENHP. TE's attempts to sink a workable shaft at Falconbridge in 1902 and 1903 were defeated by layers of quicksand. He eventually abandoned the mine. "Thomas Edison," Sudburymuseum.com.

61. *Edison: Invention of the Movies,* DVD 1.

62. Musser, *Before the Nickelodeon,* 184–90; Israel, *Edison,* 425, 405; "Wonders of New Edison Battery," *Rochester Democrat and Chronicle,* 25 Aug. 1901.

63. Thomas Armat to TE, 15 Nov. 1901, TENHP. For Armat's aggressive defense of his own patents at this time, see Musser, *Emergence of Cinema,* 333.

64. Thomas Armat to TE, 15 Nov. 1901, TENHP.

65. Ibid.

66. TE superscript on ibid; TE to T. Cushing Daniel, 29 Nov. 1901, TENHP; Daniel to TE (with TE superscript reply), 13 Dec. 1901, TENHP.

67. This account of the Marconi dinner derives from *Transactions of the American Institute of Electrical Engineers* 19 (Jan.–July 1902), 93–121.

68. MME to Mary V. Miller, 16 Jan. 1902, EFW.

69. William Edison to MME, 28 Jan. 1902, PTAE.

70. Ibid.

71. "Edison's Sons Under Arrest," *New York Evening World,* 11 Mar. 1902. See also *Chicago Tribune* and *Baltimore Sun,* same date.

72. *New York Evening World,* 11 Mar. 1902.

73. Musser, *Before the Nickelodeon,* 196; Israel, *Edison,* 400–1; Welch and Burt, *Tinfoil to Stereo,* 80ff.; MME to Mary A. Miller, 20 Apr. 1902, EFW.

74. Israel, *Edison,* 415; MME to Mary V. Miller, 25 May 1902, EFW; Millard, *Edison and Business,* 189; Charles Edison reminiscing in Venable, *Out of the Shadow,* 6–12.

75. MME to Mary V. Miller, 25 May 1902, EFW; Madeleine Edison Sloane Oral History, 16, COL.

76. TE superscripts on L. C. Weir of Adams Express Co. to TE, 11 and 22 Apr. 1902, TENHP; Israel, *Edison,* 415.

77. TE, "The Storage Battery and the Motor Car," *North American Review* 174 (July 1902).

78. Ibid.

79. Musser, *Before the Nickelodeon,* 213.

80. Ibid., 233. Eileen Bowser uses the phrase *temporal overlaps* in her commentary on the film in *Edison: Invention of the Movies,* DVD 1. The following account of the action is the author's own. See Musser, *Emergence of Cinema,* 325–29, for a professional analysis.

81. *Edison: Invention of the Movies,* DVD 1. For a full discussion of the myth arising out of this documentary, see "Did Edison Really Execute Topsy the Elephant?," *The Edisonian,* Vol. 11, at http://edison.rutgers.edu/newsletter11htm#7.

82. Joseph McCoy to Howard Hayes, 19 Dec. 1902, TENHP.

83. Memorandum of agreement, 19 Mar. 1899; McCoy to Hayes, 19 Dec. 1902, both TENHP.

84. McCoy to Hayes, 19 Dec. 1902, TENHP.

85. Thomas Edison, Jr., to TE, 20 Dec. 1902, TENHP.

86. TE to MME, 9 Feb. 1898, PTAE; Hayes to Randolph, 8 Jan. 1903, TENHP.

87. Thomas Edison, Jr., to TE, 29 Dec. 1902, TENHP.

88. Ibid.

89. Ibid.

90. *Wilmington* (DE) *Evening Journal,* 21 Jan. 1903. The background to this precautionary suit is confused. When Joseph McCoy first alerted Edison's legal department to the deal between Tom and Stilwell in December 1902, he specifically cited their formation of a phonograph company in competition to Edison's own. By the time the suit was filed in January 1903, most of McCoy's allegations were applied instead to the Thomas A. Edison, Jr., Chemical Company.

91. "Edison Contra Jungner," *Nya Dagligt Allehanda,* 3 Jan. 1903, reprinted in *Horseless Age,* 28 Jan. 1903.

92. Frank Dyer to Brandon Bros., 13 May 1904, TENHP.

93. "Two Years' Rest for Edison," New York *Sun,* 15 Feb. 1903.

94. Smoot, *Edisons of Fort Myers,* 64, 60, 61; MME to Madeleine Edison, 22 Feb. 1903, DSP.

95. *New York Times,* 3 Mar. 1903; *Philadelphia Inquirer* and *Allentown* (PA) *Leader,* 16 Mar. 1903; *Scranton* (PA) *Republican* and *Buffalo Enquirer,* 3 Mar. 1903.

96. MME to Madeleine Edison, 4 Mar. 1903, DSP; Dyer and Martin, *Edison,* 618 (photograph); Nerney, *Edison, Modern Olympian,* 242.

97. Israel, *Edison,* 391; René Rondeau, *Lost to History: Thomas A. Edison, Jr.* (2010), Edisontinfoil.com.; Thomas A. Edison, Jr., to TE, 21 July 1903, TENHP; TE superscript on Thomas Edison, Jr., to TE, 21 July 1903, TENHP.

98. William Edison to TE, 12 July 1903, TENHP; H. F. Miller memorandum, 17 July 1903, department records, TENHP. The loan eventually totaled $2,544 and was strictly divided into monthly notes payable.

99. William Edison to TE, 12 July 1903, TENHP.

100. "The Situation Regarding the Edison Storage Battery," *Electrical Review* 43 (8 Aug. 1903).

101. Ibid.

102. Dyer and Martin, *Edison,* 562–63; Vanderbilt, *Edison, Chemist,* 213.

103. Frank Dyer to Brandon Bros., 7 Dec. 1903, legal department records, TENHP.

104. One of Dyer's first assignments was to transfer the old Edison-Gilliland "grasshopper" wireless patent to Guglielmo Marconi. Edison insisted on this cashless deal, despite a preemptive offer from the Postal Telegraph & Cable Company. "Marconi is responsible for making a success of wireless telegraphy," he said. "I would be the last man to put obstacles in his way."

105. *Motography* 6, no. 1 (July 1911); TE to Frank Dyer, 10 Nov. 1903, TENHP; Musser, *Emergence of Cinema,* 238–39.

106. This paragraph is based on a survey of Dyer's densely detailed diary for 1906, preserved at TENHP.

107. *Edison: Invention of the Movies,* DVD 1.

108. DeGraaf, *Edison and Innovation,* 136.

109. Thomas Edison, Jr., to John Randolph, 17 Dec. 1903, and TE to William Edison, 13 Oct. 1903, both TENHP.

110. William Edison to TE, 17 Oct. 1903; Samuel Scoggins to TE, 27 Dec. 1903; Blanche Edison to TE, 12 Dec. 1903; all TENHP.

111. John Randolph superscript on Blanche Edison to TE, 12 Dec. 1903; William Edison to TE, 16 Dec. 1903, both TENHP.

112. Charles Stilwell to John Randolph, 18 Dec. 1903, TENHP.

113. Frank Dyer to Brandon Bros., 13 Mar. 1904, TENHP.

114. TE to Theodore Roosevelt (draft), 10 Dec. 1903, TENHP.

115. Theodore Roosevelt to Frederick I. Allen, 11 Dec. 1903, TENHP.

116. Dr. L. Sell to TE, 10 Jan. 1904, TENHP.

117. Legal File, 4 Feb. 1904, TENHP; William Edison to John Randolph, 12 Feb. 1904, TENHP; Israel, *Edison,* 416; Vanderbilt, *Edison, Chemist,* 212–13; Israel, *Edison,* 416.

118. Vanderbilt, *Edison, Chemist,* 212–19; Dyer and Martin, *Edison,* 560; Josephson, *Edison,* 414.

119. *Electrical Review* 44, no. 8 (20 Feb. 1904). The following account is taken from Jones, *Edison: Sixty Years,* 262–66. Jones misdates the visit by one month, saying it occurred in June rather than on Saturday 14 May.

120. *Washington Evening Star* and *Buffalo Morning Express,* 15 June 1904; *Indianapolis News,* 16 June 1904.

121. *Washington Evening Star,* 15 June 1904.

122. Tom was cooperating with Post Office investigators prosecuting directors of

the Thomas A. Edison, Jr. Chemical Company for mail fraud. Jeffrey, "Tom and Beatrice," 3; Gilbert King, "Clarence Dally—The Man Who Gave Thomas Edison X-Ray Vision," Smithsonian.com, 14 Mar. 2012.

123. Carolyn T. de la Pera, *Body Electric: How Strange Machines Built the Modern Era* (New York, 2005), 175. Pierre Curie asked Hammer, in return, for a sample of the tungstate of calcium that Edison was using in some private lighting experiments. "Is it prepared in some special way, and can I find it in commerce?" William J. Hammer to TE, 10 Nov. 1903, TENHP.

124. Israel, *Edison*, 422; William Hammer to TE, 10, 20 Nov. 1903, TENHP; Elizabeth Chapin (biographer of TE's gastroenterologist Max Einhorn) to Norman Speiden, ca. 1942, Edison folder, Biographical Collection, TENHP; TE quoted in New York *World*, 3 Aug. 1903 (see ibid. for a detailed account of TE's work with Dally). William S. Andrews, another of the men who worked with Edison on X-rays, died of lingering radiation burns in 1929. Jehl, *Menlo Park Reminiscences*, 684.

125. Vanderbilt, *Edison, Chemist*, 212; Israel, *Edison*, 415–17; Frank Dyer to Dr. L. Sell, 28 Nov. 1905, PTAE; Nerney, *Edison, Modern Olympian*, 180; Dyer and Martin, *Edison*, 566.

126. *New York Times*, 24 Jan. 1905; Nerney, *Edison, Modern Olympian*, 180. A recording of TE telling the quoted story in 1906 is available from Michigan State University's Vincent Voice Library at http://archive.lib.msu.edu/VVL/dbnumbers/DB500.mp3.

127. *Akron Beacon Journal*, 25 Jan. 1905; Frank Dyer to Dr. L. Sell, 28 Nov. 1905, TENHP; TE Patent 252,932.

128. TE/Aylesworth Patent 976,791.

129. Frank Dyer to Meffert and Sell, 28 Nov. 1903.

130. Thomas E. Jeffrey, "Beatrice Heyzer Edison and the Heyzer Family," unpublished research note, 2018; Jeffrey, "Tom and Beatrice," 3. According to Jeffrey, historian of the Edison family, the weight of reliable evidence indicates that Beatrice was born Matilda R. Heyzer in New York in 1874. In later life she would identify herself as Beatrice La Montagne Edison, the Kentucky-born daughter of Charles La Montagne-Hazeur, M.D. Beatrice Edison, "Brief Biography," 30 Sept. 1929, TENHP.

131. Jeffrey, "Beatrice Heyzer Edison." See also *New York Times*, 11 Jan. 1906.

132. C. Wilmot Townsend to John Randolph, 23 Sept. 1903, TENHP.

133. A. E. R. Laning to John Randolph, 27 Sept. 1905, TENHP.

134. Thomas Edison, Jr., to TE, 22 Nov. 1905, TENHP.

135. Ibid.

136. Ibid.

137. John Randolph to "Burton Willard," 11 Dec. 1905, TENHP; TE superscript on Thomas Edison, Jr., to TE, 22 Nov. 1905, TENHP.

138. William Edison to TE, 16 Dec. 1905, TENHP.

139. Ibid.

140. Frank Dyer Diary, entry for 2 Jan. 1906, TENHP.

141. Ibid., entries for 30 Nov. and 25 July 1906, TENHP. Only the 1906 volume of Dyer's diary appears to have survived.

142. Ibid., entry for 5 Jan. 1906, TENHP.

143. Ibid., entry for 8 Feb. 1906, TENHP; *Boston Post,* 23 Feb. 1906. Press reports gave the cause of death as an unidentified, two-day "illness." Jeffrey declines to accuse anyone of foul play but notes that Beatrice "allegedly once tried to poison her own sister—the allegations coming from the sister herself." Jeffrey, "Tom and Beatrice," 5.

144. *New York Times,* 22 Feb. 1906; A. G. Cottell, Undertaker and Embalmer, to TE, 20 Feb. 1906, TENHP ("The family appreciates your generosity"); Frank Dyer Diary, entry for 21 Feb. 1906.

145. TE memo to John Randolph, n.d., ca. mid–Nov. 1905, TENHP; Frank Dyer Diary, entries for 19 Feb. and 9 July 1906, TENHP. The wedding was held in Trenton and attended only by members of the Heyzer family. Tom signed the register with his real name, while Beatrice did so as "Miss Beatrice Matilda Heyzer." This did not fool reporters, who identified her as "Mrs. Thomas Montgomery." Upon returning to Burlington, the newlyweds once again became "Burton and Beatrice Willard." *St. Louis Post-Dispatch,* 11 July 1906; Jeffrey, "Beatrice Heyzer Edison."

146. Frank Dyer Diary, entries for 12 and 9 Oct. 1906, TENHP; Benjamin M. Dugger, *Mushroom Farming* (New York, 1920), *passim.*

147. Frank Dyer Diary, entry for 9 Oct. 1906, TENHP.

148. Ibid., entry for 4 Sept, 1906, TENHP; Musser, *Before the Nickelodeon,* 329–30; TE superscript on A. S. Cushman to TE, ca. 29 Oct. 1906, TENHP.

149. Dyer and Martin, *Edison,* 516–17; Vanderbilt, *Edison, Chemist,* 187; TE quoted in *Washington Evening Star,* 21 Feb. 1906.

150. Bryan, *Edison: The Man,* 254–55; Dyer and Martin, *Edison,* 937–42; "The Building Materials of the Future," *Insurance Engineering* 1, no. 3 (June 1901).

151. "Building Materials of the Future."

152. "Edison Plans a Revolution in Building Houses," *New Castle* (PA) *Herald,* 10 Aug. 1906; TE quoted in *Price* (UT) *News-Advocate,* 27 Dec. 1906. See especially Brian Charlton, "Cement City: Thomas Edison's Experiment with Worker's Housing in Donora," *Western Pennsylvania History,* Fall 2013.

153. "Edison Plans Revolution in Building Houses."

154. *Papers,* 1.652, 642–43. The device referred to was the Edison Universal Stock Printer, U.S. Patent 126,532.

155. *Papers,* 1.638, 2.784; Musser, *Before the Nickelodeon,* 377 and 433ff.; Musser, *Emergence of Cinema,* 450–51; Josephson, *Edison,* 401–2; "Frank L. Dyer," *Electrical World,* Nov. 1910; De Graaf, *Edison and Innovation,* 139. See especially Robert J. Anderson, "The Motion Picture Patents Company," Ph.D. diss., University of Wisconsin, 1983.

156. The phrase *sweet savor of wax,* along with other details in this sentence, comes from Leroy Hughbanks's memoir, *The Story of the Phonograph,* 34–35.

157. MTE Patent 713,209 (11 Nov. 1902); De Graaf, *Edison,* 106; Welch and Burt, *Tinfoil to Stereo,* 84–85; Vanderbilt, *Edison, Chemist,* 129; Millard, *Edison and Business,* 194–95; MME to Theodore Edison, 13 May 1909, PTAE.

158. *Oakland Tribune,* 26 June 1909. See, e.g., Holland, "The Edison Storage Bat-

tery"; C. W. Bennett and H. N. Gilbert, "Some Tests of the Edison Storage Battery," *Chemical and Metallurgical Engineering* 11 (1913).

159. Hutchison, *Edison Storage Battery*, 20–22.

160. Israel, *Edison,* 419; Dyer and Martin, *Edison,* 935 (magnified illustration) and 931; Hutchison, *Edison Storage Battery*, 16.

161. Dyer and Martin, *Edison,* 926–36; TE in *Oakland Tribune,* 26 June 1909; Fred H. Colvin, "The Mechanics of the Edison Battery," *American Machinist,* 10 Aug. 1911; Hutchison, *Edison Storage Battery*, 17–19.

162. TE to Robert Bachman, quoted in Hutchison, *Edison Storage Battery*, 24; TE in *Oakland Tribune*, 26 June 1909.

163. Israel, *Edison,* 420–21.

164. TE quoted in Dyer and Martin, *Edison,* 608; Elizabeth Wadsworth to TE, 16 May 1905, TENHP.

165. Thomas Edison, Jr., to Mary V. Miller, 15 Mar. 1910, EFW.

166. William Edison to Harry Miller, 3 Ang. 1911, TENHP; Frank Dyer to William Edison, 4 May 1909, TENHP.

167. "Thomas A. Edison," *Fra: A Journal of Affirmation* 5, no. 1 (Apr. 1910). In 1912 TE became a self-proclaimed "believer in the utilization of Tidal powers" after seeing the Maine inventor Thomas A. McDonald's "Tidal Power Wheel." He said however that such a technology was feasible only in areas where there was a very high demand for energy. TE superscript on Harry C. Webber to TE, 27 Jan. 1912, TENHP.

PART FOUR · MAGNETISM (1890–1899)

1. Association of Edison Illuminating Companies, *"Edisonia,"* 73; New York *Evening World* and *New York Tribune,* 2 Jan. 1890; Israel, *Edison,* 334–35, citing Henry Villard to TE, 2 Feb. 1890. See also McDonald, *Insull,* 39ff.

2. Josephson, *Edison,* 341.

3. TE to Henry Villard, 8 Feb. 1890, PTAE.

4. Ibid.

5. Henry Villard to TE, 13 Feb. 1890, PTAE.

6. Tate, *Edison's Open Door,* 247–48; Thomas V. Leidy and Donald R Shelton, "Titan in Berks: Edison's Experiments in Iron Concentration," *Historical Review of Berks County* (Fall 1958); Israel, *Edison,* 347–48; *Buffalo Evening News,* 20 Jan. 1890; Elizabeth Earl to MME, 10 Mar. 1890, PTAE.

7. Elizabeth Earl to MME, 10 Mar. 1890, PTAE; Marion Edison Öser, "Wizard of Menlo Park, *passim,* TENHP.

8. Elizabeth Earl to MME, 10 Mar. 1890. "The abscess on her back [inflicted] permanent injury to the spine, and when they lanced it . . . she bled so profusely they feared for her life." Hemorrhagic smallpox is almost always fatal.

9. Michele Albion, "Mina Miller Edison Pregnancies and Miscarriages," unpublished research note, 28 Feb. 2007 (additional research by Thomas E Jeffrey, Author's Collection; Albion memo; *Statement of St. Paul's School,* 1891, TENHP. The fee for two boarders at St. Paul's that year was $1,200 per annum, or $34,200 in 2018 dollars.

10. Sarah Brigham to MME, 7 Apr. 1890, PTAE.

11. TE quoted in Meadowcroft, *Boys' Life of Edison*.

12. Vanderbilt, *Edison, Chemist*, 138; Israel, *Edison*, 342.

13. See TE's rationale for the mining and concentration of Eastern iron ore, untitled ms. essay, ca. Oct. 1894, PTAE, henceforth TE Rationale.

14. TE quoted in *Atlanta Constitution*, 25 Feb. 1890; TE Rationale, 2; TE in *Engineering and Mining Journal*, 52 (26 Dec.1891); TE in Dyer and Martin, *Edison*, 478–79; Theodore Waters, "Edison's Revolution in Iron Mining," *McClure's Magazine*, Nov. 1897. The Ogden mine was the largest of three Edison sought to develop in 1887–90. The other two, at Humboldt, Michigan, and Bechtelsville, Pennsylvania, were unsuccessful. Israel, *Edison*, 344–48.

15. Israel, *Edison*, 345–47; Carlson, "Edison in the Mountains," 43. The New Jersey & Pennsylvania Concentrating Works had been formed in December 1888 to develop TE's preliminary mine in Bechtelsville.

16. Charles Batchelor Diary, entry for 24 Mar. 1890, PTAE; *Wall Street Journal*, 31 Mar. 1890; New York *World*, 1 Apr. 1890; *Chicago Tribune* and *Philadelphia Times*, 8 Apr. 1890; McDonald, *Insull*, 48; Israel, *Edison*, 334; Dyer and Martin, 53.

17. Parker, *Natural Philosophy*, 18.

18. *Iron Age* 46 (27 Nov. 1890).

19. Edison made this comparative remark when being quizzed on his lack of religious belief. *New York Times*, 2 Oct. 1910.

20. TE in *Telegrapher*, 8 Aug. 1868 and 16 Oct. 1869; TE, "On a Magnetic Bridge or Balance for Measuring Magnetic Conductivity," *Proceedings of the American Association for the Advancement of Science* 36 (Aug. 1887); William J. Hammer in *Electrical World*, 21 Sept. 1889; John Birkinbine and Thomas Edison, "The Concentration of Iron-Ore," *Transactions of the American Institute of Mining Engineers* 17 (Feb. 1889).

21. Hounshell, "Edison and the Pure Science Ideal." For an account of TE's explorations of electromagnetic science, see Israel, *Edison*, 30–61.

22. TE Patent 228,329, executed 3 Apr. 1880, issued 1 June 1880.

23. Birkinbine and Edison, "Concentration of Iron-Ore"; TE Patent 430,280, issued 17 June 1890. At this early stage of Edison's development of the movie camera, he used the names "Kinetoscope" and "Kinetograph" indiscriminately. They later took on more precise meaning. See, e.g., TE Patent 589,168.

24. "Edison Has an Idea," *Minneapolis Times*, 17 Apr. 1890; Paul C. Spehr, "Unaltered to Date: Developing 35mm. Film," in John Fullerton and Astrid Soderburgh Widding, *Moving Images: From Edison to the Webcam* (Bloomington, IN, 2000), loc. 378ff.

25. Spehr, *Man Who Made Movies*, 288; TE and W. K. L. Dickson Patent 434,588, issued 19 Aug. 1890.

26. Carlson, "Edison in the Mountains," 42.

27. *Hornellsville* (NY) *Weekly Tribune*, 21 Mar. 1890; Marion Edison to MME, Mar. 1890, PTAE; Israel, *Edison*, 384; Marion Edison to TE, 28 Dec. 1890, PTAE ("Do please add a line or two to one of Mina's letters I should like so much to hear from you"), and to MME, ca. Aug. 1890, PTAE.

28. Marion Edison to MME, ca. Aug. 1890, PTAE.

29. *Engineering News,* Oct. 1980; Johnson, *Edison's "Ogden Baby,"* 119–20, 144; *Engineering and Mining Journal,* 10 Oct. 1891.

30. Spehr, *Man Who Made Movies,* 191; Johnson, *Edison's "Ogden Baby,"* 145.

31. Tate to Samuel Insull, 10 July 1890, PTAE; Lathrop, "Talks with Edison"; Lathrop to TE, 10 Aug. 1890, PTAE.

32. Lathrop to Tate, 30 June 1890, PTAE.

33. Finding aid, Rose Hawthorne Lathrop Papers, Dominican Sisters of Hawthorne Archive, Hawthorne, NY.

34. TE to Lathrop, ca. early Oct. 1890, PTAE.

35. Lathrop to TE, 13 Oct. 1890 and 10 Aug. 1891; TE to Lathrop, ca. early Oct. 1890; all PTAE.

36. Lathrop to TE, 10 Aug. 1891, and TE to Lathrop, ca. early Oct. 1890, both PTAE.

37. Lathrop to TE, 10 Aug. 1891, PTAE.

38. Tate, *Edison's Open Door,* 283; Dickson and Dickson, *History of Kinetograph,* 8–12; TE caveat, 8 Oct. 1888, TENHP; Musser, *Before the Nickelodeon,* 30–31; *Minneapolis Times,* 17 April 1890.

39. Spehr, *Man Who Made Movies,* 210. A reshoot, *Monkeyshines No. 2,* already showed improvement in camera technique. See https://www.youtube.com/watch?v=9jSbExx-960.

40. Dickson, "Brief History"; John Belton, *Widescreen Cinema* (Cambridge, MA, 1992), 18. For an exhaustive account of film R&D at the Edison laboratory in 1889–90, see Spehr, "Unaltered to Date."

41. Dickson, "Brief History."

42. The complex chronology of the invention of cinema, involving simultaneous experiments and claims of precedence in France, Britain, and the United States, is a subject of continuing, unresolved debate by scholars in all three countries. Edison's relations with Étienne-Jules Marey and his pioneer work on the Kinetoscope in 1888 and 1889 will be discussed in Part Five.

43. *Poughkeepsie Daily Eagle,* 1 Jan. 1891, quoting New York *Sun*; Lathrop to TE, 10 Aug. 1891, PTAE; *Buffalo Enquirer,* 27 May 1891; *Hartford Courant,* 10 June 1891.

44. Lathrop to TE, 10 Aug. 1891, 17 Mar. and 15 Apr. 1892; Tate to Lathrop, 15 Apr. 1891; Lathrop to MME, 24 June 1891, all PTAE.

45. Spehr, *Man Who Made Movies,* 190. On 11 Apr. 1883, Edward H. Johnson specifically recommended Sprague to TE, as "of all men the very one to take charge of your Railway Experiments." TE did hire him, but assigned him to install small-town lighting systems at the very time he was forming his own Edison-Field Electric Railway Company. *Papers,* 7.61–65, 67.

46. *Buffalo Commercial,* 16 Jan. 1890.

47. Rowsome, *Birth of Electric Traction,* loc. 1420, 1461ff.

48. Frank J. Sprague to Edison General Electric Co. and Henry Villard, 2 Dec. 1890, PTAE. See also Rowsome, *Birth of Electric Traction,* loc. 1547ff.

49. See, eg., Sprague in *The New York Times,* 23 Sept. 1928, and Harriet Sprague, *Frank J. Sprague and the Edison Myth* (New York, 1947), *passim.*

50. Harry Livor to TE, 4 Mar. 1891, PTAE; Israel, *Edison,* 335; *New York Evening World,* 5 Feb. 1891; *Wall Street Journal,* 7 Feb. 1891. See also McDonald, *Insull,* 39ff.

51. Donald R. Baker and A. F. Buddington, *Geology and Magnetite Deposits of the Franklin Quadrangle and Part of the Hamburg Quadrangle, New Jersey,* U.S. Geological Survey Professional Paper 638 (Washington, DC, 1970), 38, 49, 52; TE Rationale, 5–6, PTAE; TE in *Engineering and Mining Journal,* 10 Oct. 1891.

52. "Report of Mr. Edison on Mill and Property at Ogden," NJ&PCW Board Minutes, 16 July 1890, PTAE; *Engineering and Mining Journal,* 10 Oct. 1891; Johnson, *Edison's "Ogden Baby,"* 36–37.

53. Johnson, *Edison's "Ogden Baby,"* 72.

54. Ibid., 34; Walter Mallory deposition in *Edison v. Allis Chalmers,* 640; Israel, *Edison,* 349; *Scranton Republican,* 9 Apr. 1891; Charles Batchelor Diary, entry for 8 Mar. 1891, PTAE.

55. TE to Henry Livor, 10 June 1891, and Livor to TE, 14 Apr. 1890, PTAE.

56. Israel, *Edison,* 350; Johnson, *Edison's "Ogden Baby,"* 33.

57. Carlson, "Edison in the Mountains," 48–49; Henry Livor to TE, 11 June 1891; Bethlehem Iron to W. S. Perry, 15 June 1891; Perry to Bethlehem Iron, 24 July 1891, all PTAE.

58. Israel, *Edison,* 350; TE in *Buffalo Morning Express,* 8 Nov. 1891; 47 F. 454 (SDNY, 1891), 1891 U.S. App. Lexis 1151; "Locomotion in Water Studies by Photography," *Scientific American Supplement,* 10 Jan. 1891. Edison was familiar with Marey's pioneering work and could not fail to recognize its superiority to his own. He was at least dimly aware of that of William Friese-Greene. The British inventor wrote to him on 18 Mar. 1890 to say he was sending by separate post "a paper with description of Machine Camera for taking 10 a second." There is no trace of this paper in TENHP, but receipt of the letter was acknowledged. Hendricks, *Origins of American Film,* 1.178; William Friese-Greene to TE, 18 Mar. 1890, PTAE.

59. *Reading* (PA) *Times,* 14 May 1891. See also *Chicago Tribune,* 13 May 1891.

60. Ibid.

61. *Philadelphia Inquirer,* 14 May 1891.

62. *New York Tribune,* 27 May 1891. Richard Dyer was the brother and partner of Frank Dyer, future president of Thomas A. Edison, Inc.

63. Ibid.; Allerhand, *Illustrated History,* 226ff.; *Electrical Review,* 15 Oct. 1892; *New York Evening World,* 26 May 1891.

64. Ibid. The New York *Sun,* 28 May 1891 is the source of a doubtful story that TE first demonstrated the Kinetograph to a group of women touring his laboratory a few days previously. It is true that Mina Edison hosted a lunch at Glenmont on 20 May for some 200 members of the General Federation of Women's Clubs, and that she invited them to visit the plant afterward. However, Edison was in Chicago that day. Jane C. Croly's densely detailed account of the event (which gives a good idea of the elegance of Mina's entertainments) makes no mention of a movie show. Jane C. Croly, *The History of the Woman's Club Movement in America* (New York, 1898), 110. See also Musser, *Emergence of Cinema,* 504.

65. New York *Sun*, 28 May 1891; Lathrop to TE, 29 May, 1891.

66. New York *Sun*, 28 May 1891.

67. Ibid.

68. Ibid.

69. TE Patent 493,426; Musser, *Emergence of Cinema*, 71; New York *Sun*, 28 May 1891.

70. New York *Sun*, 28 May 1891.

71. "Edison's Kinetograph," *Harper's Weekly* 35 (13 June 1891).

72. Ibid.

73. Henry Hart interviewed in *New York Morning Journal*, 26 July 1891.

74. Ibid.

75. *Phonogram*, TE's house magazine, predicted that with all other lighting companies included as liable in the infringement decision, Edison General Electric was due as much as $50 million in back damages, and $2 million a year in future royalties. *Phonogram* 2, no. 10 (Oct. 1892).

76. TE quoted in *Asheville Citizen-News*, 11 Nov. 1891.

77. Dyer and Martin, *Edison*, 661; TE U.S. Patents 589,168 and 493,426; TE quoted in *Asheville Citizen-News*, 11 Nov, 1891. TE made no attempt to patent the Kinetograph overseas, probably because he quailed at the cost and difficulty of claiming precedence over the rival inventions of Marey, Le Prince, Friese-Greene, and others. But he thereby lost millions and enabled such French competitors as Lumière and Pathé to make substantial inroads into the U.S. market. For a discussion of TE's decisions to patent, or not patent, various aspects of his motion picture devices, see Spehr, *Man Who Made Movies*, chap. 18. For an online study of the even more momentous lamp case, see Ron D. Katznelson and John Howells, "Inventing Around Edison's Lamp Patent: The Role of Patents in Stimulating Downstream Development and Competition," www.law.northwestern.edu. The authors argue that it was a victory for all parties, particularly the American consumer. For a general analysis of TE's patenting policy, see Israel, *Edison*, 316–19.

78. *New York Morning Journal*, 26 July 1891.

79. Ibid.

80. Ibid. Poe's protagonist is named "Ellison."

81. Tate to Lathrop, 27 Aug. 1891, and Lathrop to TE, 10 Aug. 1891, PTAE.

82. Lathrop to TE, 10 Aug. 1891, PTAE.

83. Lathrop to Tate, 29 Aug. 1891, PTAE; *Galveston Daily News*, 13 Dec. 1896; Greg Daugherty, "Thomas Edison's Forgotten Science Fiction Novel," *Smithsonian*, 3 Jan. 2018. For an extensive discussion of TE's relationships with science fiction writers, see Israel, *Edison*, 363–69.

84. TE to Edward Marshall, *Seattle Times*, 25 Jan. 1920.

85. TE quoted in New York *Sun*, 28 May 1891. TE later proposed that a "scientifically-kept watch for interstellar signaling should be established in Michigan, where enormous masses of ore might be expected particularly to attract magnetic signals from space if any should be sent." TE to Marshall, *Seattle Times*, 25 Jan. 1920.

86. Carlson, "Edison in the Mountains," 50; Johnson, *Edison's "Ogden Baby,"* 23, 145; Israel, *Edison,* 350.

87. *Engineering and Mining Journal,* 10 Oct. 1891.

88. Thomas Robins, "Friends in a Lifetime," ts. memoir, 1944, 3–4, Biographical Collection, TENHP.

89. Ibid., 4; Thomas Robins, "Notes on Conveyor Belts and Their Use," *Transactions of the American Institute of Mining Engineers* 26 (Apr. 1896).

90. Robins, "Friends In a Lifetime," 5; Frederic V. Hetzel, *Belt Conveyors and Belt Elevators* (New York, 1922), 10–11.

91. Minutes, Board Meetings of the NJ&PCW, 31 Aug. 1891, 20 Feb. 1892, PTAE.

92. TE in New York *Sun,* 21 Feb. 1892; McDonald, *Insull,* 45–50; Hammond, *Men and Volts,* 173, 194; Josephson, *Edison,* 363–64; Jean Strouse, *Morgan: American Financier* (New York, 1999), 313.

93. Charles Batchelor Diary, entry for 6 Feb. 1892, PTAE; Josephson, *Edison,* 364; *New York Times,* 6 Feb. 1892.

94. *Monthly Notices of the Royal Astronomical Society* 52, no. 5 (11 Mar. 1892); Annie S. Maunder and E. Walter Maunder, *The Heavens and Their Story* (Boston, 1908), 182; *New York Evening World,* 8 Feb. 1892; *New York Times,* 9 Feb. 1892.

95. Sherburne B. Eaton to TE, 8 Feb. 1892, PTAE ("I suppose Insull will be back in the morning with some interesting news from Boston"); New York *Sun,* 15 Feb. 1892. For TE's ongoing electromagnetic research at this time, see Israel, *Edison,* 306-11.

96. *New York Evening World,* 8 Feb. 1892; *New York Times,* 9 Feb. 1892.

97. New York *Sun,* 15 Feb. 1892. See the "photographic registers" of the solar storm at its peak on 13 Feb. 1892 in Maunder and Maunder, *Heavens and Their Story,* 182.

98. New York *Sun,* 15 Feb. 1892.

99. Tate, *Edison's Open Door,* 261; "Conference re Edison Museum," ts., 20 Aug. 1928, TENHP.

100. *New York Tribune,* 20 Feb. 1892. Some newspapers reported that Edison had attended a pre-merger conference with Villard and Insull. When the latter complained about Coffin's coup in reversing what had been planned as an Edison General Electric acquisition, Villard allegedly snapped, "It never would have happened except for your mismanagement." *New York Tribune,* New York *World,* and *Middletown* (NY) *Times-Press,* 20 Feb. 1892.

101. *New York Times,* 21 Feb. 1892. See also McDonald, *Insull,* 51. Insull, in his unpublished memoirs, confirmed that he and Edison resumed friendly relations, but the "spell had been broken" between them and was never restored. Israel, *Edison,* 336–37.

102. McDonald, *Insull,* 51. See, e.g., Hammer Reminiscences, Biographical Collection, TENHP. "What he did to Edison and his interests . . . would fill a book. . . . Considerable of it was crooked." For a more charitable view of Insull in 1892, see Tate, *Edison's Open Door,* 265.

103. McDonald, *Insull*, 52, 339 and *passim*; *Sterling* (IL) *Daily Gazette*, 25 June 1892; Charles Batchelor Diary, entries for 29 and 24 June 1892, PTAE.

104. Israel, *Edison*, 336; Strouse, *Morgan*, 314.

105. Tate, *Edison's Open Door*, 278. Kennelly later became a professor of electrical engineering at Harvard.

106. Johnson, *Edison's "Ogden Baby,"* 146–47, 41–42; DeGraaf, *Edison and Innovation*, 152, 156; Israel, *Edison*, 352.

107. Clipping quoted in Jones, *Edison: Sixty Years*, 320; Israel, *Edison*, 352.

108. Dan Jones interview, Nov. 1928, Mary C. Nerney Notebook, TENHP.

109. Ibid.; TE to MME, n.d. "Thursday," PTAE. TE's letters to his wife during the 1890s often bear no date other than an occasional named weekday. The author's speculative dating of some of these letters sometimes differs from that of the editors of PTAE (bracketed below).

110. TE to MME, n.d. "Tuesday," ca. 1894 [1890s]; n.d. [1895]; 9 and 12 Aug. 1895; n.d. [1896]; all PTAE.

111. TE to MME, two letters [1896], PTAE.

112. Spehr, *Man Who Made Movies*, 268; *Phonogram* 2, no. 10 (Oct. 1892); Dickson, "Brief History"; Hendricks, *Origins of American Film*, 1.140–42; Spehr, *Man Who Made Movies*, 388.

113. Dickson and Dickson, *History of Kinetograph*, 19; *Indianapolis News*, 15 Mar. 1894; Spehr, *Man Who Made Movies*, 214, 265–67; Hendricks, *Origins of American Film*, 2.26; Dickson, "Brief History." Dickson's explanatory sketch of the Black Maria is reproduced in Spehr, *Man Who Made Movies*, 266.

114. Israel, *Edison*, 384–85. Carlos Levison's courtship of Marion Edison is a frequent subject of discussion in her correspondence in PTAE.

115. *Town Topics* clip, ca. Jan. 1893, preserved by Madeleine Edison, DSP. ("I have an idea that a disgruntled companion or chaperone for Marion was responsible for this.")

116. Jana F. Brown, "Sons of the Phonograph," *Horae Scholasticae*, Winter 2009; Israel, *Edison*, 385; Jones, *Edison: Sixty Years*, 266; Marion Edison to John Randolph, 18 Jan. 1893, and Randolph to Marion Edison, 31 Jan. 1893, PTAE; TE to MME, "Thursday," ca. Mar. 1893 [1895], PTAE.

117. TE to MME, "Thursday," ca. Mar. 1893 [1895], PTAE.

118. "Mr. Dickson is seriously ill." Theodore Lehmann to Alexander Elliott, Jr., 3 Feb. 1893, PTAE.

119. J. V. Miller to Elliott Joslin, 3 Aug. 1931, PTAE; Israel, *Edison*, 461; loan certificates in H. F. Miller Legal File, 12 and 19 June 1893, PTAE; TE to Charles Kintner, 20 June 1893, quoted in Hendricks, *Origins of American Film*, 1.97; TE Patents 513,097, 567,187, 602,064, 605,475, and 607,588.

120. Allerhand, *Illustrated History*, 297–98. Although TE remained on the board of GE for a few years, he took no part in the affairs of the company.

121. John Randolph to Marion Edison Öser, 19 Nov. 1894, PTAE. The reconstruction of Ogden required another capitalization increase to $1.5 million, almost half of which came out of TE's own pocket. Israel, *Edison*, 352, 354.

122. TE to Tate, 19 Apr. 1893, quoted in Josephson, *Edison,* 374.

123. J. V. Miller to Elliott Joslin, 3 Aug. 1931, PTAE; Israel, *Edison,* 461; loan certificates in H. F. Miller Legal File, 12 and 19 June 1893, PTAE; *Review of Reviews,* July 1893; TE to Charles Kintner, 20 June 1893, quoted in Hendricks, *Origins of American Film,* 1.97. In mid-1893 Bradstreet gave TE an astonishing credit rating of $3 million. He ascribed it to property. "It did not come from my inventions." TE in *Review of Reviews,* July 1893. The modern equivalent would be $86.4 million.

124. TE Patents 513,097, 567,187, 602,064, 605,475, and 607,588.

125. Spehr, *Man Who Made Movies,* 296; *Scientific American,* 20 May 1893; TE inscription in a copy of Muybridge's *Descriptive Zoopraxigraphy* (1893), quoted by Josephson, *Edison,* 392. When Dr. Hopkins asked if the Kinetograph was going to be exhibited at the world's fair in its full with-sound form, TE replied, "No, didn't have time to perfect." Superscript on George Hopkins to TE, 25 Apr. 1893, PTAE. For an account of this public relations disaster, see Spehr, *Man Who Made Movies,* 297–99.

126. *Scientific American,* 20 May 1893.

127. Ibid. See also Musser, *Before the Nickelodeon,* 34–36 and Spehr, *Man Who Made Movies,* 296–97.

128. TE Rationale, 16–16A; Johnson, *Edison's "Ogden Baby,"* 107; Mallory deposition, 645; Waters, "Edison's Revolution"; Emil Herter deposition in *Edison v. Allis Chalmers,* 546.

129. Herter deposition, 129, 575, 554; TE Rationale, 16B; but see "Edison's Revolution in Iron Mining," *McClure's Magazine,* Nov. 1897.

130. Mallory deposition, 642–43, 646; Herter deposition, 555, 567, 548–49; Johnson, *Edison's "Ogden Baby,"* 108–10.

131. Herter deposition, 546; Mallory deposition, 655.

132. See Spehr, *Man Who Made Movies,* 254–57, for the development of the peephole player.

133. See http://www.ifbbpro.com/news/the-first-bodybuilding-movies; *Atlanta Constitution,* 9 Mar. 1894. See also Musser, *Before the Nickelodeon,* 39–40.

134. https://www.youtube.com/watch?v=HWM2ixqua3Y.

135. Tate, *Edison's Open Door,* 286–87; Spehr, *Man Who Made Movies,* 307; Millard, *Edison and Business,* 54; Musser, *Before the Nickelodeon,* 44.

136. Tate, *Edison's Open Door,* 286–87.

137. Ibid., 286.

138. Musser, *Before the Nickelodeon,* 40–44; *Poughkeepsie Eagle-News,* 11 Aug. 1884; "Annabelle," *Who's Who of Victorian Cinema,* http://www.victorian-cinema.net/annabelle; Joshua Yumibe, *Moving Color: Early Film, Mass Culture, Modernism* (New Brunswick, NJ, 2012), loc. 713ff.; http://earlysilentfilm.blogspot.co.uk/2013/08/peerless-annabelle-symphony-in-yellow.html, which contains the best biographical details.

139. Hendricks, *Origins of American Film,* 2.6off.; Musser, *Before the Nickelodeon,* 45. See W. K. L Dickson, "Edison's Invention of the Kineto-Phonograph," *Century Magazine,* June 1884.

140. The fact that Edison's statement was published in *Century Magazine* as an introduction to an article about the Kinetoscope by W. L. K. and Antonia Dickson, as well as their identical use of it later in *The Life and Inventions of Thomas Alva Edison* (1894) and *History of the Kinetograph, Kinetoscope, and Kineto-Phonograph* (1895), suggests that he may have simply copied out a promotional text that they drafted for him. See Part Five and Spehr, *Man Who Made Movies*, 250 and *passim*, for further discussions of Dickson's compulsive revisionism.

141. Millard, *America on Record*, 42–44; Israel, *Edison*, 297; Welch and Burt, *Tinfoil to Stereo*, 35.

142. Welch and Burt, *Tinfoil to Stereo*, 26, 35; *New York Times*, 10 May 1894.

143. Tate, *Edison's Open Door*, 293.

144. Ibid., 294.

145. New York *World*, 1 Apr. 1895; Welch and Burt, *Tinfoil to Stereo*, 39; Spehr, *Man Who Made Movies*, 310.

146. Israel, *Edison*, 355; Charles Batchelor Diary, entries for 15 Apr. 1892 et seq., PTAE; Waters, "Edison's Revolution"; *Iron Age*, 28 Oct. 1897.

147. Mallory deposition in *Edison v. Allis Chalmers*, 677.

148. Josephson, *Edison*, 374.

149. Madeleine Edison Oral History, TENHP. "Thomas is back again and it is good to have him near me." MME to Lewis and Mary Miller, 8 July 1894, EFW.

150. Marion Edison to TE, 24 July 1894.

151. Lewis Miller to Mary V. Miller from Glenmont, ca. mid-Mar. 1894. See also Theodore Miller to Mary V. Miller, 18 Mar. 1894: "Isn't it funny about Marion going to Europe [so suddenly]. It is probably just as well . . . she seems to treat Mina very shabbily." EFW.

152. Oscar Öser to TE, 23 July 1894, PTAE; Louise Juechzer to TE, 23 July 1894, PTAE; Marion Edison Öser to TE, 10 Apr. 1896, PTAE.

153. TE to Editor, *Cassier's Magazine*, 14 Nov. 1894, PTAE.

154. Dickson and Dickson, *Life and Inventions*, 362; "An Authentic Life of Edison," *New York Times*, 11 Nov. 1894. See also Spehr, *Man Who Made Movies*, 395–96.

155. Frank Dyer to Francise Kehl, 29 Aug. 1936, TENHP.

156. TE to William Pilling, 12 Oct. 1894, PTAE; Israel, *Edison*, 355, 352; TE Patents 465,251 and 485,840 ("Method of Bricking Fine Iron Ores"); Johnson, *Edison's "Ogden Baby,"* 120; Vanderbilt, *Edison, Chemist*, 159; *Iron Age*, 28 Oct. 1897.

157. NJ&P Concentrating Works Letterbook, 21 Feb. and 20 Mar. 1895, PTAE; Johnson, *Edison's "Ogden Baby,"* 149; "Nobody had any hopes at all that they would ever be perfected." Herter deposition in *Edison v. Allis Chalmers*, 558.

158. *Century Magazine*, June 1884; TE to Norman C. Raff, 5 February 1895, PTAE.

159. Dickson and Dickson, *History of Kinetograph*, 19–20, 54, 14.

160. TE to Norman C. Raff., 5 Feb. 1895, PTAE.

161. *New York Times* and *Rochester* (NY) *Democrat Chronicle,* 14 Mar. 1895; *New York Press,* 14 Mar. 1895.

162. *New York Times,* 14 Mar. 1895.

163. New York *Evening Sun,* 15 Mar. 1895.

164. See, e.g., *Buffalo Evening News,* 14 Mar. 1895; *Papers,* 6.821.

165. See, e.g., *New York Journal,* 22 May 1896.

166. Carlson, *Tesla,* 239–41; TE quoted in *Philadelphia Press,* 24 July 1896 ("To my mind it solves one of the most important questions associated with electrical development"); *Baltimore Herald,* 27 May 1896.

167. Musser, *Before the Nickelodeon,* 46, and *Emergence of Cinema,* 84–86; Spehr, *Man Who Made Movies,* 358. For a detailed account of the films made in the Black Maria in 1894, see Musser, *Before the Nickelodeon,* 47–51 and Spehr, *Man Who Made Movies,* chap. 22.

168. Musser, *Emergence of Cinema,* 105; Dickson and Dickson, *Life and Inventions,* 311.

169. Musser, *Emergence of Cinema,* 91, 94; Spehr, *Man Who Made Movies,* 356–58, 360–64.

170. Spehr, *Man Who Made Movies,* 366, 311–12, 283–85, 304; Musser, *Emergence of Cinema,* 92–93, 47.

171. The author is grateful to Paul Spehr for this perception. *Man Who Made Movies,* 366.

172. Spehr, *Man Who Made Movies,* 371–72.

173. Ibid., 352–56; TE to Frederick P. Fish, 1 Nov. 1895, PTAE; Spehr, *Man Who Made Movies,* 282–85; Dickson, "Brief History." "It's too depressing, heartbreaking to go unrecognized in the art, yet it was my work which was as commercialized by me, adopted in every detail by the whole million making world—ah me." Dickson in 1932, quoted in Spehr, *Man Who Made Movies,* 282–83.

174. See Le Prince's *Roundhay Garden Scene,* at https://www.youtube.com/watch?v=nR2r__ZgO5g. There is a detailed account of his *disparation* in Jean-Jacques Aulas and Jacques Pfend, "Louis Aimé Augustin Le Prince, inventeur at artiste, précurseur du cinéma," Journals.openedition.com (2000). See also Richard Howells, "Louis Le Prince: The Body of Evidence," *Screen* 47, no. 2 (July 2006), and *The First Film,* a 2013 documentary by David Nicholas Wilkinson, at https://vimeo.com/ondemand/thefirstfilm/181293064.

175. Dickson and Dickson, *Life and Inventions,* 309; Musser, *Before the Nickelodeon,* 55–56; De Graaf, *Edison and Innovation,* 133ff.

176. Musser, *Before the Nickelodeon,* 91.

177. TE Rationale, 23A, 27A, 30A, 23A–24A, PTAE.

178. TE quoted in Johnson, *Edison's "Ogden Baby,"* 120; Waters, "Edison's Revolution." See, e.g., illustration in *Harper's New Monthly Magazine,* Jan. 1894, 409.

179. Dan Smith interviewed by Mary Nerney, Nov. 1928, TENHP.

180. TE to MME, 9 Aug. 1895, PTAE.

181. Ibid., 11 and 18 Aug. 1895, PTAE.

182. Ibid., 21 Aug. 1895, PTAE.

183. Johnson, *Edison's "Ogden Baby,"* 84. For a table of the various wages and salaries TE paid at Ogden, see 85–91.

184. Walter Mallory to James C. Parrish, 9 Sept. 1895, and TE to MME, 23 Aug. 1895, both PTAE.

185. TE to MME, 23 Aug. 1895, and Mallory to James C. Parrish, 9 Sept. 1895, PTAE.

186. Mallory, "Edison Could Take It," 3.

187. Mallory deposition, 648; Dyer and Martin, *Edison,* 501; TE to MME, 18 Aug. 1895, PTAE; Israel, *Edison,* 337.

188. "Special cable despatch," 6 Jan. 1896, to New York *Sun,* 7 Jan. 1896.

189. Andrew Robinson, "Radiation's Risks and Cures," *The Lancet,* 16 Mar. 2016.

190. "Ten hours after [the] cable despatch . . . I started experimenting." TE superscript on William Bowen to TE, 7 Apr. 1898, PTAE; TE to Arthur Kennelly, 27 Jan. 1896, quoted in Israel, *Edison,* 309.

191. James Barry to TE, 4 Feb. 1896, and William Randolph Hearst to TE, 5 Feb. 1896, PTAE; David Shepherd, "Thomas Edison's Attempts at Radiography of the Brain (1896)," *Mayo Institute Proceedings* 49 (Jan. 1974); Edward P. Thompson, *Roentgen Rays and Phenomena of the Anode and Cathode* (New York, 1896), 117; TE quoted in *New York Times,* 11 Feb. 1896.

192. Musser, *Emergence of Cinema,* 115–66; Israel, *Edison,* 301–2.

193. Roland Burke Hennessy, "Edison and the Röntgen Light," *Metropolitan Magazine* (UK) 3, no. 3 (Mar. 1896).

194. Ibid.

195. *Electrical Review,* 18 Mar. 1896. TE published another information-sharing article, "Influence of Temperature on X-Ray Effects," in *Electrical Engineer* on 22 Apr. 1896. He also contributed to "Photographing the Unseen: A Symposium of the Roentgen Rays," *Century Magazine* 52 (May 1896).

196. *Electrical Review,* 18 Mar. 1896; Carlson, *Tesla,* 224; TE to Nikola Tesla, 13 Mar. 1896, PTAE. For summaries of the X-ray research respectively accomplished by TE and Tesla in 1896, see Thompson, *Roentgen Rays,* chaps. 10 and 11.

197. Until at least 1893, Tesla reportedly "had the strongest admiration" for TE. T. Commerford Martin, *The Inventions, Researches and Writings of Nikola Tesla* (1893; New York, 1995), 4; quoted in *Brooklyn Daily Eagle,* 5 Mar. 1896.

198. *New York Morning Journal,* 22 May 1896; TE quoted in *Electrical Review,* 2 May 1896; TE Patent 865,367, "Fluorescent Electric Lamp."

199. TE to Sir John Pender, 13 Mar. 1896, PTAE; TE in *New York Herald,* 28 Mar. 1896; TE in *Electrical Engineer,* 1 Apr. 1896; TE to MME [1896], PTAE; Israel, *Edison,* 310. It would have been risky for Edison to patent the fluoroscope, since three other scientists had developed similar devices earlier in the year. However, as Adam Allerhand notes, "Edison's [fluoroscope] ultimately prevailed. Edison did the most thorough research, promoted his device, and marketed it." C. C. Trowbridge, "The Use of the Fluoroscopic Screen in Connection with Röntgen Rays," *Annals of the New York Acad-*

emy of Sciences 11, no. 3 (30 Mar. 1898); Allerhand, *Illustrated History*, 462.

200. Michael Pupin to TE, 28 Mar. 1896, PTAE.

201. TE superscript on ibid.

202. *New York Tribune* and New York *Sun*, 27 Apr. 1896.

203. TE to Charles W. Price, 29 May 1896, PTAE. Regarding Tesla's current lamp, TE remarked, "He gets his results from the inductive coil and the Geissler tube. It is of a ghastly color. You cannot get a pleasant mellow yellow light without low temperature waves as well as light." *New York Morning Journal*, 26 July 1896.

204. "Claims of Moore, Tesla, and Edison," *Western Electrician*, 6 June 1896. See, e.g., *New York Morning Journal*, 22 May 1896; *New York Times*, 22 Mar. 1896.

205. Israel, *Edison*, 301–2, 374.

206. Quoted in Musser, *Before the Nickelodeon*, 63, 85, 82. A selection of early Edison movies collected by the Library of Congress can be seen at https://www.youtube.com/playlist?list=PLD28424FAA9414F49.

207. Blackton took a bow at the end of this ninety-five-second short, and went on to become a major movie producer and the father of film animation.

208. Image THF96100, HFM. TE's meticulous account of this convention in a letter to Mina (16 Aug. 1896, PTAE) corrects some of the memory slips in Ford's own several accounts, recorded many years later. While TE does not mention Ford, his letter makes plain that their meeting must have occurred between 11:30 P.M. and 12:30 A.M. that night. See also *Western Electrician*, 22 Aug. 1896.

209. Henry Ford interview, "When did I first see Mr. Edison?", 1928, TENHP; Henry Ford, "My Life and Work," *McClure's Magazine* 54 (Oct. 1922).

210. Leigh Dorrington, "The First Automobile Races in America," Prewar.com.

211. TE to MME, 16 Aug. 1896, PTAE.

212. Ibid.; TE to MME, ca. 1896, "Tuesday" [1890s], PTAE.

213. Vanderbilt, *Edison, Chemist*, 167; American Iron and Steel Association, *Statistics of the American and Foreign Iron Trades for 1896* (Philadelphia, 1897), 32; *Harrisburg Daily Independent*, 26 Dec. 1896; *Wilkes-Barre Times*, 6 Jan. 1897.

214. Musser, *Before the Nickelodeon*, 42–43; DeGraaf, *Edison and Innovation*, 106; Vanderbilt, *Edison, Chemist*, 122–23; *Buffalo News*, 30 Aug. 1896; *Louisville Courier-Journal*, 24 Nov. 1896.

215. *Louisville Courier-Journal*, 24 Nov. 1896; Nikola Tesla folder, Biographical Collection, 1896, TENHP; *New York Morning Journal*, 22 May 1896; "Tesla on the Roentgen Streams," *Electrical Review*, 2 Dec. 1896.

216. Carlson, *Tesla*, 224; TE in *Brooklyn Daily Eagle*, 6. Oct. 1912. By late November 1896, many other patients and scientists, notably Elihu Thompson, had begun to suffer radiation damage. See *New York Morning Journal*, 29 Nov. 1986, and *New York Press*, 30 Nov. 1896.

217. Francis L. Chrisman, notes from an unpublished interview with TE, ca. Nov. 1896, Articles File, PTAE.

218. Ibid. See also Percy Brown, "Clarence Madison Dally (1806–1904)," *American Journal of Radiology* 165 (Jan. 1995).
219. Leonard Peckitt Reminiscences, Biographical Collection, TENHP.
220. Ibid.
221. Ibid.
222. Ibid.
223. Ibid.
224. Leonard Peckitt to TE, 22 Jan. 1897, PTAE; Israel, *Edison*, 359.
225. Peckitt to TE, 22 Jan. 1897, PTAE; Musser, *Before the Nickelodeon*, 93; Dan Smith interviewed by Mary Nerney, Nov. 1928, TENHP.
226. Johnson, *Edison's "Ogden Baby,"* 95–96; Herter deposition, 546–59ff.; *Iron Age*, 28 Oct. 1897; *Harrisburg Daily Independent*, 28 Dec. 1896; Waters, "Edison's Revolution."
227. Mallory to Stuart Coats, 29 Jan. 1897, PTAE; Israel, *Edison*, 358.
228. *Harrisburg Daily Independent*, 28 Dec. 1896. This text of this article was widely syndicated. The price did reach $2.10 in June 1897 and increased by only five cents over the next fiscal year. American Iron and Steel Association, *Statistics of the American and Foreign Iron Trades for 1896* (Philadelphia, 1897), 26.
229. *Transactions of the American Institute of Mining Engineers* 27 (1897), 457–58.
230. Israel, *Edison*, 356, 360.
231. TE/Thomas Edison, Jr., release memorandum, 23 Feb. 1897, Legal File, TENHP; Thomas Edison, Jr., to TE, 14 Jan. 1897, PTAE.
232. Thomas Edison, Jr., to MME, 4 Jan. and 19 May, 1897, PTAE.
233. Ibid., 6 Aug. 1897, PTAE.
234. TE Patent 675,057; Meadowcroft, *Boys' Life of Edison*, 223; Johnson, *Edison's "Ogden Baby,"* 151–52; TE quoted in Israel, *Edison*, 360.
235. The following account of the operation of the Ogden plant is based on articles in *Iron Age*, 28 Oct. 1897; *McClure's Magazine*, Nov. 1897; and *Scientific American*, 22 Jan. 1898.
236. *Iron Age*, 28 Oct. 1897.
237. *McClure's Magazine*, Nov. 1897; *Engineering and Mining Journal*, 10 Oct. 1891; Israel, *Edison*, 403.
238. Vanderbilt, *Edison, Chemist*, 110; TE to E. Hubbell Hotchkiss, 25 Jan. 1899, TENHP. TE had also borrowed $9,000 from Tom's estate on 18 Jan. 1897, when he was still his son's guardian. A month later, on 23 Feb., he released the estate to Tom, at which time it was worth $17,309.91 ($533,145 in modern money). It is not clear from surviving documents whether the 30 Sept. 1987 loan was a reworking of the earlier debt, which was not paid off until Apr. 1905. TE/Thomas Edison, Jr., bond, 25 Apr. 1898, Legal File, TENHP.
239. Thomas Edison, Jr., to William Edison, 16 Dec. 1898, PTAE; Bond and mortgage memorandum, 18 Jan. 1897, Legal File, TENHP.
240. Thomas Edison, Jr., to MME, 27 Nov. 1897, PTAE.
241. Thomas Edison, Jr., to MME, 4 Nov. 1897, PTAE.
242. On 1 Oct. 1897 Tom ordered 200,000 bulbs of the type "now known as X-Ray lamps" from the Shelby Electric Company of Ohio. Memorandum in

PTAE. In New York *Sunday Herald,* 5 Dec. 1897, Tom mentions showing a sample bulb to TE, who declined to comment on it.

243. Thomas Edison, Jr., to MME, 4 Nov. 1897, PTAE; U.S. Trademark 34,806 (15 Dec. 1897); Gitelman, *Scripts, Grooves,* 161; New York *Sunday World,* 5 Dec. 1897. The New York *Sunday Herald* simultaneously published a similar story under the headline "EDISON JR., WIZARD."

244. Thomas Edison, Jr., to MME, 12 Nov. 1897, PTAE.

245. NJ&PCW Board Minutes, 12 Jan. 1898, PTAE; TE to MME, 9 Feb. 1898, PTAE.

246. NJ&PCW Board Minutes, 12 Jan. 1898, PTAE. Walter Cutting had succeeded his brother Robert L. Cutting, Jr., as a more cautious backer of the NJ&PCW.

247. Mallory in NJ&PCW Board Minutes, 12 Jan. 1898, PTAE. Later in the meeting, Mallory confirmed that Edison would lend the NJ&PCW $51,500 ($1.6 million in 2018 dollars) over the next six months. Walter Mallory in NJ&PCW Board Minutes, 12 January, 1898, PTAE.

248. Mallory to Ira Miller, 29 Jan. 1898; New York *World,* 2 Jan. 1898; *Omaha Bee,* 22 Jan. 1898; *Trenton Evening Times,* 5 Feb. 1898; New Mexico Mining and Minerals Division, "Real de Dolores Mine Safeguard Project," Emnrd. state.nm.us; Dyer and Martin, *Edison,* 583–84. The records of the Dolores mine are in the Document File Series—Mining (1899–1901), TENHP. See also Ralph E. Pray, "Edison's Folly," http://www.mine-engineer.com/mining /edison.htm.

249. NJ&PCW Board Minutes, 12 Jan. 1898, PTAE; Mallory in NJ&PCW Board Minutes, 12 Jan. 1898, PTAE; Mallory quoted in "Edison's Revolution in Iron Mining," *McClure's Magazine,* Nov. 1897.

250. TE to MME, 9 Feb. 1898, PTAE.

251. Thomas Edison, Jr., to MME, 3 Feb. 1898, DSP ("I may never come back to you all alive—but you have not lost much—for I know I have not been the son to you as I should have."); TE to MME, 9 Feb. 1898, PTAE.

252. Thomas Edison, Jr., to MME, 19 Aug. 1898, DSP; Mary V. Miller to TE and MME, 23 Jan. 1898, DSP.

253. Theodore Miller to Lewis Miller, 15 April 1898, EFW; George E. Vincent, ed., *Theodore W. Miller, Rough Rider* (Akron, OH, 1899), 64; William Edison to MME, ca. 12 Mar. 1898, DSP; Thomas Edison, Jr., to John Randolph, 19 Apr. 1898, PTAE; agreement between Thomas Edison, Jr., and Charles F. Stilwell, 19 Mar. 1898, PTAE.

254. Vincent, *Theodore Miller,* 74; William Edison to MME, ca. early June 1898, DSP.

255. "It was only as one might stand in their [giant rolls] vicinity and hear the thunderous roar accompanying the smashing and rending of the massive rocks as they disappeared from view that the mind was overwhelmed with a sense of the magnificent proportions of this operation." Dyer and Martin, *Edison,* 484.

256. Ibid., 587–88.

257. Israel, *Edison,* 360; *Iron Trade Review,* 62.1142; Carlson, "Edison in the Mountains," 42.

258. Michael Peterson, "Thomas Edison, Failure," *Invention and Technology* 6,

no. 3 (Winter 1991); Nerney, *Edison, Modern Olympian*, 149; Johnson, *Edison's "Ogden Baby,"* 76; Wayne T. McCabe, *Sussex County* (Charleston, SC, 2003), 44.

259. TE quoted in William A. Simonds, *Edison: His Life, His Work, His Genius* (London, 1935), 270; Dan Smith interviewed by Mary Nerney, Nov. 1928, TENHP, 8.

260. Vincent, *Theodore Miller*, 135–42; Theodore Roosevelt, *The Rough Riders* (1899; New York 2003).

261. William Edison to TE, ca. late Sept. 1898, DSP; Thomas Edison, Jr., to Edward Redington, 27 Aug. 1898, PTAE, and to MME, 10 and 19 Oct. 1898, DSP. William remained in the army through 25 Jan. 1899.

262. Mallory to Pilling and Crane, 7 Dec. 1898, PTAE; *Iron Age,* 5 Jan. 1899; Johnson, *Edison's "Ogden Baby,"* 72, 154.

263. TE to MME, 2, 5 and n.d. Dec. 1898, PTAE.

264. Mallory to Pilling and Crane, 7 Dec. 1898, PTAE; Israel, *Edison,* 361.

265. William Edison to Thomas Edison, Jr., ca. 15 Dec. 1898, PTAE.

266. Thomas Edison, Jr., to TE, 17 Dec. 1898, PTAE.

267. Thomas Edison, Jr., to William Edison, 16 Dec. 1898, PTAE.

268. Mallory deposition in *Edison v. Allis Chalmers,* 648; Josephson, *Edison,* 377; DeGraaf, *Edison and Innovation,* 164–65.

269. Israel, *Edison,* 403–4.

270. *Cleveland Press,* 21 Feb. 1898; *Boston Post,* 18 Feb. 1898; *Rochester* (NY) *Democrat and Chronicle*, 18 Feb. 1898.

271. William S. Bayley, *Iron Mines and Mining in New Jersey* (Trenton, NJ, 1910), 15.

272. Mallory to Josiah Reiff, 10 Apr. 1899, and to Stuart Coats, 11 May 1899, TENHP.

273. Israel, *Edison,* 399; Dyer and Martin, *Edison,* 508ff.

274. Israel, *Edison,* 404; *Iron Age,* 15 June 1899; Israel, *Edison,* 410–11; Mallory interview, 1908, William Meadowcroft Collection, TENHP; Desmond, *Innovators,* loc. 1410; Israel, *Edison,* 400.

275. *Boston Globe,* 3 July 1899; William Edison to TE, nd, ca. Oct. 1899, PTAE.

276. Johnson, *Edison's "Ogden Baby,"* 154, 291ff.; Dan Smith interviewed by Mary Nerney, Nov. 1928, TENHP; NJ&PCW Board Meeting Minutes, 10 Jan. 1900, TENHP; Mallory to Pilling & Crane, 18 Sept. 1897, PTAE.

277. Josephson, *Edison,* 377–78.

278. Robins, "Friends in a Lifetime," 6–7.

279. Mallory, "Edison Could Take It," 2. This conversation may have occurred a year earlier, but the weight of evidence indicates December 1899. See Dyer and Martin, *Edison,* 502–3, for a similar council of war leading to the organization of the Edison Portland Cement Company in June 1899.

280. Mallory, "Edison Could Take It," 1.

281. The plant did reopen for a few months in 1900, but failed to satisfy Bethlehem Iron's low-phosphorus requirement and was forced to dispose of its remaining briquettes below cost. It closed finally for dismantlement on 10 December. Carlson, "Edison in the Mountains"; Mallory to H. S. Gay, 27 Sept. 1900;

Mallory to Pilling and Crane, 8, 12, 15, 22 Oct. 1900, PTAE; Johnson, *Edison's "Ogden Baby,"* 154–55.

282. TE quoted in John Coakley Oral History, TENHP, 14. Edison may have exaggerated his estimated $3 million loss, although Bernard Carlson endorses it. An authoritative contemporary statement of what the Ogden project cost was made by Walter Mallory in 1910. He said that up till 10 December 1900, his boss contributed, in cash, $2,174,000 "out of a total expenditure of about two and a half million dollars." The modern equivalent of what Edison lost would be $66.9 million. Carlson, "Edison in the Mountains"; Walter Mallory testimony in *Edison V. Allis Chalmers,* 648.

PART FIVE · LIGHT (1880–1889)

1. TE quoted in *Engineering World,* Nov. 1922.
2. Walter P. Phillips, *Sketches Old and New* (Boston, 1897), 189. Phillips misre-members this conversation as occurring in 1876, but his context makes clear it took place four years later.
3. "Menlo Park Lighted," *Philadelphia Public Ledger,* 27 Dec. 1879; *New York Herald,* 1 Jan. 1880.
4. George W. Soren to Francis Upton, 29 Dec. 1879, PTAE; McPartland, "Almost Edison," 190; *Papers,* 5.644. See also Gall, "Edison: Managing Menlo Park," for a discussion of TE's problem of reconciling his open laboratory policy with intellectual property theft.
5. *Harper's Weekly,* 3 Feb. 1880. Thornall Avenue was the old post road to Philadelphia, later the Lincoln Highway.
6. Except where otherwise indicated, documentary details in the following account of TE's "open house" at Menlo Park in the early days of 1880 come from newspaper clippings in the Charles Batchelor Scrapbooks, PTAE, and others quoted in *Papers,* 5.539–42.
7. New York *Sun* and Philadelphia *Times,* 27 Dec. 1879; *Times* (London), 29 Dec. 1879.
8. *New York Herald,* 30 Dec. 1879 and 1 Jan. 1880; *Times,* 29 Dec. 1879; Dyer and Martin, *Edison,* 362; Strouse, *Morgan,* 181; *Philadelphia Ledger,* 28 Dec. 1879.
9. TE first used the word *filament,* hitherto applied to thread, to describe his incandescent lamp element in a draft lighting caveat, ca. 25 Feb. 1880, in *Papers,* 5.652. He made it public when executing his Patent 525,888 on 10 Mar. 1880 ("the filament of my electric lamp"). It appeared even earlier, when the *Philadelphia Ledger,* in an article about TE at Menlo Park datelined 27 Dec. 1879, referred to the "carbon filament" of his successful lightbulb. The *Oxford English Dictionary* and the Online Etymological Dictionary still cite 1881 as the year of first such usage, by the British physicist Sylvanus P. Thompson.
10. Hammer, "Edison and His Inventions," v; Martin, *Forty Years of Edison Service,* 21–22; *Papers,* 5.1025–26.
11. New York *Sun,* 12 Feb. 1880.

12. "Edison's Light," *New York Herald*, 28 Dec. 1879; *New York Herald*, 1 Jan. 1880; *Chicago Tribune*, 4 Jan. 1880. See also Friedel and Israel, *Edison's Electric Light, passim*. Friedel observes that the crowds attaching themselves to Edison at this time represented "a new relationship between advanced technology and the common man. Edison's electric light was as mystifying and awe-inspiring as any invention of the age. . . . The wizardry of scientific technology was now a source, not of distrust, but rather, of hope. This attitude toward the powers of science and technology was one of the nineteenth century's most important legacies, and no single instance exemplifies it better than the enthusiasm with which the crowds ushered in the new decade at Menlo Park" (89–90).

13. Quoted in *Bristol Mercury*, 15 Jan. 1880 [original dateline Philadelphia, 30 Dec. 1879].

14. Menlo Park farmer, quoted in *Philadelphia Times*, 2 Jan. 1880; *New York Herald*, 1 Jan. 1880; Hammer Reminscences, TENHP; *Philadelphia Public Ledger*, 7 Jan. 1880. See the photograph in Jehl, *Menlo Park Reminiscences*, 670.

15. New York *Herald*, 2 Jan. 1880; Taylor, *Mr. Edison's Lawyer*, 46.

16. *Times*, 30 December 1879.

17. *Nature*, 1 Jan. 1880. See also Joseph Swan, "The Sub-Division of the Electric Light," *Journal of the Society of Telegraphic Engineering* 9 (1880) and Allerhand, *Illustrated History*, 128–32.

18. Quoted in ibid., 131.

19. Bowers, *Lengthening the Day*, 84; *Scientific American*, 27 Nov. 1880; quoted in Allerhand, *Illustrated History*, 131; *Journal of Gas Lighting*, 20 Jan. 1880.

20. Bowers, *Lengthening the Day*, 70.

21. *Reports of Patents, Design, and Trademark Cases Decided by Courts of Law in the United Kingdom* (London, Apr. 1887), IV.4, 83.

22. *Saturday Review*, 10 Jan. 1880; *Nature*, 12 Feb. 1880.

23. *Papers*, 5.727.

24. *Le Temps*, 8 Jan. 1880 (author's translation).

25. Henry Morton letter in *Sanitary Engineer*, 1 Jan. 1880; Morton interviewed in *New York Times*, 28 Dec. 1879. See also, e.g., the criticism of "a well-known electrician" of Cleveland, Ohio (probably Charles F. Brush), in "Lighting a Great City," *New York Times*, 7 Feb. 1880.

26. *New York Times*, 28 Dec. 1879; New York *Sun*, 27 Dec. 1879.

27. See Ron D. Katznelson and John Howells, "Inventing Around Edison's Lamp Patent: The Role of Patents in Stimulating Downstream Development and Competition," www.law.northwestern.edu.

28. TE Patent 369,280.

29. Ibid., figs. 3 and 4.

30. Ibid.

31. Francis R. Upton, "Edison's Electric Light," *Scribner's Monthly*, Feb. 1880.

32. TE Patent 369,280.

33. Ibid.

34. Ibid.; Dyer and Martin, *Edison*, 389; McPartland, "Almost Edison," 206ff.

35. TE Patent 369,280.

36. TE Patents 227,226 and 369,280.

37. Friedel and Israel, *Edison's Electric Light*, 94. See *Papers*, 5.692, for the "fundamental change" in Menlo Park operations from research to manufacturing during the course of 1880.

38. *Papers*, 5.542; New York *Sun*, 12 Jan. 1880. For an eloquent description of Menlo Park at night in the summer of 1882, see *York* (PA) *Daily*, 15 July 1882.

39. The best guide to TE's inventive progress is the list of his patents by execution date compiled by the Edison Papers project. It is available online at http://edison.rutgers.edu/patente1.htm et seq.

40. Friedel and Israel, *Edison's Electric Light*, 118. "There is no record of such a large corps of trained men and so complete an establishment devoted to scientific research. Menlo Park is one of the wonders of this wonderful age." *Boston Globe,* 2 May 1880.

41. Jehl, *Menlo Park Reminiscences,* 359, 262–63, 858–59, 866; Hammer Reminiscences, TENHP; Marshall, *Recollections of Edison,* 31–32. See also Bernard S. Finn, "Working at Menlo Park," in Pretzer, *Working at Inventing,* 32–47.

42. Dyer and Martin, *Edison,* 637; Jehl, *Menlo Park Reminiscences,* 693, 241.

43. *Papers,* 5.545; Charles T. Hughes reminiscence, 19 June 1907, TENHP. See also "The Menlo Park Mystique" in Friedel and Israel, *Edison's Electric Light,* 118–20. On 2 May 1880 the *Boston Globe* commented, "There is no record of such a large corps of trained men and so complete an establishment devoted to scientific research."

44. See Friedel and Israel, *Edison's Electric Light,* 131–34, on this "most critical lamp production problem."

45. *Papers,* 5.624–26, 752, 636; TE Patent 248,418.

46. W. C. White, "Electrons and the Edison Effect," *General Electric Review,* Oct. 1943. Edison used the phrase "molecular bombardment." Jehl, *Menlo Park Reminiscences,* 618.

47. *Papers,* 5.753, 627–30; TE Patent 307,031; William H. Preece, "On a Peculiar Behaviour of Glow-Lamps when raised to High Incandescence," *Proceedings of the Royal Society of London* 38 (London, 1885); Harold G. Bowen, *The Edison Effect* (West Orange, NJ, 1951); Sungook Hong, *Wireless: From Marconi's Black Box to the Audion* (Cambridge, MA, 2001), 121–22.

48. *Papers,* 5.546, 677; Friedel and Israel, *Edison's Electric Light,* 111–20; McPartland, "Almost Edison," 201–2.

49. *Papers,* 5.671–73 and 713–14; "On the Efficiency of Edison's Light," *American Journal of Science,* Apr. 1880; McPartland, "Almost Edison," 203.

50. Grosvenor Lowrey to Kate Armor, 20 Feb. 1880, quoted in Taylor, *Mr. Edison's Lawyer,* 50.

51. *Scientific American,* 22 May 1880; Association of Edison Illuminating Companies, *"Edisonia,"* 133. See also Allerhand, *Illustrated History,* 177–81. TE was not required to provide lights for navigation, that function being performed by the much more powerful arc lamps of Hiram Maxim.

52. A fourth dynamo excited the functional three. Association of Edison Illuminating Companies, *"Edisonia,"* 129, 131; Friedel and Israel, *Edison's Electric Light*, 116–17; TE Patent 227,226; *Papers*, 5.598–99.

53. Dyer and Martin, *Edison*, 372.

54. *Papers*, 5.600, 694; *New York Herald* clipping, "April 29, 1880," Charles Batchelor Scrapbook 1878–1881, PTAE; Ray E. Kidd, "Lighting the Steamship *Columbia* with Edison's First Commercial Light Plant," General Electric release, June 1936; De Borchgrave and Cullen, *Villard*, 310.

55. Marshall, *Recollections of Edison*, 17; Israel, *Edison*, 198.

56. Marshall, *Recollections of Edison*, 18; Association of Edison Illuminating Companies, *"Edisonia,"* 124.

57. *New York Times*, 9 Aug. 1880; Association of Edison Illuminating Companies, *"Edisonia,"* 121–22; drawings in TE Patent 475,591. For an extended discussion of TE's 1880 electric railway, see Jehl, *Menlo Park Reminiscences*, 576–86; Association of Edison Illuminating Companies, *"Edisonia,"* 122; *Scientific American*, 5 June 1880; Dyer and Martin, *Edison*, 452; *Papers*, 5.231–34, 1020–21. Edison's application for a U.S. patent for his electric railway was found to be in interference with that of Werner von Siemens, and disallowed. He wrote a detailed description of the system when applying successfully for a British patent in September 1880. See *Papers*, 5.846–53.

58. *Papers*, 5.739. The phrase *odor of armature* was used by T. Commerford Martin in "Edison's Pioneer Electric Railway Work," *Scientific American*, 18 Nov. 1911.

59. Lowrey to Kate Armour, 5 June 1880, quoted in Taylor, *Mr. Edison's Lawyer*, 53. See also Dyer and Martin, *Edison*, 463.

60. *Papers*, 5.738, 735; *New York Herald*, 10 Aug. 1880. For an extended account of TE's railway work in the 1880s, see Dyer and Martin, *Edison*, 454–72.

61. *Papers*, 5.738; Taylor, *Mr. Edison's Lawyer*, 27 and *passim*.

62. Jocelyn P. Kennedy and Robert C. Koolikian, "Grosvenor Porter Lowrey," ts., Biographical Collection, TENHP.

63. *Boston Globe*, 2 May 1880.

64. *Boston Globe*, 2 May 1880. William S. Pretzer remarks on the "mystical" quasi-religious nature of the Menlo Park fraternity in Pretzer, *Working at Inventing*, 18.

65. Friedel and Israel, *Edison's Electric Light*, 95–97; *Papers*, 5.790, 764; *New York Herald*, 10 Aug. 1880; McPartland, "Almost Edison," 204; Allerhand, *Illustrated History*, 136, 135; Israel, *Edison*, 196.

66. Friedel and Israel, *Edison's Electric Light*, 107–8; *Papers*, 5.563; Association of Edison Illuminating Companies, *"Edisonia,"* 97; Israel, *Edison*, 196.

67. Friedel and Israel, *Edison's Electric Light*, 128–30; Jehl, *Menlo Park Reminiscences*, 620; *Papers*, 5.1050; Hammer Reminiscences, TENHP; Hammer, "Edison and His Inventions," vi. In Association of Edison Illuminating Companies, *"Edisonia,"* there is a photograph of this historic bulb, which TE gave to Hammer as a souvenir (106).

68. Jehl, *Menlo Park Reminiscences*, 674; *Papers*, 7.725.

69. Friedel and Israel, *Edison's Electric Light*, 164–65; Vanderbilt, *Edison, Chemist*, 348–49 McPartland, "Almost Edison," 218. Upton went so far as to deduce a general law of distribution: "The cost [of conductors] increases as the square of the distance from the central station." *Papers*, 5.573.

70. TE Patent 264,642; Hughes, *Networks of Power*, 21; TE Patent 239,147; Dyer and Martin, *Edison*, 386.

71. Josephson, *Edison*, 23.

72. TE Patent 264,642; Dyer and Martin, *Edison*, 343; *Manchester Guardian*, 6 Sept. 1882.

73. *Papers*, 5.764, 783–84, 840–41, 843; Charles L. Clarke, "Menlo Park in 1880," TENHP; Marshall, *Recollections of Edison*, 30; W. S. Andrews, "A Short History of the First Underground System Used for Edison Lamps," ts., 1907, 1–3, TENHP; *Papers*, 5.764; Jehl, *Menlo Park Reminiscences*, 723ff.

74. Andrews, "Short History," 3.

75. Charles Mott laboratory diary, entry for 3 Nov. 1880, PTAE; Andrews, "Short History," 3. TE's grand gesture nearly became an embarrassment. There was a surge back to the Democracy after midnight, and Garfield won by only 1,898 votes.

76. Friedel and Israel, *Edison's Electric Light*, 151; *Papers*, 5.908–9, 898–90. For a detailed account of lamp production in this period, see Jehl, *Menlo Park Reminiscences*, 786–816.

77. *Papers*, 5.889–90.

78. *Papers*, 5.739; Friedel and Israel, *Edison's Electric Light*, 134–46. William Hammer notes that power was delivered to the factory via a three-quarter-mile overhead cable, "[giving] to the world the first demonstration of how current could be distributed successfully." Hammer, "Edison and His Inventions," vi.

79. Friedel and Israel, *Edison's Electric Light*, 134; Pretzer, *Working at Inventing*, 44; Jehl, *Menlo Park Reminiscences*, 406–7.

80. Association of Edison Illuminating Companies, *"Edisonia,"* 149; Jehl, *Menlo Park Reminiscences*, 805. The last eight words of this sentence are taken from Friedel and Israel, *Edison's Electric Light*, 134.

81. *Papers*, 5.882, 825–26; Charles Mott laboratory diary, 4 and 6 Aug. 1880, PTAE.

82. See Dyer and Martin, *Edison*, chap. 13, "A World-hunt for Filament Material."

83. *Papers*, 5.891.

84. Allerhand, *Illustrated History*, 226; *Cincinnati Enquirer*, 13 Nov. 1880; *Papers*, 5.922. According to Francis Jehl, Maxim's visit took place a few weeks before Ludwig Boehm left Menlo Park to work for him. Jehl, *Menlo Park Reminiscences*, 612.

85. Jehl, *Menlo Park Reminiscences*, 707; Friedel and Israel, *Edison's Electric Light*, 160; "Has Edison Been Outdone?" New York *Sun*, 17 Nov. 1880.

86. *Papers*, 5.915, 923; *New York Evening Post*, 22 Nov. 1880.

87. Allerhand, *Illustrated History*, 172. But see TE to Morton in *Papers*, 5.906: "I certainly have believed that you have not treated me exactly right for reasons which I cannot fathom."

88. Pretzer, *Working at Inventing*, 119. For the early relations of TE and Barker, see Hounshell, "Edison and the Pure Science Ideal."

89. *Papers*, 5.922.

90. Stathis Arapostathis and Graeme Gooday, *Patently Contestable: Electrical Technologies and Inventor Identities on Trial in Britain* (Cambridge, MA, 2013, 182; Joseph Swan, "The Sub-Division of the Electric Light," *Journal of the Society of Telegraphic Engineers* 9 (1880), 346, 342, 362.

91. Alexander Muirhead to TE, 13 Jan. 1881, PTAE.

92. *Papers*, 5.920–21.

93. *Papers*, 5.941, 921.

94. *Papers*, 5.943–44; Hounshell, "Edison and the Pure Science Ideal."

95. *Papers*, 5.761–62, 778; Hounshell, "Edison and the Pure Science Ideal"; Israel, *Edison*, 464.

96. *Papers*, 5.944.

97. *Papers*, 5.889–91; *Edison Electric Light Company vs. United States Electric Lighting Company—On Letters Patent 223,898 (1890)*. See Allerhand, *Illustrated History*, 225–27 and 125–28.

98. *Papers*, 5.894; "Thomas Edison in Kansai," *Kansai Culture* (Japan), 10 Apr. 2018; Hammer, "Edison and His Inventions," vi; Dyer and Martin, *Edison*, 376.

99. *Papers*, 5.903–44.

100. Friedel and Israel, *Edison's Electric Light*, 157.

101. *Papers*, 5.959; Strouse, *Morgan*, 232. Except where otherwise indicated, the following account is based on Friedel and Israel, *Edison's Electric Light*, 154; Francis Jehl, "Thomas A. Edison – Menlo Park," ts. memoir, June 1908, TENHP; Jehl, *Menlo Park Reminiscences*, 779–84; *New York Times*, 21 Dec. 1880, a detailed report from which the quotations are taken. Additional material from New York *World*, 21 Dec. 1880, and *Chicago Tribune*, 24 Dec. 1880.

102. *New York Times*, 21 Dec. 1880; Friedel and Israel, *Edison's Electric Light*, 172; Dyer and Martin, *Edison*, 385.

103. *New York Times*, 21 Dec. 1880; Charles Mott laboratory diary, 20 Dec. 1880; Jehl, "Thomas A. Edison," 88. "Here breathed a little community of kindred spirits, all in young manhood, enthusiastic about their work, expectant of great results; moreover, often loudly explosive in word, emphatic in joke and vigorous in action." Charles Clarke quoted in Jehl, *Menlo Park Reminiscences*, 858.

104. *Chicago Tribune*, 1 Jan. 1881. The eclipse, "two new moons" in December, and TE's "melancholy," were all noted by Charles Mott in his laboratory diary, 31 Dec. 1880, PTAE.

105. *Chicago Tribune*, 1 Jan. 1881.

106. Arapostathis and Gooday, *Patently Contestable*, 177; Mary and Kenneth Swan, *Joseph William Swan: Inventor and Scientist* (Newcastle-upon-Tyne, UK, 1929), 58–60.

107. *Chicago Tribune*, 1 Jan. 1881.

108. Strouse, *Morgan*, 232; Charles Mott laboratory diary, 6 Jan. 1881; New York

Sun, 7 Jan. 1881. Amos J. Cummings of the *Sun* accompanied the financial party.

109. Hammer Reminiscences, TENHP.

110. Strouse, *Morgan*, 232.

111. New York *Sun* quoting an unnamed EELC director, 7 Jan. 1881; Friedel and Israel, *Edison's Electric Light*, 174–75.

112. Charles L. Clarke, "Economy Test of the Edison Electric Light at Menlo Park, 1881," in Association of Edison Illuminating Companies, *"Edisonia,"* 166. TE wanted to publish the results of the test at once, but when Henry Villard reminded him that knowledge was—especially in this case—power, he allowed it to remain secret until 1904.

113. Ibid., 169.

114. Ibid., 173, 177. See also Friedel and Israel, *Edison's Electric Light*, 144–45; TE quoted in Association of Edison Illuminating Companies, *"Edisonia,"* 85.

115. Strouse, *Morgan*, 230; *Papers*, 5.973.TE's temporary fashion consciousness was noted by Jehl, *Menlo Park Reminiscences*, 506.

116. This portrait of Mary Edison at Menlo Park is based on Öser, "Wizard of Menlo Park," and Jehl, *Menlo Park Reminiscences*, 507–14. See also Israel, *Edison*, 230–31. "She was a very nice woman, bright and vivacious, beautiful in appearance and character . . . very much devoted to [Edison.]" Statement by Charles T. Hughes, 19 June 1907, Meadowcroft Collection, TENHP.

117. In the 1880s "midtown" Manhattan centered on fashionable Union Square. *Papers*, 6.2; Israel, *Edison*, 230–31; TE to Naomi Chipman, 18 June 1881, PTAE.

118. *New York Herald* clipping, 21 Jan. 1881, in Charles Batchelor Scrapbook (1878–1881), PTAE.

119. Charles Batchelor Scrapbook, 1881, 1573, PTAE; Dyer and Martin, *Edison*, 328.

120. *Papers*, 5.968.

121. Clarke in Jehl, *Menlo Park Reminiscences*, 862.

122. Ibid.

123. *Papers*, 5.969; Suncalc.net.

124. *Papers*, 5.730ff.; *Scientific American*, 22 Jan. 1881; Jehl, *Menlo Park Reminiscences*, 714.

125. William H. Preece, "Electric Lighting at the Paris Exhibition," *Journal of the Society of Arts*, 16 Dec. 1881; *Papers*, 5.818–19.

126. The following account is taken from the recollections of Charles Clarke in a speech to the New York Illuminating Engineering Society, 14 Nov. 1907, transcript in William Meadowcroft Collection, TENHP; also Clarke in Jehl, *Menlo Park Reminiscences*, 860, and TE quoted in Jones, *Edison: Sixty Years*, 116. Slightly different RPM figures are given by other witnesses in *Papers*, 5.991.

127. Charles Clarke in Jehl, *Menlo Park Reminiscences*, 860. Wagner applied the term *Erdenton* to the deep E-flat that resounds at the beginning of his *Ring* cycle.

128. Association of Edison Illuminating Companies, *"Edisonia,"* 35; TE quoted in

Jones, *Edison: Sixty Years,* 116. This test dynamo was never used, but served as a model for TE's famous "Jumbo" generator, developed in the summer of 1881.

129. McDonald, *Insull,* 10, 17–18. Stockton Griffin, TE's previous secretary, had either resigned or been fired in February. *Papers,* 5.970.

130. "Mr. Insull's Notes, Feb. 09," ts., in Meadowcroft Collection, 5, 3, TENHP; McDonald, *Insull,* 14–17, 20–21; Dyer and Martin, *Edison,* 329–30.

131. McDonald, *Insull,* 21. At the beginning of the year TE's bank balance had stood at $64,825. TE Private Ledger 1880–81, TENHP; Dyer and Martin, *Edison,* 331. The modern equivalent would be $1.6 million.

132. McDonald, *Insull,* 18, 22, 13; *Papers,* 5.968, 990, 996–97.

133. Charles Clarke to Francis Jehl, 29 Dec. 1932, TENHP.

134. McDonald, *Insull,* 26; Israel, *Edison,* 210–11, 324; Eaton portrait in Findagrave.com.

135. "Mr Insull's Notes," 20; *Papers,* 6.659, 5.xxv; Samuel Insull to John Kingsbury, 1 May 1881, PTAE. See also *Papers,* 6.42–45.

136. Israel, *Edison,* 212; list in Dyer and Martin, *Edison,* 355.

137. Quoted in Dyer and Martin, *Edison,* 719.

138. *Papers,* 5.996, 988.

139. McDonald, *Insull,* 22.

140. *Papers,* 5.993–94; Friedel and Israel, *Edison's Electric Light,* 162, 174, 172.

141. *Papers,* 6.31, 1023, 1029; DeGraaf, *Edison and Innovation,* 55. At the time, the new location of the Edison Lamp Company was referred to as Harrison, New Jersey. Francis Upton urged transferring lamp production there because the locality offered "plenty of boys and girls at low wages." *Papers,* 5.967.

142. *Edison Monthly,* Aug. 1922; Dyer and Martin, *Edison,* 394; Friedel and Israel, *Edison's Electric Light,* 177.

143. Friedel and Israel, *Edison's Electric Light,* 177; construction drawing, 15 Mar. 1882 in *Papers,* 6.429; Dyer and Martin, *Edison,* 402, 400; Samuel Insull in *Papers,* 6.34.

144. *Papers,* 6.84.

145. *Papers,* 6.85, 84.

146. There are six references to Antoine in *The Hunchback of Notre Dame;* the wharf rats appear in the grotesque procession of the Fool's Pope. Quasimodo, whom Grolio calls "an unholy demon" at birth and wishes to "send back to Hell," carries a massive wart on his chest.

147. *Papers,* 5.103–11, 6.84–85; 7.737–38; Notebook N306103 (Spring 1881), 103–11, TENHP.

148. *Papers,* 7.632; Chris N. Alam and H. Merskey, "Neuralgia: The History of a Meaning," *Pain Research Management* (1996), 1:3.

149. Taylor, *Mr. Edison's Lawyer,* 50.

150. TE ms. reproduced in Kate Armour Reed, *A Woman's Touch: Kate Reed and Canada's Grand Hotels* (Canada, 2016), 35; Taylor, *Mr. Edison's Lawyer,* 51. See also MME warning her son Theodore not to fall in love and marry too soon: "[Your father] made such a terrible mistake that he is fearful of you." 26 May 1924, PTAE.

151. Öser, "Wizard of Menlo Park." "Everything Connected with Mrs. Edison Was Ornate in the Extreme." Marshall, *Recollections of Edison*, 28.

152. *Papers*, 6.100.

153. *Papers*, 6.103.

154. Charles Clarke in Association of Edison Illuminating Companies, *"Edisonia,"* 37.

155. Ibid.; *Papers*, 6.100, 103.

156. *Papers*, 6.100, 105–6, 168, 88, 103, 110; Charles Clarke in *"Edisonia,"* 37; TE Laboratory Notebook N-81-04-06, 126–49, TENHP.

157. TE notebook entry, 15 July 1881, *Papers*, 6.104; Charles Clarke in *"Edisonia,"* 39; Jehl, *Menlo Park Reminiscences*, 973. At 350 RPM, the machine produced an electromotive force of 110V, attainable only at 1,000 RPM on traditional Graemme dynamos.

158. *Papers*, 6.169; Charles Clarke in *"Edisonia,"* 50.

159. *Papers*, 6.170–71; Friedel and Israel, *Edison's Electric Light*, 179; Bowers, *Lengthening the Day*, 87.

160. *Papers*, 6.168, 225. Mary, accompanied by her daughter, had been visiting with TE's brother William Pitt Edison in Port Huron, Michigan. She was described as "very ill" in mid-August and "seriously ill" on 2 September. *Port Huron Times Herald*, 29 July and 14 Aug. 1881; *Minneapolis Star Tribune*, 2 Sept. 1881. She was again, or still, ill in early Oct. 1881.

161. TE in Dyer and Martin, *Edison*, 326–27, and *Papers*, 6.813; Jehl, *Menlo Park Reminiscences*, 973.

162. *Papers*, 6.175; TE to Charles Batchelor, 14 Sept. 1881, PTAE; Friedel and Israel, *Edison's Electric Light*, 180; Dyer and Martin, *Edison*, 327.

163. *Cincinnati Enquirer*, 28 Aug. 1881; *New York Times*, 5 Sept. 1881; *Manchester* (UK) *Courier*, 6 Sept. 1881; Menlo Park Scrapbooks, vol. 51A, *passim*, PTAE; *London Standard*, 24 Sept. 188; Preece, "Electric Lighting at the Paris Exhibition." The story that the liner that brought the dynamo to France had just unloaded an elephant destined for Barnum and Bailey's circus is apocryphal.

164. *L'Évènement*, 24 Sept. 1881; Bright, *Electric Lamp Industry*, 55; *Le Figaro*, 29 Oct. 1881; *Journal des Débats*, 22 Oct. 1881.

165. *Papers*, 6.96; *Expériences faites á l'Exposition International d'Électricité par Mm. Allard* . . . (Paris, 1883), 108. See also Robert Fox, "Thomas Edison's Parisian Campaign: Incandescent Lighting and the Face of Technology Transfer," *Annals of Science* 53 (1996).

166. Lowrey to TE, 22 Oct. 1881, Letterbook Series, PTAE.

167. Ibid.

168. *New York Times*, 5 Oct. 1881.

169. Ibid.

170. *Papers*, 6.90.

171. New York Edison Company, *Thirty Years of New York*, 151; *Atlanta Constitution*, 2 Oct. 1881. The arc lamps were hooked up to a small generator in Pearl Street; Friedel and Israel, *Edison's Electric Light*, 163–64; Dyer and Martin, *Edison*, 408–9. For a detailed discussion of Kruesi's insulation techniques, see Vanderbilt, *Edison, Chemist*, 79–83.

172. New York Edison Company, *Thirty Years of New York*, 35; TE quoted in *Papers*, 6.815.

173. Edouard Reményi to TE, 19 Aug. 1881, PTAE. The thirty-document series of letters between these two men tells one of the more moving personal stories in the Edison Papers. It is available online at http://edison.rutgers.edu/Names Search/NamesSearch.php.

174. Reményi to TE, 25 Apr. 1883, PTAE; *Papers*, 6.816; Hammer Reminiscences, TENHP. In 1898 Edison served as a pallbearer at Reményi's funeral in New York.

175. *Papers*, 6.252–53, 263; Israel, *Edison*, 204; *Papers*, 6.253.

176. Insull to superintendent, Pennsylvania Railroad Jersey City, 17 Dec. 1881, PTAE.

177. Jehl, *Menlo Park Reminiscences*, 641. "Its simplicity and accuracy when correctly handled were so great that it remained in use for a number of years; then it was replaced by improved mechanical types." Bright, *Electric Lamp Industry*, 69.

178. Israel, *Edison*, 495, 981–82, 208. TE had already won 170 U.S. patents before the 1880s.

179. Preece, "Electric Lighting at the Paris Exhibition." The post-lecture comments included in this transcript are indicative of the gathering respect in Britain for TE's lighting innovations.

180. Ibid.

181. *Daily News* (UK), 8 Apr. 1882.

182. *Papers*, 6.468; Otto Moses in *Papers*, 6.363. Smithsonian Image 2003-35552, William J. Hammer Collection.

183. Friedel and Israel, *Edison's Electric Light*, 183; *Papers*, 6.314, 334, 350.

184. *Cincinnati Enquirer*, 1 Jan. 1882; *Papers*, 6.701; Bowers, *Lengthening the Day*, 91; *Papers*, 6.260.

185. *Papers*, 6.348.

186. *Papers*, 6.348, 313, 417; New York dispatch to *Topeka Daily Capital*, 4 Mar. 1882; Insull to E. H. Johnson, *Papers*, 6.431. There are a number of references to TE's "recuperative" vacation in Florida in U.S. newspapers during the month.

187. *Scientific American*, 22 July 1882.

188. *Papers*, 6.446.

189. *New York Times*, 9 Mar. 1882.

190. Israel, *Edison*, 167–68.

191. *Papers*, 6.425ff.; New York Edison Company, *Thirty Years of New York*, 29; Friedel and Israel, *Edison's Electric Light*, 185.

192. TE Patent 460,122. See *Edisonian* 7 (Rutgers, NJ, 2012), online at http://edi son.rutgers.edu/newsletter7.html#2; *Papers*, 6.446–47. The first "village" to install this system was Roselle, New Jersey, on 19 Jan. 1883. Israel, *Edison*, 219.

193. Bright, *Electric Lamp Industry*, 68–69; *Papers*, 6.582, 794–95. Edison delayed executing his patent on this invention until 27 Nov. 1882, allowing John Hopkinson to file a similar application in Britain well before that date. *Papers*,

6, 582, 794–95. The Edison system was awarded priority in the United States on 20 Mar. 1883 (U.S. Patent 274,290), but by then Hopkinson already had his British patent.

194. TE to T. C. Martin, June 1909, quoted in Dyer and Martin, *Edison,* 342–43. See *Papers,* 6.609–11, for more details of the Wilber affair.

195. *Papers,* 6.447.

196. *Scientific American,* 22 July 1882. As of late August, TE had initiated three isolated lighting plants in England and three in Germany, as well as others in France, Holland, Italy, Hungary, Cuba, and Chile. Edison Electric Light Co., *Thirteenth Bulletin,* 28 Aug. 1882.

197. *New York Tribune,* 5 Sept. 1889; *Scientific American,* 26 Aug. 1882; *Papers,* 6.428.

198. *Scientific American,* 26 Aug. 1882.

199. Ibid.

200. *Janesville* (WI) *Daily Gazette,* 5 Sept. 1882.

201. New York *Sun,* 5 Sept. 1882. There is some ambiguity in the newspaper accounts covering the system activation, *The New York Times* reporting that the lights came on two hours after the dynamos. But that was still two hours before dark. Edison Electric Light Co., *Fourteenth Bulletin,* 14 Oct. 1882, states specifically, "The plant was started and the district lighted up at 3 P.M." Whatever the time, TE wrote in old age that starting up the system was "the most thrilling event of my life." TE to F. D. Hopley, 11 Apr. 1921, HFM.

202. "Edison's Electric Light," *New York Times,* 5 Sept. 1882; see also *New York Herald,* New York *World,* and *New York Tribune,* same date.

203. *New York Times,* 5 Sept. 1882.

204. Ibid. and New York *Sun,* 5 Sept. 1882. Francis Jehl, writing more than fifty years later in *Menlo Park Reminiscences,* 1065, had TE lighting up his system in the offices of Drexel, Morgan, but most contemporary accounts put him in the Pearl Street station. See, however, *Papers,* 6.539.

205. New York *Sun* and *New York Tribune,* 5 Sept. 1882.

206. The *New York Herald* actually had its own (isolated) Edison light system. For the pleased reaction of one group of night reporters, see New York Edison Company, *Thirty Years of New York,* 26–27.

207. See, e.g., the London *Standard* and *Daily News,* 6 Sept. 1882; *Boston Globe,* 6 Sept. 1882.

208. New York Edison Company, *Thirty Years of New York,* 44. The date of the following incident is uncertain, except that it occurred on a Sunday, probably 10 September. See *Papers,* 6.670 and 676–77.

209. *Papers,* 814–15 (TE dictating in June 1909).

210. Clarke in Association of Edison Illuminating Companies, *"Edisonia,"* 49; New York Edison Company, *Thirty Years of New York,* 30; Dyer and Martin, *Edison,* 404. See also Hughes, *Networks of Power,* 43.

211. *Papers,* 6.815.

212. Edison Electric Light Co., *Fifteenth Bulletin,* 20 Dec. 1882. See also *Scientific American,* 30 Dec. 1882; Clarke in *"Edisonia,"* 47, 49; *Papers,* 6.671, 815,

676–77; *Scientific American,* 30 Dec. 1882; New York Edison Company, *Thirty Years of New York,* 46.

213. Stephen Garmey, *Gramercy Park: An Illustrated History of a New York Neighborhood* (New York, 1984), 74; Israel, *Edison,* 201; *Papers,* 6.675.

214. *Papers,* 6.199, 680, 683. The phrase "the most fashionable quarter of the city" is TE's, *Papers,* 7.724.

215. *Papers,* 6.675.

216. *Papers,* 7.745, 724. The diary must have been left behind by Morse's widow, a prior renter. Garmey, *Gramercy Park,* 154.

217. *Papers,* 5.909–10; Tate, *Edison Open Door,* 35. TE's total debt to Mrs. Seyfert was actually around $7,000 ($181,300 in today's money) comprising interest and another note later dropped from her suit. The case grew to rival *Jarndyce v. Jarndyce* in complexity but is ably summarized in *Papers,* 7.603-4 and 8.328–29.

218. Edison Electric Light Co., *Sixteenth Bulletin,* 2 Feb. 1883; *Times,* 5 Jan. 1883.

219. Edison Electric Light Co., *Sixteenth Bulletin,* 2 Feb. 1883; *Papers,* 6.668.

220. Edison Electric Light Co., *Sixteenth Bulletin,* 2 Feb. 1883; Insull in *Papers,* 6.669.

221. Edward Johnson looked over the system at Morgan's request and said, "If it was my own, I would throw the whole damned thing into the street." "That's just what Mrs. Morgan says," the financier replied. However, he was too aware of the moneymaking potential of domestic incandescent light to jettison either the system or his own investment in it. Strouse, *Morgan,* 233–34; *Papers,* 6.750–51.

222. Edison Electric Light Co., *Sixteenth Bulletin,* 2 Feb. 1883; *North British Daily Mail,* 4 Dec. 1882.

223. *Papers,* 6.736.

224. *Papers,* 7.727.

225. Snow Removal File (1922), TENHP; *Papers,* 6.802, 23ff., 824–25; TE Patent 228,329 (described at length by TE in *New York Evening Post,* 24 May 1881; *Chicago Tribune,* 24 June 1881. TE did not personally discover the Quogue deposit, as some sources suggest. His only recorded visit to the site took place with Insull in early June 1881. *Papers,* 6.76.

226. Edison Ore-Milling Co. Minutes, 2 June and 28 Oct. 1881, 17 Jan. 1882, PTAE; Sherburne Eaton in *Papers,* 6.765–70; *Engineering and Mining Journal* 52 (1891). TE won a contract to supply two hundred tons of beach magnetite ore to the Poughkeepsie Iron & Steel Co., but that concern got into difficulties and canceled its order. TE closed down his Rhode Island operation in December 1882. See Eaton in *Papers,* 6.765–70.

227. *Papers,* 6.756, 793, 754, 773.

228. *Papers,* 6.772–73; Samuel Johnson to Lord Chesterfield, 7 Feb. 1755, quoted in Boswell's *Life of Johnson,* chap. 13.

229. Edouard Reményi to TE, 25 Apr. 1883, PTAE.

230. *Papers,* 7.73, 6.809; Israel, *Edison,* 218, 221.

231. *Papers,* 6.793–94.

232. Israel, *Edison*, 218–19; Edward Johnson to TE, *Papers*, 7.129–34; Bowers, *Lengthening the Day*, 105.

233. *Papers*, 7.609–10, 75; Israel, *Edison*, 221.

234. Israel, *Edison*, 223–24; *Papers*, 6.798.

235. *Papers*, 7.74, 6.809; DeGraaf, *Edison and Innovation*, 67.

236. *Brooklyn Daily Eagle*, 29 July 1883.

237. Öser, "Wizard of Menlo Park."

238. *Papers*, 7.217–18.

239. William Pitt Edison to TE, 12 Aug. 1883, PTAE.

240. "TE" to William Pitt Edison, 14 Aug. 1883, PTAE. This letter's salutation, "Friend Pitt," and sign-off, "With kind regards," suggest it was written by Insull.

241. Reprinted in *Science*, 2.29 (24 Aug. 1883).

242. Ibid. If any individual other than TE was impugned in Rowland's address, it was the publicity-seeking physicist George Barker of the University of Pennsylvania. See Hounshell, "Edison and the Pure Science Ideal."

243. Bowers, "Edison and Early Electric Engineering in Britain," in Graham Hollister-Short and Frank James, eds., *History of Technology*, vol. 13 (New York, 1991).

244. Israel, *Edison*, 217; *Papers*, 7.190–91.

245. TE to Theodore Waterhouse, 24 July 1883, *Papers*, 7.191.

246. Ibid.

247. *Chicago Tribune*, 19 June, 1883; TE, "Instructions and Directions, Central Stations," 1883, PTAE; Israel, *Edison*, 225, 224; A. Stuart to Samuel Insull, 28 May 1884, PTAE. See also TE's detailed description of his canvassing method in *Papers*, 7.203–8.

248. *Papers*, 7.76; Israel, *Edison*, 225. See, e.g., TE's struggle to get fully paid for the village system he installed in Sunbury, Pennsylvania, in July 1883. Ibid., 201–3.

249. James Pryor to Samuel Insull, 6 Sept. 1883; TE superscript on Pryor to Insull, 6 Sept. 1883; Pryor to TE, 19 Sept. 1883; TE to Pryor, 20 Sept. 1883; all PTAE.

250. Pryor to TE, 24 Sept. 1883, PTAE; *Papers*, 7.268. The Gramercy Park house cost TE $400 a month, whereas his hotel suite, with incidental expenses, cost $200 a week. Pryor-Edison Lease, 23 Sept. 1882, PTAE; *Papers*, 7.309.

251. Öser, "Wizard of Menlo Park."

252. McPartland, "Almost Edison," 252; Bright, *Electric Lamp Industry*, 71.

253. Dickson to TE, 23 May 1883, PTAE. In old age Dickson claimed that TE hired him, but his work application in March 1883 was acted on by Insull, and his *billet-doux* to TE in May made clear that he had not yet been noticed by the Old Man. TE probably took an interest in him after W. S. Andrews, the chief engineer at the Machine Works, praised his skill at the end of the year. Dickson to Insull, and Raymond Sayer to TE, both 28 Mar. 1883, PTAE; Dickson to TE, 23 May 1883, PTAE; *Papers*, 7.101–2; Andrews to TE, 16 Dec. 1883, PTAE.

254. TE Patent 307,031, executed 15 Nov. 1883, issued 21 Oct. 1884.

255. Dickson to TE, 23 Jan. 1924, TENHP.

256. *Papers*, 7.369; U.S. Patent 307,031.

257. *Papers*, 7.102.

258. *Papers*, 7.369; TE interviewed in New York *Sun*, 27 Aug. 1884.

259. *Papers*, 7.370, 482, 481.

260. DeGraaf, *Edison and Innovation*, 67; McDonald, *Insull*, 31; Israel, *Edison*, 228. "If you want money wire me." Insull to TE, 22 Feb. 1884, *Papers*, 7.434. See also Mary Edison to Insull, 27 Feb. 1884, *Papers*, 443.

261. Insull to A. O. Tate, 27 Oct. 1884, PTAE; McDonald, *Insull*, 31–32.

262. S. B. Eaton to Insull, 18 Feb. 1884, PTAE.

263. *Papers*, 7.436, 454.

264. Ibid.

265. TE pocket notebook 84-02-85, 2, TENHP. The editors of the Edison Papers identify this sketch as "a telephone and resonator in a Wheatstone bridge arrangement." *Papers*, 7.434.

266. Insull to TE, 22 Feb. 1884, *Papers*, 7.433.

267. *Papers*, 7.480.

268. "Edison's Electric Shark Hunt," *Nebraska Daily State Journal*, 4 April 1884.

269. Ibid. The Vedder Museum, maintained by the St. Augustine Historical Society, was a local attraction for many years. It burned down in 1914. "The Vedder Museum," n.d., http://lostparks.com/vedder.html.

270. *Papers*, 7.346, 480–81.

271. *Papers*, 7.480; De Borchgrave and Cullen, *Villard*.

272. TE to Eaton, 24 Apr. 1884, PTAE.

273. Rowsome, *Birth of Electric Traction*, loc. 661, 692.

274. *Papers*, 7.482–83; TE to Eaton, 24 Apr. and 9 May 1884, PTAE.

275. *Papers*, 7.494. Nicholas Stilwell suffered from dementia, and Mary had been keeping him in her house under nursing care.

276. *Papers*, 8.328.

277. *Papers*, 7.518–19, 536–57; Mary Edison to Middlesex County Sheriff, 15 May 1884, PTAE.

278. *Papers*, 7.561; James W. Pryor House Inventory, 1882, PTAE; Mary Edison to Insull, 30 Apr. 1884, PTAE; Olive Harper, "In the Wizard's Home: How Thomas A. Edison's Residence Is Fitted Up," New York *World*, 1 June 1884. The article was reprinted or excerpted in several major newspapers. See also *Papers*, 7.562–70.

279. *Washington Post*, 26 November 1878.

280. Harper, "In the Wizard's Home."

281. *Papers*, 7.632; Öser, "Wizard of Menlo Park."

282. Mary Edison described her children in Harper, "In the Wizard's Home."

283. *Papers*, 7.483, 575.

284. Carlson, *Tesla*, 69–70; *Papers*, 6.821. See also Jehl, "Thomas A. Edison," 90. ("He used to order every dish twice"). According to T. C. Martin, writing in the February 1894 issue of *Century Magazine,* Edison also boggled at Tesla's appetite and asked if he was a cannibal.

285. Carlson, *Tesla*, 68–70. Tesla began to work for TE on 8 June 1884. He was paid $100 a month, a high wage by TE's standards, and the equivalent of $2,690 in 2018. Salary List, Edison Machine Works, PTAE.

286. TE Edison Medal acceptance speech, 18 May 1917, *Electrical Review and Western Electrician* 70 (26 May 1917).

287. Mary Edison to Andrew Disbrow, 13 May 1884, PTAE; *Papers*, 7.632.

288. *Papers*, 8.328.

289. TE to Eaton, 22 July 1884, PTAE; Israel, *Edison*, 234–35.

290. *Papers*, 7.620–21.

291. Ibid., 7.622; Öser, "Wizard of Menlo Park," 5.

292. Alice Stilwell Holzer to William A. Symonds, 2 July 1932, HFM. In old age, Marion Edison Öser said the same thing, very likely repeating what she had been told as a child. Israel, *Edison*, 233.

293. William R. Gowers, *A Manual of Diseases of the Nervous System* (London, 1893), 2.375–76. See also the editorial discussion of Mary Edison's death in *Papers*, 7.620–24. It includes the pertinent information that in late 1883 the Edison household ordered "two one-half ounce bottles of sulphate of morphia, a form suitable for hypodermic injection."

294. New York *World Supplement*, 17 Aug. 1884; *Papers*, 7.630–35.

295. The full article is reprinted with commentary in *Papers*, 7.630–35. It states that TE tried for two hours after his wife's death to revive her with shocks from an electric "cabinet." Electrotherapy in overdose cases was not unknown in the nineteenth century, but later that year TE said he did not believe in it. *Papers*, 7.633.

296. TE quoted in New York *Sun*, 27 Aug. 1884; Mary Edison Holzer to Francis Jehl, 27 Apr. 1935, TENHP.

297. Insull to Harriet Clarke, 17 Sept. 1884, PTAE.

298. *Papers*, 7.590, 573; *San Francisco Examiner*, 21 Sept. 1884; Smithsonian Institution, SI neg. 85-8773. The column's 2,100 bulbs represented one day's output of the Edison Lamp Works.

299. TE in *Buffalo Courier*, 8 Sept. 1884.

300. *Edison and Gilliland v. Phelps*, Testimony on Behalf of Edison (2 June 1886), 3; Wile, "Edison and Growing Hostilities"; *Papers*, 7.658.

301. *Edison and Gilliland*, 3; TE Patent 438,304. TE executed eight more sonic-communication patents before the end of 1884.

302. Israel, *Edison*, 227–29, 322, 228; *Papers*, 7.685–88, 7.687–98; Lowrey to TE, 19 Oct. 1884, PTAE; McDonald, *Insull*, 32ff.; De Graaf, *Edison and Innovation*, 67; Insull to Tate, 27 Oct. 1884. In 1909 TE boasted to his biographer T. C. Martin, "I am the only man that ever beat Drexel & Morgan Company over an election of directors and officers." *Papers*, 7.731.

303. U.S. Patent 422,577, filed 1 Dec. 1884. The idea of inductive signaling from moving trains was not original to TE or Gilliland. It appears to have been first proposed by A. C. Brown of the Eastern Telegraph Co., in 1881, and more fully articulated by the British inventor Willoughby Smith in his paper, "Voltaic-Electric Induction," read before the Institution of Electrical Engineers on 8 Nov. 1883. Fahie, *History of Wireless Telegraphy*, 100–111.

304. *Edison and Gilliland*, 5; *Papers*, 7.681, MME interviewed by Milton Marmor of AP, 10 Jan. 1947, TENHP.
305. TE to Richard Dyer from Adrian, ca. 22 Feb. 1885. *Papers*, 8.38–40.
306. *Papers*, 7.48–49, 8.22; Fritz, *Bamboo and Sailing Ships*, 5.
307. Fritz, *Bamboo and Sailing Ships*, 5.
308. *Papers*, 8.64–65. The price was later reduced to $2,750.
309. *Papers*, 8.64, 179–80.
310. Israel, *Edison*, 237–38.
311. *Papers*, 8.163–64; MME/Marmor interview, TENHP; TE quoted in Nerney, *Edison, Modern Olympian*, 273. The date of this first encounter between TE and MME is not known, but it likely occurred early in 1885, when MME was in school and TE was often visiting Boston. Some sources suggest that they met, improbably, in New Orleans at the beginning of March. However, they both recalled meeting in the Gilliland apartment, and each mentioned the musical incident. Marion Edison Öser stated that they already knew each other when they met again at the beach house that summer. "Wizard of Menlo Park." See also *Papers*, 8.163–65.
312. MME interview, 10 Jan. 1947, TENHP.
313. Israel, *Edison*, 244; Ellwood Hendrick, *Lewis Miller: A Biographical Essay* (New York, 1925); *Papers*, 8.246–47. On 25 Feb. 1886 the New York *Sun* estimated Miller's fortune at $2.5 million, or $68.7 million in today's money.
314. Israel, *Edison*, 244; Hendrick, *Lewis Miller*; *Papers*, 8.246–47.
315. "We all set around the table to write up our diaries." TE Diary, entry for 15 July 1885, PTAE. This seven-day journal—the only personal diary TE ever kept—can be read in its calligraphed entirety online at edison.rutgers.edu /NamesSearch/SingleDoc.php?Docid=MA001. It has also been published in Runes, *Diary and Sundry Observations of Thomas Alva Edison*, and in *Papers*, 8.162ff.
316. Ibid., entries for 14 and 15 July 1885.
317. Ibid., entries for 17 and 19 July 1885.
318. Ibid., entry for 21 July 1885.
319. Ibid.
320. Ibid., entry for 12 July 1885.
321. Ibid., entries for 12 and 15 July 1885; *Papers*, 8.189; TE Diary, entry for 20 July 1885.
322. TE Diary, entries for 12–21 July 1885, *passim*. See also the editorial annotations in *Papers*, 8.170–89.
323. TE Diary, entry for 17 July 1885.
324. Israel, *Edison*, 246–47.
325. MME interview, 10 Jan. 1947, TENHP; *Papers*, 8.217. Louise Igou married Robert A. Miller in 1887.
326. MME interview, 10 Jan. 1947, TENHP.
327. AP clipping, 2 Feb. 1947, PTAE.
328. Öser, "Wizard of Menlo Park," 10. See also Israel, *Edison*, 233, 253.
329. TE to Lewis Miller, 30 Sept. 1885, PTAE.
330. Edward Johnson to Uriah Painter, 12 Oct. 1885, PTAE.

331. The female members of the Miller family were more doubtful about TE than the patriarch. Israel, *Edison,* 248.

332. TE Diary, entry for 19 July 1885, PTAE; "The Most Difficult Husband in America," *Collier's Magazine,* 18 July 1925. See *Papers,* 8.256–58 for TE's current plans for the development of his Florida estate.

333. Kristin Herron, *The House at Glenmont: Edison National Historic Site* (West Orange, NJ, 1998). See *New York Tribune,* 19 July 1884, for details of the Pedder case.

334. Glenmont was valued at $400,200 at the time Pedder's creditor, Arnold, Constable & Co., took it over in part payment of his debt. *New York Tribune,* 19 July 1884.

335. *Papers,* 8.315; Hughes, *Networks of Power,* 45; Bright, *Electric Lamp Industry,* 71, 75.

336. *Papers,* 8.328–29, 261, 257–58, 319; TE to William Mawer, 9 Feb. 1886, PTAE; *Fort Myers Press,* 9 Nov. 1885 and 13 Feb. 1886; *Philadelphia Inquirer,* 3 Feb. 1886.

337. *Papers,* 8.348, 344; Charles Batchelor Diary, entries for 20 and 23 Feb. 1886, PTAE; New York *Sun,* 21 Feb. 1886.

338. "Under the Wish-Bone," *Akron Daily Beacon,* 25 Feb. 1886; *Philadelphia Inquirer,* 25 Feb. 1886; Nerney, *Edison, Modern Olympian,* 273–74. Despite the attendance at the ceremony of TE's eldest associates, Edward Johnson and Charles Batchelor, TE chose Lt. Frank Toppan, USN, a friend of Ezra Gilliland's, to be his best man. *Papers,* 8.339–40.

339. *Atlanta Constitution,* 27 Feb. 1886; *Papers,* 8.423–24, 429.

340. MME to Mary V. Miller, 28 Feb. 1886, PTAE; *Doctrines and Discipline of the Methodist Episcopal Church* (Cincinnati, 1854).

341. MME to Mary V. Miller, 28 Feb. 1886, PTAE.

342. TE executed only thirty-one patents in the eighteen months following Mary Edison's death, compared to 164 in the equivalent period preceding.

343. *Papers,* 8.348, 427; TE Florida Notebook, entry for 16 Apr. 1886, 178, PTAE.

344. Author's count of devices or experimental innovations, in TE's Florida notebooks, that are not merely whimsical.

345. *Papers,* 8.483–87, 493–95, 105–11, 375, 475; George P. Lathrop, "An Interview with the Wizard of Menlo Park," *New York Union and Advertiser,* 22 May 1885.

346. This observation was first made by the editors of the Edison Papers. *Papers,* 8.105.

347. *Fort Myers Press,* 20 Feb. 1886; MME to Mary V. Miller, 28 Feb. 1886, PTAE.

348. This marital problem, and MME's general unhappiness on honeymoon, may be clearly deduced from Lewis Miller's letter to her of 26 Apr. 1887. See *Papers,* 8.695–97.

349. TE memorandum to Eli Thompson, ca. Apr. 1886, *Papers,* 8.517–24.

350. *Papers,* 8.362, 377.

351. TE quoted in Nerney, *Edison, Modern Olympian,* 277.

352. Hammer, "Edison and His Inventions," ii.

353. TE quoted in *Chautauqua Assembly Magazine,* Oct. 1886; *Papers,* 8.549–50; McDonald, *Insull,* 38. For a short discussion of TE's attitude to labor, see Israel, *Edison,* 444.

354. See Israel, *Edison,* 238–39, 241–42; TE Patent 333,291 ("Way-Station Quadruplex Telegraph") issued 29 Dec. 1885, and 422,072, issued 25 Feb. 1890.

355. Israel, *Edison,* 243.

356. Ibid., 243, 255; *Chicago Tribune,* 14 Aug. 1886; *Papers,* 8.527; Nerney, *Edison, Modern Olympian,* 274.

357. Quoted in Israel, *Edison,* 267.

358. "Mina Miller Edison Pregnancies and Miscarriages/Stillbirths," unpublished research note by Michele Albion, 28 Feb. 2007, with additional research by Thomas E. Jeffrey, author's collection; *Papers,* 8.682–84.

359. U.S. Patent 352,105; Allerhand, *Illustrated History,* 273–78.

360. For an analysis of the internet myth that TE "stole" technology from Tesla, see *The Edisonian,* vol. 11 at http://edison.rutgers.edu/newsletter11.htm#7.

361. Frank Sprague to Edward H. Johnson, 13 Sept. 1886, quoted in *Papers,* 8.621; Israel, *Edison,* 324–25. See also *Papers,* 8.625.

362. The actual purchase was made by Francis Upton, who took advantage of its availability in Paris on 25 Nov. 1886. *Papers,* 8.655–56.

363. Allerhand, *Illustrated History,* 279–83; *Papers,* 8.637–38.

364. *Papers,*7.729, 8.672, and 667.

365. *Papers,* 8.675–76; New York *Sun,* 15 Feb. 1887; *Fort Myers Press,* 14 Apr. 1887.

366. *Papers,* 8.696.

367. Lewis Miller to MME, 26 Apr. 1887, *Papers,* 8.695.

368. Jane Miller to MME, 19 May 1887, TENHP.

369. Welch and Burt, *Tinfoil to Stereo,* 20.

370. See *Papers,* 8.714; Israel, *Edison,* 281. TE was not afraid of an infringement suit regarding wax incision, since he had experimented with it himself in the 1870s.

371. Welch and Burt, *Tinfoil to Stereo,* 23; Israel, *Edison,* 281.

372. TE quoted in Israel, *Edison,* 281.

373. TE to George Gouraud, 21 July 1887, *Papers,* 8.768. Gouraud was taken aback by TE's ferocity on this subject. Gouraud to TE, 6 Aug. 1887, PTAE. For a detailed account of the contention between the parties, see Wile, "Edison and Growing Hostilities."

374. Israel, *Edison,* 282, 280; Wile, "Edison and Growing Hostilities," 13; TE superscript, 2 Dec. 1887, on Uriah Painter to TE, 30 Nov. 1887, PTAE; Gardiner Hubbard to Edward Johnson, 13 Oct. 1887, PTAE.

375. TE to James Hood Wright, ca. Aug. 1887, PTAE.

376. Wile, "Edison and Growing Hostilities," 13.

377. Edward Johnson superscript on Uriah Painter to Johnson, 12 Feb. 1888, PTAE.

378. TE to Edward Johnson, 12 Feb. 1888, PTAE.

379. *Scientific American,* 31 Dec. 1887.

380. Wile, "Edison and Growing Hostilities," 16; Israel, *Edison,* 282.

381. National Park Service map of 1888 plant, TENHP; Arthur Kennelly interview,

Biographical Collection, TENHP; National Park Service, *Edison Laboratory,* 1.13–15; Vanderbilt, *Edison, Chemist,* 44–45.

382. National Park Service, *Edison Laboratory,* 17, 19; TE in *Evansville Courier,* 27 Aug. 1928: DeGraaf, *Edison and Innovation,* 85.

383. Israel, *Edison,* 271; Harry F. Miller Reminiscences, Biographical Collection, TENHP.

384. Israel, *Edison,* 292; Spehr, *Man Who Made the Movies,* 76–77. TE planned to experiment with high-speed photography in his new laboratory as early as Nov. 1887, three months before meeting Muybridge (79).

385. Israel, *Edison,* 292–93; Spehr, *Man Who Made the Movies,* 76–77.

386. Wile, "Edison and Growing Hostilities," 21; Israel, *Edison,* 289.

387. Ibid., 286.

388. Insull to Alfred Tate, 23 May 1888, PTAE.

389. Tate, *Edison's Open Door,* 153–55.

390. Israel, *Edison,* 293; Wile, "Edison and Growing Hostilities," 24–25.

391. Wile, "Edison and Growing Hostilities," 25.

392. Madeleine Edison was born on 31 May 1888.

393. Quoted in Nerney, *Edison, Modern Olympian,* 132.

394. TE to Ezra Gilliland, 11 Sept. 1888, quoted in Israel, *Edison,* 289.

395. Gilliland to TE, 13 Sept. 1888, PTAE; Albion, *Florida Life of Edison,* 43–44.

396. TE Caveat 110, 8 Oct. 1888, PTAE. Facsimile reproduction in DeGraaf, *Edison and Innovation,* xviii.

397. Richard Howells, "Louis Le Prince: The Body of Evidence," *Screen* 47, no. 2 (July 2006). See especially *The First Film,* a 2013 documentary by David Nicholas Wilkinson, available at https://vimeo.com/ondemand/thefirstfilm/181293064.

398. TE to Dyer and Seely, 8 Oct. 1888, PTAE. For a detailed account of the development of the Kinetograph by TE and Dickson, see Spehr, *The Man Who Made the Movies,* 82 ff.

399. TE quoted by Lucile Erskine in *St. Louis Post-Dispatch,* 10 Mar. 1912.

400. For accounts of the "War of the Currents," critical of Edison's and the Light Company's cynical misuse of the electrocution issue to attack Westinghouse, see Richard Moran, *Executioner's Current: Thomas Edison, George Westinghouse, and the Invention of the Electric Chair* (New York, 2002); Mark Essig, *Edison and the Electric Chair: A Story of Life and Death* (New York, 2003); Jill Jonnes, *Empires of Light: Edison, Tesla, Westinghouse, and the Race to Electrify the World* (New York, 2004).

401. Israel, *Edison,* 329; Charles Batchelor account in *Cassier's Magazine* 5, Nov. 1893; Allerhand, *Illustrated History,* 284 ff.

402. *New York Times,* 13 Dec. 1889.

403. Moran, *Executioner's Current,* chapter 4.

404. Essig, *Edison and the Electric Chair,* 196–97; TE quoted in Nerney, *Edison, Modern Olympian,* 124; Moran, *Executioner's Current,* loc. 2071.

405. TE's moral opposition prevented the Light Company and its successor, Edison General Electric, from entering the AC market until the spring of 1890. Israel, *Edison,* 332–33.

406. *New York Times,* 6, 13, and 18 Dec. 1889.

407. Moran, *Executioner's Current,* loc. 2147; *New York Times,* 6 Dec. 1888.

408. See *A Warning from the Edison Electric Light Co.* (privately printed, 1888) and McPartland, "Almost Edison," 328–30, for Johnson's early involvement in the AC/DC rivalry.

409. Hendricks, *Origins of American Film,* 1.29–30.

410. Leroy Hughbanks, *Talking Wax: The Story of The Phonograph* (New York, 1945), chap. 3.

411. De Graaf, *Edison,* 80; Stephan Puille, "Prince Bismarck and Count Moltke Before the Recording Horn: The Edison Phonograph in Europe, 1889–1890," translated by Patrick Feaster, 1912, PTAE. The Goethe quotation is from *Faust,* Part One. Some of these unique cylinders, including Mark Twain's, melted away in the Edison Works fire of December 1914. One that survived was recorded by TE himself. It was vocally addressed to James G. Blaine, the 1888 Republican presidential candidate, and took him on an imaginary tour of the world. *Schenectady Gazette,* 28 January 1996.

412. Jones, *Edison: Sixty Years,* 147; Julius Block Edison cylinder recording, ca. 14 Oct. 1889, Marston.records.com.; Julius Block, "Edison Album" 1889, New York Public Library Digital Collections.

413. Israel, *Edison,* 370; Lewis Miller to Mary V. Miller, ca. mid-Apr. 1889, EFW. Hammer's 7-part retrospective series on TE's inventions was serialized in *Electrical World* between 31 Aug. and 12 Oct. 1889.

414. Alan Walker, *Hans von Bülow: A Life and Times* (NY 2010), 409; *New Brunswick Home News,* 2 Mar. 1888; *New York Times,* 21 Apr. 1888; Israel, *Edison,* 321–23. Villard had previously and successfully combined all Edison's European lighting companies (321).

415. Israel, *Edison,* 322; Josephson, *Edison,* 353; TE to Henry Villard, 1 Apr. 1889, quoted in Israel, *Edison,* 324.

416. Ibid.

417. The word *apotheosis* was used to describe TE's stay in Paris by *Figaro* on 30 Aug. 1889. Except where otherwise indicated, the following account is based on daily accounts in *Le Figaro,* 12 Aug.–12 Sept. 1889, with additional details from Annegret Fauser, *Musical Encounters at the 1889 Paris World's Fair* (New York, 2005); Tate, *Edison's Open Door,* 233–44 and Israel, *Edison,* 370–71. TE's own account is in Dyer and Martin, *Edison,* 747–50.

418. Francis Upton Scrapbook, 22 Aug. 1889, PTAE, *Boston Globe,* 8 Sept. 1889.

419. *Le Figaro,* 28 Aug. 1889; *New Albany Evening Tribune,* 10 Sept. 1889; Étienne-Jules Marey, *La Chronophotographie* (Paris, 1829), 26; Marta Braun, *Picturing Time: The Work of Étienne-Jules Marey, 1830–1904* (Chicago, 1995), 189–90.

420. William J. Hammer, "Edison's Display at the Paris Exposition," trilingual booklet in Beinecke Library, Yale University; Israel, *Edison,* 371; K. G. Beauchamp, *Exhibiting Electricity* (London, 1997), 182–84; TE in Dyer and Martin, *Edison,* 747ff.; New York *World,* 16 Sept. 188.

421. Auguste Villiers de l'Isle-Adam, *L'Ève-Future* (Paris, 1886). The complex story of Villiers's twelve-year obsession with TE, culminating in book publica-

tion of his seminal science fiction novel, is well told in Gaby Wood, *Edison's Eve: A Magical History of the Quest for Mechanical Life* (New York 2002); A. W. Raitt, *The Life of Villiers de l'Isle-Adam* (New York, 1981); Carol de Dobray-Rifelj, "La Machine Humaine: Villiers' *Éve-Future* and the Problem of Personal Identity," *Nineteenth-Century French Studies* 20 (Spring 1992); and Ritch Calvin, "The French Dick: Villiers de l'Isle-Adam, Philip K. Dick, and the Android," *Extrapolation,* 22 June 2007 (www.thefreelibrary.com). TE seems to have remained unaware of the novel until the dying Villiers sent him a copy just before his arrival in Paris. In 1910 he contributed $25 toward the erection of a statue in Villiers's memory.

422. *Le Figaro,* 3 Sept. 1889, translated by the author.

423. TE in Dyer and Martin, *Edison,* 750–53; *Scientific American,* 21 Sept. 1889; Israel, *Edison,* 371; *Le Figaro,* 12 Aug. 1889.

424. Puille, "Prince Bismarck"; Israel, *Edison,* 371–72; *Tageblatt der 62. Versammlung deutscher Naturforscher und Ärzte in Heidelberg vom 18–23 Sept.* (Heidelberg, 1890); 141; *New York Herald,* 21 Sept. 1889; *Le Figaro,* 12 Aug. 1889.

425. See TE to Henry Villard, 8 Feb. 1890, PTAE: "I would now ask you not to oppose my gradual retirement from the lighting business, which will enable me to enter fresh and more congenial fields of work."

426. This fantasy of TE's was ventured, shortly after his return to the United States, on the writer George Parsons Lathrop, and published in "Talks with Edison." TE might possibly have been channeling two passages by Tennyson: "And the soul of the rose went into my blood," from *Maud,* and "I am a part of all that I have met," from *Ulysses.*

PART SIX · SOUND (1870–1879)

1. *Papers,* 1.151–56; Israel, *Edison,* 52.

2. *Papers,* 1.146, 151.

3. Nerney, *Edison, Modern Olympian,* 53–54 [misdated as 1869]. John Ott was the elder brother of Frederick Ott, who also worked for TE. He suffered a crippling stroke in 1895, but TE continued to employ and support him for life. His crutches and wheelchair were displayed near TE's coffin. Jeffrey, *Phonographs to U-Boats,* 53.

4. William Ford, *Industrial Interests of Newark, N.J.* (New York 1874), 231; Israel, *Edison,* 52–55; Nerney, *Edison, Modern Olympian,* 53.

5. TE Patent 128,608; *Papers,* 1.147, 151–54. This instrument, which substituted a single print wheel for Calahan's two, never went into production.

6. Israel, *Edison,* 52.

7. *Papers,* 1.161–70. The Edison-Pope gold printer remained in common use throughout the 1880s.

8. *Papers,* 1.172–73.

9. TE to Sam and Nancy Edison, 9 May 1870, HFM.

10. The "Family" folders in TENHP offer abundant evidence of the pecuniary consequences of worldly success.

11. J. J. Anger to TE, 28 Oct. 1929, TENHP.

12. Israel, *Edison,* 54–55.

13. *Papers,* 1.196–207.

14. *Papers,* 1.182; Prescott, *Electricity and Telegraph,* 688–89.

15. Prescott, *Electricity and Telegraph,* 725; TE Patent 114,656; *Papers,* 1.173–75. TE also briefly experimented with automatic telegraphy in Boston in May 1868; Israel, *Edison,* 60ff.; *Papers,* 1.242–43; *Papers,* 1.246.

16. This document (*Papers,* 1.208–9) is apparently the source of the tallest of TE's autobiographical tales, first published in Dyer and Martin, *Edison,* 132–33—that he asked Lefferts to name a price for his universal stock ticker, hoping for $3,000 to $5,000, and was flabbergasted to be given a check for $40,000, or $796,000 in today's money. The story may be a mélange of memories, and is impossible to substantiate from the available records. However, Lefferts's draft did result in a contract with TE on 26 May 1871, conditionally worth much more than $40,000. See *Papers,* 1.283–87.

17. TE to his parents, 30 Oct. 1870, HFM.

18. Walter L. Welch, *Charles Batchelor* (Syracuse, NY, 1972), *passim.* Welch accurately describes Batchelor as "the balance wheel of an organization of which Edison was the mainspring" (5).

19. Jehl, *Menlo Park Reminiscences,* 506.

20. Israel, *Edison,* 61; *Papers,* 1.218–19, 222–23.

21. *Papers,* 1.218, 220.

22. Daniel Craig to TE, 12 Jan. 1871, *Papers,* 1.235.

23. Israel, *Edison,* 62; *Papers,* 1.232.

24. *New York Times,* 23 Apr. 1878; *The Independent,* 25 Apr. 1878.

25. *Papers,* 1.237; William Orton to Anson Stager, 20 and 24 Jan. 1871, quoted in Israel, *Edison,* 54.

26. *Papers,* 1.270.

27. TE quoted in *New York Herald Tribune,* 19 Oct. 1931.

28. *Papers,* 1.277–78, 226.

29. David Hochfelder, *The Telegraph in America, 1832–1920* (Baltimore, MD, 2012), 109; *Papers,* 283–92.

30. Ibid., 1.283–94.

31. *Papers,* 1.225–26. The purchase deal benefited Pope and Ashley at 510 Gold & Stock shares each, as opposed to only 180 for TE. Israel, *Edison,* 54–55; *Papers,* 1.226.

32. Mary Edison interview in New York *World,* 1 June 1884, quoted in *Papers,* 1.563–64.

33. Mary Edison interview, *Papers,* 1.564.

34. Dated personal documents are scanty for this period of TE's life. The author infers his chronology from the reminiscences of Mary Edison, cited above, and Edward Johnson, cited below. Mary says she was "going home from school" when she took shelter in the factory, which suggests June at the latest. If she was speaking precisely in saying she was "fifteen and a half years old" at the time, the encounter could have been as early as April, but her memory of a five-month courtship ties in better with the later date. It

implies a proposal from TE in November, followed by their confirmed marriage on 25 December. Johnson's memories of TE being almost penniless when *they* first met suggests a rough coincidence with the latter's begging letter to Harrington on 22 July. Finally, Mary's denial that she ever worked for TE in a "factory" does not conflict with evidence that she was employed in a short-lived news reporting business he established that fall. See *Papers,* 7.566–67.

35. *Papers,* 1.295–96. TE gives a breezy account of his business method ("which certainly was new") in ibid., 1.644.

36. TE to George Harrington, fragment, 22 July 1871, PTAE.

37. *Papers,* 1.308–9, 644.

38. *Papers,* 1.346; TE Patents 123,984 and 121,601; *Edison,* 69, 67.

39. *Papers,* 1.264–65. The Automatic Telegraph Company building stood at 66 Broadway.

40. Edward Johnson to T. C. Martin, 21 Nov. 1908, TENHP. He remembered this first spell of association with TE as lasting about three months.

41. Johnson quoted in *Executive Intelligence Review,* 9 Feb. 1896; Dyer and Martin, *Edison,* 148.

42. Johnson quoted in *Electrical World,* undated clipping, ca. Mar. 1899, 1899 General File, TENHP.

43. *Electrical World,* undated clipping, ca. Mar. 1899, 1899 General File, TENHP.

44. *Papers,* 7.564.

45. Ibid.

46. Israel, *Edison,* 74–75; *Papers,* 1.346. Until the publication of volume 7 of the *The Papers of Thomas A. Edison,* which contained Mary Edison's own account of her wedding, a nineteenth-century legend that TE returned to laboratory after the ceremony and forgot to come home that night was widely accepted by Edison biographers. That story is now discredited. It is true, though, that he went back to work briefly, with Mary's permission, to deal with a problem to do with his stock-ticker delivery. See *Papers,* 7.560–62.

47. Description based on early photographs of Mary Stilwell Edison. Nicholas Stilwell's occupation was probably the reason Edison included, in his last notebook entry before the wedding, a double-tooth design to prevent band saws from running out of line. *Papers,* 1.376.

48. *Newark Daily Advertiser,* 5 Jan. 1872; *Papers,* 1.385, 7.635; TE quoted in Israel, *Edison,* 75.

49. Ibid.

50. *Papers,* 1.377, 429, 430.

51. *Papers,* 1.429–31.

52. *Papers,* 1.496, 645. See, e.g., S. A. Woods Co. to Edison and Murray, *Papers,* 1.499.

53. Israel, *Edison,* 77; *Papers,* 1.493, 496–97.

54. Ibid.

55. Facsimile in *Papers,* 1.437.

56. *Papers,* 1.506–7.

57. See, e.g., *Papers*, 1.508–12; William Orton testimony, *Atlantic & Pacific Telephone Company v. George B. Prescott* [et al.], vol. 71, 117–18, 125; Josephson, *Edison*, 109.

58. Orton testimony, 129–32; Israel, *Edison*, 79. TE's specific mandate was to develop duplex or diplex designs that would amplify but not conflict with the Stearns patent, which Western Union owned. Israel, *Edison*, 79; Orton testimony, 118ff. Orton and TE made their agreement verbally—a mistake on the former's part that later involved them both in tormented lawsuit over quadruplex rights. Litigation Series, Quadruplex Case, vols. 70ff., PTAE.

59. Phillips, *Sketches Old and New*, 183.

60. Ibid.

61. "Affidavit of Thomas A. Edison in Regard to his Inventions of Duplex and Quadruplex Telegraphy," 27 Apr. 1875, reprinted in *Papers*, 2.810; TE, "Testimony of invention" to Lemuel Serrell, 15 Feb. 1873, Quadruplex Case, 71.2, PTAE. See also *Papers*, 1.527, 529.

62. Dyer and Martin, *Edison*, 148–49; Israel, *Edison*, 83; *Papers*, 1.591.

63. Ibid. Dyer and Martin, *Edison*, 149.

64. Israel, *Edison*, 83; TE in Dyer and Martin, *Edison*, 150.

65. TE in Dyer and Martin, *Edison*, 150.

66. Francis B. Keene, U.S. consul in Geneva, quoted in David Lindsay, *Madness in the Making: The Triumphant Rise and Untimely Fall of America's Show Inventors* (1997; New York, 2005), 229. See also Leah Burt, "George Edward Gouraud," ts., Biographical Collection, TENHP.

67. TE in Dyer and Martin, *Edison*, 151.

68. Ibid., 150.

69. *Papers*, 1.501–2; TE in Dyer and Martin, *Edison*, 152.

70. TE in Dyer and Martin, *Edison*, 151; Israel, *Edison*, 87.

71. TE in Dyer and Martin, *Edison*, 152; London *Echo*, 22 Aug. 1889.

72. Joseph Murray to TE, 12 June 1873.

73. Orton quoted in Maury Klein, *The Life and Legend of Jay Gould* (Baltimore, MD, 1986), 198; Israel, *Edison*, 99; *Papers*, 2.235.

74. Israel, *Edison*, 93–95; *Papers*, 2.3 ff.

75. TE in *Golden Book*, Apr. 1931, quoted in Nerney, *Edison, Modern Olympian*, 232. The rest of this paragraph is closely based on Israel, *Edison*, 87–95.

76. TE quoted in Dyer and Martin, *Edison*, 158; *Papers*, 2.239. The article erroneously listed George Prescott of Western Union as co-inventor of the quadruplex. There were legal repercussions. See Israel, *Edison*, 98.

77. Dyer and Martin, *Edison*, 156; TE quoted in Dyer and Martin, *Edison*, 156. For another indication of the quadruplex's complexity, see *Papers*, 2.314.

78. Prescott, *Electricity and Telegraph*, 843–44. "In essence," Paul Israel writes, "Edison used a cascade of electromagnets to bridge over the time during which the reversed current regenerated the magnetic field in the main relay magnet." Israel, *Edison*, 98. See also TE Patent 207,724, executed on 14 Dec. 1874.

79. Quoted in Israel, *Edison*, 99.

80. TE Patent 158,787, filed 13 Aug. 1874; TE in *Scientific American*, 5 Sept.

1874; TE in Dyer and Martin, *Edison*, 183. TE discovered the motograph principle on 10 Apr. 1874. *Papers*, 2.178–79.

81. TE Patent 158,787.

82. *Papers*, 2.178; Jehl, *Menlo Park Reminiscences*, 69; *Scientific American*, 5 Sept. 1874 (italics added). TE's letter to the magazine on this date is reprinted in its entirety in *Papers*, 2.282–83.

83. *Papers*, 2.315–20, 281–82, 301; TE in *Operator*, 25 Nov. 1874.

84. George Barker to TE, 3 Nov. 1874, PTAE; "Edison and *The Telegrapher*," *Papers*, 2.305–7. Ashley's attacks on TE continued through May 1875.

85. *Papers*, 2.331, 369. The editors speculate that TE might have been musing invective to get back at James Ashley.

86. *Papers*, 2.360.

87. *Papers*, 2.364, 813, 361. This was the note that later became the property of Lucy Seyfert, and the cause of TE's protracted legal squabble with her, described in Part Five.

88. *Papers*, 2.813, 801, 366, 341, 365; Israel, *Edison*, 102; William Orton to Joseph Stearns, 2 Dec. 1874. TE did, however, file four precautionary caveats on 4 Dec. 1874. See *Papers*, 2.347–60.

89. *Papers*, 2.364, 813, 801.

90. TE reminiscence, *Papers*, 2.780. The date of this visit is uncertain, but it probably occurred on or just before 30 Dec. 1874, the day Gould concluded his acquisition of the Automatic Company.

91. Phillips, *Sketches Old and New*, 186; Klein, *Life of Gould*, 216, 197–200; Israel, *Edison*, 102.

92. TE reminiscence, *Papers*, 2.780.

93. *Papers*, 2.788. The details of the transaction were more complex than TE chose to remember. See *Papers*, 2.378–79.

94. Israel, *Edison*, 102; *Papers*, 2.801.

95. *Papers*, 2.405, 407. Orton died on 22 April 1878, aged fifty-one.

96. *Papers*, 2.375, 382. There is a facsimile of Mary's elaborate invitation card in *Papers*, 2.418.

97. *Papers*, 2.463. Harrington, ailing, sold TE's automatic patents to Gould in Apr. 1875 and moved to England, leaving his partners much distressed. The territorial and patent wars between Western Union and A&P were resolved in a merger of both companies in 1877. Israel, *Edison*, 104.

98. *Papers*, 2.488; Israel, *Edison*, 104–5.

99. *Papers*, 2.493–94, 2.495, 500–502.

100. *Papers*, 2.502.

101. Jehl, *Menlo Park Reminiscences*, 10ff., 99; *Papers*, 2.561–62, 582; TE "Autographic Press" draft caveat in *Papers*, 586ff.; Israel, *Edison*, 106. DeGraaf, *Edison and Innovation*, notes the electric pen's ancestry of today's tattoo needle (16–17).

102. Charles to Tom Batchelor, 1 Sept. 1875, PTAE.

103. U.S. Patent 141,777; TE testimony in *Speaking Telephone Interferences*, 1.5, Litigation Series, PTAE, hereafter *Telephone Interferences*. A year later Elisha Gray adopted the same principle in his "musical telegraph." It was the feature

that distinguished his telephone design from (and above) Bell's, when they both filed for patents on 14 Feb. 1876. See Seth Shulman, *The Telephone Gambit: Chasing Alexander Graham Bell's Secret* (New York, 2009), for the possibly criminal consequences of this coincidence.

104. *Papers,* 2.581; TE caveat "Acoustic Telegraphy," 22 Nov. 1875, *Papers,* 2.645.

105. *Papers,* 2.647; "Edison's Discovery of a Supposed New Force," *Operator,* Jan. 1876. TE sketched some of these scintillations, even recording their colors and "scents." *Papers,* 2.689.

106. *Papers,* 2.494, 648.

107. Israel, *Edison,* 112; Arthur Kennelly interview, 19 May 1936, Biographical File, TENHP.

108. Addresses of William J. Hammer and Arthur Kennelly, *Minutes of the 38th Meeting of the Association of Edison Illuminating Cos.* (Oct. 1922), 397. See also the comments of Oliver Lodge in Francis Jehl, "Edison's Contributions to Wireless," *Edison Monthly,* Dec. 28, and Arthur Kennelly interview, 19 May 1936, Biographical File, TENHP. The latter described TE's dark box as "the first piece of wireless apparatus in the world." TE's etheric force experiments in November and December 1875 are detailed in *Papers,* 2.646–702.

109. Israel, *Edison,* 130, states that TE studied Hermann von Helmholtz's seminal *On the Sensations of Tone as a Physiological Basis for the Theory of Music* around this time. He cites as evidence TE's copious marginalia in a copy of the July, 1875, English translation preserved as a family relic in ENHP. Contemporary references in the marginalia, however (to "Marconi the Receiver," e.g.,) indicate that they were written in the 1890s, and possibly not until TE's "Insomnia Squad" experiments in 1912. Patrick Feaster notes that as late as the spring of 1877, TE had little knowledge of Helmholtzian acoustic theory. "Speech Acoustics and the Keyboard Telephone: Rethinking Edison's Discovery of the Phonograph Principle," *ARSC Journal* 38.1 (2007), 16–17.

110. See Gall, "Edison: Managing Menlo Park"; quote in DeGraaf, *Edison and Innovation,* 22; Israel, *Edison,* 130.

111. *Papers,* 2.524, 526, 720, 723; TE Patent 141,777; Dyer and Martin, *Edison,* 176.

112. *Papers,* 2.629.

113. Israel, *Edison,* 123. There is a reproduction of the invitation to this party (calligraphed with an electric pen) in *Papers,* 2.769.

114. Gall, "Edison: Managing Menlo Park"; Josephson, *Edison,* 96.

115. *Papers,* 3.3.

116. Samuel Edison fathered three children by Mary Sharlow, whom he never married. "Edison, Miller, and Affiliated Families" in Jeffrey, *Phonographs to U-Boats,* 163.

117. George P. Lathrop, "Edison's Father," unidentified newspaper article, 20 Jan. 1894, TENHP.

118. Israel, *Edison,* 123; Vanderbilt, *Edison, Chemist,* 31.

119. "A Visit to Edison," *Philadelphia Weekly Times,* 29 Apr. 1878; Josiah Reiff testified to TE's need for ambient silence in *Telephone Interferences,* 1.274–78. Bell's historic "Mr. Watson" telephone call had been made on 10 Mar. 1876.

120. See, e.g., TE Patent 198,087, "Telephonic Telegraph."

121. U.S. Centennial Commission, *List of Awards Made to the American Exhibitors, International Exhibition 1876* (Philadelphia, 1876), 79, 113; Israel, *Edison,* 125. Thomson was equally impressed with TE's American automatic telegraph, which could send more than one thousand words a minute, ten times the speed "attained by the best of the other systems hitherto in use in America or any other part of the world." *Papers, 3.55.*

122. Shulman, *Telephone Gambit,* 189–90; Edward Johnson to T. C. Martin, 21 Nov. 1908, TENHP.

123. *Bucks County* (PA) *Gazette,* 1 June 1876; TE to Frederick Royce, 29 May 1876, PTAE. There are no records in PTAE of TE attending the centennial, but the collection is sparse in its coverage of the early Menlo Park period.

124. *Papers,* 3.27–53, 46.

125. Israel, *Edison,* 131, 127; *Papers,* 3.64, 82–83; Israel, *Edison,* 131.

126. *Papers,* 3.229, 344, 172–73, 120, 359, 168, 312, 254–55; TE quoted 255.

127. *Papers,* 1.659 ("It was this instrument which gave me the idea for the phonograph") and 3.250. Edison meant it to improve on the already impressive performance of his automatic telegraph, which on 5 Dec. 1876 transmitted President Grant's 12,600-word annual message from Washington to New York in just over an hour. Jehl, *Menlo Park Reminiscences,* 77.

128. TE technical note NS7704ZC11, 8 Sept. 1877, PTAE; TE in Dyer and Martin, *Edison,* 206–7; TE Patent 213,554, "Automatic Telegraph." See also *Papers,* 3.248–50.

129. Pretzer, *Working at Inventing,* 88; *Papers,* 3.257; Feaster, "Speech Acoustics," 12–13. TE's first official notice that he was intent on transmitting "spoken words regardless of musical key" was in his application for U.S. Patent 474,230 (27 Apr. 1877). He called this metal-diaphragm device a "phonetic or speaking telegraph."

130. *Papers,* 3.257–88.

131. *Papers,* 3.300 ("We can get everything perfect except the lisps & hissing parts of speech such as 'Sh' in shall = get only .o in coach"); TE quoted in Israel, *Edison,* 132–33. See also Pretzer, *Working at Inventing,* 91.

132. Pretzer, *Working at Inventing,* 92–93; *Papers,* 3.518.

133. *Papers,* 3.361; Feaster, "Speech Acoustics," 19. The author is indebted in the sections that follow to Feaster's important article.

134. TE unbound notebook no. 11, 26 May 1877, 109 (signed and witnessed by Batchelor).

135. The circuit-breaking tonewheel had been an essential part of TE's acoustic telegraph designs in 1875 and 1876. *Papers,* 3.361–62; Feaster, "Speech Acoustics," 20–21.

136. Feaster argues that at this stage, TE was still misinformed as to the oscillographic nature of acoustic recording, and that ironically, his ignorance allowed him to proceed without prejudice into development of the phonograph. "Speech Acoustics," 36.

137. Ibid., 19–23.

138. Israel, *Edison,* 137; *Papers,* 3.435, 363; *Journal of the Telegraph,* 1 June 1877.

139. *Papers,* 3.379, 427–28; TE Patents 474,231 and 474,232; Charles to Tom Batchelor, 11 June 1877, PTAE; *Operator,* 15 June 1877.

140. TE Patent 203,014; Elisha Andrew to Gardiner Hubbard (Bell's backer), 16 Jul. 1877, *Papers,* 3.435.

141. Charles Batchelor to Ezra Gilliland, 26 July 1877, *Papers,* 3.463.

142. Marshall, *Recollections of Edison,* 56.

143. *Papers,* 3.444.

144. TE interviewed in *Talking Machine Journal,* Sept. 1927; Feaster, "Speech Acoustics," 22, 28, 24–25; TE, "The Perfected Phonograph," *North American Review,* June 1888; *Buffalo Evening News,* 2 Aug. 1904; TE quoted in Minneapolis *Star Tribune,* 4 Mar. 1878; Dickson and Dickson, *Life and Inventions,* 122–23; TE in Dyer and Martin, *Edison,* 207; Tate, *Edison's Open Door,* 115.

145. TE in Dyer and Martin, *Edison,* 207.

146. Ibid.

147. Transcript in George Gouraud Biographical Collection, TENHP.

148. *Papers,* 3.444. Previously, the word *phonograph* had meant a shorthand stenographer.

149. Patrick Feaster, "Perfectly Reproduced Slow or Fast": A New Take on Edison's First Playbook of Sound," *The Sound Box,* Mar. 2011; *Papers,* 3.439 (facsimile).

150. *Papers,* 3.440. The first scholar to deduce the full significance of this document was Patrick Feaster, in "Speech Acoustics," 25–26. See also Israel, *Edison,* 143 (facsimile). Feaster points out that the phonographic device, as sketched, would not have worked, because TE imagined the cylinder could be played back at a slow speed for a copyist to transcribe, making the sound unintelligible. "Neither Edison nor anyone else would yet have been in a position to know this."

151. Charles Batchelor testimony, *American Graphophone Company v. U.S. Phonograph Company et al.,* 586. For another version of this reminiscence, see *Papers,* 3.699–700.

152. *Papers,* 3.495. Coincidentally, the "phonograph" TE did design on 12 Aug. 1877 differed totally from the Kruesi model. It consisted of a resonating box with two diaphragm-needle attachments, one labled "spk" and the other "listen," following the groove of a roll of paper tape unrolling beneath them. TE Technical Drawing NS7703B, Unbound Notes 1877, PTAE.

153. Allen Koenigsberg, *Edison's Cylinder Records, 1889–1912* (New York, 1987), xiii–xiv; Pretzer, *Working at Inventing,* 110; *Papers,* 3.446–47, 449–50. Israel, *Edison,* points out that TE devoted himself entirely to telephone improvements through October 1877, and did not return to phonograph development until November (139). But see also the cogent observation in Dyer and Martin, *Edison,* that around this time all TE's "phonic" researches tended to blend in his head (198).

154. Charles Batchelor "The Invention of the Phonograph," in *Papers,* 3.699.

155. See, e.g., *Wilmington* (DE) *News-Journal,* 1 Oct. 1877 ("This is no joke").

156. Reprinted in *Papers,* 3.670–74.

157. *Papers*, 3.671 (facsimile), 672.

158. *Papers*, 3.674.

159. According to Newspapers.com, TE was called "Professor" five times in 1877, and 823 times in 1878.

160. *Papers*, 3.628, 4.140, 58–59, 133, 140; *New York Tribune*, 25 Mar. 1877; George Bliss to TE, 13 Apr. 1878, PTAE; *St. Paul* (MN) *Globe*, 3 Mar. 1878 (a widely reprinted article from the New York *Sun*); *New York Daily Graphic*, 10 Apr. 1878.

161. *Papers*, 4.197, 191.

162. TE's letter was reproduced in facsimile in *Daily Graphic* on 16 May 1878.

163. *Washington Evening Star*, 19 Apr. 1878; *Papers*, 4.242–43; *New York Tribune*, 20 Apr. 1878.

164. *Washington Evening Star*, 19 Apr. 1878.

165. *New York Tribune*, 20 Apr. 1878.

166. *Washington Evening Star*, 19 Apr. 1878.

167. John Brisben Walker to TE, 26 Feb. 1924, TENHP; *Philadelphia Inquirer*, 20 Apr. 1878; *Papers*, 4.243, 863.

168. *Feather River* (Quincey, CA) *Bulletin*, 29 June 1878. This article is misdated as 29 Oct. 1877 at Newspapers.com. See also TE's comments on Scott in Tate, *Edison's Open Door*, 114–15.

169. Reproduced above.

170. TE Patent 200,521 was issued on 19 Feb. 1878.

171. *Papers*, 4.134, 3–4; Raymond E. Wile, "The Rise and Fall of the Edison Speaking Phonograph Company, 1877–1880," ts., ca. 1970s, TENHP.

172. Adrian Hope, "A Century of Recorded Sound," *New Scientist* 22 (24 Dec. 1977).

173. *Papers*, 4.133, 862; Pretzer, *Working at Inventing*, 94–95.

174. TE in New York *World*, 23 Apr. 1878 ("Whenever I get to love a man, he dies right away. Lefferts went first, and now Orton's gone too"); Israel, *Edison*, 157, 140; *Papers*, 4.78, 862.

175. *Papers*, 4.181. Orton died on 22 April.

176. *Papers*, 4.243.

177. Israel, *Edison*, 141; Pretzer, *Working at Inventing*, 84.

178. *Papers*, 4.862. Edward Johnson remarked that Edison, "the Dam fool," could have gotten £12,000 ($233,000) if he sold his carbon telephone to a British consortium. As it was, he included all his telephone patents in his deal with Western Union. The telegraph giant then tried to use them to batter the Bell Telephone Company into insolvency. Although this effort failed after many years of litigation back and forth, Western Union at last profited greatly by selling out to Bell. *Papers*, 4.178; Pretzer, *Working at Inventing*, 96.

179. DeGraaf, *Edison and Innovation*, 40–41; *Papers*, 4.259, 55, 260–62.

180. *Papers*, 4.260, 270. Although TE patented the tasimeter in Britain, he left it unprotected in the United States, on the grounds that he sought no royalties from "instruments of a purely scientific character." *Papers*, 4.458.

181. Pretzer, *Working at Inventing*, 96–97; *Papers*, 4.88, 135, 282, 352ff., 265; TE interviewed in *Daily Graphic*, 4 May 1878.

182. Israel, *Edison,* 157; TE to William Preece, 19 May 1878, PTAE.

183. *Telegraphic Journal and Electrical Review,* 1 July 1878; TE to Preece, 4 May 1878, PTAE; Israel, *Edison,* 158; Preece to TE, 22 May 1878, PTAE.

184. Israel, *Edison,* 158.

185. Ibid., 159; William Thomson to Preece, 12 June 1878, PTAE; TE to Henry Edmunds, Jr., 26 May 1878.

186. For a detailed account of the microphone controversy, see Israel, *Edison,* 158–60. It is extensively covered in *Papers* vol. 4, *passim.* Although at least one British journal, *Engineering,* deliberately suppressed data that supported Edison's case, his reputation in Britain suffered as a result of bringing it. Late in the year Thomson criticized him for failing to acknowledge that he had overreacted. "There is no doubt he is an exceedingly ingenious inventor, and I should have thought he had it in him to rise above . . . the kind of puffing of which there has been so much." Thomson to Preece, quoted in *Papers,* 4.677–78.

187. *Papers,* 4.368–72, 347. For the interest of scientific officials in the tasimeter, see *Papers,* 4.270.

188. *Papers,* 4.401.

189. *Papers,* 4.373

190. *Papers,* 4.376; New York *World,* 27 Aug. 1878; N. N. Craig (Western Union telegraph operator), "Looking for Thrills," ts. memoir, 25, TENHP; TE reminiscing in *Papers,* 4.856. "Texas Jack," alias John B. Omohundro, was a former army scout pursuing a theatrical career in 1878. See David Baron, *American Eclipse: A Nation's Epic Race to Catch the Shadow of the Moon and Win the Glory of the World* (New York, 2017), loc. 1216. This book offers an excellent account of the total eclipse of 1878.

191. *New York Herald,* 29 July 1878.

192. *Laramie Daily Sentinel,* 30 July 1878.

193. *New York Herald,* 29 July 1878; John A. Eddy, "Thomas A. Edison and Infra-Red Astronomy," *Journal of the History of Astronomy* 3 (1972); *Telegraphic Journal,* 1 Aug. 1878; *Papers,* 4.435.

194. *Laramie Daily Sentinel,* 30 July 1878. The following account is largely based on the reporting of Edwin Fox, TE's companion, in *New York Herald,* 30 July 1878.

195. Rebecca Hein, "Moon Shadows over Wyoming: The Solar Eclipse of 1878, 1889, and 1918," Wyohistory.org, 29 June 2017; TE reminiscence in Dyer and Martin, *Edison,* 230; *New York Herald,* 30 July 1878; *Papers,* 4.435.

196. TE interviewed in New York *World,* 27 Aug.1878; TE, "On the Use of the Tasimeter for Measuring the Heat of the Stars and of the Sun's Corona," paper read to the American Association for the Advancement of Science, 15 Aug. 1878, in *Papers,* 4.432–36; *New York Herald,* 30 July 1878.

197. Phil Roberts, "Edison, the Light Bulb, and the Eclipse of 1878," 8 Nov. 2014, Wyohistory.org.

198. See Allerhand, *Illustrated History,* 40, 107, and *passim.*

199. TE, "The Beginnings of the Incandescent Lamp and Lighting System," ts., 1926, HFM.

200. Dyer and Martin, *Edison*, 450; *Papers*, 5.232–34.

201. See Baron, *American Eclipse* for the mystical effect of a total eclipse on many observers (chap. 16).

202. *Papers*, 4.432–35; Stockton Griffin to TE, 5 Aug. 1878, PTAE.

203. Stockton Griffin to E. H. Brown, 20 Aug. 1878, PTAE; *Papers*, 4.441. TE declined to read three other papers, on his carbon button, carbon telephone, and a new "sonorous voltameter." Barker presented them in his stead. *Papers*, 4.375.

204. New York *Sun*, 29 Aug. 1878.

205. *Papers*, 4.445.

206. TE notebook entry, 2 Feb. 1877, in *Papers*, 3.246; Israel, *Edison*, 165; *Papers*, 4.325.

207. *Papers*, 4.468, 867, 868.

208. "An acquaintance," almost certainly Charles Frederick Chandler, another member of Barker's party, quoted in *Denver* (CO) *News*, 5 July 1891.

209. Ibid.; New York *Sun*, 10 Sept. 1878; *Papers*, 4.469.

210. Jehl, *Menlo Park Reminiscences*, 211.

211. *Papers*, 4.469; New York *Sun*, 20 Oct. 1878.

212. *Papers*, 4.470, 473–487; TE to William Wallace, 13 Sept. 1878, PTAE.

213. "Edison's Newest Marvel," New York *Sun*, 16 Sept, 1878. The entire interview is printed in *Papers*, 4.503–5.

214. "Edison's Newest Marvel."

215. Friedel and Israel, *Edison's Electric Light*, 16; Jehl, *Menlo Park Reminiscences*, 241, 235; TE Patent 214,636; TE Patent 214,637, "Thermal Regulation for Electric-Lights."

216. Strouse, *Morgan*, 183.

217. Hughes, *Networks of Power*, 48–49; *Papers*, 4.551; DeGraaf, *Edison and Innovation*, 50.

218. *Papers*, 4.586.

219. New York *Sun*, 20 Oct. 1878.

220. Ibid.

221. *Papers*, 4.648, 642–43. *New York Herald* announced that "A New Talking Machine Makes Its Appearance in Menlo Park."

222. *Papers*, 4.664.

223. McPartland, "Almost Edison," 117; *Papers*, 4.657–58, 664.

224. Lowrey to TE, 10 Dec. 1878, PTAE; Gall, "Edison: Managing Menlo Park," 40.

225. Francis Upton, 12 Dec. 1878, PTAE.

226. Strouse, *Morgan*, 230–31; TE in New York *Sun*, 19 Dec. 1878; Friedel and Israel, *Edison's Electric Light*, 19–21. The boilers were fed with piped water from a nearby brook. Jehl, "TAE – MP," 26.

227. Friedel and Israel, *Edison's Electric Light*, 29–30; Bright, *Electric Lamp Industry*, 47–53.

228. Lowrey to TE, 25 Jan. 1879, quoted in Friedel and Israel, *Edison's Electric Light*, 31.

229. TE quoted in Jehl, *Menlo Park Reminiscences,* 197; *Papers,* 4.756.

230. *Papers,* 4.756; TE reminiscence in *Papers,* 4.864.

231. *Papers,* 4.507–8, 561–62, 235, 125; *Scientific American,* 26 Apr. 1878.

232. *Papers,* 5.149, 124.

233. London *Times,* 22 Mar. 1879.

234. *Papers,* 5.126.

235. *Papers,* 5.126, 157; Francis Upton to Elijah Upton, 23 Feb. 1879, PTAE.

236. "Edison's Electric-Light Inventions," *Appleton's Cyclopedia 1879* (New York, 1880); *Papers,* 5.53, 5.55.

237. Quoted in Israel, *Edison,* 172. McPartland, "Almost Edison," 218, points out that TE was unconsciously restating Joule's law of resistance, W=V2 / R.

238. Friedel and Israel, *Edison's Electric Light,* 33, 53–57; Israel, *Edison,* 182.

239. Upton addressing the Edison Pioneers, 11 Feb. 1918, Biographical File, PTAE.

240. W. S. Andrews to E. H. Mullin, 4 Apr. 1898, Meadowcroft Collection, TENHP; Friedel and Israel, *Edison's Electric Light,* 58; Israel, *Edison,* 182; Association of Edison Illuminating Companies, *"Edisonia,"* 28–29. See also TE Patent 227,229.

241. *Journal of Gas Lighting,* 20 Jan. 1880.

242. TE quoted in Friedel and Israel, *Edison's Electric Light,* 58; TE in Mar. 1882, quoted in Israel, *Edison,* 182.

243. Francis to Elijah Upton, 6 July 1878, PTAE.

244. *Papers,* 5.161, 227, 242–43.

245. Jehl, *Menlo Park Reminiscences,* 325–27; New York *Herald,* 10 Dec. 1880; Öser, "Wizard of Menlo Park."

246. Jehl, *Menlo Park Reminiscences,* 250–51. See also Friedel and Israel, *Edison's Electric Light,* 45ff.; *Papers,* 5.4.

247. *Papers,* 5.277, 195–96; Friedel and Israel, *Edison's Electric Light,* 63. Seven of the five hundred volumes Edison ordered for his library in 1879 were mineralogy studies.

248. Friedel and Israel, *Edison's Electric Light,* 46; *Proceedings of the American Association for the Advancement of Science* (1880), 173ff. See also *Papers,* 5.347–54; Francis to Elijah Upton, 31 Aug. 1878, PTAE.

249. Friedel and Israel, *Edison's Electric Light,* 62–63.

250. "Edison and the Electric Incandescent Lamp," *The Street Railway Bulletin,* 14 (Oct. 1914).

251. Jones, *Edison: Sixty Years,* 252; Edwin M. Fox, "A Night with Edison," New York *Herald,* 31 Dec. 1879.

252. Fox, "Night with Edison."

253. Ibid.

254. Ibid.

255. Ibid.

256. Ibid.

257. Ibid.

258. Ibid.

259. *Brooklyn Daily Eagle,* 30 Dec. 1879; *New York Tribune,* 1 Jan. 1880.

PART SEVEN · TELEGRAPHY (1860–1869)

1. "Sunday 1 Mr. Edison's Interview" galley, E.Bio, TENHP. See also Dyer and Martin, *Edison,* 40–41. Note: The most important source of information about TE's teenage years remains his series of dictated or jotted reminiscences, prepared in 1908–9 at the request of his official biographers, Dyer and T. C. Martin. Volume 1 of *The Papers of Thomas A. Edison* reprints all the memoranda relevant to TE's childhood and early youth as Appendix 1, pp. 627ff. Except where otherwise indicated, the following narrative through 1869 is based on this appendix. TE's memory often has to be taken on trust, for lack of supporting evidence, throughout the 1860s. Whenever possible, slips in his recall have been corrected in text. For example, when he recalls driving the locomotive "62 and a half miles," he is remembering the full length of the Grand Trunk Railway from Port Huron to Detroit. It was more likely 47 ½ miles, because he also says that the engineer drove "fifteen miles" first.

2. Runes, *Diary and Sundry Observations of Thomas Alva Edison,* 50.

3. TE's working day on the Grand Trunk Railway in 1860 was at first shorter than he remembered. The train left from Port Huron at eight A.M. and returned from Detroit at 3:50 P.M. Each journey took about three hours. By the spring of 1862, schedule changes had lengthened his hours from seven A.M. to about ten-thirty P.M. *Detroit Free Press,* 9 Aug. and 21 Nov. 1859; TE, *Weekly Herald,* 3 Feb. 1862, HFM; *Papers,* 1.27.

4. *Papers,* 1.6–7, 630; *Detroit Free Press,* 15 Sept. 1859; George P. Lathrop, "Edison's Father"; TE reminiscing in Lucius Hitchcock to Aunt Sade, 3 Dec. 1930, TENHP.

5. Ballentine, "Early Life of Edison"; map drawn and marked "1860" by TE, reproduced in Stamps, Hawkins, and Wright, *Search for the House,* 16; *Papers,* 1.632.

6. *Papers,* 1.629, 631; Dyer and Martin, *Edison,* 32; Nancy Wright, "The House in the Grove," in Stamps, Hawkins, and Wright, *Search for the House,* 41. This essay is an important source of information about TE's youth in Port Huron.

7. *Detroit Free Press,* 19 May 1860; Paul Taylor, *"Old Slow Town": Detroit During the Civil War* (Detroit, MI, 2013), loc. 77; *Detroit Free Press,* 9 Sept. 1860, e.g. Young Al also did a good trade in Lincoln campaign buttons. TE memo, ca. 1929, Biographical Collection, TENHP.

8. Despite the fanatical anti-Lincoln bias of its leading newspaper, Michigan strongly supported the Union cause. Taylor, *"Old Slow Town,"* chap. 2.

9. Jones, *Edison: Sixty Years,* 28; *Weekly Herald,* 3 Feb. 1862.

10. *Weekly Herald,* 3 Feb. 1862.

11. Ibid.

12. Dyer and Martin, *Edison,* 33. The story, in this and other biographies, that the British engineer George Stephenson traveled on the Grand Trunk and publicized TE's newspaper in the London *Times* is apocryphal. Only one other issue of the *Weekly Herald* has survived, as a facsimile in *Magazine of Michigan* 1 (Oct. 1929). It is datelined "June [1862], Published by the Newsboy on the Mixed Train." See also *Battle Creek Evening News,* 21 Oct. 1959.

13. Jones, *Edison: Sixty Years*, 16–17.
14. *Papers*, 1.629; *Detroit Free Press*, 9 Apr. 1862.
15. *Detroit Free Press*, 9 Apr. 1862; *Papers*, 1.629; TE, *Weekly Herald*, 3 Feb. 1862. The generally accepted casualty total for Shiloh is 17,854.
16. Lathrop, "Talks with Edison."
17. Ibid. TE mistakenly recalled that he dealt with the editor's predecessor, Wilbur F. Storey. *Papers*, 1.630.
18. *Papers*, 1.630.
19. TE to Willis Engle, 10 Aug. 1862. *Papers*, 1.27 ("My time is taken up with my business in the cars. . . . I don't get home until ten in the evening"). For TE's hearing loss around this time, see Dyer and Martin, *Edison*, 37; TE to William J. Curtis, 7 May 1920 ("I have been deaf for 60 years"); Ford, *Edison As I Knew Him*, 20, 24–25; TE Diary; *Papers*, 1.670; and Israel, *Edison*, 17. For the scarlet fever theory, see Robert Traynor, "The Deafness of Edison," *Hearing International*, 19 Feb. 2013, hearinghealthmatters.org.
20. Dyer and Martin, *Edison*, 49–51.
21. *Telegraph* 6, no. 1 (28 Aug. 1869).
22. Dyer and Martin, *Edison*, 51. Note: It is not clear from the sparse records of this part of TE's life just when his service with the Grand Trunk ended, or when exactly he took up telegraphy. He seems to have worked the Mount Clemens–Port Huron segment for at least a while in the fall and winter of 1862, delegating the rest of the trip to another boy. Several operators along the line recalled him stopping by to practice on their instruments. According to his own recollection, he sent a few primitive messages along a half-mile stovepipe wire rigged between his house and a friend's in Fort Gratiot. His serious study of the medium, however, began under Mackenzie. Israel, *Edison*, 18; Dyer and Martin, *Edison*, 48–51.
23. *Papers*, 1.10–11.
24. *Papers*, 1.6, reproduces TE's society membership card, dated 15 Sept. 1862. Library regulations did not permit the registration of boys younger than eighteen. *Papers*, 1.7; Detroit—Young Men's Society, *Catalogue of the Library, with a Historical Sketch* (Detroit, 1865). All the volumes TE claimed to have read in youth are listed in this catalog with the exception of Fresenius's *Chemical Analysis*.
25. Norman Speiden to George E. Probst, 4 Mar. 1959, TENHP; Dyer and Martin, *Edison*, 51; *Papers*, 1.631.
26. *Papers*, 1.10; William D. Wright to TE, 29 June 1900, TENHP; Israel, *Edison*, 21.
27. Ibid., 3; Henry Hartsuff on 3 Apr. 1863, quoted in "House in the Grove," 44.
28. Ibid., 44–45; *Papers*, 1.631; "Edison, Miller, and Affiliated Families," in Jeffrey, *Phonographs to U-Boats*, 151ff.
29. Dyer and Martin, *Edison*, 54–55.
30. *Papers*, 1.631.
31. TE reminiscing in *Papers*, 1.632. Note: A number of unverifiable or conflicting stories about TE's teenage wanderings were published during his lifetime. Francis Arthur Jones, e.g., writing in 1908, has TE working as a railroad night

operator in Port Huron before doing so in Canada, and locates the train-convergence anecdote in Sarnia rather than Stratford, Ontario. Jones, *Edison: Sixty Years,* 39–44. Since so few actual records of the period survive, this biography as far as possible confines itself to accounts TE himself authorized.

32. *Papers,* 1.661.

33. David Hochfelder, *The Telegraph in America, 1832–1920* (Baltimore, MD, 2012), 3.

34. Banner motto of *Telegraph* in the mid–1860s. See, e.g., *Papers,* 1.13. TE refers nostalgically to corn whiskey in a letter to Charles Mixer, 24 July 1920, TENHP.

35. *Papers,* 1.655. TE recalled that only "3 percent" of these jokes were publishable (1.661).

36. Dee Alexander Brown, *The Bold Cavaliers* (Philadelphia, 1959), 80. Coincidentally, George Ellsworth, the telegrapher whom Brown cites as having this sixth sense, demonstrated it when working with TE in Cincinnati in 1865. *Papers,* 1.663.

37. W. P. Phillips (a former TE telegraph associate), *Oakum Pickings* (New York, 1876), 138–39 and *passim* for stories of life on the telegraph trail in the 1860s.

38. Dyer and Martin, *Edison,* 65; Ezra Gilliland interview, *Cincinnati Commercial,* 19 Mar. 1879. This is the only source that specifically states TE moved to Adrian "in 1863." An old-timer quoted in *Papers,* 4.452 recalls him arriving there in an "old straw hat, linen coat and pants."

39. Josephson, *Edison,* 44; Dyer and Martin, *Edison,* 65–66; *Papers,* 4.873.

40. *Papers,* 1.659.

41. *Papers,* 1.36–37; Dyer and Martin, *Edison,* 67–68.

42. Israel, *Edison,* 29; *Papers,* 1.661. "This peculiar state of the brain doing intellectual work unconsciously should be investigated," TE wrote in 1908. It has been, in modern times. See Iwan Morus, " 'The Nervous System of Britain': Space, Time, and the Electric Telegraph in the Victorian Age," *British Journal for the History of Science* 33, no. 4 (2000).

43. Dyer and Martin, *Edison,* 70.

44. Ibid.; *Papers,* 1.660. See also TE memo, ca. 1929, Biographical Collection, TENHP.

45. Quoted in Phillips, *Sketches Old and New,* 65; Dyer and Martin, *Edison,* 68–69.

46. "Tom Edison's Operating Days," *Operator* 9 (1 Apr. 1878); Israel, *Edison,* 28–29.

47. *Papers,* 1.637, 660–61; Dyer and Martin, *Edison,* 73.

48. *Papers,* 1.661.

49. There is uncertainty about the chronology of TE's stays in these cities in late 1865 and early 1866. His movements have to be inferred from fragmentary sources, best summarized in Israel, *Edison,* 30–33.

50. *Papers,* 1.28.

51. TE to Sam and Nancy Edison, ca. spring 1866, PTAE; *Papers,* 1.29, 657; Israel, *Edison,* 32.

52. *Papers,* 1.650; TE reminiscing in Frank L. Dyer Diary, entry for 20 Feb. 1906, TENHP.

53. Wright, "House in the Grove," 48, 26, 49.

54. Ibid., 47–48; *Papers*, 1.17.

55. Dyer and Martin, *Edison*, 80; *Papers*, 1.75–76.

56. George Bliss quoted in *Papers*, 4.875; TE reminiscing in *Papers*, 1.653, 650; TE quoted in Dyer and Martin, *Edison*, 80.

57. *Papers*, 1.653–54; *Louisville Courier-Journal*, 4 Dec. 1866 and 8 Jan. 1867. The earlier issue contains an editorial reference to "the extraordinary and late rush of telegraphic copy this morning," whereby space needed for Johnson's message "has compelled us to exclude many important dispatches." *Papers*, 1.75.

58. *Papers*, 1.30–31, 36–49, 662; Israel, *Edison*, 36. Faraday died on 25 Aug. 1867.

59. Z. T. Underwood in *Oakland Tribune*, 19 Oct. 1931; *Papers*, 4.874; Phillips, *Sketches Old and New*, 179; Eugene Baker in *Cincinnati Post*, 22 July 1929, clipping in Biographical Collection, TENHP.

60. See Israel, *Edison*, 35–36, for the sophistication of Cincinnati's telegraph elite at this time.

61. *Papers*, 1.56.

62. "Edison's Double Transmitter," *Telegrapher*, 11 Apr. 1868, *Papers*, 1.56–58. Little is known of the approximately six months TE spent in Port Huron between October 1867 and March 1868. It is possible he returned there to help his parents adjust to their new straitened circumstances. He appears to have spent the winter working on telegraph designs and speculative articles. In March, short of money, restless, and hopeful for a job with Western Union in the East, he wrangled a free pass from the Grand Trunk Railway and made a snowbound journey via Montreal to Boston, arriving there around the end of the month. *Papers*, 1.635–36.

63. *Papers*, 1.51.

64. TE reminiscing in *Papers*, 1.636; Dyer and Martin, *Edison*, 99.

65. Dyer and Martin, *Edison*, 99–100; *Papers*, 1.637.

66. *Telegrapher*, 1 Aug. 1868.

67. *Papers*, 1.633, 38.

68. *Papers*, 1.635.

69. Ibid.

70. Dyer and Martin, *Edison*, 100–1; Josephson, *Edison*, 62.

71. *Journal of the Telegraph*, 11 Apr. 1868; *Papers*, 1.98–100.

72. *Papers*, 1.61–62.

73. Ibid., 1.77–81, 67. In August 1870 Daniel Craig wrote a letter introducing TE to Farmer, praising him as "a genius only second to yourself" and proposing a collaborative venture on a printing machine. Neither the meeting nor the venture seem to have occurred. *Papers*, 1.182–84.

74. *Papers*, 1.67, 83; TE to John Van Duzer, 5 Sept. 1868. TE's proposed facsimile telegraph was intended for the Asian export market. See Israel, *Edison*, 43–44.

75. TE Patent 90,646, issued 1 June 1969; *Papers*, 1.84–86. His whittling marks are still visible on the preserved model of the vote recorder in HFM. Jehl, *Menlo Park Reminiscences*, 38.

76. Dyer and Martin, *Edison*, 103. TE vaguely implied in 1908 that he took the recordograph to Washington and "exhibited [it] before a committee that had something to do with the Capitol," but there was no contemporary evidence of his doing so.

77. *Papers*, 1.60; Israel, *Edison*, 42–43.

78. *Papers*, 1.111.

79. TE Patent 91,527.

80. *Papers*, 1.102–3; Israel, *Edison*, 45–46. Laws abandoned his Boston venture later in 1869, when the two Gold & Stock companies merged in New York.

81. *Papers*, 1.113; Israel, *Edison*, 44–45.

82. *Papers*, 1.121.

83. Josephson, *Edison*, 698; *Papers*, 1.116, 121, 118; Israel, *Edison*, 48; Charles A. Barnes, former Rochester telegraph executive, in *Plainfield* (NJ) *Courier-News*, 25 Oct. 1929; TE to Louis Wiley, 9 Feb. 1925, TENHP; TE to Frank Hanaford, 26 July 1869, PTAE; TE phonograph reminiscence quoted in John W. Lieb, *Dinner in Honor of Thomas A. Edison and in Commemoration of Forty Years of Edison Service in New York, 11th September 1922* (New York, 1923), copy in TENHP.

84. Israel, *Edison*, 48; *Papers*, 1.121.

85. Dyer and Martin, *Edison*, 122; Jones, *Edison: Sixty Years*, 62–63.

86. TE reminiscing, quoted in Dyer and Martin, *Edison*, 123.

87. James Laws Ricketts to Charles Edison, 3 Feb. 1947, Biographical Collection, TENHP. TE's memory of being at once put in charge of Laws's "whole plant" accelerates chronology. That promotion, and consequent "violent jump" in salary, did not occur until the summer.

88. *Papers*, 1.128; Dyer and Martin, *Edison*, 123.

89. TE Patent 96,567; *Papers*, 1.179, 126; Israel, *Edison*, 50.

90. *Papers*, 1.114.

91. *Papers*, 1.163.

92. *Papers*, 1.136, 138; TE in *New York Morning Journal*, 26 July 1891; Klein, *Life of Gould*, 69–70, 109, 112–13; Cornwallis, *Gold Room and the Stock Exchange*, 15–16; TE reminiscing in Dyer and Martin, *Edison*, 126.

93. Ashley quoted in Israel, *Edison*, 51; *Papers*, 1.132–34; advertisement facsimile in Dyer and Martin, *Edison*, 128.

94. Ibid., 129.

PART EIGHT · NATURAL PHILOSOPHY (1847–1859)

1. This Part, like Part Seven, is based on TE's reminiscences recorded in 1908 and 1909 for his biographers Frank Dyer and T. C. Martin, and republished as Appendix 1 in vol. 1 of *The Papers of Thomas A. Edison*, pp. 627ff. Supplementary or corrective details from other sources are documented below.

2. Sam Edison stood 6′2″ tall, and with his muscular frame had the effect of filling every room he entered. According to his son, he was deemed at age 69 to have had "the highest chest expansion of anyone except one" in the annals of the Mutual Life Insurance Co. of New York. *Papers*, 4.870; Sheldon Wood to

TE, 5 Sept. 1929, TENHP; TE superscript on his copy of *DeLemar Lectures 1925–1929* 127, TENHP.

3. Henry Ford, "The Greatest of Americans," *Hearst's International / Cosmopolitan*, July 1930; TE superscript on W. H. Raymenton to TE, 14 Nov. 1921, TENHP; Marion Estelle Edison (1829–1900) was married to Homer Page on 19 Dec. 1849.

4. TE's other living siblings were William Pitt Edison (1831–91) and Harriet "Tannie" Edison (1833–63). For a compact genealogy of all his relatives, see Thomas E. Jeffrey, "Edison, Miller, and Affiliated Families," in Jeffrey, *Phonographs to U-Boats*, 151ff.

5. Samuel Ogden Edison, Jr., was born on 16 Aug. 1804. Nancy Elliot Edison's birthdate is the subject of some dispute, being either 4 Jan. 1808 or 4 Jan. 1810. Jeffrey, "Edison, Miller, and Affiliated Families."

6. Israel, *Edison*, 2–4; Nerney, *Edison, Modern Olympian*, 25.

7. "The Birthplace of Edison Dreams of Her Fallen Greatness," *Firelands Pioneer* 13 (1900), 716ff.; Wallace B. White, *Milan Township and Village: One Hundred and Fifty Years* (Milan O., 1959) 15–19.

8. "Birthplace of Edison," 440; milanarea.com.; TE to *Sandusky Register*, 31 Dec. 1922, TENHP.

9. Dyer and Martin, *Edison*, 14; TE to Ruth Thompson, 12 June 1922, TENHP; George Minard, TE schoolmate, in *Sandusky Star-Journal*, 13 Aug. 1923; R. F. McLaughlin to Mary C. Nerney, 4 Nov. 1929, Biographical Collection, TENHP.

10. Maria Cooke via Isoline Minty to Mary C. Nerney, 14 Dec. 1929, Biographical Collection, TENHP. See also Rev. C. Emmons in *Pigott* (AL) *Banner*, 25 Oct. 1929.

11. Samuel Edison, Jr., aged ninety, interviewed in Port Huron by George P. Lathrop, ca. 16 Aug. 1894.

12. TE quoted in Josephson, *Edison*, 13; *Papers*, 2.786. See also Dyer and Martin, *Edison*, 18.

13. "Second reader . . . as they used to term it." Cooke/Minty to Mary C. Nerney, 14 Dec. 1929, TENHP; TE quoted in Jones, *Edison: Sixty Years*, 6–7.

14. Israel, *Edison*, 4; Ballentine, "Early Life of Edison," 1; oldtowneporthuron.com.

15. Wright, "House in the Grove," 13ff.

16. Terry Pepper, "Seeing the Light: The Lighthouses of Lake Huron," Terrypepper.com; Dyer and Martin, *Edison*, 24–25; Jones, *Edison: Sixty Years*, 12–13.

17. Israel, *Edison*, 7; Dyer and Martin, *Edison*, 23–24; Israel, *Edison*, 6.

18. Taylor, *"Old Slow Town,"* loc. 5322; A. T. Andreas Co., *St. Clair County, Michigan* (Chicago, 1883), 550; Israel, *Edison*, 6.

19. Dyer and Martin, *Edison*, 25; *Papers*, 1.23–24, 27; Wright, "House in the Grove," 30–31.

20. TE quoted in Jones, *Edison: Sixty Years*, 7; M. G. Bridges to TE, 19 June 1911, TENHP; Nerney, *Edison, Modern Olympian*, 24; illustration in DeGraaf, *Edison and Innovation*, xxii; Israel, *Edison*, 2; Dyer and Martin, *Edison*, 15; TE quoted in Israel, *Edison*, 14; reminiscence of a Port Huron contemporary of TE in Jones, *Edison: Sixty Years*, 11–12.

21. A. R. Ogden to Arthur E. Bestor, 23 July 1879, PTAE; Israel, *Edison*, 7; Jones, *Edison: Sixty Years*, 11; John F. Talbot to TE, 26 Nov. 1920, TENHP.

22. *Papers*, 4.870; TE superscript on Charles Gaston to TE, 23 July 1927, TENHP; Dyer and Martin, *Edison*, 26.

23. Ford, *Edison As I Know Him*, 20; *Papers*, 1.24. It is possible that TE first encountered Parker's book in its abridged junior reader form (*First Lessons in Natural Philosophy*, 1848). The Union School curriculum carried the full version, as did many other American schools. *Papers*, 1.24.

24. Henceforth Parker, *Natural Philosophy*.

25. Ibid., xiii.

26. Ibid., xiv.

27. Ibid., xiv–xv.

28. Ibid., 20, 31–32.

29. Ibid., 81, 60.

30. Ibid., 98, 155, 168, 143.

31. Ibid., 98, 155, 168, 143.

32. Ibid., 258, 258.

33. Ibid., 259.

34. Apparently TE tried electrotherapy on Mattie Wilson, a young Fort Gratiot woman who concussed herself while ice skating. "My wife . . . says she well remembers the shock you gave her in the chair in the old Edison store." George A. Fritz to TE, 27 Oct. 1927, TENHP.

35. Parker, *Natural Philosophy*, 294.

36. Ibid., 298ff., 304, 310.

37. Ibid., 173, 176, 174, 178–79.

38. Ibid., 179, 210–12, 220.

39. Ibid., 245.

40. Ibid., 233, 237–42.

41. Ibid., 396.

42. Nerney, *Edison, Modern Olympian*, 6; *Papers*, 1.628–29; TE quoted in Josephson, *Edison*, 27.

43. *Papers*, 1.629.

44. Parker, *Natural Philosophy*, xv; Dyer and Martin, *Edison*, 28.

45. Dyer and Martin, *Edison*, 28; Israel, *Edison*, 11. An archeological excavation of the site of the long-razed Edison home at Fort Gratiot in the late twentieth century disclosed many bottles, beakers, crucibles, test tubes, funnels, and other chemical paraphernalia. See Stamps, Hawkins, and Wright, *Search for the House*.

46. Ballentine, "Early Life of Edison," 1; Israel, *Edison*, 11; Nerney, *Edison, Modern Olympian*, 290; Runes, *Diary and Sundry Observations of Thomas Alva Edison*, 50. "The ear boxing incident never happened," TE told Henry Ford in 1929, when they passed through Smiths Creek Station. But on the same journey he repeated his less believable story of being hauled aboard a departing train by the ears. Ford, *Edison As I Know Him*, 20, 24–25. Robert Traynor, "The Deafness of Edison," February 9, 2013, *Hearing International*,

Hearinghealthmatters.org, focuses on the theory that his hearing loss was caused by scarlet fever.

EPILOGUE (1931)

1. AP report, *Sandusky* (OH) *Register,* 21 Oct. 1931; Herbert Hoover, "Statement in a National Tribute to Thomas Alva Edison," 20 Oct. 1931, *American Presidency Project,* http://www.presidency.ucsb.edu.
2. Hoover, "Statement."
3. AP news release, 21 Oct. 1931; *Hartford Courant,* 22 Oct. 1931; New York *Daily News,* 22 Oct. 1931; *Chicago Tribune,* 22 Oct. 1931. In 1963 TE's coffin was reinterred in the grounds of Glenmont, where Mina is also buried.
4. AP news dispatch, 22 Oct. 1931; *Montreal Gazette* and *Hartford Courant,* same date.
5. *Tampa Tribune,* 22 Oct. 1931; *Chicago Tribune.*
6. As noted above, TE left an estate of $12 million ($198 million in today's money), and there was some bickering among his children as to its fair disposition.
7. The three illegitimate children Sam fathered after the death of Mary Edison in 1871 were all girls. See Jeffrey, "Edison, Miller, and Affiliated Families."
8. The following catalogue of local sound effects is taken from Ballentine, "Early Life of Edison" (Ballentine lived in the Edison house in Fort Gratiot after TE's departure), as well as from John F. Talbot to TE, 26 Nov. 1920, and TE reminiscing to Ruth Thompson, 12 June 1922, both in TENHP.

ILLUSTRATION CREDITS

Unless otherwise credited, all images are courtesy of National Park Service, Thomas Edison National Historical Park.

Frontispiece: Edison in his laboratory library, under Aurelio Bordiga's *Genius of Electricity*, 1911. Library of Congress. **13:** Edison collecting botanical specimens, circa 1927. **19:** Charles Edison, circa 1920. **21:** Aerial photograph of Thomas A. Edison, Inc., 1920s. **28:** Edison listening to phonograph records at home, 1920s. **35:** Edison napping in front of Harvey Firestone and President Harding at Vagabond camp, 23 July 1921. **38:** Thomas and Mina Edison on his seventy-fifth birthday, 11 February 1922. **45:** Theodore Edison, 1924. **60:** Edison letter regarding invention of phonograph, 1927. **62:** Henry Ford, Edison, and Harvey Firestone in Florida, circa 1928. **63:** Edison botanical sketch, 1920s. **70:** Thomas Edison brooding in the chem lab. **80:** Menlo Park reconstructed at Greenfield Village, Dearborn, Michigan, 1929. **89:** Edison and Secretary of the Navy Josephus Daniels aboard USS *New York*, 1915. **92:** Edison film studio in the Bronx, circa 1910. **107:** Madeleine Edison, circa 1911. **114:** Edison asleep in his laboratory, 1911. Photograph by Miller R. Hutchison **128:** Edison receiving Morse signals from Hutchison, circa 1912. **130:** Edison and the Insomnia Squad at midnight "lunch," fall 1912. **149:** Diamond Disc retail advertisement, 1913. **150:** Edison A-100 "Moderne" Diamond Disc phonograph, 1915. **167:** The great fire of 9 December 1914. **203:** Commodore Edison and the crew of USS *Sachem*, 1917. **209:** Commodore Edison at Key West Naval Station, 1918. **211:** Edison in his chemistry laboratory, 1902. Courtesy of the Library of Congress. **215:** Edison's chair and lamp in the sitting room at Glenmont, circa 1900s. **217:** Thomas Alva Edison, Jr., circa 1900. **225:** Illustration from Edison's cadmium-copper storage battery patent application, 15 October 1900. **243:** Edison's trademark signature, 1902. **248:** Edison at Seminole Lodge, early 1900s. **265:** Edison and model cement house, circa 1906. **269:** Edison's A-12 storage battery, 1909. **270:** Edison's A-12 storage battery, 1909. **275:** Edison at his Ogden mine, 1895. **279:** Marion Edison as a teenager. **297:** Edison's sketch of his tabletop Kinetograph, 28 May 1891. **312:** The Black Maria, circa 1893. **318:** Eugene Sandow models for W. K. Dickson's camera, March 1894. **333:** The Ogden mine workforce, circa 1895. **347:** Edison sketched by William Dodge Stevens, Ogden, 1897. **351**: William Edison, circa 1898. **358**: The Ogden mill under snow, late 1890s. **359**: Thomas Edison refracted. Courtesy of Edmund Morris. **363**: Edison's house in Menlo Park, January 1880. **375**: Members of the Menlo Park laboratory team, 1880. **382**: The Edison electric train, Charles Batchelor driving. **385**: Stages of splitting and shearing a splint of madake bamboo into filaments ready for carbonization. **397**: Menlo Park in the winter of 1880–81. Painting by Richard F. Outcault. **412**: Shell winding for Edison's large magneto dynamo, 1879. **414**: Edison's "Jumbo" dynamo at the Paris Electrical Exposition, 1881. **423**: Power monitor panel, Edison Pearl Street station, 1882. **433**: Mary Edison and feathered friends, 1883. **454**: A page of Edison's diary, summer 1885. **456**: Mina Miller, at about the time Edison first met her. **461**: Glenmont in Llewellyn Park, soon after Edison's purchase of it. **463**: Edison's Magritte-like sketch of Mina as an airborne clock. **465**: Notebook pages, 1886. **467**: Edison's plan for his Fort Myers estate, spring 1886. **478**: Edison's new laboratory in West Orange. Phonograph Works in background. Original source unknown. **484**: Edison's first Kinetoscope caveat, 8 October 1888. **489**: Edison with his microphotographic camera. Photograph by W. K. Dickson, 1888. **497**: Edison with his phonograph in Washington, April 1878. Original source unknown. **512**: Mary Stilwell Edison, circa 1871. **513**: Edison chemical printer spelling out the word "BOSTON," January 1872. **515**: Edison & Murray workforce, Ward Street, Newark, 1873. **516**: A freehand sketch by Edison of his quadruplex system, 25 November 1875. **528**: The Edison Electric Pen with batteries and press, 1875. **537**: Edison sketch instruments, 1876. **538**: Edison sketch voices, 1876. **540**: Edison's embossing recorder-repeater, February 1877. **547**: Edison's sketch of his first phonograph, circa November 1877. **551**: Edison's letter of thanks to *The Daily Graphic,* 16 May 1878. Thomas A. Edison Papers, Rutgers University. **561**: The total eclipse seen from Creston, Wyoming, 29 July 1878. Astronomical drawing by E. L Trouvelot. Original source unknown. **578**: Charles Batchelor in the Menlo Park laboratory. The first photograph ever taken by incandescent light, 22 December 1879. **580**: Edison's "New Year's Eve Lamp," 1879. Division of Work and Industry, National Museum of American History, Smithsonian Institution. **581**: Edison as a young telegrapher, circa 1863. **585**: Al Edison, newsboy, circa 1860. **588**: *The Detroit Free Press* reports the Battle of Shiloh, 10 April 1862. Original source unknown. **609**: Gold price postings annotated by future president James A. Garfield. Black Friday, 1869. Original source unknown. **613**: Alva Edison as a child, circa 1850. **616**: Edison's birthplace in Milan, Ohio. Original source unknown. **617**: "Milan from near the Sandusky City Road," by J. Brainerd, 1847. Original source unknown. **622**: Nancy Elliott Edison, circa 1854. **625**: Locomotive, from Richard Green Parker's *Natural Philosophy.* **628**: Sam Edison tilling his field at Fort Gratiot, date unknown.

INDEX

ABOUT THE AUTHOR

EDMUND MORRIS was born and educated in Kenya and attended college in South Africa. He worked as an advertising copywriter in London before immigrating to the United States in 1968. His first book, *The Rise of Theodore Roosevelt*, won the Pulitzer Prize and the National Book Award in 1980. Its sequel, *Theodore Rex*, won the *Los Angeles Times* Book Prize for Biography in 2002. In between these two books, Morris became President Reagan's authorized biographer, and wrote the national bestseller *Dutch: A Memoir of Ronald Reagan*. He then completed his trilogy on the life of the twenty-sixth president with *Colonel Roosevelt*, also a bestseller, and published *Beethoven: The Universal Composer* and *This Living Hand: And Other Essays*. He was married to fellow biographer Sylvia Jukes Morris for fifty-two years. Edmund Morris died in 2019.

ABOUT THE TYPE

This book was set in Sabon, a typeface designed by the well-known German typographer Jan Tschichold (1902–74). Sabon's design is based upon the original letterforms of sixteenth-century French type designer Claude Garamond and was created specifically to be used for three sources: foundry type for hand composition, Linotype, and Monotype. Tschichold named his typeface for the famous Frankfurt typefounder Jacques Sabon (c. 1520–80).